Lecture Notes in Logistics

Series editors

Uwe Clausen, Dortmund, Germany
Michael ten Hompel, Dortmund, Germany
Robert de Souza, Singapore, Singapore

More information about this series at http://www.springer.com/series/11220

Michael Freitag · Herbert Kotzab
Jürgen Pannek
Editors

Dynamics in Logistics

Proceedings of the 5th International
Conference LDIC, 2016 Bremen, Germany

Editors
Michael Freitag
BIBA-Bremer Institut für Produktion
  und Logistik GmbH
Bremen
Germany

Jürgen Pannek
BIBA-Bremer Institut für Produktion
  und Logistik GmbH
Bremen
Germany

Herbert Kotzab
University of Bremen
Bremen
Germany

ISSN 2194-8917   ISSN 2194-8925 (electronic)
Lecture Notes in Logistics
ISBN 978-3-319-83214-2   ISBN 978-3-319-45117-6 (eBook)
DOI 10.1007/978-3-319-45117-6

© Springer International Publishing Switzerland 2017
Softcover reprint of the hardcover 1st edition 2016
This work is subject to copyright. All rights are reserved by the Publisher, whether the whole or part of the material is concerned, specifically the rights of translation, reprinting, reuse of illustrations, recitation, broadcasting, reproduction on microfilms or in any other physical way, and transmission or information storage and retrieval, electronic adaptation, computer software, or by similar or dissimilar methodology now known or hereafter developed.
The use of general descriptive names, registered names, trademarks, service marks, etc. in this publication does not imply, even in the absence of a specific statement, that such names are exempt from the relevant protective laws and regulations and therefore free for general use.
The publisher, the authors and the editors are safe to assume that the advice and information in this book are believed to be true and accurate at the date of publication. Neither the publisher nor the authors or the editors give a warranty, express or implied, with respect to the material contained herein or for any errors or omissions that may have been made.

Printed on acid-free paper

This Springer imprint is published by Springer Nature
The registered company is Springer International Publishing AG
The registered company address is: Gewerbestrasse 11, 6330 Cham, Switzerland

Preface

Continuing in the footsteps of the four previous International Conferences on Dynamics in Logistics, LDIC 2016 was the fifth event in this series to be held in Bremen (Germany) from February 22 to February 25, 2016, this year in conjunction with the 7th IFAC Conference on Management and Control of Production and Logistics—IFAC MCPL 2016. The LDIC 2016 was accompanied by an Internet of Things (IoT) Workshop as well as by the 1st LogDynamics Summer School (LOGISS) 2016 as satellite events. Similar to its predecessors LDIC 2007, LDIC 2009, LDIC 2012, and LDIC 2014, the Bremen Research Cluster for Dynamics in Logistics (LogDynamics) of the University of Bremen organized the conference in cooperation with the Bremer Institut für Produktion und Logistik (BIBA), which is a scientific research institute affiliated to the University of Bremen.

The conference was concerned with dynamic aspects of logistic processes and networks. The spectrum of topics reaches from modeling, planning and control of processes over supply chain management and maritime logistics to innovative technologies and robotic applications for cyber-physical production and logistic systems. The growing dynamic confronts the area of logistics with completely new challenges: it must become possible to describe, identify and analyze the process changes. Moreover, logistic processes and networks must be redevised to be rapidly and flexibly adaptable to continuously changing conditions. LDIC 2016 addressed scientists in logistics, operations research, production and industrial engineering, and computer science and aimed at bringing together researchers and practitioners interested in dynamics in logistics.

The LDIC 2016 proceedings consist of 43 papers selected by a strong reviewing process. The volume is organized into the main areas

- Distributed and Collaborative Planning and Control,
- Predictive Analytics and Internet of Things and Services,
- Technology Application in Logistics,
- Transport, Maritime and Humanitarian Logistics,
- New Business Models, and
- Frameworks, Methodologies and Tools.

There are many people whom we have to thank for their help in one or the other way. For pleasant and fruitful collaboration we are grateful to the members of the program and organization committees

Michael Bourlakis, Cranfield (UK)
Sergey Dashkovskiy, Erfurt (Germany)
Neil A. Duffie, Madison (Wisconsin, USA)
Enzo M. Frazzon, Florianópolis (Brazil)
Kai Furmans, Karlsruhe (Germany)
David B. Grant, Hull, Yorkshire (UK)
Axel Hahn, Oldenburg (Germany)
Kap Hwan Kim, Pusan (Korea)
Aseem Kinra, Copenhagen (Denmark)
Matthias Klumpp, Essen (Germany)
Antônio G.N. Novaes, Florianópolis (Brazil)
Kulwant S. Pawar, Nottingham (UK)
Alexander Smirnov, St. Petersburg (Russia)
Gyan Bahadur Thapa, Kathmandu (Nepal)
Dieter Uckelmann, Stuttgart (Germany)
Karl Worthmann, Ilmenau (Germany)

Carrying the burden of countless reviewing hours, we would like to thank our secondary reviewers Till Becker, Michael Beetz, Julia Bendul, Matthias Busse, Tobias Buer, Carmelita Görg, Hans-Dietrich Haasis, Otthein Herzog, Aleksandra Himstedt, Michael Hülsmann, Frank Kirchner, Herbert Kopfer, Hans-Jörg Kreowski, Walter Lang, Michael Lawo, Indah Lengkong, Marco Lewandowski, Rainer Malaka, Jens Pöppelbuß, Ingrid Rügge, Klaus-Dieter Thoben, Katja Windt for their help in the selection process. We are also grateful to Indah Lengkong, Marco Lewandowski, Ingrid Rügge, Tobias Sprodowski, Marit Hoff-Hoffmeyer-Zlotnik, Aleksandra Himstedt, and countless other colleagues and students for their support in the local organization and the technical assistance during the conference. Special thanks go to Tobias Buer and Till Becker for organizing the 1st LogDynamics Summer School as well as to Marco Lewandowski for organizing the Internet of Things (IoT) Workshop. Moreover, we would like to acknowledge the financial support by the BIBA, the Research Cluster for Dynamics in Logistics (LogDynamics), and the University of Bremen. Finally, we appreciate the excellent cooperation with the Springer-Verlag, which continuously supported us regarding the proceedings of all LDIC conferences.

Bremen, Germany
April 2016

Michael Freitag
Herbert Kotzab
Jürgen Pannek

Contents

Part I Distributed and Collaborative Planning and Control

1 Model-Based Specification and Refinement for Cyber-Physical Systems. 3
Rolf Drechsler, Serge Autexier and Christoph Lüth

2 Routing and Scheduling Transporters in a Rail-Guided Container Transport System 19
Xuefeng Jin, Hans-Otto Guenther and Kap Hwan Kim

3 The Influence of Manufacturing System Characteristics on the Emergence of Logistics Synchronization: A Simulation Study. 29
Stanislav M. Chankov, Giovanni Malloy and Julia Bendul

4 Safety Requirements in Collaborative Human–Robot Cyber-Physical System 41
Azfar Khalid, Pierre Kirisci, Zied Ghrairi, Jürgen Pannek and Klaus-Dieter Thoben

5 A Trust Framework for Agents' Interactions in Collaborative Logistics. 53
Morice Daudi, Jannicke Baalsrud Hauge and Klaus-Dieter Thoben

6 Quality-Aware Predictive Scheduling of Raw Perishable Material Transports 65
Xiao Lin, Rudy R. Negenborn and Gabriël Lodewijks

7 Process Maintenance of Heterogeneous Logistic Systems—A Process Mining Approach 77
Till Becker, Michael Lütjen and Robert Porzel

| 8 | A Synergistic Effect in Logistics Network | 87 |

Sergey Dashkovskiy, Petro Feketa, Christian Kattenberg and Bernd Nieberding

| 9 | A Step Toward Automated Simulation in Industry | 99 |

Himangshu Sarma, Robert Porzel and Rainer Malaka

Part II Predictive Analytics and Internet of Things and Services

| 10 | Application Potential of Multidimensional Scaling for the Design of DSS in Transport Insurance | 109 |

Victor Vican, Ciprian Blindu, Alexey Fofonov, Marta Ucinska, Julia Bendul and Lars Linsen

| 11 | Methodological Demonstration of a Text Analytics Approach to Country Logistics System Assessments | 119 |

Aseem Kinra, Raghava Rao Mukkamala and Ravi Vatrapu

| 12 | Big Data Analytics in the Maintenance of Off-Shore Wind Turbines: A Study on Data Characteristics | 131 |

Elaheh Gholamzadeh Nabati and Klaus Dieter Thoben

| 13 | Mitigating Supply Chain Tardiness Risks in OEM Milk-Run Operations | 141 |

Antonio G.N. Novaes, Orlando F. Lima Jr., Monica M.M. Luna and Edson T. Bez

| 14 | What Hinders the Implementation of the Supply Chain Risk Management Process into Practice Organizations? | 151 |

Pauline Gredal, Zsófia Panyi, Aseem Kinra and Herbert Kotzab

| 15 | The Fashion Trend Concept and Its Applicability to Fashion Markets and Supply Chains | 163 |

Samaneh Beheshti-Kashi

| 16 | Intelligent Packaging and the Internet of Things in Brazilian Food Supply Chains: The Current State and Challenges | 173 |

Ana Paula Reis Noletto, Sérgio Adriano Loureiro, Rodrigo Barros Castro and Orlando Fontes Lima Júnior

| 17 | Methodology for Development of Logistics Information and Safety System Using Vehicular Adhoc Networks | 185 |

Kishwer Abdul Khaliq, Amir Qayyum and Jürgen Pannek

| 18 | Power Management of Smartphones Based on Device Usage Patterns | 197 |

Lin-Tao Duan, Michael Lawo, Ingrid Rügge and Xi Yu

| 19 | Integration of Wireless Sensor Networks into Industrial Control Systems... | 209 |

T. Raza, W. Lang and R. Jedermann

| 20 | Advantages of Sub-GHz Communication in Food Logistics and DASH7 Implementation................................. | 219 |

Chanaka Lloyd, Sang-Hwa Chung, Walter Lang and Reiner Jedermann

Part III Transport, Maritime and Humanitarian Logistics

| 21 | Pre-selection Strategies for the Collaborative Vehicle Routing Problem with Time Windows............................... | 231 |

Kristian Schopka and Herbert Kopfer

| 22 | Transportation Planning with Forwarding Limitations .... | 243 |

Mario Ziebuhr and Herbert Kopfer

| 23 | Sustainable Urban Freight Transport: Analysis of Factors Affecting the Employment of Electric Commercial Vehicles ...... | 255 |

Molin Wang and Klaus-Dieter Thoben

| 24 | Evaluation of Practically Oriented Approaches for Operational Transportation Planning..................................... | 267 |

Heiko Kopfer, Herbert Kopfer, Benedikt Vornhusen and Dong-Won Jang

| 25 | Intelligent Transport Systems for Road Freight Transport—An Overview................................... | 279 |

Ilja Bäumler and Herbert Kotzab

| 26 | Imposing Emission Trading Scheme on Supply Chain........... | 291 |

Fang Li and Hans-Dietrich Haasis

| 27 | Planning of Maintenance Resources for the Service of Offshore Wind Turbines by Means of Simulation .... | 303 |

Stephan Oelker, Abderrahim Ait Alla, Marco Lewandowski and Michael Freitag

| 28 | Shopper Logistics Processes in a Store-Based Grocery-Shopping Environment .... | 313 |

Jon Meyer, Herbert Kotzab and Christoph Teller

| 29 | Empty Container Repositioning from a Theoretical Point of View... | 325 |

Stephanie Finke

| 30 | Frugal and Lean Engineering: A Critical Comparison and Implications for Logistics Processes .... | 335 |

Eugenia Rosca and Julia Bendul

31 Logistics Dynamics and Demographic Change 347
    Matthias Klumpp, Hella Abidi, Sascha Bioly, Rüdiger Buchkremer,
    Stefan Ebener and Gregor Sandhaus

Part IV New Business Models

32 Future Logistics: What to Expect, How to Adapt 365
    Henk Zijm and Matthias Klumpp

33 Concept and Diffusion-Factors of Industry 4.0 in the Supply
    Chain 381
    Hans-Christian Pfohl, Burak Yahsi and Tamer Kurnaz

34 On Upper Bound for the Bottleneck Product Rate Variation
    Problem 391
    Shree Ram Khadka and Till Becker

35 Crowdsourcing in Logistics: An Evaluation Scheme 401
    Matthias Klumpp

36 Modularization of Logistics Services—An Investigation
    of the Status Quo 413
    Aleksander Lubarski and Jens Pöppelbuß

Part V Frameworks, Methodologies and Tools

37 Toward a Unified Logistics Modeling Language: Constraints
    and Objectives 425
    Michael Freitag, Martin Gogolla, Hans-Jörg Kreowski,
    Michael Lütjen, Robert Porzel and Klaus-Dieter Thoben

38 Potential of Improving Truck-Based Drayage Operations
    of Marine Terminals Through Street Turns 433
    Niklas Nordsieck, Tobias Buer and Jörn Schönberger

39 Inventory Routing Problem for Perishable Products by
    Considering Customer Satisfaction and Green Criteria 445
    Mohammad Rahimi, Armand Baboli and Yacine Rekik

40 Decentralized Routing of Automated Guided Vehicles
    by Means of Graph-Transformational Swarms 457
    Larbi Abdenebaoui and Hans-Jörg Kreowski

41 Interoperability of Logistics Artifacts: An Approach for
    Information Exchange Through Transformation Mechanisms 469
    Marco Franke, Till Becker, Martin Gogolla, Karl A. Hribernik
    and Klaus-Dieter Thoben

42	Airflow Behavior Under Different Loading Schemes and Its Correspondence to Temperature in Perishables Transported in Refrigerated Containers .... Chanaka Lloyd, Reiner Jedermann and Walter Lang	481
43	VANET Security Analysis on the Basis of Attacks in Authentication... Nimra Rehman Siddiqui, Kishwer Abdul Khaliq and Jürgen Pannek	491

Part I
Distributed and Collaborative Planning and Control

Chapter 1
Model-Based Specification and Refinement for Cyber-Physical Systems

Rolf Drechsler, Serge Autexier and Christoph Lüth

Abstract *Cyber-physical systems* are small yet powerful systems which are embedded into their environment, adapting to its changes and at the same controlling it, and often operating autonomously. These systems have reached a level of complexity that opens up new application areas, but at the same time strains the existing design flows in system development. To ameliorate this problem, we propose a novel design flow for cyber-physical systems by adapting model-based specification and refinement methods known from software development. The design flow allows to start with a system specification and its essential properties at a high level of abstraction, and gradually refines it down to an electronic system level. Properties of higher levels can be inherited during refinements to lower levels by relying on local proof obligations only, which results in a design flow capable to keep up with the incre asing complexity of cyber-physical systems.

1.1 Introduction

Embedded systems have become powerful devices which play an increasingly important role in many areas such as production, transport, medicine and logistics: they control autonomous vehicles, aeroplanes, trains and other transport systems, they run production systems up to whole industrial plants, or they can be found in medical implants. We call these *cyber-physical systems*: embedded systems which are

---

Research supported by BMBF grant 01IW13001 (SPECifIC) and by the German Research Foundation (DFG) within the Reinhart Koselleck project grant DR 287/23-1.

R. Drechsler (✉) · S. Autexier
Deutsches Forschungszentrum Für Künstliche Intelligenz (DFKI), Bremen, Germany
e-mail: drechsler@uni-bremen.de

S. Autexier
e-mail: serge.autexier@dfki.de

C. Lüth
Universität Bremen, FB 3—Mathematics and Computer Science, Bremen, Germany
e-mail: christoph.lueth@dfki.de

© Springer International Publishing Switzerland 2017
M. Freitag et al. (eds.), *Dynamics in Logistics*, Lecture Notes in Logistics,
DOI 10.1007/978-3-319-45117-6_1

connected to the Internet and thus merge the boundaries of the virtual and physical world, which are autonomous, and adapt to and control their environment.

Cyber-physical systems are often employed in safety-critical situations, where failure is not an option and their correct functioning is of paramount importance; however, their complexity strains the currently existing design flows in system development, and makes this correctness hard to guarantee. This paper presents first steps towards a novel design flow for cyber-physical systems by applying methods from model-based software engineering.

Existing design flows model the system on the so-called Electronic System Level (ESL) using languages such as SystemC The IEEE Computer Society (2011) or System Verilog. These system level descriptions hide details of the precise realisation in hardware and software while still allowing the execution and simulation of the design. They are executable, concrete models which do not allow to state the desired properties of the system abstractly and formally tractable. However, the initial system specification is mostly given informally in natural language which is typically far away from the ESL. To bridge this gap in expressiveness, the Formal Specification Level (FSL) has been proposed (Drechsler et al. 2012; Soeken and Drechsler 2015). Its aim is to close this gap by employing formal descriptions means such as the UML to give an abstract specification of the system. It has been used for error-detection in early design phases Seiter et al. (2014) and to provide a notion of refinement for operations at the FSL Przigoda et al. (2015).

The contributions of this paper are twofold: first, to develop a semantics of the FSL which was lacking so far and enables to formulate system properties at the FSL. Second, it provides provably correct notions of refinement guaranteeing that system properties are preserved under refinement.

This paper is structured as follows: Illustrated by a simple access control system example, Sect. 1.2 introduces the specification formalisms and semantics for the central Formal Specification Level (FSL) as well as the kind of correctness properties that can be expressed and proven on that basis. Section 1.3 defines the methods to refine an abstract system specification to more concrete specifications. The refinements are defined such that more detailed specifications inherit properties already established at higher levels of abstraction and are illustrated by the running example. They allow to gradually add more details to the specification until it eventually contains all details of an executable implementation at the Electronic System Level (Sect. 1.4). Section 1.5 compares the contributions with related work and concludes the paper.

1.2 Introducing the FSL

As a running example, we consider an access control system which governs the access of people to buildings connected by gates, originally modelled by Abrial (1999) using the B method Abrial et al. (2005). We start with very high-level specifications describing the overall behaviour.

An FSL model consists of classes and operations. The operations can be restricted by OCL constraints, which are either invariants constraining all operations of the class, or pre-/postconditions for specific operations.

In the initial specification (*ACS-1*), the model contains only persons and buildings (given by the two classes). Each person is authorised to enter a number of buildings (attribute `aut`), and will always be in exactly one building (attribute `sit`). The class invariant P5 models the requirement specification that a person must only be in a building where they are allowed to be. The pre- and postcondition on the `enter` operation specify that to enter a building, a person has to be authorised for the building, and that he is not allowed to enter the building he is currently in.

In a first refinement (*ACS-2*), still on a very abstract level we introduce connections between buildings (given by a UML association). This constrains the `pass` operation: persons can only go from one building in which they are into another if they are authorised to do so, and if the buildings are connected.

Figure 1.1 shows the first two steps of our example. It gives a good feel on how system development can start at a very abstract level. At this level, it is clear that the central safety invariant is adequate. We cannot prove it yet, but it gives a good idea on how it could be proven: if we can show that all operations preserve it, and that it holds in the initial states.

```
class Building {}

class Person {
  op void pass(Building b);
  ref Building [*] aut;
  ref Building [1] sit ;
}

context Person
  inv P5: self.aut→includes(self.sit)

context Person::pass(b: Building):
  pre pass_pre1: self.aut→ includes(b)
  pre pass_pre2: self.sit ≠b
  post pass_post: self.sit =b
```

ACS-1

```
class Building {
  ref Building [*] #building gate;
  ref Building [*] #gate building;
}

class Person {
  op void pass(Building b);
  ref Building [*] aut;
  ref Building [1] sit ;
}

context Person
  inv P5: self.aut→includes(self.sit)
context Building
  inv P7: not (self.gate→includes(self))
context Person
  inv P10: self.aut→ forAll(b|self.aut.building→includes(b))
context Person::pass(b: Building):
  pre pass_pre1: self.aut→ includes(b)
  pre pass_pre2: self.sit.gate→ includes(b)
  post pass_post: self.sit =b
```

ACS-2

Fig. 1.1 Initial specification (*ACS-1*), and first refinement step (*ACS-2*), of our running example. For UML class diagrams, we use the textual EMFatic The Eclipse (2012) notion from the Eclipse Modelling Framework (EMF)

1.2.1 Semantics

Technically, an FSL specification is given by a tuple

$$SP = \langle \mathcal{M}, \mathit{init}, \textsc{Opn}, \mathit{inv}, \mathit{pre}, \mathit{post}, \mathit{st} \rangle$$

where $\mathcal{M}$ are the classes (specifically, the *object model* Richters and Gogolla 2002), *init* is a specification of the initial states of the system, OPN the operations, *inv* the class invariants, *pre* the preconditions and *post* the postconditions of the operations, and *st* the state diagrams. We do not consider state diagrams in their full UML generality (in particularly, we do not allow hierarchical states and concurrent regions). Instead, we allow state diagrams which can be encoded into pre- and postconditions on the class operations: if the operation f of some class corresponds to a transition from state d to a state r (denoted $f : d \to r$) in the state diagram associated to that class, we encode this in the pre- and postconditions using an additional attribute m_state as follows: pre s: **self** .m_state = d and **post** ps: **self** .m_state = r.

The semantics of *SP* consists of the states given by the object model $\mathcal{M}$, and state transitions given by the operations. We add a special state *STOP*, which corresponds to the system state from which no further transitions are possible, and thus models deadlock. This amounts to a *Kripke structure* $[\![SP]\!] = \langle S, I, \to \rangle$ where S is the set of states and I the initial set of states that satisfy *init* and the invariants *inv*. The transition relation $\to = \{\to_o\}_{o \in \textsc{Opn}}$ is a family of transitions labelled by operations; there is a transition $\sigma_1 \to_o \sigma_2$ iff

(i) all invariants hold in σ_1 and σ_2,
(ii) the preconditions of o are satisfied in σ_1, and
(iii) the postconditions of o are satisfied in σ_1 and σ_2;
(iv) if there is no outgoing transition from σ_1 by the previous clauses for any σ_2, we add a transition $\sigma_1 \to STOP$.

The invariants and pre-/postconditions of an operation are the *constraints* of o and denoted by $cons(o)$. For a specification *SP*, an execution trace is a infinite sequence $s = \langle s_i \rangle_{i \in \mathbb{N}}$ of states such that $s_i \to s_{i+1}$.

1.2.2 Verification Properties

Given a specification such as above, we want to exhibit certain properties of the modelled system. For example, the invariant P5 is a property we want the system to have. We call these properties *verification properties*. In its simplest form, a verification property is an OCL property, which is required to hold in all states of the execution trace of a system. This corresponds to the temporal $\square$ operator; we typically specify safety properties this way ('Something bad never happens'). We might require that other properties only hold at some point in the future, corresponding to the temporal

◊ operator; these are typically liveness properties ('Something good will eventually happen'). Generalising slightly further, we define verification properties as follows:

Definition 1 (*Verification Property*) The set of *verification properties* is defined as follows. Let ϕ be an OCL formula without the `@pre` postfix, then

(i) ϕ is a verification property,
(ii) $\Box\phi$ is a verification property ('safety'),
(iii) $\Diamond\phi$ is a verification property ('liveness'),
(iv) $\Box\Diamond\phi$ is a verification property ('fairness'),
(v) $\Diamond\Box\phi$ is a verification property ('persistence').

The two additional verification properties are 'fairness' ($\Box\Diamond\phi$), which specifies that at every point in the system, ϕ will eventually hold, and 'persistence' ($\Diamond\Box\phi$), which requires that ϕ will at some point start to hold.

Example 1 Using our running example, we illustrate below the different kinds of verification properties. ∀b:Building . ∃b':Building . b.gate-> includes(b') is a verification property requiring for the initial state, that each building is connected to at least one other building. To express that each person eventually enters each building for which he has access is achieved by ∀p:Person . ◊(p.aut->includes(p.sit). To express that in each building there are always at least two persons is achieved by ∀b:Building.□∃ p,p':Person. p!= p' **and** p.sit=b **and** p'.sit=b.

Example 2 In the *STOP* state, every OCL expression (including True) evaluates to *False*. Thus, the verification formula □ True holds for all systems which do not reach the *STOP* state, i.e. it expresses freedom from deadlock.

1.3 Refinement in the FSL

In general terms, refinement is a property-preserving mapping from an 'abstract' model to a more 'concrete' one. Semantically, a refinement should restrict the possible implementations of the specification. As properties are defined in terms of traces in the Kripke structure, to preserve properties for each trace in the concrete model there needs to be a trace in the abstract one (Fig. 1.2).

1.3.1 Data Refinement

Data refinement allows to change the system state of our models. It can be constructed by mapping all classes of operations of the abstract model to the concrete one. Such

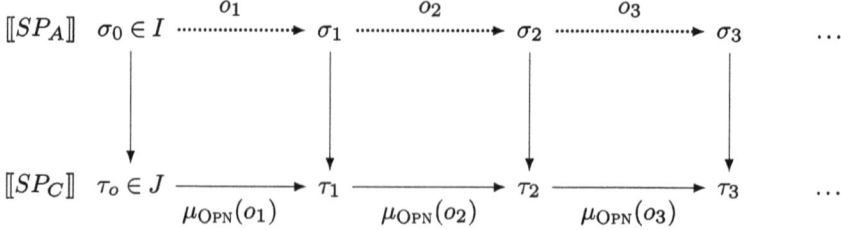

Fig. 1.2 Data refinement: for each trace in $[\![SP_C]\!]$, we can construct one in $[\![SP_A]\!]$

a mapping is given by a *model morphism*. Given two object models $\mathcal{M}, \mathcal{M}'$, a model morphism is a tuple of maps $\mu = \langle \mu_C, \mu_A, \mu_O, \mu_{\text{Assoc}} \rangle$ with $\mu_C, \mu_A, \mu_O, \mu_{\text{Assoc}}$ maps between the class names, attribute names and association names respectively, which preserve the type hierarchy, the types of attributes and operations, and the associations and their cardinality. Given such a model morphism, the *homomorphic extension* $\mu^{\#}$ maps OCL expressions ϕ in $\mathcal{M}$ to OCL expressions in $\mathcal{M}'$, by replacing all classes, attributes and associations in ϕ with their image under μ. A model morphism becomes a map between FSL specifications if it also preserves the initial states and, crucially, the invariants on states as well as the pre- and postconditions of the operations in the abstract model (translated appropriately) are implied by the ones in the concrete model:

Definition 2 (*Specification Morphism*) Given two FSL specifications $SP_A = \langle \mathcal{A}, I_A, \text{OPN}_A, inv_A, pre_A, post_A, st_A \rangle$ and $SP_C = \langle \mathcal{C}, I_C, \text{OPN}_C, inv_C, pre_C, post_C, st_C \rangle$, a *specification morphism* is a tuple of maps $\mu = \langle \mu_M, \mu_I, \mu_{\text{OPN}} \rangle$ where μ_M is a model morphism, and μ_I and μ_{OPN} are maps from the initial states and operations of SP_A to those of SP_C, satisfying:

$$\tau \in I_A \implies \mu_I(\tau) \in I_C \tag{1.1}$$

$$cons_C(\mu_{\text{OPN}}(o)) \implies \mu_M^{\#}(cons_A(o)) \tag{1.2}$$

If all initial states and operations in SP_C are in the image of μ, and the model morphism is injective on the state, we have a *data refinement* from SP_A to SP_C. The following lemma shows why data refinements are useful tools: they preserve all verification properties (in fact, all LTL properties).

Lemma 1 *Given two FSL specifications* $SP_A = \langle \mathcal{A}, I_A, \text{OPN}_A, inv_A, pre_A, post_A, st_A \rangle$ *and* $SP_C = \langle \mathcal{C}, I_C, \text{OPN}_C, inv_C, pre_C, post_C, st_C \rangle$, *and a specification morphism* $\mu : SP_A \to SP_C$. *If* μ_M *is injective, and* μ_I *and* μ_{OPN} *are surjective, then all verification properties which hold in* SP_A *hold in* SP_C *as well.*

Proof We first show that for any execution trace t in $[\![SP_C]\!]$, the Kripke structure induced by SP_C, there is a trace in $[\![SP_A]\!]$. This corresponds to the fact that for any transition $\sigma \to_o \sigma'$ in $[\![SP_C]\!]$ there is a transition $\tau \to_o \tau'$ in $[\![SP_A]\!]$. There is a transition $\sigma \to_o \sigma'$ in $[\![SP_C]\!]$ only if $cons_C(o')$ holds in σ and σ' for some o'.

Because μ_{OPN} is surjective, $o' = \mu_{\text{OPN}}(o)$ and by (1.2), $\mu_{\mathcal{M}}^{\#}(cons_A(o))$ holds in σ and σ'. Hence there are states τ, τ' such that $cons_A(o)$ holds, and there is a transition $\tau \rightarrow_o \tau'$ as required.

Now given any LTL property ϕ, ϕ holds for SP_C if it holds for all traces in $[\![SP_C]\!]$. Assume ϕ holds for SP_A. To show ϕ holds for SP_C, we have to show ϕ holds for all traces in SP_C. But for any such trace we have shown there exists a trace in $[\![SP_A]\!]$, for which ϕ holds by assumption. □

Example 3 (Running Example) Continuing our running example, Fig. 1.1 shows a simple example of data refinement. The model morphism is the injection from *ACS-1* to *ACS-2*, and thus by construction injective; it is easy to see that it is surjective on the operations. It remains to show that the model morphism is a specification morphism to be able to apply Lemma 1. For demonstration, we will show here in detail how syntactic proof obligations are derived, and how they are proven.

From Definition 2 it follows that we need to show that $cons_{ACS-1}(o) \implies cons_{ACS-2}(o))$ for all operations. Since there is only one operation pass, we have to show[1]

ACS2.P5 **and** ACS2.P7 **and** ACS2.P10 **and**
ACS2.pass_pre1 **and** ACS2.pass_pre2 **and** ACS2.pass_post
implies ACS1.P5 **and** ACS1.pass_pre1 **and** ACS1.pass_pre2 **and** ACS1.pass_post

To show this proof obligation, we break it down into four obligations using conjunction introduction. We note that except for pass_pre2 all the other axioms are the same in *ACS-1* and *ACS-2*, hence three of the resulting proof obligations are trivial, like the following:

P5 **and** P7 **and** P10 **and** pass_pre1 **and** ACS2.pass_pre2 **and** pass_post
implies P5

The remaining one we need to show is ACS1.pass_pre2. To do so, we need to unfold the axioms P7 and ACS2.pass_pre2. (In order to avoid overcrowding, we drop the unneeded axioms P5, P10, ACS2.pass_pre1 and pass_post).[2]

∀**self@pre**: Building. **not** (**self@pre**.gate→ includes(**self@pre**))
and self@pre.sit.gate→ includes(b)
implies self@pre.sit.gate $\neq$ b

Note that the invariant is universally quantified over all buildings, because that is its context (we omit the outer universal quantifiers in the conclusion). Eliminating the quantifier in the premise and instantiating **self@pre** with b results in

not (b.gate→ includes(b)) **and self@pre**.sit.gate→ includes(b)
implies self@pre.sit.gate $\neq$ b

---

[1] In this ad-hoc notation we replace axioms by their name, and use qualified notation s.a to refer to axiom a from specification s.
[2] For readability, we use the notation ∀a:C. p instead of the correct allInstances(C)->forAll(a|p).

This is proven by reductio ad absurdum: assume **self@pre.sit.gate=b**, then the second premise reduces to b.gate->includes(b), and together with the first premise we can derive False, thus proving **self@pre**.sit.gate ≠ b.

1.3.2 Operation Refinement

Data refinement allows changes in the state space; the typical use cases are introduction of new classes or attributes. If we want to replace an abstract operation by more concrete ones, or if we want to move an operation from one class to another, the restrictions of Definition 2 are too prohibitive. In this case, we need *operation refinement*.

Operation refinement decomposes the state space of the refining model SP_C such that the additional states are used for the refining operations (Fig. 1.3). This requires some restrictions on the refining operations, in particular in how they are allowed to be composed. We express these restrictions as a particular form of UML state diagrams.

When refining an abstract operation $f : d \to r$ of an FSL specification, the following LTL properties that must be proven about the refining state machine $\mathcal{S}$:

(A) the invariants *inv* of the FSL specification must also hold for $\mathcal{S}$, hence we need to prove $\Box inv$ for $\mathcal{S}$ (see also the condition of case (2) in Lemma 2);
(B) the preconditions $pre(f)$ must hold for the initial states;
(C) the postconditions $post(f)$ relate values in the state before and the state after f. This can be reformulated as an LTL property for $\mathcal{S}$ by introducing variables in a straight forward, though technical manner. Instead of providing a formal definition, we illustrate the construction by the following example postcondition:

self@pre.p < **self@pre**.q and **self@pre**.q =**self**.q * 2

This postcondition can be reformulated to the LTL formula

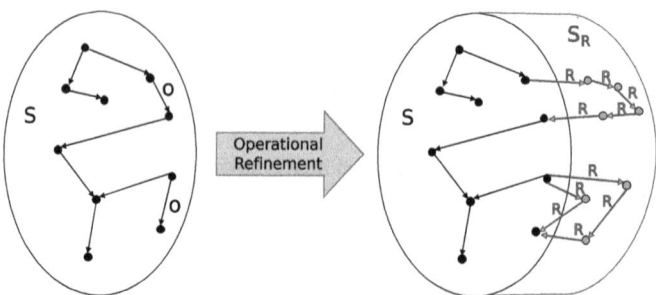

Fig. 1.3 Operation refinement of the Kripke structure $\mathcal{K}_A$ to $\mathcal{K}_C$: in each trace in $\mathcal{K}_C$, sequences of steps starting in I_R, going through S'_R and ending in F_R can be mapped to one step in S

$$\forall x{:}\mathsf{Int} \ . \ (x = \textbf{self}.q \ \textbf{and} \ \textbf{self}.p < \textbf{self}.q) \ \textbf{implies} \ \Diamond\Box \ x = \textbf{self}.q * 2$$

Requiring S to have that property transfers the postconditions to the refining state machine. Essentially, it means the refining state machine S must at some point make the postcondition true such that from that on it holds until S terminates.

We denote by $cons_\Box(f)$ the conjunction of the above LTL properties (A)–(C). Additionally, the refining state machine S must be always terminating, which can be formulated as a proof obligation about S and be proven by respective methods from termination analysis. Finally, all traces in S must start in d and terminate in r, which we formulate as an additional requirement to be proven.

Definition 3 (*Operation Refinement*) Given an FSL specification $SP_A = \langle \mathcal{M}, I, \{f : d \to r\} \uplus \text{OPN}_A, inv_A, pre_A, post_A, st_A \rangle$ an *operation refinement* of SP_A wrt. f is given by a FSL specification $SP_C = \langle \mathcal{M}, I, \text{OPN}_c, inv_C, pre_C, post_C, st_C \rangle$, new operations OPN_{C0} over the states T_{C0} such that:

(i) The states T_{C0} are disjoint from the states of SP_A,
(ii) $\text{OPN}_C = \text{OPN}_{C0} \uplus \text{OPN}_A$, and for all $o \in \text{OPN}_A$, $cons_C(o) = cons_A(o)$
(iii) the state machine $S_{\text{OPN}_{C0}}$ induced by OPN_{C0} is with initial state d terminating with final state r, and $S \models cons_\Box(f)$

Remark 1 This includes a specific case where an abstract operation f is refined by one concrete operation g which does not preserve the types and hence cannot be expressed by specification morphisms. In that case the proof obligation simplifies to $cons_C(g) \implies cons_A(f)$.

Lemma 2 *Let SP_C be an operational refinement of an operation o of an FSL specification SP_A using the operations OPN and P, Q OCL-formulas without @pre. The following verification properties of SP_A are preserved by operation refinement to SP_C:*

(i) $\Diamond P$
(ii) $\Box P$ *if, and only if, $S_{\text{OPN}} \models \Box P$, i.e. the refining structure satisfies $\Box P$*
(iii) $\Box \Diamond P$
(iv) $\Diamond \Box P$ *if $S_{\text{OPN}} \models P \Rightarrow \Box P$ (note that this is not an equivalence).*

Proof First we observe, that for each trace of τ_C of the concrete FSL specification SP_C we can construct a trace of SP_A by replacing all maximal finite traces of the refining state machine S_{OPN} which occur in τ_C by a single transition $\to_o$ of the refined operation o of SP_A. As a consequence all states in τ_A occur in the same order in τ_C, except that they may be interspersed with states from S_{OPN}. Thus, if $\Diamond P$ or $\Box \Diamond P$ hold for τ_A, they also hold for τ_C, which proves (i) and (iii). Assume $\Box P$ holds for τ_A we need to know that P also holds for all inserted states from S_{OPN}, which is ensured by the proof obligation $\Box P$ for the S_{OPN}, proving (ii). Finally, assume $\Diamond \Box P$ holds for τ_A; this means that there is an infinite suffix of the trace τ_A for which $\Box P$ holds. Now consider a state s in τ_C at which a trace from S_{OPN} starts: either it is before the suffix, then $\Box P$ will be reached later in the trace; or it is in the suffix, then P holds

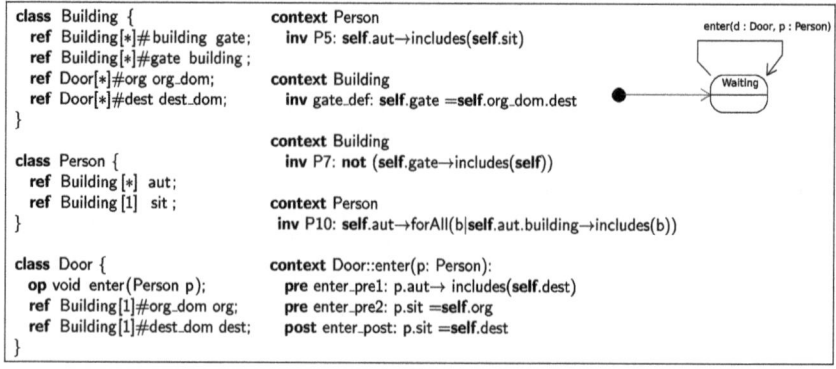

Fig. 1.4 Third refinement: the class Door is introduced to connect buildings, and the pass operation is implemented by the enter operation from the new class

for s and because $P \Rightarrow \Box P$ holds for $\mathcal{S}_{\text{OPN}}$, the rest of the sequence in $\mathcal{S}_{\text{OPN}}$ satisfies $\Box P$, hence $\Box P$ holds, proving (iv). □

Example 4 (Running Example) We now consider the refinement step from *ACS-2* to *ACS-3* (see Fig. 1.4), which consists of moving the pass operation from the class Person to the class Door, where it is called enter. This refinement cannot be expressed with a specification morphism, because the refinement step actually consists of two refinements: first, a data refinement which introduces the class Door, and then an operational refinement which moves the operation pass from the class Person to the operation enter in Door. It is thus an example of a simple operational refinement mentioned in Remark 1. For this refinement, we have to show that the pre- and postconditions on enter imply those on pass. This means we have to prove that

(∀**self**: Door ∀p: Person. p@pre.aut→ includes(**self**@pre.dest) **and**
p@pre.sit= **self**@pre.org **and** p.sit= **self**.dest)
implies (∀**self**: Person ∀b: Building. **self**@pre.aut→ includes(b@pre) **and**
self@pre.sit.gate→includes(b@pre) **and**
self.sit= b)

This proof is a bit more delicate. We start by expanding the universal quantification of the goal by substituting **self**: Person and b: Building with indefinite constants p0 and b0. Conversely, in the premise the universally quantified variable **self**: Door is instantiated with b0.dest_dom and p:Person with p0[3] to obtain the following three subgoals:

---

[3] Note that substituting **self** by b0.dest_dom in **self@pre**.dest moves the suffix **@pre** inside to result in b0@pre.dest_dom.dest.

(a) `p0@pre.aut-> includes(b0@pre.dest_dom.dest)`
    and `p0@pre.sit= b0@pre.dest_dom.org` **and** `p0.sit= b0.dest_dom.dest)`
    implies `p0@pre.aut-> includes(b0)`
(b) `p0@pre.aut-> includes(b0@pre.dest_dom.dest)`
    and `p0@pre.sit= b0@pre.dest_dom.org` **and** `p0.sit= b0.dest_dom.dest)`
    implies `p0@pre.sit.gate->includes(b0)`
(c) `p0@pre.aut-> includes(b0@pre.dest_dom.dest)`
    and `p0@pre.sit= b0@pre.dest_dom.org` **and** `p0.sit= b0.dest_dom.dest)`
    implies `p0.sit= b0`

To show the first conjoint (4), knowing that `b0@pre = b0`, `p0@pre.aut= p0.aut`, `b0@pre.dest_dom=b0.dest_dom`, and `b0.dest_dom.dest= b0` this simplifies to

(p0.aut→ includes(b0) **and** p0@pre.sit= b0.dest_dom.org **and** p0.sit= b0) **implies** p0.aut→ includes(b0)

which holds trivially.

Applying the same simplifications to (4) results in

(p0.aut→ includes(b0) **and** p0@pre.sit= b0.dest_dom.org **and** p0.sit= b0) **implies** p0@pre.sit.gate→includes(b0)

By applying `p0@pre.sit= b0.dest_dom.org` we obtain the goal

(p0.aut→ includes(b0) **and** p0@pre.sit= b0.dest_dom.org **and** p0.sit= b0) **implies** b0.dest_dom.org.gate→includes(b0)

Now using the invariant `gate_def` we can further transform the goal to

(p0.aut→ includes(b0) **and** p0@pre.sit= b0.dest_dom.org **and** p0.sit= b0) **implies** b0.dest_dom.org.org_dom.dest→includes(b0)

This follows because `org` is inverse to `org_dom` and further `dest` is inverse to `dest_dom`, and some further technical reasoning about the OCL semantics of collections.[4]

Finally, to show the third conjoint (c) we also apply the same simplifications to obtain

(p0.aut→ includes(b0) **and** p0@pre.sit= b0.dest_dom.org **and** p0.sit= b0) **implies** p0.sit= b0

which holds trivially.

---

[4]The relevant properties are ∀d:Door . d.org.org_dom->includes(d) and ∀b:Building . b.dest_dom.dest->includes(b).

The final step within the FSL towards the ESL as in Fig. 1.6 is to refine the `enter` operation in the `Door` class. If the person entering the door is allowed to pass, a green light should indicate this; after the person has passed, or at most after 30 s, the green light should be off again. If the person entering is not allowed to pass, a red light should be lit for 2 s, and the door should stay locked.

To model the relevant aspects of this behaviour (we do not model the timing aspects here), we need to add attributes for the green and red lights, and introduce operations such as `accept`, `pass_thru`, `green_off` which model the desired state transitions. The state diagram modelling the desired behaviour is on the right of Fig. 1.5. As new state names, we introduce `Waiting`, `Refusing` and `Accepting`. We incur three proof obligations, corresponding to the pre- and postconditions of the `enter` operation which have to be derived from the conjunction of pre- and postconditions of the `accept` and `refuse` operations and the invariants. The proofs are more elaborate than the ones we have seen so far, but mathematically routine. We further have to prove that the refining operations induce a state machine, which starting in state `Waiting` always terminates in state `Waiting`.

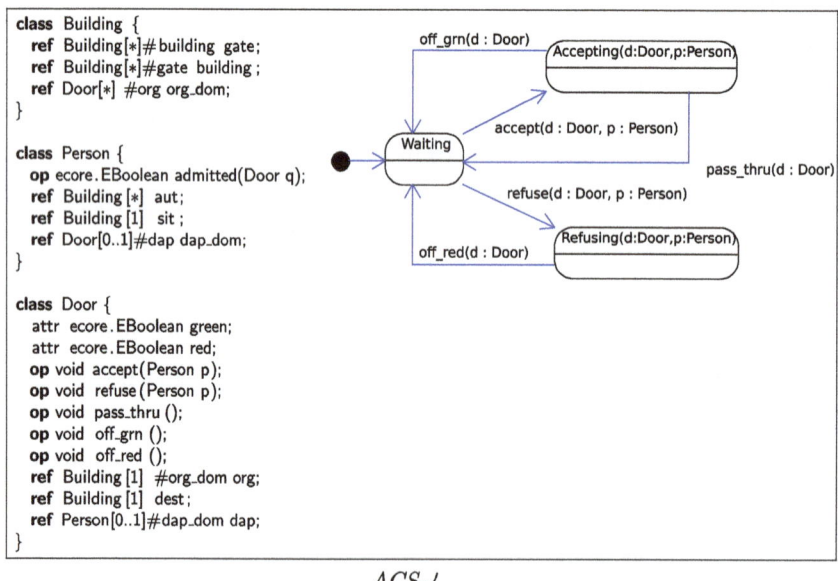

Fig. 1.5 In this step, we refine the `enter` operation by introducing several new operations, and describe their behaviour in a UML state machine (*right*)

1 Model-Based Specification and Refinement for ...   15

```
class Users {                          SC_MODULE(Door)              void operate()
private:                               {                            {
  static std::unordered_set<Building>    sc_in<Person> card;          while (true) {
       aut[NUM_PERSONS];                 sc_out<bool> green;            wait(card.value_changed_event());
  static Building sit[NUM_PERSONS];      sc_out<bool> red;              dap= card.read();
                                                                       if (Users::admitted(dap, dest))
public:                                public:                         {
  static void init()                     Building org;                   accept();
  {                                      Building dest;                  wait(sc_time(TIMEOUT_GREEN, TIME_UNIT),
    for (Person p= 0;                                                               passed.posedge_event());
         p< NUM_PERSONS; p++)           private:                         if (passed.read())
    {                                    Person dap= NO_PERSON;            pass_thru();
      aut[p]=                                                            else
         std::unordered_set<Building>(); public:                           off_grn();
      sit[p]= DEFAULT_BUILDING;          SC_CTOR(Door)                  }
    }                                    {                              else
  }                                        dap= NO_PERSON;              {
                                           SC_THREAD(operate);            refuse();
  static bool admitted(Person p,         }                                wait(TIMEOUT_RED, TIME_UNIT);
                       Building b)     }                                  off_red();
  {                                                                     }
    return (aut[p].count(b) > 0);                                     }
  }                                                                 }
};
```

Fig. 1.6 Implementation of the example in the ESL. We show the relevant excerpts, the actual SystemC implementation about 360 loc

1.4 Refinement from the FSL to the ESL

The ESL language we are using here is SystemC, a C++ class library which allows a cycle-accurate model of the hardware. It provides classes to simulate the hardware constructs; these can be mixed in with usual C++ to allow hardware–software co-design.

In brief, the class `sc_module` models the basic building blocks of the hardware. The signals going in and out of such a block are modelled by generic datatypes `sc_in<T>` and `sc_out<T>`.

To map FSL specifications to the ESL, we map classes in the FSL to either instances of the `sc_module` class (if they are implemented in hardware), or to usual C++ classes (if they are implemented in software). For the former, public attributes are mapped to signals, corresponding to the fact that they can be read or written. Methods are mapped to function members of type **void** `f()`; all parameter and result passing must be performed by reading and writing to attributes (i.e. signals). An instance of `sc_module` must have a main thread, which must implement the state diagram of the FSL specification.[5]

Example 5 (Running Example: Finale) Fig. 1.6 shows an excerpt of the ESL implementation. The operation `admitted` is implemented in software, by the `Users` class (left of Fig. 1.6). The class `Door` is implemented by the `Door` instance of `sc_module`; Fig. 1.6 shows the declaration of the instance, and the implementation of the main method `operate`.

At this point, we do not aim to verify formally that the SystemC code satisfies the pre- and postconditions and preserves invariants, as formal verification of C++ is still very much a research problem in its own right. There are various methods

---

[5] As we currently do not consider concurrency, there can be only one state diagram, and hence only one main thread.

by which we can *validate* the pre- and postconditions and invariant preservation to a certain degree, i.e. check for obvious violations or contradictions (Przigoda et al. 2015; Stoppe et al. 2014).

1.5 Conclusions and Outlook

We have presented the first steps towards a new design flow for cyber-physical systems in order to keep up with the rapidly increasing complexity of these systems. It applies methods from model-based software engineering and allows to start from an abstract specification which states the essential properties down to an implementation in SystemC using refinement steps which have been proven correct with respect to a comprehensive semantics based on Kripke structures. It thus bridges the gap in expressiveness between system-level modelling languages such as SystemC and initial specifications in natural language leveraging the advantages of the UML and its existing tools without forcing system engineers to give up their existing design flow.

We have developed two refinement operations in Sect. 1.3, but these should be seen as representative. There are many more possible refinement operations (e.g. an obvious one that comes to mind is to remove attributes), which may be uncovered by further case studies. Also, the current approach does not allow to handle concurrency, which is another major area for future work.

In related work, there are a number of so-called wide-spectrum languages which cover the whole of the design flow. For example, our running example was originally conceived by Abrial for the B language Abrial et al. (2005). Atelier B, the tool supporting B, covers the whole design flow (and there is a connection to VHDL called BHDL), but it does not easily allow designers to keep their existing work flow, and has a steeper learning curve than our UML-based approach. The advantage of using UML is that engineers can start with a light-weight modelling, using just a few UML diagrams (class diagrams with OCL pre- and postconditions and invariants as in our example). Another relevant language is Event-B, an extension of B with events, which is supported by the Rodin tool chain Abrial et al. (2010). There is work on UML-B Snook and Butler (2006), a UML front-end for Event-B which even supports a notion of refinement Said et al. (2009), which essentially has the same aim as our approach, namely allowing the engineer to use well-known UML concepts rather than having to learn a new specification language, except it uses Event-B as the semantic 'back-end'; our use of Kripke structures makes it easier to connect to a completely separate ESL language such as SystemC.

We have implemented a prototypical tool support for our approach using tools from the Eclipse Modelling Framework (EMF) for the UML, and the LLVM toolset (clang) for SystemC. Our tool can automatically find the obvious mappings between the refinement steps, but has no proof support for proof obligations arising in the development. Its aim is also to be able to track the impact of changes in the development e.g. if we find during validation that the initial specification is not adequate.

In closing, we are confident this work will bring the benefits of model-based software engineering into the systems development community by combining the advantages of existing industrial-scale system design flows with well-known model-based specification and refinement concepts.

References

Abrial JR (1999) System study: method and example. http://atelierb.eu/ressources/PORTES/Texte/porte.anglais.ps.gz

Abrial JR, Hoare A (2005) The B-book: assigning programs to meanings. Cambridge University Press

Abrial JR, Butler M, Hallerstede S, Hoang TS, Mehta F, Voisin L (2010) Rodin: an open toolset for modelling and reasoning in event-B. International J Soft Tools Technol Transf 12(6):447–466

Drechsler R, Soeken M, Wille R (2012) Formal specification level: towards verification-driven design based on natural language processing. In: Forum on specification and design languages (FDL) 2012, IEEE, pp 53–58

Przigoda N, Stoppe J, Seiter J, Wille R, Drechsler R (2015) Verification-driven design across abstraction levels—a case study. In: Euromicro conference on digital system design (DSD)

Przigoda N, Wille R, Drechsler R (2015) Contradiction analysis for inconsistent formal Models. In: International symposium on design and diagnostics of electronic circuits and systems (DDECS)

Richters M, Gogolla M (2002) OCL: syntax, semantics, and tools. In: Clark T, Warmer J (eds) Object Modeling with the OCL, no. 2263 in Lecture Notes in Computer Science. Springer, pp 42–68

Said MY, Butler M, Snook C (2009) Class and state machine refinement in UML-B. In: Proceedings of workshop on integration of model-based formal methods and tools (associated with iFM 2009)

Seiter J, Wille R, Kühne U, Drechsler R (2014) Automatic refinement checking for formal system models. In: Forum on specification and design languages (FDL), pp 1–8

Snook C, Butler M (2006) UML-B: formal modeling and design aided by UML. ACM Trans Softw Eng Methodol 15(1):92–122

Soeken M, Drechsler R (2015) Formal specification level. Springer

Stoppe J, Wille R, Drechsler R (2014) Validating systemc implementations against their formal specifications. In: Symposium on integrated circuits and system design (SBCCI)

The Eclipse Foundation (2012) Emfatic: a textual syntax for EMF ecore (meta-)models. http://www.eclipse.org/emfatic

The IEEE Computer Society (2011) Standard systemc language reference manual. In: IEEE standard 1666–2011

Chapter 2
Routing and Scheduling Transporters in a Rail-Guided Container Transport System

Xuefeng Jin, Hans-Otto Guenther and Kap Hwan Kim

Abstract One of the important issues in container terminals for the efficient operation of a rail-based transport system, called "flat car system (FCS)," is how to determine the route and travel schedule of each flat car (FC). It is assumed that a flat car may have to reserve multiple resources simultaneously at a moment during its travel. Example of the resources are transfer points (TPs) and intersections (ISs) on guide-path network. A travel-scheduling algorithm based on Dijkstra's algorithm is suggested using the concept of time-windows.

Keywords Transporters · Container terminal · Routing and scheduling

2.1 Introduction

When applying an FCS to container deliveries, the following problems must be studied: (1) guide-path design; (2) a method for dispatching FCs; (3) an FC routing or routing-scheduling method for FCs; and (4) traffic management. The routing and travel-scheduling methods for FCs are the issue of this research.

Routing implies the selection of a path for an FC to travel from its origin to its destination, while the travel-scheduling of an FC is the determination of not only the path for the FC from the origin to the destination but also the time-points for an FC to enter and exit from every segment on the path.

X. Jin (✉) · K.H. Kim
Department of Industrial Engineering, Pusan National University, Jangjeon-dong 2, Geumjeong-gu, Busan 609-735, Korea, South Korea
e-mail: hbkim@pusan.ac.kr

K.H. Kim
e-mail: kapkim@pusan.ac.kr

H.-O. Guenther
Institute of Logistics Innovation and Networking, Pusan National University, Jangjeon-dong 2, Geumjeong-gu, Busan 609-735, Korea, South Korea
e-mail: hans-otto.guenther@hotmail.de

Usually, FCs travel on the floor, which has a network consisting of TPs and ISs. Designers of an FCS usually predetermine allowable paths on the floor and thus, the allowable paths can be represented by a guide-path network in which nodes represent TPs and ISs and arcs represent path segments between nodes. When a guide-path network and the origin and destination of a delivery order are provided, the travel schedule for an FC with the delivery order must be constructed so that the travel time is minimized. In the process of constructing the schedule, collisions and interferences between moving FCs must be considered.

Contrary to static routing, the dynamic-routing method considers the travel schedules of other FCs when the supervisory computer constructs a travel schedule for the given FC. Since the travel schedules of other FCs are respected, the congestion or the interference of the FC with other FCs can be considered for reducing the travel time as much as possible.

The concepts of conflict-free shortest-time vehicle routing have been introduced by some researchers (Broadbent and Besant 1985; Egbelu and Tanchoco 1983; Huang and Palekar 1989; Walker and Premi 1985). However, Kim and Tanchoco (1990, 1997) were the first to present an optimizing algorithm for finding conflict-free, shortest-time routes for AGVs by utilizing time-window concepts. They introduced the concept of a time-window graph in which the node-set represents the free (uncommitted or unreserved) time-windows and the arc-set represents reachability between the free-time-windows. The algorithm routes the vehicles through the free-time-windows of the time-window graph instead of the physical nodes of the flow-path network.

Rajotia et al. (1998) proposed another method for dynamic routing, where the main algorithm is similar to the path-planning method of Kim et al. (2006). The study maintained time-window data not only at nodes but also at arcs for applying Dijkstra's algorithm for finding the shortest-time routes for a vehicle. With the same concept of time-windows, Lim et al. (2002) suggested a dynamic-routing algorithm under the assumption that the schedules of previous vehicles can be altered when the schedule of a new vehicle is constructed.

Maza and Castagna (2005) improved the dynamic-routing algorithm so that the algorithm could handle changing situations such as vehicle delays and failures. His study proposed a two-stage approach: one stage for finding the shortest-time routes for AGVs and the other stage for avoiding conflicts and deadlocks in real time while maintaining the determined routes of AGVs.

2.2 Problem Definition and a Scheduling Algorithm

This paper assumes that three types of equipment are used for ship operations: QCs (Quay Cranes), FCs (Flat cars), and OSS (Overhead Shuttle System). For example, unloading operations are performed in the following three steps. The first step involves QCs, which transfer containers from a ship to FCs. The second step is to deliver containers from the quay side to the yard side through FCs. The last step is to transfer containers from FCs to the storage positions through OSs.

Figure 2.1 illustrates a new concept of an automated container terminal and the equipment used for the unloading operation. The loading operation is performed in the reverse order to that of the unloading operation.

Figure 2.2 illustrates a layout of the rail-based driving area. In this example, the driving area consists of three main parts: berth lanes, parking lanes, and block lanes. On berth lanes, an FC travels to the transfer point of a QC and receives (or transfers) a container from (or to) the QC. On parking lanes, FCs may park to await the next delivery order or pass over to travel between berths and blocks. On block lanes, FCs either move to (from) transfer points at blocks (TPs) for transferring containers to (or from) yard cranes or travel parallel to the quay across different blocks. We define physical nodes as TPs or ISs.

The following notation will be used for describing the algorithm.

p_i ID for physical node i
n_i state node i, which is defined by a state of an FC at a physical node p_i, which are TPs or ISs, as illustrated in Fig. 2.3
S_i state of n_i, which is represented by (physical node ID, movement direction, speed)
Q_i set of resources associated with n_i, including TPs or ISs on which an FC cannot stay when an FC is on the physical node corresponding to n_i
F_i^n set of time-intervals corresponding to free (unreserved) time-windows of n_i
R_i^n set of time-intervals corresponding to reserved time-windows of n_i
f_j free-time-window (FTW) node j

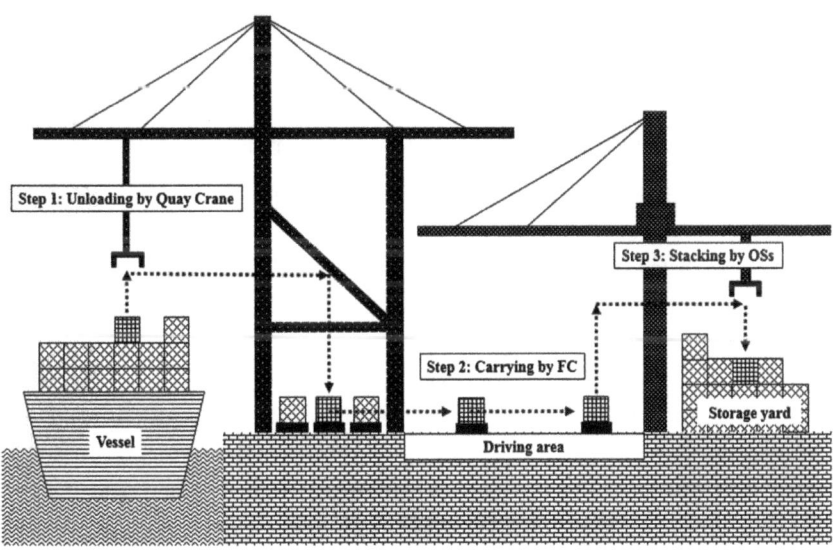

Fig. 2.1 An illustration of unloading operations and the terminal equipment

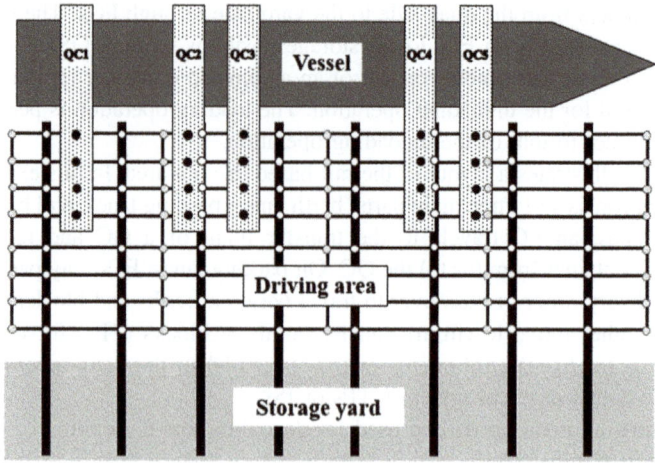

Fig. 2.2 Layout of the terminal

Fig. 2.3 An example of a node

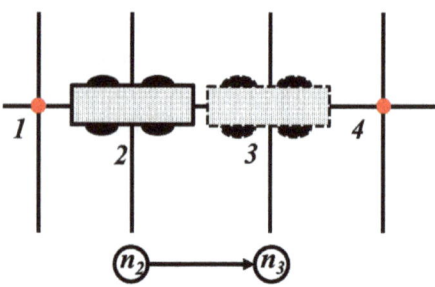

$\hat{f}_j$	node in the state-feasible graph from which f_j originated
t_j^l	lower bound of the free-time-window (FTW) node j
t_j^u	upper bound of FTW node j
L	set of labeled FTW nodes, according to Dijkstra's algorithm
L_{max}	maximum allowed number of labeled FTW nodes in L

2.2.1 Constructing a State-Feasible Graph

A state node n_i is defined by the state S_i of an FC. An example of the state of an FC is the triple consisting of the physical node ID, the moving direction, and the speed.

Suppose state node n_2 is defined by the state S_2, $(p_2, R, 3)$ (which means that the FC is located at physical node 2 and moves from left to right at a speed of 3 m/s) and node n_3 is defined by $(p_3, R, 0)$. Then, an arc can be connected from n_2 to n_3 only if an FC can reduce its speed from 3 m/s to 0 m/s during the travel from p_2 to p_3.

Let $G(V,A)$ be the graph constructed by $V = \{n_i\}$ and $A = \{a_{ij}\}$, where a_{ij} stems from n_i and is directed to n_j and is defined only when the state change is feasible (considering their relative speeds and positions) from n_i to n_j. Graph $G(V,A)$ is called a "state-feasible graph." In Fig. 2.3, node n_3 can be connected to node n_2 if the state change is feasible (considering their relative speeds and positions).

When the number of attributes defining the state increases, the number of nodes also increases. As a result, the size of the problem for Dijkstra's algorithm for finding the shortest-time route may become large. Previous approaches maintain information about free-time-windows only for nodes, while the algorithm in this paper needs to maintain those not only for nodes but also for resources. The time segment for a node is available for a vehicle only if the same time segment is available for all the resources related to the node. This relationship will be maintained by a propagation process to be explained later.

2.2.2 Free-Time-Window Graph

For an FC to travel from an origin to a destination, it must reserve the required time segment of each state node on the travel route. Each state node has both free and reserved time-windows. The reserved time-windows represent the time-segments that other FCs have already reserved for their travel. The free-time-windows are the remaining time segments after the reserved time-windows are removed from the entire planning-horizon.

In a free-time-window (FTW) graph, each FTW $\left(F_i^n\right)$ of a state node in the state-feasible graph becomes an FTW node. For an FTW node (f_i), to have a directed arc to another FTW node (f_j), the following two conditions must be satisfied: (1) in the state-feasible graph, there is an arc from node n_i, from which f_i originates, to node n_j, from which f_j originates; and (2) the travel from f_i to f_j is time-feasible. The time feasibility can be checked as follows: suppose that the earliest departure time from node n_i is t and the travel time from n_i to n_j be t_{ij}. It is time-feasible if there exists (t_i, t_j) where (C1) $t \leq t_i$; (C2) $t_j = t_i + t_{ij}$; and (C3) $t_j \leq t_j^u - t(n_j)$, and where $t(n_j)$ is the shortest time for the FC to move to the position where there is no interference another FC at the physical position of n_j. Among all the possible values of t_i, we will select the earliest time.

2.2.3 Propagation Process

For an FC to travel from n_i to n_j during (t_i, t_j), applying a conservative strategy, the time-window (t_i, t_j) should be reserved for n_i and $(t_i, t_j + t(n_j))$ should be reserved for n_j. Then, all the related resources in Q_i and Q_j, respectively, should be reserved for the same corresponding time-window.

2.2.4 Constructing a Shortest-Time Route for an FC

This paper assumes that a travel schedule is constructed whenever an FC starts its travel from an origin to a destination. In this process, the travel schedules of other FCs, which were constructed before, are respected and are therefore unchanged. Using the FTW graph, the algorithm in this study attempts to find the shortest-time path for an FC from an origin to a destination. This algorithm is based on Dijkstra's algorithm combined with a heuristic rule for restricting the number of labeled nodes during the search process. The details of the algorithm for constructing a travel schedule are explained below.

- Step 0: $L := \emptyset$. Include the source FTW node (the FTW node that includes time zero of the starting state) in L.
- Step 1: List all the unlabeled FTW nodes by (1) selecting all the unlabeled nodes directly connected to one of the nodes in L; (2) and then selecting one node with the connecting arc of the shortest travel time among all unlabeled nodes connected to each node in L.
- Step 2: Select the listed FTW node that has the shortest travel time from the source FTW node. Check if the selected node can be connected to any other FTW node that is not in L. If yes, then go to step 3. Otherwise, remove this FTW node from the set of listed FTW nodes and repeat step 2.
- Step 3: If the selected FTW node is the destination node, then go to step 4. Otherwise, add the selected FTW node to L and calculate the departure time from n_i (t_i) and the arrival time at n_j (t_j). Add the selected node to the current FTW graph. If the number of elements in L is larger than L_{max}, then remove an element from L that has the longest expected travel time from the source FTW node to the destination node. Go to step 1.
- Step 4: List the shortest-time route. Reserve the corresponding time-windows of all the nodes in the shortest-time route and perform the propagation process. Stop.

2.2.5 A Heuristic Method for Reducing the Computational Time

Since the travel schedule must be constructed in real time, the computational time of the algorithm is very critical. When the number of nodes in the driving area becomes large, the size of the FTW graph may become too large for finding the shortest-time route in real time. For reducing the computational time, the maximum number of labeled FTW nodes is not allowed to exceed L_{max}. If the number of labeled FTW nodes already exceeds L_{max}, the FTW nodes with the longest estimated travel times to the destination are removed from L until the cardinality of L is at most L_{max}. Note that each labeled FTW node of the form f_j has the shortest travel time from the origin to f_j and thus, the expected time from the origin to the destination can be calculated by adding the shortest travel time up to f_j and the expected travel time from $\hat{f_j}$ to the destination.

The initial value of the expected travel time from one node to another node is calculated by dividing the corresponding shortest travel distance by the maximum speed of an FC. Subsequently, through the following formula, the expected travel time from a node O to another node D is revised whenever a vehicle arrives at the destination of a route on which both nodes O and D are included: $t_{OD}^{rev} = \alpha \cdot t_{OD}^{new} + (1-\alpha) t_{OD}^{old}$, where t_{OD}^{rev} and t_{OD}^{old} are the revised and prior estimates of the expected travel time, respectively. t_{OD}^{new} is the travel time that has just been realized by the FC upon arriving at node D. α is the smoothing parameter of exponential smoothing.

Additional strategies for improving performance will be tested in the future.

(S1) **Delaying the travel start**: FCs should not arrive at the destination earlier than necessary. If it is too early for an FC to start its travel considering its due time, then the FC should intentionally delay its departure. The intention of this logic is to reduce the time of stay of an FC at congested areas. To delay the departure, the FC may be routed to a prespecified parking area before it starts its travel.

(S2) **Backward scheduling**: another scheduling strategy is to back schedule an activity for just-in-time arrival at the destination.

(S3) **Changing schedules of FCs which started their travel earlier**: the algorithm in this study gives higher priorities to the schedules being implemented. This strategy is for the convenience of the scheduling process, which may limit the optimality of the schedules. One minor modification is to track the total slack at each node, which represents the maximum time allowed to be delayed without delaying the scheduled arrival time at the destination. Then, when an FC incurs interference during the schedule at a node, then it is checked whether previous schedules may be delayed to make a way to the new schedule by utilizing the slacks.

The above strategies are subject to intensive testing for their effectiveness.

2.3 Conclusions

FCs routing and travel-scheduling are key issues of operation control of an FCS. In this paper, we focus on both, the problem of FC routing and that of travel-scheduling. A travel-scheduling algorithm based on Dijkstra's shortest-path algorithm is suggested using the concept of time-windows.

For future research, it is necessary to develop a simulation model for testing the performance of the developed heuristic and compare the method in this paper with existing heuristic algorithms. For reducing the computational time, further studies on various strategies for restricting the search space are necessary.

Acknowledgments This work was supported by the Technological Development of Low-carbon Automated Container Terminals funded by the Ministry of Oceans and Fisheries, Korea (Project Number: 201309550003). This work was also supported by the Korean Federation of Science and Technology Societies (KOFST) grant funded by the Korean government (MSIP: Ministry of Science, ICT and Future Planning).

References

Ashayeri J, Gelders LF, Van Looy PM (1985) Micro-computer simulation in design of automated guided vehicle systems. Mater Flow 2(1):37–48
Broadbent AJ, Besant CB (1985) Free-ranging AGV systems: promises, problems and pathways. In: Proceeding of the 2nd international conference on automated material handling, Birmingham UK, pp 221–237
Egbelu PJ, Tanchoco JMA (1983) Designing the operations of automated guided vehicle system using AGVSim. In: Proceeding of the 2nd international conference on automated guided vehicle systems, Stuttgart Germany, pp 21–30
Em-plant software. http://www.emplant.de/english/index.html
Haines CL (1985) An algorithm for carrier routing in a flexible material-handling system. IBM J Res Dev 29(4):356–362
Huang J, Palekar W (1989) A labeling algorithm for the navigating. In: Advanced manufacturing system engineering: proceedings of the ASME Winter meeting, San Francisco, California, pp 181–193
Kim KH, Jeon SM, Ryu KR (2006) Deadlock prevention for automated guided vehicles in automated container terminals. OR Spectrum 28:659–679
Kim CW, Tanchoco JMA (1990) Prototyping the integration requirements of a free-path AGV system. In: Proceedings of the 1990 material handling research colloquium, Hebron KY, pp 355–363
Kim CW, Tanchoco JMA (1997) Conflict-free shortest-time bidirectional AGV routeing. Intern J Prod Res 29(12):2377–2391
Lim JK, Lim JM, Yoshimoto K, Kim KH (2002) A construction algorithm for designing guide paths of automated guided vehicle systems. Intern J Prod Res 40(15):3981–3994
Lim JK, Kim KH (2002) Dynamic routing in automated guided vehicle systems. JSME Intern J 45 (1):323–332
Liu C-I, Jula H (2004) Automated guided vehicle system for two container yard layouts. Transp Res Part C 12:349–368
Maza S, Castagna P (2005) A performance-based structural policy for conflict-free routing bi-directional automated guided vehicles. Comput Ind 56:719–733
Majety SV, Wang MH (1995) Terminal location and guide path design in terminal based AGV systems. Intern J Prod Res 33(7):1925–1938
Rajotia S, Shanker K, Batra JL (1998) A semi-dynamic time window constrained routeing strategy in an AGV system. Intern J Prod Res 36(1):35–50
Rajeeva LM, Wee HG (2003) Cyclic deadlock prediction and avoidance for zone-controlled AGV system. Intern J Prod Econ 83:309–324

Walker SP, Premi SK (1985) The imperial college free-ranging AGV (ICAGV) and scheduling system. In: Proceeding of the 3rd international conference on automated guided vehicle systems, Stockholm Sweden, pp 189–198

Zeng J, Hsu WJ (2008) Conflict-free container routing in mesh yard layouts. Robot Auton Syst 56:451–460

Chapter 3
The Influence of Manufacturing System Characteristics on the Emergence of Logistics Synchronization: A Simulation Study

Stanislav M. Chankov, Giovanni Malloy and Julia Bendul

Abstract The term "synchronization" in manufacturing refers to the provision of the right components to the subsequent production steps at the right moment in time. It is still unclear how manufacturing system characteristics impact synchronization. Thus, the purpose of this paper is to investigate the effect of manufacturing systems' characteristics on the emergence of logistics synchronization in them. We conduct a discrete-event simulation study to examine the effect of three system characteristics: (1) material flow network architecture, (2) work content variation, and (3) order arrival pattern. Our findings suggest that the material flow network architecture and the work content variation are related to logistics synchronization. Linear manufacturing systems with stable processing times such as flow shops operate at high logistics synchronization levels, while highly connected systems with high variability of processing times such as job shops exhibit lower synchronization levels.

Keywords Synchronization · Manufacturing system · Discrete-event simulation

---

S.M. Chankov (✉) · J. Bendul
Department of Mathematics and Logistics, Jacobs University Bremen,
Campus Ring 1, 28759 Bremen, Germany
e-mail: s.chankov@jacobs-university.de

J. Bendul
e-mail: j.bendul@jacobs-university.de

G. Malloy
School of Industrial Engineering, Purdue University, 315 N. Grant Street,
West Lafayette, IN 47907, USA
e-mail: malloyg@purdue.edu

© Springer International Publishing Switzerland 2017
M. Freitag et al. (eds.), *Dynamics in Logistics*, Lecture Notes in Logistics,
DOI 10.1007/978-3-319-45117-6_3

3.1 Introduction

Synchronization phenomena from various scientific fields have been intensively studied as part of the theory of dynamical systems (Pikovsky et al. 2003). In the context of manufacturing systems, the term "synchronization" refers to the provision of the right components to the subsequent production steps at the right moment in time. These just-in-time material flows are believed to lead to higher efficiency for manufacturing systems (Miller and Davis 1989). Previous work has focused on defining this form of logistics synchronization (Chankov et al. 2014), identifying if it occurs in job shop manufacturing environments (Becker et al. 2013) and on developing quantifying synchronization measures for it (Chankov et al. 2015). However, it remains unclear if synchronization can occur in any manufacturing system type, how manufacturing system characteristics impact synchronization and how the emergence of synchronization affects logistics performance.

The purpose of this paper is to contribute to closing this gap by investigating the effect of manufacturing systems' characteristics on the emergence of synchronization in them. We study three main system characteristics: (1) material flow network architecture, (2) work content variation, and (3) order arrival pattern. A discrete-event simulation study is applied in order to study types of manufacturing systems with diverse network architectures and varying work content distributions (line production, flow shop production, job shop production and cellular manufacturing) in order to compare the synchronization phenomena occurring in them. The paper is organized as follows. Section 3.2 presents different types of synchronization phenomena occurring in manufacturing and appropriate measures for them. Section 3.3 explains the methodology used in this study. We present and discuss our results in Sect. 3.4. Finally, Sect. 3.5 provides a brief summary of the investigation, its limitations and outlook for further research.

3.2 Synchronization in Manufacturing Systems

There are two views of synchronization: flow-focused and system-focused (Chankov et al. 2015). Within the manufacturing and logistics domain, synchronization is seen as the flow-oriented coordination of materials between systems (Wiendahl 1998) and thus closely related to the just-in-time philosophy, while within the natural science domain synchronization is defined as the adjustment of rhythms of systems due to interaction (Pikovsky et al. 2003). Chankov et al. (2015) term the two separate views logistics and physics synchronization. Based on the flow-focused view, they define logistics synchronization as "the coupling of work systems (WSs) that are linked by material flows," while physics synchronization is derived from the system-focused view as "the rhythm and repetitive behavior of production processes in a manufacturing system." In addition, they observe differences in the synchronization behavior of job shops and flow shops, and thus suggest that the type of manufacturing system influences its synchronization behavior. However, they do not examine which manufacturing system characteristics lead to those differences. Both network connectivity (Becker et al. 2012) and the variability in processing times (Bondi and Whitt 1986) have been suggested as distinctive system characteristics and found to impact manufacturing systems. Moreover, a study on the synchronization in railway timetables by Fretter et al. (2010) indicates that the type of arrival events in their avalanche model affects synchronization. Transferring this to the manufacturing context, we suggest that the order arrival pattern has similar effects on the synchronization level of manufacturing systems. Accordingly, we hypothesize that the following three manufacturing system characteristics affect the emerging in manufacturing systems synchronization: (1) material flow network architecture, (2) work content variation, and (3) order arrival pattern.

Chankov et al. (2015) suggest quantitative measures for both synchronization types. Our paper aims at understanding what triggers the emergence of logistics synchronization within manufacturing systems, therefore we only consider their logistics synchronization measure. It is based on cross-correlation, which is a standard measure of linear synchronization (Becker et al. 2013). The cross-correlation of two discrete univariate time series x_t and y_t spanning over a time period $t = 1 \ldots N$ is:

$$c_{x,y}(\tau) = \frac{1}{N-\tau} \sum_{t=1}^{N-\tau} \left(\frac{x_t - \bar{x}}{\sigma_x} \right) \left(\frac{y_{t+\tau} - \bar{y}}{\sigma_y} \right) \qquad (1)$$

where $\bar{x}$ and σ_x represent the mean and the standard deviation of the time series, respectively, while the parameter τ is a time lag. Thus, a value of zero represents zero synchronization and values of ± 1 indicate perfect (anti-)correlation.

The cross-correlation of the WIP development of two work systems provides information about their synchronization for a specific time lag. Obtaining a global quantification index for the whole manufacturing system requires using the maximal correlation independent of the time delay at which it occurs given by

$$c^*_{x,y} = \max_{\tau > 0} \left| c_{x,y}(\tau) \right| \qquad (2)$$

Chankov et al. (2015) hypothesize that "in manufacturing systems, which exhibit logistics synchronization, the maximum cross-correlations of the linked by material flows WS pairs will be higher than the maximum cross-correlations of the non-linked pairs". Thus, a logistics synchronization index is formulated as

$$I_{LS} = \frac{\frac{1}{L} \sum_{x \to y} c^*_{x,y}}{\frac{1}{M} \sum_{i,j} c^*_{i,j}} \qquad (3)$$

Where $x \rightarrow y$ stands for a material flow from WS x to WS y, L is the number of linked WS pairs and M is the total number of WS pairs. A value of 1 for the index shows that linked WSs are equally synchronized to the non-linked ones, while values above 1 show that they are more synchronized and values below 1 that they are less synchronized than the non-linked ones. The comparability of results across systems with different characteristics requires the use of a z-score:

$$z_{LS} = \frac{I_{LS} - \mu_{I_{LS}}^{(R)}}{\sigma_{I_{LS}}^{(R)}} \quad (4)$$

where $\mu_{I_{LS}}^{(R)}$ and $\sigma_{I_{LS}}^{(R)}$ denote the mean and standard deviation of the logistics synchronization index for given number of random scenarios (obtained by shuffling the maximal cross-correlations values randomly among the WS pairs).

3.3 Methodology

3.3.1 Simulation Model

Discrete-event simulation is a widely used simulation method for manufacturing systems, in which components are modeled as objects that have certain attributes representing the object states. Changes in those states are triggered by events. For manufacturing systems, the objects can be machines or workers, for example, the corresponding attributes can be the time needed for a task or the object's availability, while events can be the arrival of a new order or a machine breakdown (Kelton and Law 2000). We use discrete-event simulation because of its versatility and reliability in representing manufacturing processes.

The simulation model presented in this paper was created in FlexSim 7.3. It consists of fifty work systems that can be arranged into different manufacturing system designs (see Fig. 3.1a). Each WS is composed of a pre-process buffer, a processing machine and a post-process buffer (see Fig. 3.1b). Moreover, the transport from one WS to the next is considered to be part of the subsequent WS and is modeled with the use of a processor. Thus, the model does not involve any predetermined material flows and can be used to model manufacturing systems with diverse material flows. Besides, the model allows for utilizing different priority rules, transport modes as well as the introduction of transportation times and set up times. For matters of simplicity, for this study we have used the standard first-in-first-out (FIFO) priority rule and have kept the transport and set up times at zero.

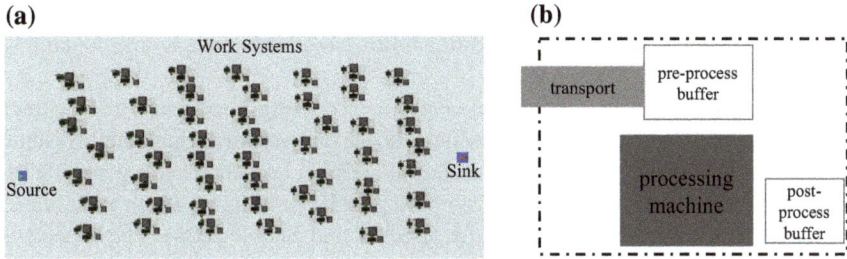

Fig. 3.1 Simulation model: **a** model overview and **b** single work system

3.3.2 Experiment Setup

To study the effect of manufacturing systems' characteristics on the emergence of synchronization, we design six cases representing different types of manufacturing systems (see Table 3.1). Although several manufacturing system classifications exist, we follow a widely used one suggested by Chryssolouris (2006), who distinguishes five manufacturing system types: flow line, cellular system, job shop, project shop, and continuous system. The first three types are different from the latter two as they represent systems in which discrete products move from WS to WS, while a project shop is used for products whose position is fixed and a continuous system produces liquids or gases. Hence, our study focuses on the first three types.

A flow line, also known as a flow shop, is a manufacturing system in which "the machines and other equipment are ordered according to the process sequences of the parts to be manufactured" (Chryssolouris 2006). Thus, a sequence of work systems is dedicated to one particular product or product family. The simplest form of a flow shop is the transfer or assembly line, which only contains a single

Table 3.1 Overview of selected manufacturing systems

No	Networks	Work systems	Orders	Operations (k)	Average operations per order
I	Line production	50	400	20	50
II	Flow shop production 1	50	2000	20	10
III	Flow shop production 2	50	2000	20	10
IV	Cellular manufacturing	15	5320	20	4
V	Job shop production 1	50	3532	20	6
VI	Job shop production 2	50	1851	20	11

sequence of WSs that take in material flow one at a time. Network I in our study represents such a flow line. It is a traditional flow shop with a single line containing 50 WSs.

More sophisticated flow shops involve not just a single sequence of machines but several ones that run in parallel, which allows for manufacturing a high volume of a limited variety of goods (Chryssolouris 2006). Flow shops can generally be organized in two fashions. The first is a pure parallel fashion, in which each line is dedicated to a different product family (Becker and Scholl 2006). The second is parallel fashion with crossovers, in which materials can be transferred from one line to another (Freiheit et al. 2004). Networks II and III represent those two flow shop types. Both have 50 WSs grouped in 5 parallel lines, but while Network II is a pure parallel flow shop, network III allows crossovers at all stages.

Further, a cellular system is similar to a parallel flow-shop since it can manufacture several product families. In cellular manufacturing, work systems are grouped into cells and each cell is dedicated to a particular product family (Chryssolouris 2006). Network IV represents such a cellular manufacturing system with 15 WSs split in two cells and is based on the system presented by Witte (1980).

Finally, job shops group machines with similar functions together and can produce products with largely differing process sequences (Chryssolouris 2006). As a result their material flow networks are rather complex and involve numerous production path options, which makes them difficult to model. We suggest modelling job shops with random graph networks (Erdős and Rényi 1959). Random graphs are networks in which an edge between two nodes exists with a given probability p. Since the material flow networks of job shops contain a large variety of links between the WSs, we argue that it is appropriate to model them with random graphs. Accordingly, we generate two job shop production networks. The first one (network V) is a directed random graph of 50 WS nodes in which the edges occur with a probability $p = 0.10$ and the second one (network VI) also has 50 WS nodes but this time the first 5 WSs are connected in a production line, which is subsequently followed by a directed random graph of the remaining 45 WSs, in which the edges occur again with a probability $p = 0.10$. The logic behind network VI is that even though job shops involve largely differing process sequences, some job shops have process sequences that always have the same start of the process sequence (for example, quality control of parts). This can also be observed in the material flow networks of five real-world job shop manufacturers presented in Chankov et al. (2015). The material flow networks for all six cases of our study are depicted on Fig. 3.2.

For each of the six networks, we generate production orders for a total of 20000 operations and run simulations. The equal number of operations ensures the comparability of the results. Figure 3.3 presents examples of orders from the six networks illustrating two key components: (1) WS sequence and (2) WS work content (WC).

First, it has to be noted that for some networks the WS sequence is fixed, while for others it varies. For example, orders on network I always have to go through the

3 The Influence of Manufacturing System Characteristics ...

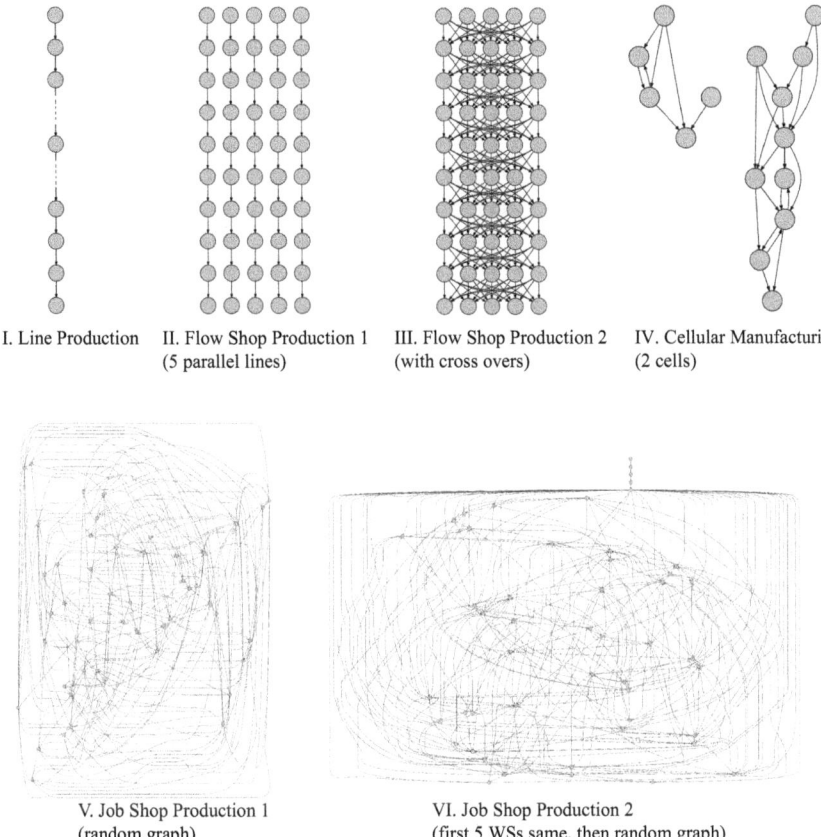

Fig. 3.2 Material flow networks of selected manufacturing systems

entire line of the 50 WSs it contains (thus each order has 50 operations) and orders on network II always go on one of the five lines with 10 WSs (thus each order has 10 operations). Further, network III involves orders with 10 operations but the sequence of the 10 WSs depends on the crossovers between its 5 lines (each crossover is chosen at random). Network IV has orders that are dedicated to one of its cells and the exact material flows are based on the example of Witte (1980). Network V starts at a random WS and goes through the system by selecting at random each of the possible subsequent WSs. The orders of network VI always go through the same five WSs and then proceed in the same way as network V.

Second, some networks have stable WC distributions while others don't. Flow line production normally utilizes cycle time (Becker and Scholl 2006) and thus we have assumed that all WSs belonging to network I have constant WC of 0.1 days. The parallel flow shops without crossovers normally have a fixed cycle time per line, accordingly we have assigned constant WC per line for network II (0.2 days for the chosen example of Fig. 3.3). The flow shop with crossovers requires the

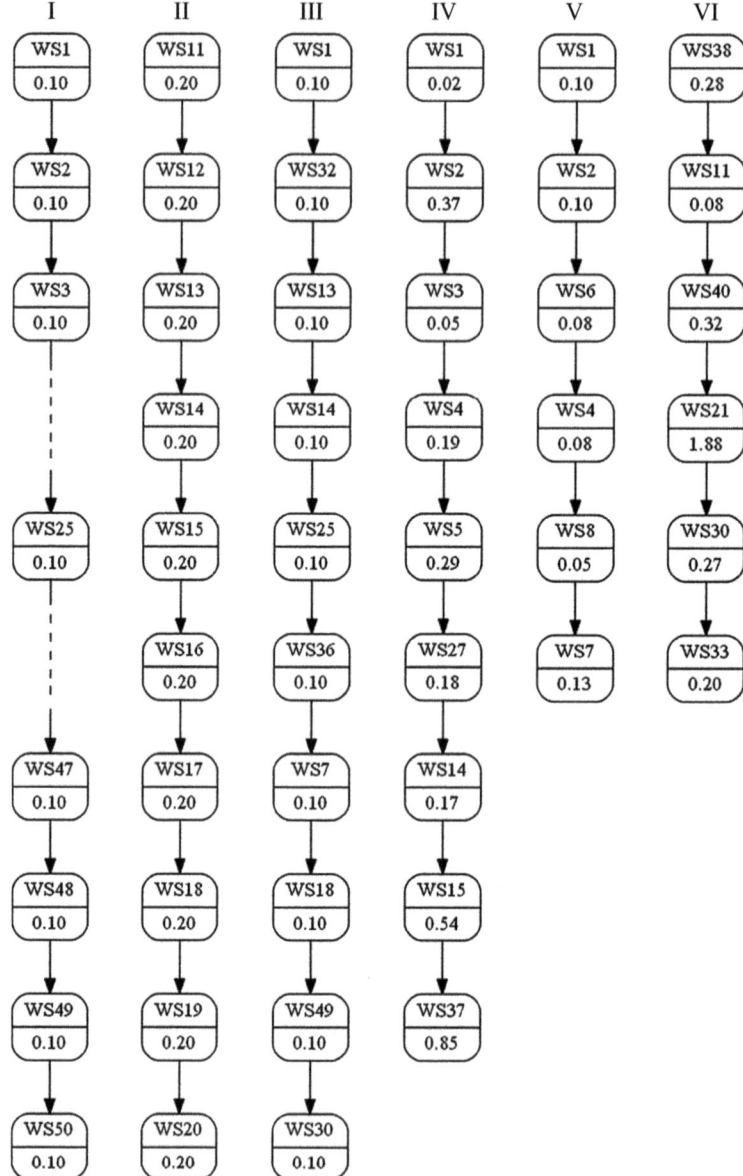

Fig. 3.3 Examples of orders for selected manufacturing systems

different lines to operate at the same cycle time, hence all WSs part of network III have constant WC of 0.1 days. Further, cellular systems do not utilize cycle times and the WC can vary among the WSs. In our study, the WC of network IV is based on Witte (1980). Finally, job shops have largely differing work contents. Consequently, the WC of networks V and VI is not fixed. Instead, each WS is assigned an average WC between 0.05 and 0.5 days and the WC for each operation is drawn from a normal distribution with that average and a corresponding standard deviation ensuring a coefficient of variation (CV) of 100 % (log-normal distribution is used in order to avoid negative numbers for WC as suggested by Mood et al. (1974)).

In order to control for effects of the order arrival, we run simulations using three order arrival patterns: (1) fixed-interval, (2) Poisson-process, and (3) batch. In the fixed-interval case, an order arrives every 0.10 days, in the Poisson-process case, the inter-arrival times between orders follow an exponential distribution with $\beta = 0.05$, and in the batch case 20 orders arrive in the beginning of every day (with the exception of network IV, where the daily batch size is based on Witte (1980)).

3.4 Results and Discussion

After running experiments on the described above model, we are able to calculate the emerging in every scenario logistics synchronization (Eqs. 1–4). The obtained logistics synchronization z-scores are shown on Fig. 3.4. It can be seen that the *line* and *flow shop production 1* scenarios exhibit the highest synchronization levels (z-scores reaching values of 15), while the two *job shop production* cases exhibit no synchronization with z-score values close to zero. Moreover, the *flow shop production 2* and the *cellular manufacturing* scenarios have average positive z-scores, with the exception of the fixed-interval order arrival case of *flow shop production 2*, which shows a negative z-score of −5.

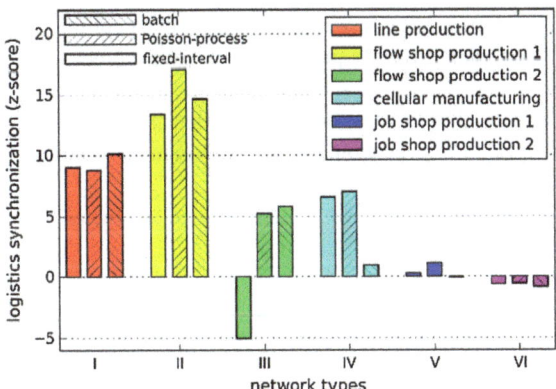

Fig. 3.4 Logistics synchronization results

To examine the relation between manufacturing system characteristics and emerging synchronization, we study the influence of three parameters (1) material flow network architecture, (2) work content variation, and (3) order arrival pattern.

First, the material flow networks of the different manufacturing system types are differently connected. The average node degree (ratio between number of nodes and links in a network) has been suggested as an indicative measure for the connectivity of material flow networks (Becker et al. 2012). Accordingly, to study the relation between material flow network architecture and logistics synchronization, we perform a Pearson's correlation analysis with the hypothesis that the average node degree of the material flow network of a manufacturing system is related to logistics synchronization. The results shown on Fig. 3.5a are significant on the 1 % level and indicate that more connected networks show lower synchronization levels.

Second, the level of variability of processing times differs across manufacturing system types. The CV (ratio of the standard deviation and the mean) of the work content of all operations performed by a WS has been suggested as practical measure of this variability (Bondi and Whitt 1986). Hence, to examine the relation between WC variation and logistics synchronization, we perform a Pearson's correlation analysis with the hypothesis that the average CV of WC among all WSs part of a manufacturing system is related to logistics synchronization. Figure 3.5b shows the results, which are significant on the 1 % level and indicate that manufacturing systems with higher variability of processing times have lower synchronization.

Third, to investigate if there are differences in synchronization across the different order arrival patterns, we perform a repeated measures analysis of variance (ANOVA) (Field 2013) with the null hypothesis that the synchronization measurements across the three arrival patterns in our study have the same means. The results show that logistics synchronization is not significantly affected by the order arrival pattern, $F(2, 10) = 1.09, p > 0.05$. Hence, we accept the null hypothesis and can conclude that the arrival pattern does not affect the emergence of synchronization.

Our results suggest that two of the three studied manufacturing system characteristics are related to synchronization emergence: material flow network

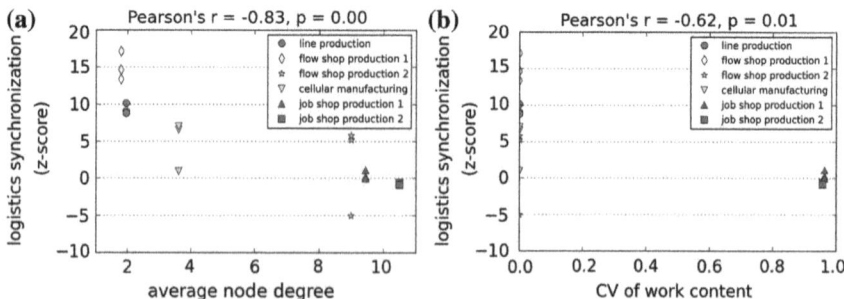

Fig. 3.5 Relation between manufacturing system characteristics and synchronization: **a** average node degree and **b** coefficient of variation of work content

architecture and WC variation. To begin with, systems with lower connectivity and WC variations exhibit high logistics synchronization. The examined *line production* and *flow shops production 1* are linear systems that operate at defined cycle times. Thus, a high coupling level between connected WSs emerges. Further, despite the presence of some variation in terms of available production sequences, the *cellular manufacturing* system and *flow shop production 2*, also show relatively high synchronization that can be explained by the stable processing times in those systems. Finally, it is not surprising that highly connected systems with high variability of processing times exhibit low logistics synchronization. *Job shops* are such systems which manufacture a high variety of products that undergo diverse production se-quences and involve varying processing times at each WS. As a result the coupling between connected WSs is weak, leading to low logistics synchronization. The last studied parameter, order arrival pattern, did not have a significant relation to synchronization, which could be due to the fact that inherent system characteristics play a more important role for the synchronization emergence than varying conditions.

3.5 Conclusion

In this paper, we conducted a discrete-event simulation study to investigate the effect of manufacturing system characteristics on the emergence of logistics synchronization. Our findings suggest that the material flow network architecture and the work content variation are two features of manufacturing systems that are related to logistics synchronization. Linear manufacturing systems with stable processing times such as flow shops operate at high logistics synchronization levels, while highly connected systems with high variability of processing times such as job shops exhibit lower synchronization levels. However, our simulation study does not consider several manufacturing system parameters, such as setup and transport times, priority rules, applied production planning and control methods, and machine breakdowns. Further research is required to investigate if these parameters also affect the emergence of logistics synchronization. Besides, studying factors that trigger the emergence of physics synchronization in manufacturing is also suggested.

References

Becker C, Scholl A (2006) A survey on problems and methods in generalized assembly line balancing. Eur J Oper Res 168(3):694–715
Becker T et al (2012) The impact of network connectivity on performance in production logistic networks. CIRP J Manuf Sci Technol 5(4):309–318
Becker T, Chankov SM, Windt K (2013) Synchronization measures in job shop manufacturing environments. Procedia CIRP 7:157–162

Bondi A, Whitt W (1986) The influence of service-time variability in a closed network of queues. Perform Eval 6(3):219–234
Chankov SM, Becker T, Windt K (2014) Towards definition of synchronization in logistics systems. Procedia CIRP 17:594–599
Chankov SM, Huett M-T, Bendul J (2015) Synchronization in manufacturing systems: quantification and relation to logistics performance. Submitted to International Journal of Production Research
Chryssolouris G (2006) Manufacturing systems: theory and practice, 2nd ed. Springer, New York
Erdős P, Rényi A (1959) On random graphs I. Publ. Math. Debrecen 6:290–297
Field A (2013) Discovering statistics using ibm spss statistics, 4th edn. SAGE Publications, London
Freiheit T, Shpitalni M, Hu SJ (2004) Productivity of paced parallel-serial manufacturing lines with and without crossover. J Manuf Sci Eng 126(2):361
Fretter C et al (2010) Phase synchronization in railway timetables. Eur Phys J B 77(2):281–289
Kelton WD, Law AM (2000) Simulation modeling and analysis. McGraw Hill, Boston
Miller WA, Davis RP (1989) Synchronization of material flow to aid production planning in a job shop. Eng Manage Int 5(3):179–184
Mood AMF, Graybill FA, Boes DC (1974) Introduction to the theory of statistics, 3rd edn. McGraw-Hill, New York
Pikovsky A, Rosenblum M, Kurths J (2003) Synchronization: a universal concept in nonlinear sciences. Cambridge University Press, Cambridge
Wiendahl H-H (1998) Zentralistische Planung in dezentralen Strukturen?—Orientierungshilfen für die Praxis. In: Westkämper E, Schraft RD (eds) Auftrags- und Informationsmanagement in Produktionsnetzwerken—Konzepte und Erfahrungsberichte. Fraunhofer IPA, Stuttgart, pp 79–107
Witte J De (1980) The use of similarity coefficients in production flow analysis. Int J Prod Res 18(4):503–514

Chapter 4
Safety Requirements in Collaborative Human–Robot Cyber-Physical System

Azfar Khalid, Pierre Kirisci, Zied Ghrairi, Jürgen Pannek
and Klaus-Dieter Thoben

Abstract The paper identifies the basic technology and functional requirements of a cyber-physical system to control human–robot collaboration in an industrial environment. The paper defines the collaboration grading of human–robot co-working environment based on the prevailing safety concepts in workspace sharing. Detailed requirements are generated for each interaction mode and few collaboration indices are established. Different indices are found to be useful for the purpose of categorization of collaboration levels. A specific case is discussed later for a detailed Cyber-Physical System solution in a smart production or logistical context. The paper ends with a general guideline that is formulated to cater for various industrial level human–robot collaborative scenarios. An important aspect of the collaboration guideline is a sensor catalogue to meet cyber-physical system design requirements.

Keywords Cyber-physical system · Human–robot collaboration · Safety requirements

4.1 Introduction

The manufacturing horizon for industry 4.0 comprises a paradigm shift from the automated manufacturing toward an intelligent manufacturing concept. The exclusive feature in industry 4.0 is to fulfil the real-time customer demand of variation in product in a very small lot size. This will enable the manufacturing system to meet individual customer requirement without wasting for re-configuration of assembly line or set-up times. The intelligent manufacturing

---

K.-D. Thoben
University of Bremen, Bibliothekstraße 1, 28359 Bremen, Germany

A. Khalid (✉) · P. Kirisci · Z. Ghrairi · J. Pannek
BIBA-Bremer Institut für Produktion und Logistik GmbH (BIBA),
Hochschulring 20, 28359 Bremen, Germany
e-mail: kad@biba.uni-bremen.de

implementation will take place though the concept of internet of things (IoT), in which each participating component has its known IP address. In this context, the smart manufacturing and logistics systems should not only fulfil the real-time customer demands with product variation in very small lot sizes, but also include characteristics such as better predictive maintenance, robustness in product design and adaptive logistics. For a smart robotic factory to work in the context of industry 4.0 and (IoT), where high productivity is the demand of the future so the robots will take most of the workshare in the future manufacturing but the human worker has to stay in the work area either in supervision role or for the jobs for which the robots are not trained. The constant human presence in or near the robot's work area develops a shift about how the robot work areas are fenced and prohibited for the humans. The futuristic approach is to implement robotic applications where robot and human workers can co-exist and collaborate safely.

In this setting, the robots share the same workspace with human counterparts and do industrial activities such as raw material handling, assembling and industrial goods transfer. The conventional approach is to expose human worker up to a limited extent with the robot and with appropriate safety control that leads to full stoppage (safe hold) of machine in case of worker violation of robot workspace. This causes interruptions and resetting procedures to be activated and hampers productive time. The proposed approach is to exhibit safe intermediate human–robot collaboration (HRC) without any fencing. In order to realize, extra safety and protection measures needs to be implemented for a collaborative robotic cyber-physical system (CPS). These safety and security requirements are based on the level of interaction between humans and robots on the shop floor to increase productivity. In fact, the approach in the design of collaborative robotic CPS is to merge the safety and security concerns just like designing industrial facility, control and risk assessment that consider both aspects (Kriaa et al. 2015). However, in this paper, only the safety aspects are considered for collaborative robotic CPS development (specific to large size robots). An existing example of such a system is Robonaut 2 (NASA) which is designed to work in a dexterous manner with humans to perform difficult jobs in and around International Space Station. Another important work (Lasota et al. 2014) for conversion of present day industrial robots to become interactive with humans are reported in (Knight 2014) in addition to the contributions from other robot developers.

A CPS is a smart system in which the computational and physical systems are integrated to control and sense the changing state of real-world variables (NIST 2013). The success of such CPS relies on the sensor network and communication technologies that are reliable, safe and secure. In CPS, all the functional components are in modules and interconnected wirelessly in the production line or in smart factory. Even, raw materials and machines are connected on network cooperating with human workers through human–machine interactive (HMI) systems. Hence, the CPS platform evolves its architecture to engineer across the digital-physical divide and removing the borders among the key technologies. In particular, the robotic CPS consists of electronics, computing, communications, sensing, actuation, embedded systems and sensor network (Zamfirescu et al. 2013).

For this application, the deployment of a full-scale CPS accounts for the human worker as an inherent part of the system. To state the robotic CPS definition, the three components are clearly evident in the model (See Fig. 4.1).

The human component (HC), the physical component (PC) and the computational component (CC) are the three modules integrated together. There is an increasing interaction among the three components as the enabling technologies are reducing the borders. The human component is well connected through different adaptor technologies, e.g. human position tracking and safety distance parameter are important considerations for worker safety in the robotic CPS. The robotic CPS is a highly automated system as it removes the boundaries between the composite elements and preferring their operational interactions. There are various HMI technologies based on human senses of vision, audition, and touch. The proposed robotic CPS can use vision system for detection, tracking and gesture recognition of human worker. The robots can also be commanded using audio signals from human. Additionally, variety of sensors and actuators can provide the interaction between HC, CC and PC.

In order to maintain the human–robot co-existence in the context of robotic CPS, such systems must exhibit properties such as integrality, sociability, locality and irreversibility. Moreover, it must be adaptive, autonomous and highly automated (Wang et al. 2015). The ability of CPS to interact with other CPS through different communication technologies defines the sociability. It will encompass not only devices but also integrates humans as well. Locality term introduces the computational, human and physical capabilities of a CPS, as bounded by spatial properties of the environment. Irreversibility of the CPS makes it self-referential in timescale and state-space. The adaptive characteristic makes the system self-organized and evolving. The autonomy (Pirvu et al. 2015) characterizes the control loop must close over the lifecycle of a CPS.

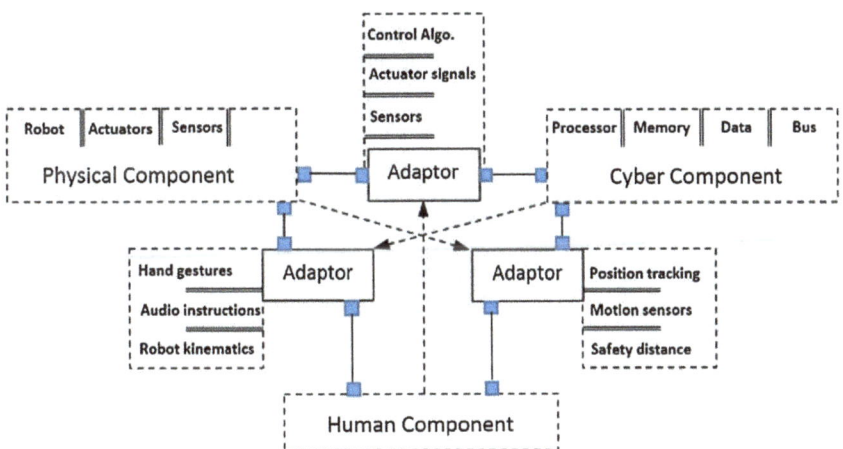

Fig. 4.1 Robotic CPS modules

4.2 Collaboration Classification

In robotic CPS industrial environment, the smooth overlapping of workspace zones of robot and human is sought in which both can collaborate. The formal grading of the human–robot collaboration involves the level of interaction between the two entities. The level of interaction can be formalized based on the distance between the two entities, workspace share level and the complexity of collaborative tasks which both are doing mutually. There is also an issue of mental strain on humans in addition to the physical interaction of robot and human. It is discussed in (Arai et al. 2010) that by restricting the moving area and moving speed of robots, the mental strain of a human operator remains low. Also, the prior accurate information of robot motion is essential to decrease the strain on human operator.

To formally grade the HRC, the safety approaches in practice must be known first. These approaches are based on the available sensor technologies. The conventional approach is to (only with small robots) provide guidance manually or reduce the robot speed as per the requirement. This manual approach is open loop, without sensing, has high HRC level, restricted to small size robots and depends on the defined risk assessment. The second safety approach is to specify a work zone that is covered with sensors like laser scanner or proximity sensor. In this case, the robots must stop at the human access to the work area. These systems are sensor dependent, closed loop and has almost no HRC level attainment (See Fig. 4.2 for the collaboration schemes).

The third approach is the speed or distance monitoring through vision-based system or other possible techniques. Speed reduction of robot applies with a possible stop in case of worker entry into the dangerous zone. This safety concept uses multiple integrated sensors and sensor fusion techniques. High HRC level attainment is possible but pose challenges to the risk assessment in case of a failure of a monitoring function. The last concept is the force monitoring through the use of force sensors. This system will also work with the help of a vision field which will guide the robot about the presence of human. The robot speed and acceleration reduction will take place according to the level of force allowed to hit the worker's body part. The force magnitude will vary for different body parts. This scheme

Fig. 4.2 Collaboration classification. **a** Robot on safe hold against human violation. **b** Speed reduction even the worker is in the robot work zone. **c** Robot touching the human with a pre-defined calibrated force

provides highest level of HRC attainment but also demands integration of multiple types of sensors, sensor fusion and pose challenge to risk assessment in case of failure of monitoring function.

By looking at different collaboration techniques and the corresponding HRC attainment level, it is feasible to formalize the HRC grading scheme. Figure 4.3 shows the grading pattern of HRC from low to high. The start point to gauge the HRC is first to count the number of sensors installed in the CPS system (S). In the case of large CPS, this variable represents the number of types of sensors. The next necessary variable is the data rate (D_i) of individual sensor or group of sensors of the same type. Data rates (ms) are important as the delay time from every sensor counts on the overall system's reaction time to respond. Thus, the system's overall delay time is the key performance indicator (*KPI*) enabling the robot to initiate the safety protocol in time to meet any hazard. Larger delay time will affect the robot's reaction time adversely and hence reduce the HRC attainment. Further, the use of sensor fusion technique can improve the *KPI* and result in better HRC attainment level. On the right side in Fig. 4.3, the HRC grades are specified. Here, a_1 grade shows the highest level of HRC attainment followed by a_2 and so on.

In addition to the HRC measurement through *KPI* computation, there is another indicator in the case of high collaboration level techniques and that is the safety distance calculation. The safe distance (*SD*) formula for the safety of human working with industrial robots is given in EN ISO 13855 (See Fig. 4.3). *SD* computes the minimum safety distance from the risk zone. K is the speed of the man approaching to collision with the robot (mm/s). T is the robot's follow-up time to stop completely, once the brakes are applied in seconds. C is the additional distance (mm) for safety compliance that depends on the sensor's capability or resolution.

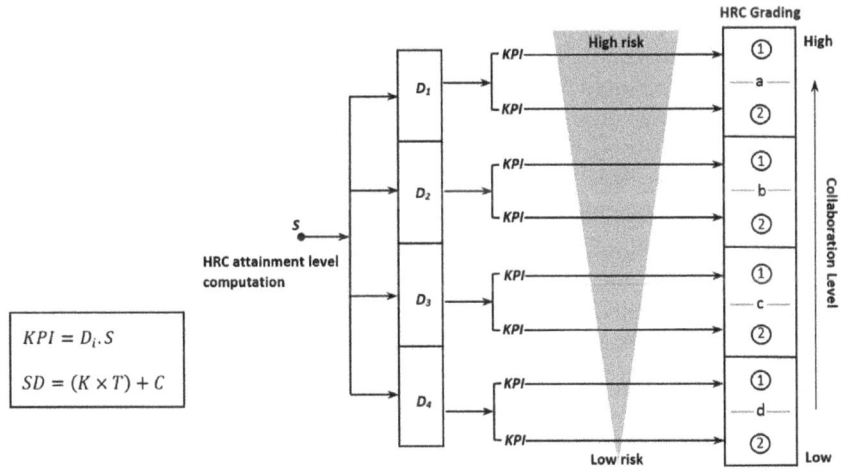

Fig. 4.3 Human–robot collaboration grading scheme

In case of multiple sensors used in a system, the sensor with lowest resolution will decide the resolution of the overall system.

4.3 Sensors Catalogue

To assess the HRC attainment level, it is necessary to compute the *KPI* in a given collaboration context. For this purpose, the safety schemes mentioned above are further explained at the sensor level and the possible risk reduction approach. These HRC contexts address the requirements of the robotic CPS and generate sensor catalogue for each type of collaboration. The basic condition for CPS implementation is that the position of the human worker and the robot must be known in real time. In strict terms, the detailed positions of the worker's body organs needs to be established. The vision sensors that can be employed for the position information associated with other sensors must work in all the lighting conditions, e.g. in low visibility or in a rough industrial environment. Further, the communication must be faster for an immediate and accurate response of the robot preferably through low distance, safe wireless network. Overall, the system must comply with the relevant safety standards such as EN ISO 13849-Part 1 and 2 and EN ISO 13855 etc. These standards provide principles, safety requirements and guidance for the design and integration of safety-related parts of control systems.

Table 4.1 defines the collaboration level of different safety approaches, employable risk reduction scheme and the basic sensor pack to implement the scheme. The optimized solution can be found based on the industrial scenario and HRC level sought. For the speed and distance monitoring case, inertial measurement units (IMU) are employed in addition to the basic area and position monitoring sensor systems. Similarly for force monitoring-based HRC system, the basic area and position monitoring will be a requirement for implementation of the robotic CPS in addition to the force sensors. In force monitoring, different types of geometry-adapted tactile sensors are available to instal on robots or in the robot joints, with shock-absorbing properties for safe collision detection and touch-based interaction. Force sensors (normal data rate of 1 kHz) of different force ranges can be used for assessment of force exposure limits for separate human body organs.

Table 4.2 shows the computation of some indices that are checked for different employable sensors in the robotic CPS. The number of sensors are selected according to the practical requirement for such a system. It is noted that the safety distance is significant in case of camera system as compared to other sensors. Moreover, around 1 m safety distance is required in any case for the deployment of safety speed reduction scheme, e.g. if a worker is coming towards a robot with a speed of 1600 mm/s and the robot's follow-up time is 0.42 s, then the robot must exhibit safety speed reduction when the worker distance remains 1 m. For the HRC calculation, ultrasonic sensors shows the best result.

Table 4.1 Sensors against the collaboration levels

Safety concept used	Collaboration level	Risk reduction approach	Technology (sensors employed)
Manual operation (hand guided)	High HRC but for small robots only	Physical ergonomics-based assessment	No sensors, passive protection guards
Complete isolation (HRC = 0)	Robot held due to line crossing	Robot workspace or path calibration	Laser scanner, proximity sensor
Speed and distance monitoring**	High level interaction	Robot workspace, human-robot distance and speed calibration	Distance monitoring: e.g. ultrasonic speed monitoring: e.g. IMU
Force monitoring**	High level interaction	Force calibration	Force monitoring: force sensors

**Basic area and position monitoring sensors are required additionally. Area monitoring (Proximity sensors, Laser scanner), position monitoring (Vision system) as an example

Table 4.2 Indices computation for sensors

Indices	Sensors			
	Laser scanner	Camera system	IMU	Ultrasonic
Data rate (D_i) (ms)	62.5	50	16.6	20
Sensor detection capability (d) (mm)	70	145	38	40
Number of sensors (S)	1	2	4	2
Additional distance based on sensor resolution (C) (mm)	448	1048	192	208
Safety distance (SD)* (mm)	1120	1720	864	880
HRC	62.5	100	66.64	40

*$K = 1600$ mm/s, $T = 0.42$ s, $C = 8(d - 14)$

4.4 Collaboration Case Study

The core of the robotic factory CPS development is the integration of dynamic characteristics of the individual components. The individual protection components register context, situation and status of worker, machine, plant and process and activate protective mechanisms before a hazard, e.g. collision, can occur. The production process will run without threats and interruptions and this will achieve the level of security and safety meeting worker safety legal requirements on industrial floor. Symbiotic human–robot collaboration (Wang et al. 2015) is defined for a fenceless environment in which productivity and resource effectiveness can be improved by combining the flexibility of humans and the accuracy of machines. Robotic CPS can enable such human–robot collaboration with the characteristics of dynamic task planning, active collision avoidance and adaptive robot control. Humans are part of the CPS design in which human instructions to robots by speech, signs or hand gestures are possible during collaborative handling, assembly, packaging, food processing or other tasks. All of these industrial tasks bring the focus of human–robot collaboration current research on heavy payload robots.

Figure 4.4 reveals a monitored area in which human and robot are interacting for completion of an industrial task. The vision system can be established through overhead 2D cameras or a 3D stereo vision camera and an additional aid of a laser scanner in case of any violation of robot workspace by human worker. The vision system is providing the worker real-time location information to the system. The robot system is normally programmed to reveal its end-effector position in all six DOFs. The next module is a sensor-mounted worker suit, which the worker will wear all the time. The suit contains multiple IMU fitted at various organ locations of the human worker thereby providing position and rate information to the CPS. The same can be proposed for an IMU fitted helmet for accurate head positioning information. These IMUs typically contain six sensors, i.e. three gyros for the three angular deflections and three accelerometers for linear acceleration measurement.

A pre-defined safe distance margin enables the robot to identify the worker near itself and speed or acceleration reduction can be started suddenly. This mode will feature speed and acceleration reduction upon identification and may lead to full stoppage of the robot till the time worker leaves the safe distance limit in the workspace. The robot will continue its job from the point it went into the full stop. The third module can be an interaction mode in which either the hands or the worker voice can be utilized to train the robot. For this, different hand gestures can

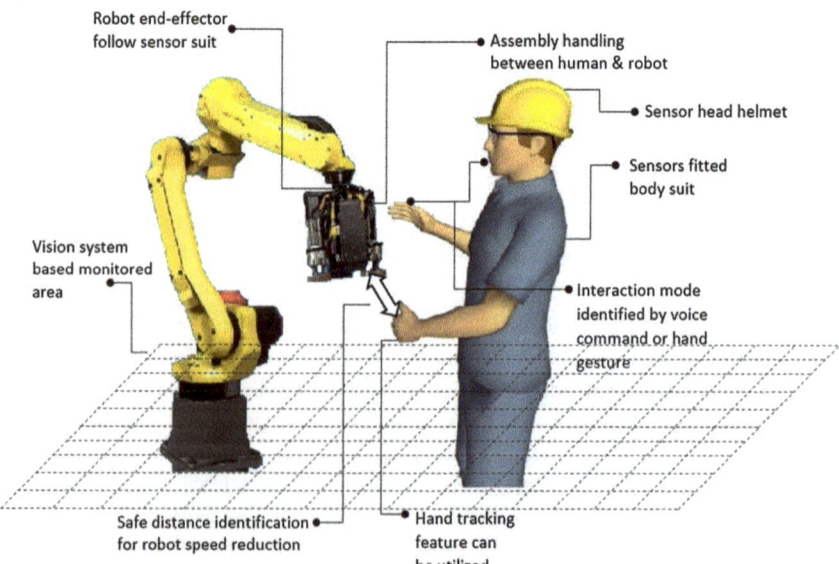

Fig. 4.4 Human–robot collaboration in CPS design

be used to enable interaction modes. For force monitoring system, the force reduction approach applies suddenly, once the safe distance margin is reached. Force sensors can provide an additional feature in the case of touching the human worker. Force calibration for different body organs is a must requirement in order to design such systems.

4.5 Guideline for Industrial Scenarios

In addition to the safety concepts, HRC attainment level and the sensor technology employed for a particular solution, there may be multiple industrial scenarios for which a generalized solution or a guideline can be established. These industrial scenarios range from single robot to multiple robots working together with multiple human workers constitutes HRC system. The possible approaches in order to build HRC system may include options like the use of inertial sensors, vision, radar or any hybrid approach for the human position monitoring. The hybrid option may consider any of the two approaches in a combined way. The real benefit of hybrid approach is the execution of task with high precision as the positioning information from two separate sensor systems will work mutually and compensate for the errors. Additionally, one technology area might be more practical in a given scenario, like vision system will not function in poor lighting condition or while meeting vision obstacle. In that scenario, the other technology sensors will keep the system functioning.

Figure 4.5 shows the guideline of a CPS design for HRC in which the industrial scenarios are given initially. Several collaboration indices can be evaluated based

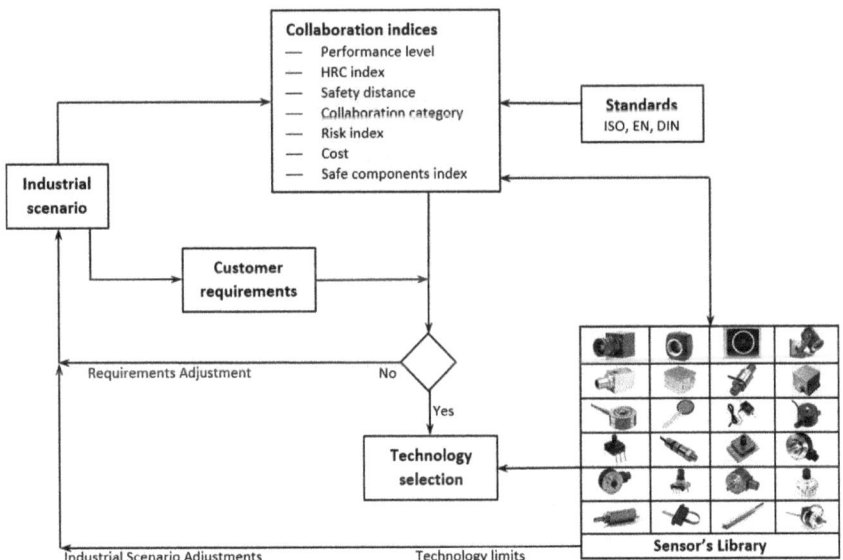

Fig. 4.5 Guideline to establish CPS for industrial HRC

on the sensor level information from the sensor library that can be established on the basis of prevailing state-of-the-art technology. The indices evaluated with the help of International Standards and sensor library catalogue are considered with customer specific requirements for the CPS design. The requirements are then adjusted according to the collaboration indices data. Once finalized with the requirement details, the technology selection process begins on the basis of sensor catalogue and requirements dossier. The technology limits will dictate the industrial scenario adjustments.

4.6 Conclusion

The paper identifies the requirements of HRC in an industrial context. A controlling CPS structure for HRC suggests the human worker to be an integrated part for which various interactive technologies can be employed. Different safety approaches for heavy payload robots are discussed. Few performance indices are established for collaboration. A very significant index is the safety distance and the calculation for different sensors reveals that 1 m distance is an essential requirement for the deployment of speed reduction approach. Safety schemes are graded for HRC attainment level and the sensor level requirements for each collaborative mode are identified. It is found that the high HRC level approaches have bounds from sensor data rates and the number and types of possible sensors used in the implementation of scheme. This highlights the technology limits and real-time issues for the achievement of high HRC. A case study revealed the detailed sensor requirements of an industrial human–robot collaboration example in which IMU's, camera system and laser scanner are the minimum required technological components. For CPS design in HRC, a generalized guideline is established to cater for various industrial scenarios. According to the guideline, many collaboration indices need to be evaluated to categorize the collaboration level. Moreover, the guideline shows that there is a necessary requirement to build a sensor catalogue that will help in industrial scenario adjustment and technology selection.

Acknowledgments The authors would like to acknowledge the support of the cLINK (Centre of excellence for Learning, Innovation, Networking and Knowledge), Erasmus Mundus Programme and the InSA Project, for this work.

References

Arai T, Kato R, Fujita M (2010) Assessment of operator stress induced by robot collaboration in assembly. CIRP Ann—Manuf Technol 59(1): 5–8. http://dx.doi.org/10.1016/j.cirp.2010.03.043
Kriaa S et al. (2015) A survey of approaches combining safety and security for industrial control systems. Reliab Eng Syst Saf 139: 156–178. http://linkinghub.elsevier.com/retrieve/pii/S0951832015000538

Knight W (2014) How human-robot teamwork will upend manufacturing. Business Report, Breakthrough Factories, MIT Technology Review

Lasota PA, Rossano GF, Shah JA (2014) Toward safe close-proximity human-robot interaction with standard industrial robots. In: The 10th IEEE international conference on automation science and engineering (CASE), Taipei, Taiwan

NASA, Robonaut R2. http://robonaut.jsc.nasa.gov. Accessed 21 Dec 2015

NIST (2013) Foundations for innovation in cyber-physical systems, workshop report. http://www.nist.gov/el/upload/CPS-WorkshopReport-1-30-13-Final.pdf

Pirvu B-C, Zamfirescu C-B, Gorecky D (2015) Engineering insights from an anthropocentric cyber-physical system: a case study for an assembly station. Mechatronics. http://linkinghub.elsevier.com/retrieve/pii/S095741581500152X

Wang L, Törngren M, Onori M (2015) Current status and advancement of cyber-physical systems in manufacturing. J Manuf Syst. http://linkinghub.elsevier.com/retrieve/pii/S0278612515000400

Zamfirescu CB, Pirvu BC, Schlick J (2013) Preliminary Insides for an Anthropocentric Cyber-physical Reference Architecture of the Smart Factory. Stud Inf Control 22:269–278

Chapter 5
A Trust Framework for Agents' Interactions in Collaborative Logistics

Morice Daudi, Jannicke Baalsrud Hauge and Klaus-Dieter Thoben

Abstract Trust is an essential factor for successful resource sharing in logistics. To build and sustain trust among collaborating partners in logistics requires, amongst others, conceptualizing on various aspects constituting underlying mistrusts. The conception is achieved by setting up a framework describing trust-based collaborative interactions of these partnering entities, referred to as *agents*. This research establishes a trust framework addressing agents' trustworthy interactions and thus aims at overcoming a knowledge gap identified in the literature. The framework depicts trust-based interactions concentrating to sharing of vehicle capacities. The trust framework is conceived on a foundation of theoretical body of knowledge in the literature. It engages knowledge on collaborative networks, logistics and transportation, agent behavior as well as trust. This research contributes by identifying key agents together with their roles, characteristics, tasks, information exchange as well as perceptions; all of which linked to agent trust. The framework is reusable in many ways, including formal conception of models aspiring to empirically investigate trust amongst agents sharing logistics resources. It also provides more understanding to practitioners, especially on issues relating to compromising differences resulting from agent's perspectives.

M. Daudi (✉) · K.-D. Thoben
International Graduate School for Dynamics in Logistics (IGS),
University of Bremen, Bremen, German
e-mail: mdaudi@uni-bremen.de

K.-D. Thoben
e-mail: tho@biba.uni-bremen.de

J.B. Hauge · K.-D. Thoben
Bremer Institut für Produktion und Logistik at the University of Bremen,
Bremen, Germany
e-mail: baa@biba.uni-bremen.de

J.B. Hauge
School of Technology and Health, KTH, Royal Institute of Technology,
Stockholm, Sweden

Keywords Trust · Trust framework · Logistics collaboration · Agent trustworthiness

5.1 Introduction

Logistics resource sharing employing collaboration strategy is increasingly becoming imperative nowadays. Through collaboration, companies reduce costs, improve quality of service, gain market position, minimize investment, as well as reducing emissions and congestion (Xu 2013). In Less-Than-Truckload (LTL), collaboration provides opportunities to exploit synergies in excess capacity, decrease lead times, and increase asset utilization (Peeta and Hernandez 2011). Despite of realized advantages, practical resource sharing under logistics collaboration is difficult due to many barriers, including the lack of trust (Xu 2013). Trust appears a primary obstacle limiting transport collaboration (Graham 2011; Hu et al. 2011) and its lack jeopardizes collaborative efforts (Hu et al. 2011). This lack of trust results into companies hesitating to share logistics resources like vehicle capacities due to uncertain outcomes of the perceived collaboration. While attempts to address trust problem in creation phase exist, there are limitations to investigation of trust in the operational phase of the collaborative organizations.

Further to performance-based trust employed in creation phase, trust in operational phase involves actions and interactions of the partners. For this reason, trust-building approaches employing only partner's "performance" (Seifert 2007; Asawasakulsorn 2009) is appropriate for partner selection and insufficient to daily execution of collaboration. Instead, such approaches must integrate partners' trusting actions and interactions occurring in operational phase. A pre-condition to this trust-building strategy comprises a feasible representation of trustworthy interactions among partnering entities. This representation is achieved by establishing a fundamental framework of trust-based interactions.

Such trust framework supposes to analyze as well as describe collaborative interactions of the partnering entities. Pursuant to the required trust framework, Collaborative Transportation Management (CTM) (VICS Logistics Committee 2004) and the conceptual framework of behavioral multi-agent model (Okdinawati et al. 2014) have provided some insights. They altogether identify CTM partners (agents), characteristics, interactions, and information exchange as well as required operating environment. Similarly, beyond identifying forms of collaboration within logistics network, Peeta and Hernandez (2011) have identified shippers and carriers as potential partners for exploiting LTL synergies. However, these works are unintended to represent trust relationships in collaborative supply networks. They lack an aspect of trust-based interactions purported to building trust supporting logistics and transportation collaboration. Moreover, Laeequddin et al. (2012) present an integrated conceptual model for trust building. The model addresses trust building from risk perspective in dyadic relationships. However, it does not support collaborations exceeding dyadic relationships and it lacks approaches to measure and assess trust. Thus, extending and integrating the model beyond its current purpose is highly needful.

The objective of this research therefore is to establish a trust framework. The framework is established by considering agents' collaborative interactions contextualized to sharing of vehicle capacities in logistics and transportation. In particular, the research identifies and specifies key partnering entities, their trustworthy characteristics, roles, tasks, exchanged information as well as perceptions. The trust framework is imperative to designing of trust models aimed to empirically investigate trust among collaborating entities.

5.2 Trust in Collaborative Networked Organizations

Emerged cloud computing platforms have lifted up the concept of collaboration with new push. It has enabled configuration of easy and feasible supply networks globally. In logistics and transportation, cloud computing platforms facilitate management of underlying collaborative processes to enable efficient flows of materials and information. There are improvements in demand forecasts and capacity planning. However, besides introducing new business models, cloud enabled collaborations pose challenges to building trust in new arisen partner relationships. Thus, this section reviews trust requirements under the settings of cloud-enabled collaboration, followed by prospective integration of underlying models.

5.2.1 Building Trust in Cloud-Enabled Collaborative Environments

Cloud computing platforms have brought in new collaborative environments. They have limited physical interactions by introducing networked society in which online trust outweighs offline trust. As online trust lacks face-to-face interactions, cloud-enabled collaboration needs appropriate trust mechanisms. To overcome this deficiency, most of the online trading partners apply "reputation systems and social graphs" (Jøsang et al. 2007; Kwan and Ramachandran 2009) to build trust in peer-to-peer business relationships. These reputations systems, however, can suffer unfair ratings by buyers and sellers, but additionally, discriminatory behaviors of the seller (Dellarocas 2003). On top of these mechanisms, investigating partners' trustworthy behaviors comprising of actions and interactions within cloud-enabled collaboration is crucial. A prerequisite to this investigation is a contextual specification of trust inherent in partners.

While trust is a context-specific construct, it challenges studying trust in all contexts (Nielsen 2011). With this view, trust contexts this paper applies are specified, considering: trust definition, trust embeddedness, and trust-building approach, as well as level of trust analysis. First, as this paper addresses trust problem in inter-firm collaboration, a trust definition in Mayer et al. (1995) is adapted. That, trust is a level of confidence trustor-party develops into trustee-party based on the expectation that trustee-party will perform a particular action necessary to trustor-party, irrespective of the ability to monitor or control trustee-party. This definition aligns to a notion of trust-as-choice or decision because it involves trusting behaviors in trust-related exchanges (Li 2012). A second context refers to trust embeddedness within actors. Laeequddin et al. (2012) highlight that trust is embedded within both actors, trustee and trustor. Trust embedded within trustee relates to trustee's competence, while the latter is conceptualized as feelings and emotions (Laeequddin et al. 2012). In this paper, trust as embedded within trustee is specified, because such trust is rational, measurable and least of psychology. Third, while the literature proposes building trust from perspectives of "performance" (Seifert 2007; Asawasakulsorn 2009) and risk (Laeequddin et al. 2012) the latter perspective matters the most. The preference grounds in reliability-based trust, a co-approach to trust building in the perspective of risk-worthy relationships. Fourth, the literature (Delbufalo 2012; Nguyen and Liem 2013) identify three levels to trust analysis: inter-personal, inter-group, and inter-organizational. Subject to trust inherent to inter-firm collaboration, this paper analyzes trust at inter-organizational level. Considering these arguments together, trust is contextualized as embedded within trustee, under risk-worthy perspective, and analyzed at inter-organizational level.

5.2.2 Collaborative Logistics and Transportation

Literature refers to Collaborative Transportation Management (CTM) as a holistic process that brings together supply chain trading partners and service providers to drive inefficiencies out of the transport planning and execution process (VICS Logistics Committee 2004). It desires fulfilling inability of Collaborative Planning, Forecasting, and Replenishment (CFPR) in executing order processes (VICS Logistics Committee 2004). In place of CFPR, CTM improves performance by reducing dwell time, dead miles, and optimizing transport asset utilization.

There are similarities among "development phases of collaborative enterprise" (Thoben and Jagdev 2001), "integrated CTM business process" (VICS Logistics Committee 2004), and "planning and management decisions within transport logistics" (Gonzalez-Feliu and Morana 2011). While collaborative enterprises develop in a life-cycle of four phases: preparation, creation, operation, and decomposition; the remaining develop in three phases. Illustratively, members prepare to collaborate by selecting trustworthy partners (creation) to start sharing resources (operation), and their alliance dissolves at the end. Within CTM business process of the transport logistics, decision-making and management by partners occur at three levels: strategic, tactical, and operational. To comprehensively understand partners' trustworthy behaviors; preparation, creation, and operation phases are proposedly merged with strategic, tactical, and operational phases respectively. To that effect, this paper establishes and formalizes three phases/stages appropriate to model agents' trustworthy actions and interactions. They are: (1) selection and front-end agreement occurring in strategic level; (2) planning/order forecast occurring in tactical level, and (3) operational stage whereby actual delivery are executed.

Nevertheless, literature describes three categories of partnering entities, which this paper refers to as "*partners*", "*actors*," or "*agents*." VICS Logistics Committee (2004) and Okdinawati et al. (2014) have identified shippers, carriers, and receivers as CTM potential partners. Gonzalez-Feliu and Morana (2011) have identified loaders (sender and receiver), logistics providers, and transporters as typical actors. Although other stakeholders like representatives of governing units exist, pursuant to trust building, shipper, carrier, and receiver are relevant partners.

5.3 Research Methodology

A trust framework depicting agents' interactions to logistics and transport collaboration is conceived on a foundation of theoretical body of knowledge in the literature. The conception engages knowledge of collaborative networks, logistics and transportation, agent behavior as well as trust (Fig. 5.1). Collaborative network contributes base concepts functional to identify life-cycle and essential components constituting collaborative enterprises. Being an application domain, logistics and transportation are necessary to defining business requirements processes. Similarly, trust concepts in supply chain and logistics as well as agents' behavior concentrating on transport collaboration are utilized to capture domain-specific connotations.

Development of the framework proceeds by integrating strengths of the existing frameworks and models to fill limitations previously outlined. This integration is constituted by: (1) CTM model and its business processes (VICS Logistics Committee 2004); (2) behavioral multiagent model for collaborative transportation (Okdinawati et al. 2014); (3) an integrated conceptual model for trust building in supply chain partners relationship (Laeequddin et al. 2012), and (4) enterprise development phases (Thoben and Jagdev 2001) (Fig. 5.1). These models constitute prominent literature whose critical analysis prior to selection is unpresented because

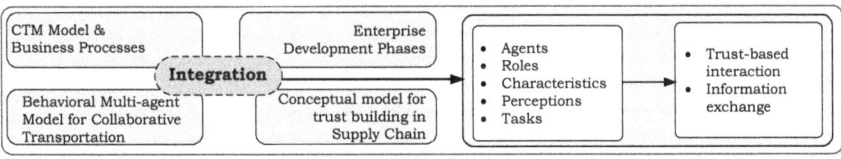

Fig. 5.1 Methodology applied to conceive the trust framework

of space limitation. Thereafter integration, resulting composition is extended to provide trust-building requirements comprising of features, actions, and interactions of agents. Moreover, regarding vehicle sharing, the trust framework establishes crucial trust relationships among partnering agents like shippers, carriers and receivers.

5.4 Conception of the Trust Framework

To conceive trust framework, the term "agent" in place of "partner" is formalized to co-establish a linkage with prospective agent trust models whose development will apply this framework. This conception falls in three stages (Fig. 5.2). Three agents: *shipper, receiver*, and *carrier* are formalized as key partnering entities, with roles of producing, ordering/receiving, and moving of goods respectively. Nevertheless, depending on collaboration and trust building requirements, each agent can consists of more than one actor servicing similar role. These agents possess distinct perspectives with respect to production, distribution and demand of goods. This paper refers to this perspective as *perception*. Perception represents agent position to synchronized decision-making in relation to underlying issues require deliberation.

Diverse interpretations, understandings, and implementations by partnering agents on production, distribution, and demand of goods create difficulties and mistrusts. The mistrusts arise whenever a satisfactory compromise is unreached among shippers, receivers, and carriers. Although there are many perceptions creating mistrusts, seven perceptions in (Okdinawati et al. 2014) are adapted and customized to needs of trust building. These perceptions (Fig. 5.2) are distinctly inherent in shippers, receivers, and carriers. Illustratively, shippers perceive *producing goods based on capabilities and capacity (P1), as well as fixing production quantities (P5)*. Receivers have perceptions on *uncertain market demand (P2), and unguaranteed on-time delivery (P8)*. Similarly, carriers perceive: *increasing profit*

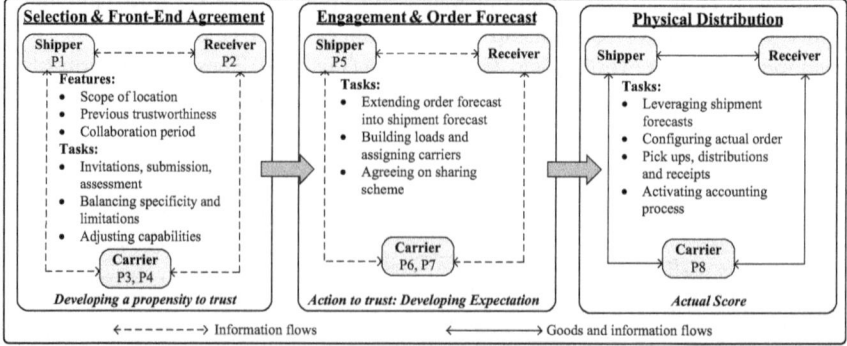

Fig. 5.2 Conceptual trust framework depicting agents' interactions in logistics collaboration

by increasing transportation rates (P3); offering limited carriage capacity (P4); fixing delivery quantities (P6); and on their full carriage capacity (P7). These perceptions are further described in subsequent paragraphs and subsections.

These perceptional differences create conflicts or "dilemmas" (Bryant 2004; Bennett 2004; Okdinawati et al. 2014) and consequently multiplying mistrusts. Aiming to model trust building, this paper emphasizes on "two dilemmas belong to collaboration: co-operation and trust dilemma" (Bryant 2004; Okdinawati et al. 2014). Bennett (2004) defines these dilemmas as follows: X has a *cooperation dilemma* if he has an improvement from his position, and; X has *trust dilemma* if Y can improve by moving away from X's position. By thoroughly analyzing these dilemmas as well as extending them to requirements of trust building, arguably, they describe similar context. This context roots from the concept of trust embeddedness. The dilemmas depict a mistrust situation whereby trustor-agent doubts whether trustee-agent will comply with established agreements.

To resolve perceptional conflicts, one has to synchronize decisions toward deliberate compromise. For example, in P5: shipper and receiver have a conflict because receiver make orders based on market demand. Therefore, a compromise is required to negotiate, whereby one agent convinces other agents to own perceptional position or follow them otherwise. The ability to negotiate stems from agent's internal ability, to either release or hold its perceptional position. Subject to which perception prevails or is compromised to, it commonly result into variant trust effect. The variance in trust effect depends on the extent of realized fairness with comparison to prior developed expectation.

5.4.1 Selection and Front-End Agreement

In creation stage (Fig. 5.2) agents establish strategic agreements thereby developing an intention to trust. According to Laeequddin et al. (2012), the intention to trust employing characteristics trust building is referred to as a propensity to trust. It is an initial willingness of partnering agents to collaborate. To develop propensity to trust, trustor-agent determines trustee's characteristics as to whether they satisfy the extent of its intention. These characteristics are unlimited to scope of geographical location, previous trustworthy relationships, production and carriage capacities, as well as market demand (Fig. 5.2). Initially, development of the propensity to trust involve three tasks: invitation, submission, and assessment. The agent taking initiative to collaborate, called *broker*, invites prospective agents to a collaborative opportunity. Interested agents apply by submitting required information. Information describing previous trust relationships is directly accessed by broker from the community in which the agent resides and operates. On submission, broker assesses capabilities of the agents as to whether they satisfy specificity of the prospective consortium, and publicize results. On receipt of results, applicant agents can: accept the proposal, suggest adjustments, or reject it. If the proposal is rejected or suggested to adjustments, broker adjusts strategic structures (capabilities) until acceptable or terminates it. Besides characteristics trust building, the propensity to trust develops as well from the compromise reached by agents.

Alongside previous highlights, prior to passing strategic agreement, four perceptions must be deliberated. These perceptions are briefly presented together with the position of each agent. First, in *P1*: as shipper perceives producing based on its capabilities, receiver perceives the shipper to produce based on demand, while carrier remains in neutral position. Second, in *P2:* whereby receiver positions itself to uncertain market demand, both shipper and carrier perceive certainty in market demand. Third, in *P3:* as carrier perceives increasing profit by increasing transport rates, both shipper and receiver perceive him to increase profit by reducing transportation cost. Fourth, in *P4:* while carrier perceives limited carriage capacity, both shipper and receiver are in a position that carrier should offer unlimited carriage capacity. A detailed representation of the deliberate compromise on agents' position will constitute a corresponding trust model, which is not part of this work.

5.4.2 Engagement and Order Forecast

Agents engage in collaborative actions thereby forecasting orders and developing rational trust (Fig. 5.2). Rational trusting is a deliberate action to trust, expressed as the degree of trust the trustor-agent develops in trustee. It is established depending on existing facts. To develop rational trust, agents execute many tasks. Specific tasks include order extension, load building, agreement on sharing scheme, and establishing expectation. Hereby a belief is, collaborating agents have more accurate information about demand and production quantities. Additionally, relying on shipment forecasts, carrier begins to build load and assigns to available trucks to early determine carriage needs. Furthermore, considering proportional scheme and Shapley values incentive schemes, agents urge deliberating which scheme they apply. On successful executions of these tasks, agents negotiate to compromise on arising perception. This stage comprises of three perceptions. They are briefly described to illustrate how specific agent perceives, and what are positions of the rival agents. First, in *P5:* as shipper perceives producing fixed quantities, receiver perceives the shipper to produce consistent with demand, while carrier is unaffected by the perception. Second, in *P6:* the carrier perceives to deliver fixed quantities, while shipper and receiver perceive that receiver should fluctuate the delivery consistent with demand. Third, in *P7:* carrier perceives a delivery to full capacity of its carriage, while receiver and shipper are reluctant to whether vehicle is full or not. Again, a resolution to these conflicts is a synchronized decision applying negotiating power to persuade other agents to own perception or follow them otherwise.

5.4.3 Physical Distribution

This is an execution stage, whereby goods are moved to indeed realize collaborative expectations, also termed as *"actual score"*. Actual scores realized constitute a crucial feedback, signaling the trustor-agent about the confidence, trust, it developed into trustee. This feedback is either below, within or above the expectation developed in prior. To realize this score, several tasks are executed. Firstly, carrier agent leverages shipment forecast depending on offered orders, and it checks for carriage constraints (if any) and leverage them accordingly. It secondly configures actual order and prepare it ready for distribution. Thirdly, carrier picks up and delivers inventories to their point of utility. Finally, the broker activates accounting process to reward partner agents. Hereafter, each agent is provided with records about operating performance resulting from executed collaboration that it compares with its prior expectations. Moreover, agents have further perceptions relating to collaboration performance. They include visibility information, lead time, responsiveness and on-time delivery. Fortunately, to these perceptions, all agents have same position except for on-time delivery *(P8)* for which carrier perceives a delay possibility.

5.5 Conclusions, Implications and Outlook

Trust building is long-term multi-disciplinary agenda in collaborative networked organizations including those in supply chain and logistics. It is however that majority of literature address trust in view of history, context, process, role, measurement as well as evaluation. Beyond these views, it is known that mistrusts root also from collaborative actions and interactions of the partnering agents. While comprehending on this view, empirical studies to investigate agents' trustworthy interactions and respective behaviors in real world of collaborative scenarios are scarce. This paper fulfills this literature gap by initially establishing the conceptual trust framework. The framework depicts trust-based interactions of the partnering agents to address the question of trust building amongst entities sharing logistics resources. It depicts key partnering agents together with their trust-based roles, characteristics, tasks, perceptions as well as information exchange. Additionally, propositions on how agents develop the propensity to trust, rational trust, as well as benchmarking expectation against the reality are established. Moreover, agents' trustworthy interactions this framework represents, inspire a significant trust building approach, which employs analysis of character behaviors and operating performance.

The proposed framework is substantially imperative in guiding the design of trust models purported to empirical investigation of trust amongst agents sharing logistics resources. To the large extent, this framework represents collaborative interactions aimed to elicit trust-building approaches in operational phase. It provides a clear understanding on mistrusts situation as well as trust scenarios agents undertake. A basis to validity of the proposed trust framework is largely determined by employed methodology. The framework is developed by integrating and extending prominent frameworks and models. Future research works are unlimited to design and evaluation of trust model resulting from this framework.

Acknowledgments The corresponding author would like to acknowledge the support of the Ministry of Education and Vocation Training (MoEVT – Tanzania), and the German Academic Exchange Service (DAAD) under the Tanzania-Germany Postgraduate Training Programme.

References

Asawasakulsorn A (2009) Transportation collaboration: partner selection criteria and IOS design issues for supporting trust. Int J Bus Inf 4(2):199–220

Bennett P (2004) Confrontation analysis: prediction, interpretation or diagnosis? In: Analyzing conflict and its resolution. Proceedings of a conference of the Institute of Mathematics and its Application, IMA, Southend-on-Sea

Bryant J (2004) Drama theory as the behavioral rationale in agent-based models. In: Analyzing conflict and its resolution, conference of the Institute of Mathematics and its Applications. Oxford

Dellarocas C (2003) Building trust online: The design of robust reputation reporting mechanisms in online trading communities. In: Information society or information economy? A combined perspective on the digital era. Idea Book Publishing, pp 95–113

Delbufalo E (2012) Outcomes of inter-organizational trust in supply chain relationships: a systematic literature review and a meta-analysis of the empirical evidence. Supply Chain Manage 17(4):377–402

Gonzalez-Feliu J, Morana J (2011) Collaborative transportation sharing: from theory to practice via a case study from France. In: Yearwood JL, Stranieri A (eds) Technologies for supporting reasoning communities and collaborative decision making: cooperative approaches, information science reference, Hershey, PA, pp 252–271

Graham L (2011) Transport collaboration in Europe. White Paper, ProLogis Research

Hu J, Bian W, Han S, Ju S (2011) An empirical research of the influence factors of the action from MF network collaboration to logistics organization performance. AISS 3(6):259–265

Jøsang A, Ismail R, Boyd C (2007) A survey of trust and reputation systems for online service provision. Decis Support Syst 43(2):618–644

Kwan M, Ramachandran D (2009) Trust and online reputation systems. In: Golbeck J (ed) Computing with social trust. Springer, pp 287–311

Laeequddin M, Sahay BS, Sahay V, Waheed KA (2012) Trust building in supply chain partners' relationship: an integrated conceptual model. J Manage Dev 31(6)550–564

Li PP (2012) When trust matters the most: The imperatives for contextualizing trust research. J Trust Res 2(2):101–106

Mayer RC, Davis J, Schoorman FD (1995) An integrative model of organizational trust. Acad Manage Rev 20(3):709–734

Nielsen BB (2011) Trust in strategic alliances: toward a co-evolutionary research model. J Trust Res 1(2):159–176

Nguyen NP, Liem NT (2013) Inter-firm trust production: theoretical perspectives. Int J Bus Manage 8(7):46–54

Okdinawati L, Simatupang TM, Sunitiyoso Y (2014) A behavioral multi-agent model for collaborative transportation management (CTM). In: Proceedings of T-LOG

Peeta S, Hernandez S (2011) Modeling of collaborative less-than-truckload carrier freight networks, USDOT Region V Regional University Transportation

Seifert M (2007) Collaboration formation in virtual organizations by applying prospective performance measurement. Dissertation, University of Bremen

Thoben KD, Jagdev HS (2001) Typological issues in enterprise networks. Manage Oper 12 (5):421–436

VICS Logistics Committee (2004) Collaborative transportation management. Accessed 8 Apr 2015

Xu X (2013) Collaboration mechanism in the horizontal logistics collaboration. Dissertation, Paris Institute of Technology

Chapter 6
Quality-Aware Predictive Scheduling of Raw Perishable Material Transports

Xiao Lin, Rudy R. Negenborn and Gabriël Lodewijks

Abstract This paper proposes a new mathematical model for predictive scheduling of perishable material transports with the aim of reducing losses of perishable goods. The model is particularly designed for allocation of potatoes from several farms to a nearby starch mill, which produces starch from a limited amount of potatoes each day. Scheduling should determine how much amount of potatoes be sent from which farm to the mill on each day. It is known that the quality of potatoes decreases over time and as a result less starch is produced. A model predictive control approach is proposed to maximize the production of starch. Simulation experiments indicate that predictive scheduling can yield higher starch production compared to non-predictive approaches.

Keywords Transport scheduling · Predictive control · Perishable products · Kinetics modeling

6.1 Introduction

Potatoes are one of the important sources of high quality starch, suitable for most industrial and food engineering (Grommers and van der Krogt 2009). It is reported that 6.9 million tonnes of potatoes are processed for starch in the year of 2014 in Europe (Starch Europe 2016). However, potatoes cannot be stored for a long time as the starch concentration drops and starch producing campaign is short in each year, lasting only a few months. Due to the high water content, the transport of starch potatoes is preferably limited from the field to nearby starch mills (DG Agriculture and Rural Development 2010).

X. Lin (✉) · R.R. Negenborn · G. Lodewijks
Department of Maritime and Transport Technology,
Delft University of Technology, Delft, The Netherlands
e-mail: x.lin@tudelft.nl

R.R. Negenborn
e-mail: r.r.negenborn@tudelft.nl

G. Lodewijks
e-mail: g.lodewijks@tudelft.nl

The extraction of starch from potatoes is an important step in starch supply chains. Usually a starch mill is located near potato farms, producing starch from potatoes during the harvest time of each year. When a starch producing campaign starts, farmers harvest potatoes and store them in their chambers, waiting to transport to the mill for processing. During this waiting time, potatoes start to deteriorate, and starch concentration starts to decrease. Therefore the scheduling of potato processing can affect the total amount of starch produced. However, limited research has been done on how to schedule the transport between farms and starch mills more effectively.

For other products, earlier research has been carried out. Grape harvesting scheduling is investigated for decision-making on which vineyards should be harvested (Ferrer et al. 2008). A model is developed to schedule harvesting time for grapes considering operation costs as well as quality of the grapes. Similarly, an orange harvesting scheduling problem (Caixeta-Filho 2006) also considers the quality of the products when deciding harvesting moments. In both of the studies, changes in quality are described using historical data. Studies are also carried out for sugarcane supply scheduling on mill sugar production (Gal et al. 2008; López-Milán and Plà-Aragonés 2015). The first study takes into account variability of sugarcane quality between areas. Historical data is used to identify regions based on different quality of yields. The second study investigates sugarcane processing supply considering an index of ripeness for deciding on harvesting moments, which affect the quality of sugar in products. The definition of this quality index is however, not explained explicitly.

Historical data can be useful in the scheduling of production processes. However, it does not thoroughly reflect real-time quality. The objective of this research is to show how real-time quality can impact the decisions for allocation and processing of raw perishable material. In this paper we consider a case in which a starch mill locates amongst several potato farms. Potatoes from the farms are sent to the mill and from which starch is extracted. Starch degrades as potatoes are stored in chambers. Therefore, the mill has to run all the time to maximize the production of starch. The key feature of the approach we propose is that it considers real-time quality of perishable goods instead of historical data to assist decision-making in fresh product logistics scheduling. This paper illustrates that with the information of "quality," fewer losses can be achieved and producers can benefit more from effective scheduling.

We adopt a system and control perspective to solve the scheduling problem. As illustrated in Fig. 6.1, there are two parts of the model. The system part is our representation of the real world, consisting of the amount and quality of potatoes in each chamber of each farm. The controller part takes measurements from the system and makes decisions on when and which farm shall be called to send potatoes to the mill. This perspective enables us to demonstrate the performance of controllers with different scheduling strategies.

The remainder of this paper is organized as follows. Section 6.2 describes the perishable material system and its dynamics. Section 6.3 discusses different strategies for controller design. Section 6.4 focuses on a case study, its results, and discussions. Section 6.5 concludes this paper and provides directions for future research.

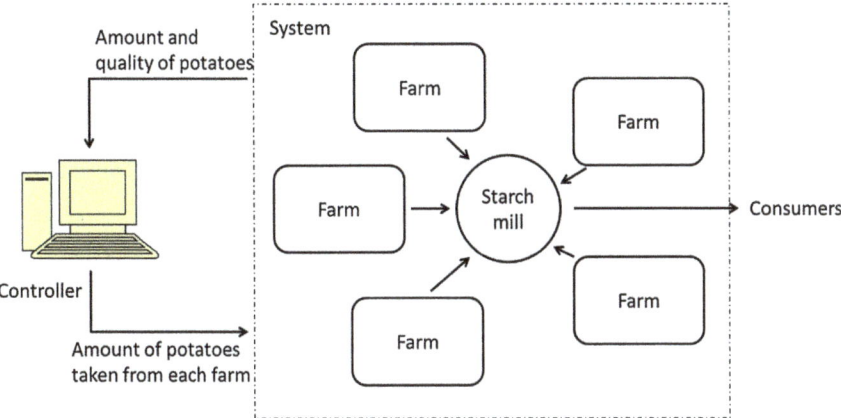

Fig. 6.1 System and control perspective

6.2 System Design for Perishable Material Distribution

This section describes the details of the system, consisting of assumptions, farm and quality dynamics and constraints.

6.2.1 Assumptions

In this study a few assumptions are made

- the farmers and transport of potatoes are assumed to be contracted; cost for potatoes and transport are therefore not considered explicitly in this study;
- the quality of potatoes decreases monotonically, and can be accurately measured and predicted; the transport process does not affect quality of potatoes;
- quality measurements, decision-making, transport, arrival, and processing of potatoes happen during the same day.

6.2.2 System Dynamics

6.2.2.1 Farm Dynamics

The system consists of a set of farms $\mathscr{F} = \{1, \ldots, i, \ldots, F\}$ with S_i chambers $\mathscr{S}_i = \{1, \ldots, j, \ldots, S_i\}$ on each farm, and a starch mill as components of the system. Each day in a chamber j of farm i, $h_{i,j}(k)$ represents the amount of potatoes brought in with the harvest, and $u_{i,j}(k)$ represents the amount of potatoes moved out for the

Fig. 6.2 A chamber

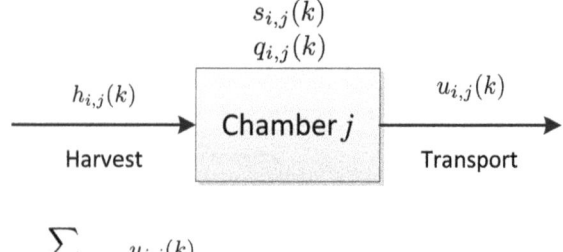

Fig. 6.3 The mill

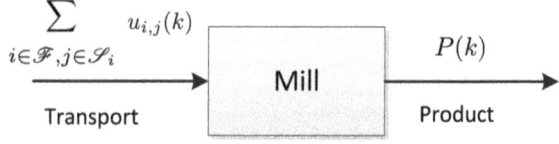

mill, where $k \in \mathcal{T} = \{1, \ldots, k, \ldots, T\}$. The amount of potatoes remaining in the chamber is $s_{i,j}(k)$, and $q_{i,j}(k)$ is the quality of them, as shown in Fig. 6.2. Each day during the campaign potatoes from different chambers are transported to the mill, and $P(k)$ is the total starch production on that day, as shown in Fig. 6.3. Therefore we have

$$s_{i,j}(k+1) = s_{i,j}(k) - u_{i,j}(k) + h_{i,j}(k), i \in \mathcal{F}, j \in \mathcal{S}_i, k \in \mathcal{T} \quad (6.1)$$

$$P(k) = \sum_{i \in \mathcal{F}, j \in \mathcal{S}_i} u_{i,j}(k) q_{i,j}(k). \quad (6.2)$$

6.2.2.2 Quality Dynamics

In food engineering, kinetic models are widely used to describe food quality (Van Boekel 2008). The following equation represents the evolution of the quality $q(t)$:

$$\frac{dq(t)}{dt} = -rq^n(t), \quad (6.3)$$

where variable n is determined by the type of reaction, and the coefficient r is determined by Arrhenius' law, which describes the temperature dependence of chemical reactions (Van Boekel 2008). The starch concentration of stored potatoes decreases over time. According to an experimental study (Nourian et al. 2003), the degradation of the starch concentration follows a first-order kinetic model, with $n = 1$ and r ranging from 0.02074 to 0.04735 in different storage temperatures. With q_0 the initial quality when harvested, the remaining quality at continuous time t can then be denoted as

$$q(t) = q_0 \exp(-rt). \quad (6.4)$$

In this study, we use the discrete kinetic model as decisions are made at discrete time steps. Let τ be the time interval between time step k and $k+1$. Note that quality at any time step k can be seen as an "initial quality" for time $k+1$ and that r is time-relevant. Let $R(k) = \exp(-r(k)\tau)$, we then have

$$\begin{aligned} q(k+1) &= q_0 \exp(-r(k+1)\tau) = q(k)\exp(-r(k)\tau) \\ &= q(k)R(k). \end{aligned} \quad (6.5)$$

6.2.2.3 System State-Space Representation

To formulate a compact state-space representation of the system dynamics, let vector $\mathbf{x}(k)$ denote the state of the system, consisting of the amount of potatoes and the quality of potatoes in all farms and chambers. In this research, the states of the system are assumed measurable, so $\mathbf{y}(k) = \mathbf{x}(k)$. Decision variables are collected in vector $\mathbf{u}(k)$, denoting the amount of potatoes to be transported from each chamber to the mill on day k. The amount of remaining potatoes in the chambers is denoted by $\mathbf{s}(k)$. The qualities of the potatoes are given by $\mathbf{q}(k)$. Vector $\mathbf{h}(k)$ represents the amount of potatoes harvested on day k, which is considered as disturbances of the system $\mathbf{d}(k)$. The element of matrix $\mathbf{H}$ (with the size of $F * T$) on row i, column k is 1 if there is harvesting for farm i, day k; and 0 otherwise. The state-space form of this system is

$$\begin{aligned} \mathbf{x}(k+1) &= \mathbf{A}(k)\mathbf{x}(k) + \mathbf{B}\mathbf{u}(k) + \mathbf{C}\mathbf{d}(k) \\ &= \begin{bmatrix} \mathbf{I} & 0 \\ 0 & \mathbf{R}(k) \end{bmatrix} \begin{bmatrix} \mathbf{s}(k) \\ \mathbf{q}(k) \end{bmatrix} + \begin{bmatrix} -\mathbf{I} & 0 \\ 0 & 0 \end{bmatrix} \begin{bmatrix} \mathbf{u}(k) \\ 0 \end{bmatrix} + \begin{bmatrix} \mathbf{I} & 0 \\ 0 & 0 \end{bmatrix} \begin{bmatrix} \mathbf{h}(k) \\ 0 \end{bmatrix} \end{aligned} \quad (6.6)$$

$$\mathbf{y}(k) = \mathbf{x}(k) \quad (6.7)$$

where

$$\mathbf{A}(k) = \begin{bmatrix} \mathbf{I} & 0 \\ 0 & \mathbf{R}(k) \end{bmatrix} \quad (6.8)$$

$$\mathbf{R}(k) = \begin{bmatrix} R_{1,1}(k) & \cdots & 0 \\ \vdots & \ddots & \vdots \\ 0 & \cdots & R_{F,S}(k) \end{bmatrix} \quad (6.9)$$

$$\mathbf{x}(k) = [\mathbf{s}^{\mathrm{T}}(k), \mathbf{q}^{\mathrm{T}}(k)]^{\mathrm{T}} \quad (6.10)$$

$$\mathbf{s}(k) = [s_{1,1}(k), s_{1,2}(k), \ldots, s_{i,j}(k), \ldots, s_{F,S}(k)]^{\mathrm{T}} \quad (6.11)$$

$$\mathbf{q}(k) = [q_{1,1}(k), q_{1,2}(k), \ldots, s_{i,j}(k), \ldots, s_{F,S}(k)]^{\mathrm{T}} \quad (6.12)$$

$$\mathbf{u}(k) = [u_{1,1}(k), u_{1,2}(k), \ldots, u_{i,j}(k), \ldots, u_{F,S}(k)]^{\mathrm{T}} \quad (6.13)$$

$$\mathbf{h}(k) = [h_{1,1}(k), h_{1,2}(k), \ldots, h_{i,j}(k), \ldots, h_{F,S}(k)]^{\mathrm{T}} \quad (6.14)$$

$$\mathbf{d}(k) = \mathbf{h}(k) \quad (6.15)$$

6.2.3 Constraints

There are a number of constraints among the system variables that need to be satisfied. First, the amount of potatoes $u_{i,j}(k)$ transported is nonnegative and no more than potatoes in the chamber. Second, the mill limits the daily processing capacity. Therefore we have

$$0 \leq u_{i,j}(k) \leq s_{i,j}(k), \tag{6.16}$$

$$\sum_{i \in \mathcal{F}} \sum_{j \in \mathcal{S}_i} u_{i,j}(k) \leq c(k). \tag{6.17}$$

6.3 Controller Design

The controller takes measurements of the system and makes decisions on what actions to take to control the system, which, in this paper, is when and from which farm potatoes shall be transported to the mill for processing (as illustrated in Fig. 6.1). We consider three different control strategies

- C_1: Quality-unaware controller: scheduling without any knowledge of quality;
- C_2: Quality-aware controller: with information of quality of the day;
- C_3: Predictive quality-aware controller: considering quality over a prediction horizon.

The difference between these strategies is the amount of information regarding quality of the goods the controller uses. Because the degradation rate of quality can easily be affected and thus differs from place to place, it is expected that controllers that use the knowledge of quality can yield better performance.

6.3.1 Quality-Unaware Controller

The quality-unaware controller does not apply any form of optimization and randomly, repeatedly picks a farm on each day, and starts the transport with the most recently harvested potatoes on that farm until the daily capacity is reached, as is shown in Algorithm 1.

Algorithm 1 Quality-unaware controller C_1

1: $u_{i,j}(k) \leftarrow 0$ for each $i \in \mathcal{F}$ and $j \in \mathcal{S}_i$
2: **repeat**
3:     Randomly pick a farm $i \in \mathcal{F}$
4:     Take away potatoes most recently harvested, $u_{i,j}(k) \leftarrow u_{i,j}(k) + s_{i,j}(k)$
5: **until** $s_{i,j}(k) \geq c(k) - u_{i,j(k)}$ (the daily capacity reached)

Algorithm 2 Quality-aware controller C_2

1: Take measurements $\mathbf{q}(k)$ for all the stored potatoes
2: Solve the optimization problem (6.18)–(6.22) to get values for $\mathbf{u}(k)$, that maximize the starch production of the day

6.3.2 Quality-Aware Controller

The quality-aware controller takes daily measurements of the quality and amount of potatoes in each chamber. It then chooses to transport potatoes that have the best quality first as shown in Algorithm 2. This controller maximizes the daily production of starch by solving the following optimization problem:

$$\max_{\mathbf{u}(k)} J(\mathbf{x}(k), \mathbf{u}(k), \mathbf{d}(k)) = \sum_{i \in \mathscr{F}} \sum_{j \in \mathscr{S}_i} q_{i,j}(k) u_{i,j}(k) \qquad (6.18)$$

subject to

$$\mathbf{x}(k+1) = \mathbf{A}\mathbf{x}(k) + \mathbf{B}\mathbf{u}(k) + \mathbf{C}\mathbf{d}(k) \qquad (6.19)$$
$$\mathbf{y}(k) = \mathbf{x}(k) \qquad (6.20)$$
$$0 \le u_{i,j}(k) \le s_{i,j}(k) \qquad (6.21)$$
$$\sum_{i \in \mathscr{F}} \sum_{j \in \mathscr{S}_i} u_{i,j}(k) \le c(k) \qquad (6.22)$$

6.3.3 Predictive Quality-Aware Controller

A predictive quality-aware controller considers predicted quality over a prediction horizon. Therefore the optimal decision for these days can be made at the beginning of each day to maximize the total production. The controller is designed in a model predictive control fashion (Camacho and Bordons 2013; Rawlings and Mayne 2013): only the decisions for the first day are implemented and rest discarded. New optimization is carried out at the next day. These iterations go on until the end of the experiment. The optimization problem solved by the controller is then[1]

$$\max_{\tilde{\mathbf{u}}(k)} J(\tilde{\mathbf{u}}(k), \tilde{\mathbf{u}}(k), \tilde{\mathbf{d}}(k)) = \sum_{l=0}^{N_p - 1} \sum_{i \in \mathscr{F}} \sum_{j \in \mathscr{S}_i} q_{i,j}(k+l) u_{i,j}(k+l) \qquad (6.23)$$

---

[1] Because the simulation is finite, on the last several days (from the day $T - N_p + 2$ to day T), the prediction horizon based on current date exceeds the length of the simulation, resulting in errors. To solve this, the controller reduces the prediction horizon if it exceeds the length of the simulation.

Algorithm 3 Predictive quality-aware controller C_3
―――――――――――――――――――――――――――――――――――――
1: Take measurements and predictions for all the stored potatoes within the prediction horizon l
2: Solve the problem (6.23)–(6.27) to find the optimal solutions for $\tilde{\mathbf{u}}(k)$
3: Implement the decision $\mathbf{u}(k)$
―――――――――――――――――――――――――――――――――――――

subject to

$$\mathbf{x}(k+l+1) = \mathbf{A}\mathbf{x}(k+l) + \mathbf{B}\mathbf{u}(k+l) + \mathbf{C}\mathbf{d}(k+l) \quad (6.24)$$

$$\mathbf{y}(k+l) = \mathbf{x}(k+l) \quad (6.25)$$

$$0 \le u_{i,j}(k+l) \le s_{i,j}(k+l) \quad (6.26)$$

$$\sum_{i \in \mathscr{F}} \sum_{j \in \mathscr{S}_i} u_{i,j}(k+l) \le c(k+l) \quad (6.27)$$

where $\tilde{\mathbf{x}}(k) = [\mathbf{x}^T(k+1), \ldots, \mathbf{x}^T(k+N_p)]^T$ is the vector consisting of all system states over the prediction horizon. Similarly, $\tilde{\mathbf{u}}(k)$ and $\tilde{\mathbf{d}}(k)$ are $[\mathbf{u}^T(k+1), \ldots, \mathbf{u}^T(k+N_p)]^T$ and $[\mathbf{d}^T(k+1), \ldots, \mathbf{d}^T(k+N_p)]^T$, respectively. Using this controller, only the first step of the decisions are implemented on the system, i.e. $u(k)$. The procedure of this controller is described in Algorithm 3.

6.4 Experiments

In this section, the potential of the three different control strategies is illustrated on scenarios randomly generated to simulate the variability of potato quality and the changing rate of it. Potatoes stored in one farm are assumed to have the same deterioration rate, which may be affected by local temperature and other conditions in the chambers. To simulate the impact from the variability of external conditions, we generate the deterioration rate from normal distributions. Results of the quality-aware controller and predictive quality-aware controller with different prediction horizons are compared to the quality-unaware controller to prove the improvements of considering quality information in scheduling.

6.4.1 Simulation Setup

Each simulation considers one scenario, in which all three strategies are implemented. We consider 10 simulations altogether. In each simulation, different strategies are performed. In order to set up different scenarios, some of the variables are nondeterministic, following some distributions. Other variables remain deterministic.

6 Quality-Aware Predictive Scheduling of Raw Perishable Material Transports

Deterministic variables The sets of farms and chambers, the length of the simulation in days, initial amount of potato storage, amounts and dates of harvesting on each farm remain the same. Variable $s_{i,1}(1)$ represents the initial storage of farm i, which is placed in the first chamber of each farm. Potatoes of new harvest are placed in other empty chambers for simplification. We consider three times of harvest for each farm. Together with the original storage the number of chambers S_i for each farm i is therefore 4. The maximum daily capacity $c(k)$ is assumed constant. Moreover, we let

$$
\begin{aligned}
F &= 5, \\
T &= 60, \\
s_{1,1}(1) &= 4400, s_{2,1}(1) = 4300, s_{3,1}(1) = 4400, s_{4,1}(1) = 4500, s_{5,1}(1) = 4200, \\
h_{i,j}(k) &= 400\mathbf{H}(i,k), i \in \mathcal{F}, k \in \mathcal{T} \\
c(k) &= 800, k \in \mathcal{T}.
\end{aligned}
$$

Nondeterministic variables Qualities of potatoes differ from farm to farm. The deteriorating rate of the potatoes is influenced by time and temperature. In each scenario these variables are randomly set from certain distributions. On the first day of the experiment, the qualities of potatoes already in first chambers of each farm $q_{i,1}$ follow normal distributions with E_i and d_i as the expectations and standard deviations. In the other chambers, the qualities are 0 until newly harvested potatoes come in. The kinetics variables also follow normal distributions with μ_i and $\bar{\sigma}_i$ as their expectations and standard deviations. The kinetics variables after the first day of simulation follow normal distributions with the kinetics value of the previous day as the expectation, and σ_i as the variables. So the deterioration rate of potatoes in one chamber does not change largely in a short period of time. The quality of potatoes harvested on farm i and stored in chamber j have the quality $q_{i,j}$ on that day, and starts deteriorating afterwards. We then have

$$
\begin{aligned}
q_{i,j} &\sim N(E_{i,j}, d_{i,j}), \\
r_{i,j}(1) &\sim N(\mu_{i,j}, \bar{\sigma}_{i,j}), \\
r_{i,j}(k+1) &\sim N(r_{i,j}(k), \sigma_{i,j}).
\end{aligned}
$$

We consider the length of simulation 60 days for the length of starch production campaign is around 2 months. The simulations consider controller C_1, C_2, and C_3 with prediction horizons varying from 2 to 60 days, so that we can compare the performance of controllers with different amount of quality information. An example of the quality deterioration on one farm is shown in Fig. 6.4. The harvesting of this farm happens on day 3, 13, and 23.

Fig. 6.4 An example of the deterioration of the quality in different chambers of one farm

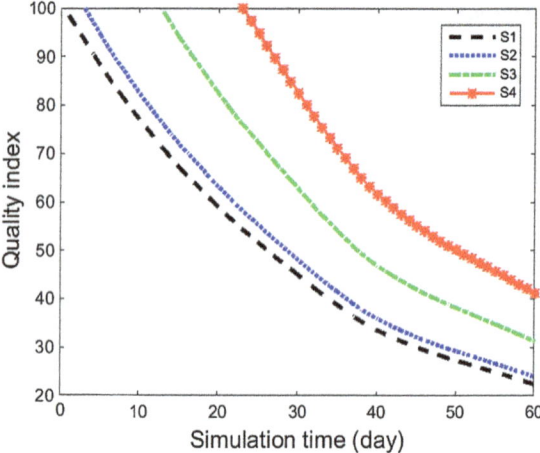

Table 6.1 Performance of different controllers

Controller	C_1	C_2	C_3 5-day	C_3 10-day	C_3 30-day	C_3 60-day
Mean (%)	100.0	109.9	111.6	112.3	114.1	114.9
Deviation	0	0.0324	0.0360	0.0387	0.0460	0.0574

6.4.2 Results

In this paper, we take the average of starch production of 10 scenarios. We normalize the results so that they are shown in percentages. Table 6.1 compares the performance of the different controllers regarding the mean and standard deviation of the results. Controller C_1 yields 100 % production, while C_2 has an increase of production by 9.9 %. The predictive controller with 5 day's prediction produces 11.6 % more starch compared with C_1. With quality prediction of 60 days, the controller has the highest production by an increase up to 14.9 % compare with C_1.

Figure 6.5 shows that as the prediction horizon increases, the controller yields higher production. The horizontal axis is different prediction horizons used by the controller, and the vertical axis is the relative production of starch compared to controller C_1. Note that the first and second point in the figure represent controller C_1 and C_2. We can immediately see the impact that information of quality brings. Besides, the predictive controller is able to obtain higher production from information of future quality. Moreover, the performance of the controller is better when using longer prediction horizons. When the prediction horizon covers the length of the campaign, the scheduling is optimal.

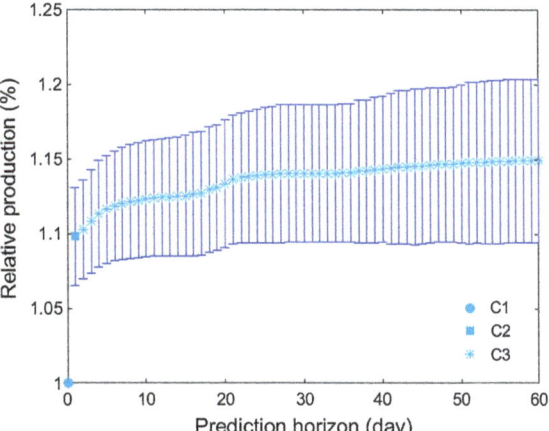

Fig. 6.5 Profit gained compared to quality-unaware controller

6.5 Conclusions and Future Research

This paper investigates the impact of accurate quality information on starch production control. We consider a transport scheduling problem with perishable items. A predictive scheduling approach is developed for raw material distribution considering the perishing nature. A model is established to describe the system of several potato farms and a starch mill. Quality decreasing is described using kinetic models and is assumed to be accurately predicted over a certain amount of days. The objective is to maximize starch production. The predictive, quality-aware scheduling strategy is performed using a model predictive control approach. The performance of the controller is compared with the performance of a quality-unaware controller and a nonpredictive, quality-aware controller. It is illustrated that the information of current as well as predicted quality can be used to improve decision-making with the model predictive control method.

Further research will include uncertainties and accuracy of quality information. Different commodities as well as other stages of supply chains will be investigated. Our detailed vision for the future research can be found in previous publication (Lin et al. 2015).

Acknowledgments The authors thank Dr. Jaap Ottjes for his valuable comments and discussions. By the China Scholarship Council under grant 201406950004 this research is supported.

References

1. Caixeta-Filho JV (2006) Orange harvesting scheduling management: a case study. J Oper Res Soc 57(6):637–642
2. Camacho E, Bordons C (2013) Model predictive control. Springer Science & Business Media

3. DG Agriculture and Rural Development (2010) Evaluation of common agricultural policy measures applied to the starch sector: final report
4. Ferrer J, Mac Cawley A, Maturana S, Toloza S, Vera J (2008) An optimization approach for scheduling wine grape harvest operations. Int J Prod Econ 112(2):985–999
5. Gal P, Lyne P, Meyer E, Soler L (2008) Impact of sugarcane supply scheduling on mill sugar production: a South Africa case study. Agric Syst 64–74
6. Grommers H, van der Krogt D (2009) Potato starch: production, modifications and uses. In: BeMiller J, Whistler R (eds) Starch, food science and technology, 3rd edn. Academic Press, San Diego, pp 511–539
7. Lin X, Negenborn RR, Lodewijks G (2015) Survey on operational perishables quality control and logistics. Proceedings of the 6th international conference on computational logistics. Delft, The Netherlands, pp 398–421
8. López-Milán E, Plà-Aragonés LM (2015) Optimization of the supply chain management of sugarcane in Cuba. In: Handbook of operations research in agriculture and the agri-food industry. Springer, pp 107–127
9. Nourian F, Ramaswamy HS, Kushalappa AC (2003) Kinetics of quality change associated with potatoes stored at different temperatures. LWT-Food Sci Technol 36(1):49–65
10. Rawlings J, Mayne D (2013) Model predictive control: theory and design. Nob Hill Publishing, LLC, Madison, Wisconsin
11. Starch Europe (2016) EU starch market data. http://www.starch.eu/european-starch-industry/. Accessed 04 Jan 2016
12. Van Boekel MAJS (2008) Kinetic modeling of food quality: a critical review. Compr Rev Food Sci Food Saf 7(1):144–158

Chapter 7
Process Maintenance of Heterogeneous Logistic Systems—A Process Mining Approach

Till Becker, Michael Lütjen and Robert Porzel

Abstract Processes in manufacturing and logistics are characterized by a high frequency of changes and fluctuations, caused by the high number of participants in logistic processes. The heterogeneous landscape of data formats for information storage further complicates efforts to automatically extract process models from this data with the tools from Process Mining. This article introduces a concept for constantly updating process models in logistics, called Process Maintenance, collects requirements for a common view on information in logistics, and shows that Process Mining with logistic data is possible, but still needs improvement to become a regular practice.

Keywords Process Mining · Logistics · Process maintenance · DOLCE

7.1 Introduction

Logistic activities in manufacturing and distribution have developed from a mere support function to key processes in the recent decades. A major cause of this is their rising influence on the overall efficiency in manufacturing and in the delivery of goods and services to customers. Just-in-Time delivery of modules and com-

---

T. Becker (✉)
Production Systems and Logistic Systems, Department of Production Engineering,
Universität Bremen, Bremen, Germany
e-mail: tbe@biba.uni-bremen.de

T. Becker · M. Lütjen
BIBA – Bremer Institut für Produktion und Logistik,
Universität Bremen, Bremen, Germany
e-mail: ltj@biba.uni-bremen.de

R. Porzel
Research Group Digital Media, Department of Mathematics
and Computer Science, Universität Bremen, Bremen, Germany
e-mail: porzel@tzi.de

© Springer International Publishing Switzerland 2017
M. Freitag et al. (eds.), *Dynamics in Logistics*, Lecture Notes in Logistics,
DOI 10.1007/978-3-319-45117-6_7

ponents in automobile industry is an essential factor for keeping production costs low (Monden 2011), and next-day delivery or even same-day delivery is taken for granted by consumers. By 'logistic activities,' we understand all activities that cover the "organization, planning, realization of the forward and reverse flow and storage of goods, data, and control along the entire product life cycle." (Schönsleben 2012) This also includes all logistic activities closely related to manufacturing (e.g., the provision of material at the assembly line or the replenishment of material buffers on the shop floor), but not the value-adding activities of manufacturing itself, i.e., the physical treatment of the product (e.g., drilling or assembly). A second phenomenon that affects management and operation of logistics in companies is the increasing level of dynamics within logistic systems. There is a trend toward a higher frequency of product innovation and shorter product life cycles (which can be clearly observed, e.g., in consumer electronics), mainly driven by the accelerated technological development and the globalization of manufacturing (Scholz-Reiter et al. 2004). As a consequence, the processes behind manufacturing and distribution of products change at a higher frequency as well.

Process models play an important role in the design, operation, and maintenance of processes (van der Aalst 2011). The logistics community has developed and adapted numerous process models over the recent decades, e.g., the Supply Chain Operations Reference (SCOR) model (Huan et al. 2004), the event-driven process chain EPC (Scheer et al. 2005), or the Business Process Modeling Notation BPMN (Zor et al. 2010). Ideally, a new process, e.g., in manufacturing, is first designed and modeled by process designers before it is implemented on the shop floor and finally carried out every day as part of the daily operations. Sometimes, legacy processes have never been modeled before and these processes are documented and modeled a posteriori to ensure that a complete and valid documentation of all processes in the companies exists. These process models can be used for, e.g., the training of new employees, variance analyses to ensure adherence to instructions, the detection of mistakes, or the creation of simulation models. However, in the light of constantly changing processes in companies caused by the high level of dynamics, the existing process models face the risk of being outdated quickly and thus not being able to serve their intended purpose. Consequently, there is a need for a systematic and periodical procedure to keep the documentation of processes up to date. This iterative approach of renewing the process documentation and adapting the existing processes we call Process Maintenance.

The goal of this article is to investigate how Process Maintenance in Logistics can be facilitated by the use of advanced data integration and data mining techniques. We propose to create a Process Maintenance System, which replaces the cycle of manual modeling and optimization of single process instances. We define the requirements of such a system with a focus on the need for a common information base in terms of an ontology. To illustrate the approach, a Process Maintenance procedure using simulated data is demonstrated.

The remainder of the article is structured as follows. In the next section, we give an overview on how processes are currently documented and modeled in logistics.

After that, we show how the technique of process mining can be utilized to create process models in an automated fashion. We also point out what the specific characteristics of data in logistics are and what the requirements the modeling procedure needs to fulfill to build enable automated modeling. Finally, we demonstrate the automated model generation based on simulated data.

7.2 Toward Business Process Management Systems in Logistics

The history of business process management (BPM) shows some overlaps with business process reengineering (BPR), Six Sigma, and workflow management (WfM). In contrast to BPR and Six Sigma, the concept of BPM can utilize radical just as incremental process improvement methodologies (Ko et al. 2009). Thereby, BPM can be described as an umbrella term for all aspects of process-focused business improvements (Verma 2009). When focusing on IT integration, the terms 'business process management systems' (BPMS) and 'workflow management systems' (WfMS) have to be mentioned. At the moment, WfMS are mostly rebranded as BPMS because BPMS add more powerful diagnostic capabilities to existing WfMS. This means improving processes by use of simulation and business intelligence in the context of process design and analysis. In general, BPMS are process-centric information technology solutions, which posses the following key capabilities (Verma 2009):

1. Closer business involvement in designing IT-enabled business process solutions
2. Ability to integrate people and systems that participate in business processes.
3. Ability to simulate business process to design the most optimal processes for implementation
4. Ability to monitor, control, and improve business processes in real time
5. Ability to detect changes in existing business processes in real time without an elaborate process conversion effort

Obviously, the idea of BPMS is not limited to business processes and can be transferred also to the domain of logistics, but at the moment only few approaches and initiatives can be found in logistics (Cabanillas 2014). The highest activity can be observed in supply chain management. Dealing with different companies, vendors, and clients, many heterogeneous processes interact with the management as well as the operational level. Existing supply chain management (SCM) solutions contain warehouse management, supply management, and inventory management components, but lack process design and analysis components. This means that with existing SCM solutions only the symptoms can be monitored, but there is no deeper understanding of the causes. For example, high inventory levels in a warehouse can be observed, but possible causes, such as delays in downstream processes, are not revealed by solely monitoring inventory figures. BPMS give the ability to close this gap and help to analyze, understand, and change processes in logistic systems more effective and in real time (Linden et al. 2014).

7.3 Process Mining—BPM Is Getting Smart

Process Mining is a young branch of research. It aims at providing the tools and techniques to automatically gain knowledge about processes from existing data sources and convert this knowledge into functioning models (van der Aalst 2011). Its main activities are the discovery of previously not documented processes, the checking of conformance of a documented process with the reality, and the enhancement of processes and process models. Figure 7.1 illustrates the general approach of Process Mining.

The novelty that Process Mining brings to BPM is the solid foundation on data. It needs to be emphasized that the data used in Process Mining is not intentionally collected for process modeling and BPM activities, but usually comes from various other processes and sources, e.g., production planning, billing, or WfMS. It uses the methods from data mining and machine learning to retrieve new information from these data sources. There are software tools and frameworks available, which support the mining process and deliver various model formats as output, e.g., BPMN or Petri Nets (van Dongen et al. 2005). However, the heterogeneity of processes, models, and data formats in logistics creates new challenges for a unified

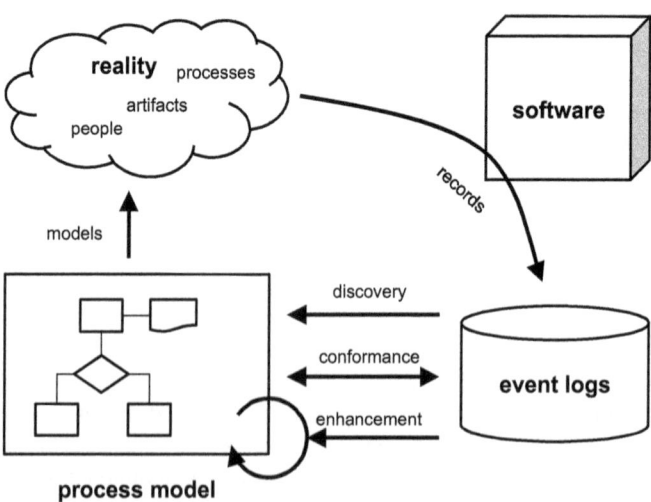

Fig. 7.1 Process Mining approach: traces of real processes are recorded by software and stored in event logs. The logs are mined to discover models, check the conformance of existing models, and to enhance models. Figure adapted from van der Aalst (2011)

modeling approach. Therefore, we claim that existing Process Mining procedures need to be accompanied by efforts to find a more general modeling approach, which serves as a connecting link between the various pieces of information from all stakeholders of a logistic process. In the subsequent section, we define the requirements for such a solution.

7.4 Requirements of Process Maintenance in Logistics

One of the fundamental requirements for Process Maintenance Systems is to devise a target representational format in which processes can be defined. Given the multitude of individual representational formats that model processes on different logistic levels, we propose a modeling approach that can subsume level-specific ones without losing their respective expressive capabilities. A suitable target representation can, therefore, be devised using description logics in the form of a formal ontology. An additional advantage of such a format is the opportunity to employ well-established modeling frameworks and principles as well as fast and robust reasoners for carrying out the targeted conformance checks.

Modern ontology engineering has departed from crafting individual domain-specific models that can neither be scaled nor reused outside of their original context. Therefore, foundational ontologies have come into play where scalability and interoperability are needed requirements. The quasi-standard in foundational ontologies is DOLCE (Masolo et al. 2002), which is used to align individual domain ontologies according to the established foundational distinctions using first-order logic. Furthermore, it provides critical design patterns, which establish a modeling coherence across different models. Employing a framework, such as DOLCE, implies the assumption of certain ontological modeling choices, which are also called ontology meta-criteria, which come with it. They are descriptive versus revisionary, multiplicative versus reductionist, possibilism versus actualism, endurantism versus perdurantism (Masolo et al. 2003). A target representation for process modeling should be descriptive in order to differentiate entities, such as machines, from processes, such as deliveries. Multiplicism is a crucial requirement because it is necessary to include different views on the ontological entities, e.g., based on the logistic level of description at hand, which is not possible with a reductionist modeling. Regarding the choice between possibilism and actualism, we claim that the latter will be more practicable for the ontology engineers involved, because modalities would raise the complexity of modeling.

An additional advantage of using a foundational framework, such as DOLCE, is that numerous individual modules have been developed that serve a domain-independent layer. Most relevant in our case is the so-called Ontology of Plans, which contains a domain-independent blueprint for our targeted representational framework. This module contains a detailed model of tasks that can be either physical or mental tasks, branching or non-branching—just to name a few distinctions—each task or sequence thereof serves to satisfy some plan. In conjunction with other pertinent modules this framework has already been successfully employed to model and drive logistic simulations (Porzel and Warden 2010; Warden et al. 2010).

In our minds the need to have an interoperable and scalable target representation for logistic processes is mainly motivated by the multitude of existing formalizations and the subsequent need to accommodate them. As mentioned in numerous publications (Oberle et al. 2007), the initial investment in a robust foundational model will provide ample revenue when the resulting model comes to its deployment in real-time applications. Modeling shortcuts and hacks will, in the long run, not scale up to reality and are, consequently, short lived or turn into idiosyncratic patch works. In the light of a future overarching logistics modeling language, we see the employment of a sound representational format as a key requirement for process modeling and maintenance as well as the subsequent conformance checking.

7.5 Toward a Process Maintenance System for Logistics

Although the need for Process Maintenance is prevalent in almost every industry, there is a specific property of logistic processes: logistic processes are characterized by a high amount of data that is generated along with the execution of the process. This data is extremely heterogeneous in terms of its structure and its purpose, because logistic processes usually run across departmental and organizational boundaries. For example, the shipping of a turbine from the manufacturer to the customer via multiple modes of transportation, such as truck, train, and vessel, induces the creation of a variety of transaction data and documents. These can be bills, delivery notes, quality certificates, manuals, shipping routes, schedules, etc. Every piece of data is collected for different purposes, e.g., billing, routing, or installation of the product. The challenge is to combine this data in heterogeneous formats from different sources in different granularity and make it usable for Process Maintenance.

Figure 7.2 illustrates the 'old' and the 'new' to automated model generation in logistics. The data generated by logistic processes are mined for process data, which is used to create a single instance of a process model, which in turn is the basis for reengineering or improvement of this process. We propose to extend this procedure and to advance it into a Process Maintenance System by installing the process mining step as an iterative, repetitive action to constantly model and check the changing logistic processes. In this way, companies create a systematic and regular procedure to keep their process documentation up to date and in alignment with possible changes in the real processes.

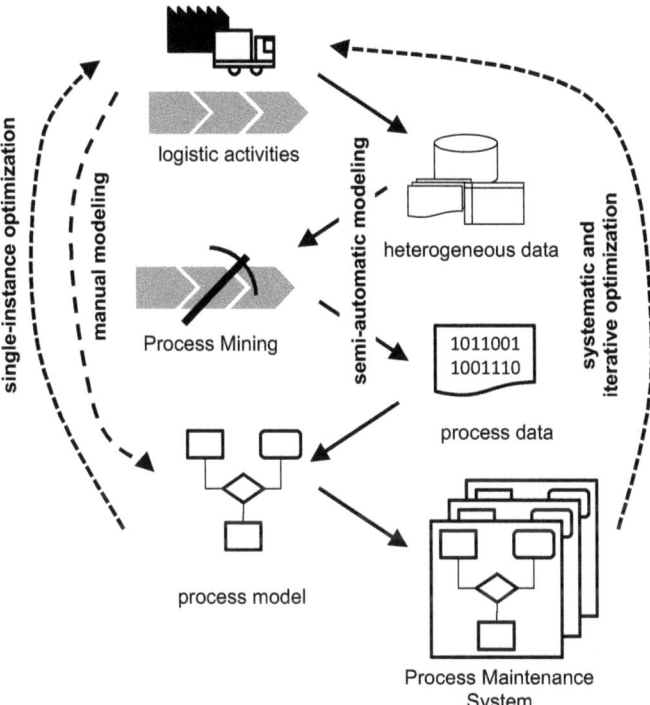

Fig. 7.2 The classical approach to process optimization in logistics is to create process models for single instances of processes and design target processes. We propose to create a Process Maintenance System, which includes semi-automatic and iterative creation of process models from heterogeneous data in highly dynamic logistic processes

7.6 Demonstration of a Process Mining Case in Logistics

This section aims to illustrate the procedure of process mining in a logistics scenario. In order to be able to compare the mining results to the real process, we demonstrate the process mining approach using the example of a simulated manufacturing process. The advantage of this approach is complete transparency of the underlying 'real' process, because the simulation model has been developed and documented by the authors (Lütjen et al. 2015). The simulation model is based on the manufacturing process of micro valves. It consists of several consecutive and partially parallel process steps, such as swaging, laser melting, quality inspection, cleaning, transport, and heat treating. To ensure a realistic picture, the simulated process is not error-free. Some of the products are not processed properly and need rework, so that the reality captured by the simulation is not fully aligned with ideal model situation. Therefore, in case of quality defects, reentrant backward flows in terms of loops are possible. Figure 7.3a depicts the intended material flow in the simulation model.

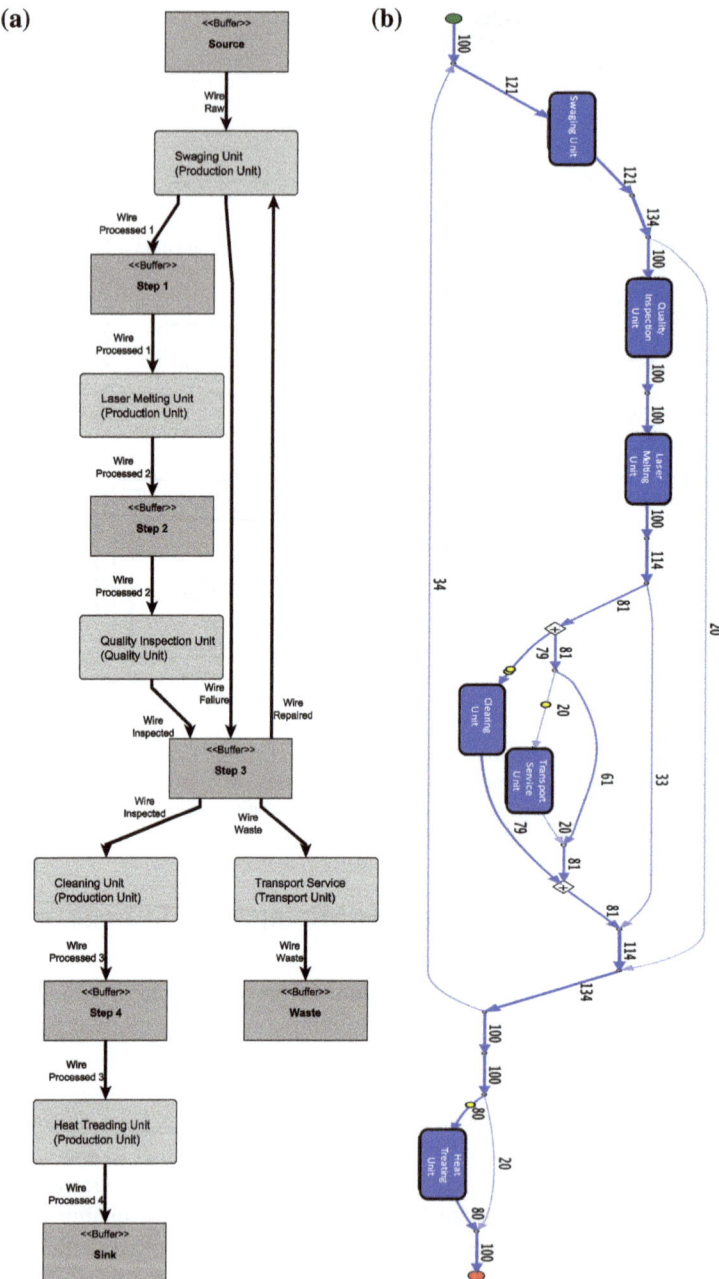

Fig. 7.3 a The original designed process model (Lütjen et al. 2015) and **b** the model gained by Process Mining from the recorded event log

The simulation is realized in TECNOMATIX PLANTSIMULATION, a discrete event simulation (DES) software for industrial scenarios. Running the simulation allow us to generate an event log in a similar manner as a manufacturing execution system (MES) would do. The resulting table of events consists of 500 records. Each record logs the start and end of an operation on the shop floor. Consequently, each record contains amongst other information an order ID, the work system that has been used, and two timestamps for start and completion of the operation.

We use the PROM tool (van Dongen et al. 2005) in its current version 6.5.1 to mine a process model from the event log. The result is displayed in Fig. 7.3b. The figure shows that the mined model corresponds to the designed model in Fig. 7.3a. However, the unplanned deviations from the designed model are also included in the new model, illustrated by the additional paths in the mined model, e.g., from 'Swaging Unit' to 'Heat Treating.'

Our application of the standard Process Mining technique shows that process in manufacturing and logistics are suitable to be the foundation for automated process modeling. However, the presented case is based on a single log file that contained all necessary information in a relatively structured fashion. In reality, there can be numerous heterogeneous data sources, and the existing approaches reach their limits.

7.7 Conclusion and Outlook

The efficiency of logistic systems is more and more important for the profitability of companies in many industries. The integral function of logistics is the glue between all other business as well as manufacturing activities. Therefore, the modeling of logistic processes and systems is important, yet challenging. The state of the art of business process management systems (BPMS) shows that there is demand to bring the two worlds of business and manufacturing together in one process-oriented representation. We followed this idea and analyzed how Process Mining can help to deal with heterogeneous logistic processes. Thereby, the requirements of such process monitoring systems describe a substantial need for the integration of foundational ontologies, which allow domain-specific as well as generic formulations of logistic systems. We defined a concept for such a system and demonstrate the use of Process Mining in a simulated scenario. We showed that the Process Mining approach was able to detect deviations, but would not be able to handle heterogeneous data sources. Further work will concentrate on this issue.

References

Cabanillas C et al (2014) Towards the enhancement of business process monitoring for complex logistics chains. Business process management workshops. Springer International Publishing
Huan SH, Sheoran SK, Wang G (2004) A review and analysis of supply chain operations reference (SCOR) model. Supply Chain Manage: Int J 9:23–29

Ko, RKL, Stephen SGL Eng WL (2009) Business process management (BPM) standards: a survey. Bus Process Manage J 15(5):744–791

Linden et al (2014) Supporting business exception management by dynamically building processes using the BEM framework. In: Dargam et al (2014) Decision support systems III-impact of decision support systems for global environments: Euro working group workshops, EWG-DSS 2013, vol 184. Springer

Lütjen M, Rippel D, Freitag M (2015) Automatic simulation model generation in the context of micro manufacturing. In: Proceedings SpringSim'15. Mod4Sim workshop, society for modeling and simulation international (SCS), Alexandria, VA, USA, pp 1024–1029

Masolo C, Borgo S, Gangemi A, Guarino N, Oltramari, A (2003) Ontology Library (final). WonderWeb Deliverable D18. http://wonderweb.semanticweb.org. Accessed Dec 2003

Masolo C, Borgo S, Gangemi S, Guarino N, Oltramari A, Schneider L (2002) The WonderWeb library of foundational ontologies. WonderWeb deliverable D17. http://wonderweb.semanticweb.org. Accessed Aug 2002

Monden Y (2011) Toyota production system: an integrated approach to just-in-time. CRC Press

Oberle D, Ankolekar A, Hitzler P, Cimiano P, Sintek M, Kiesel M, Mougouie B, Baumann S, Vembu S, Romanelli M, Buitelaar P, Engel R, Sonntag D, Reithinger N, Loos B, Zorn H-P, Micelli V, Porzel R, Schmidt C, Weiten M, Burkhardt F, Zhou J, DOLCE ergo SUMO (2007) On foundational and domain models in the SmartWeb Integrated Ontology (SWIntO). J Web Semant: Sci Serv Agents World Wide Web 5:156–174

Porzel R, Warden T (2010) Working simulations with a foundational ontology. In: Schill K, Scholz-Reiter B, Frommberger L (eds) Proceedings of the workshop on artificial intelligence and logistics at the 19th European conference on artificial intelligence, Lisbon

Scheer A-W, Thomas O, Adam O (2005) Process modeling using event-driven process chains. In: Dumas M et al Process-aware information systems. Wiley, Hoboken, New Jersey, pp 119–146

Scholz-Reiter B, Windt K, Freitag M (2004) Autonomous logistic processes: new demands and first approaches. In: Monostori L (ed) Proceedings of 37th CIRP international seminar on manufacturing systems. Hungarian Academy of Science, pp 357–362

Schönsleben P (2012) Integral logistics management: operations and supply chain management within and across companies. Auerbach Publications

van der Aalst W (2011) Process mining: discovery, conformance and enhancement of business processes. Springer Science & Business Media

van Dongen B, de Medeiros A, Verbeek H, Weijters A, van der Aalst W (2005) The ProM framework: a new era in process mining tool support. In: Ciardo G, Darondeau P (eds) Applications and theory of Petri Nets, vol 3536. Springer, Berlin, Heidelberg, pp 444–454

Verma N (2009) Business process management: profiting from process. Global India Publications PVT LTD

Warden T, Porzel R, Gehrke JD, Herzog O, Langer H, Malaka, R (2010) Towards ontology-based multiagent simulations: the PlaSMA approach. In: Proceedings of the 24th European conference on modelling and simulation, 1–4 June, Kuala Lumpur, Malaysia

Zor S, Görlach K, Leymann F (2010) Using BPMN for modeling manufacturing processes. In: Proceedings of 43rd CIRP international conference on manufacturing systems, pp 515–522

Chapter 8
A Synergistic Effect in Logistics Network

Sergey Dashkovskiy, Petro Feketa, Christian Kattenberg and Bernd Nieberding

Abstract The objective of this study is to investigate supply chain performance with a focus on transport logistics. The expected effects and capacity of potential changes in supply chain performance should be taken into account while developing a management decision on logistics network functioning. The paper proposes an analysis and a quantitative estimation of positive/negative effects and supply chain performance caused by a change of average speed and loading/unloading time in logistics network of trucks with full load. It is shown that a combination of these two factors leads to the overall synergistic effect with increased output.

Keywords Logistics network · Supply chain performance · Mathematical modelling

8.1 Introduction

At present, Germany's logistics industry shows a rapid development and faces a number of challenges. There is a certain potential to increase the efficiency and productivity in this branch as well there is a number of problems to be solved (Hagenlocher et al. 2013). For example currently there is a serious problem in recruiting junior staff to have enough drivers in Germany (Dashkovskiy and

---

S. Dashkovskiy (✉) · P. Feketa · C. Kattenberg · B. Nieberding
University of Applied Sciences Erfurt, Altonaer Str. 25, 99085 Erfurt, Germany
e-mail: sergey.dashkovskiy@fh-erfurt.de

P. Feketa
e-mail: petro.feketa@fh-erfurt.de

C. Kattenberg
e-mail: christian.kattenberg@fh-erfurt.de

B. Nieberding
e-mail: bernd.nieberding@fh-erfurt.de

Nieberding 2014). Moreover, there is a number of regulations and restriction that affect the performance of the haulers (Meyer 2006).

Every management decision on logistics network operation should be equipped with an appropriate measurement instrument for supply chain performance (Lai et al. 2002). Moreover, the effects of these decisions should be carefully analyzed and estimated in advance. The motivation for such approach is twofold. First, it helps to avoid undesirable effects which may appear as a result of management decisions. Second, it provides information about the capacity of potential changes in supply chain performance. This information can be taken into account while estimating the cost of implementing of new technologies and management solutions.

The aim of this paper is to analyze how the changes in some parameters of logistics network (like average speed, loading/unloading time, etc.) can influence overall productivity and additional revenue. Moreover as it will be shown, some adjustments of the parameters of logistics network may not lead to an additional revenue of the freight companies, but have a large influence on the work conditions of the truck drivers. It is especially important since for the past several years a profession of a truck driver becomes more and more unattractive (Bretzke and Barkawi 2010). So the hypothetical changes proposed in this paper may potentially improve the attractiveness of truck driver profession. For sure it will involve more responsible and professional people to the industry and become an additional driver of industry's growth.

The results of this paper are based on the data provided by our industrial partners. From this data we have elaborated the most common weekly route of a truck with full load. Throughout the paper we will refer to this route as a "case scenario". We believe that it reflects the actual situation of the most freight companies and allows to quantitatively estimate positive/negative effects properly.

8.2 Notation

Throughout the paper, we will use the following notations to describe a case scenario for one truck during the whole week. Let t_D be the driving time, t_P be the time that a truck driver spends on the preparation of his vehicle for a run, t_L be the time for loading/unloading procedures including a change of pallets, t_Q be the time of a truck spending in a queue waiting for a ramp to unload, t_W be the time that driver spends on paperwork and signing documents with the customers. Let an average number of orders be N_L customers per week and the whole covered distance over a week be d. The actual values of the parameters are presented in Table 8.1.

Table 8.1 Actual values of the main parameters of a case scenario

Type of parameter	Notation	Value
Driving time	t_D	40:00 h/week
Preparation for a run	t_P	2:03 h/week
Time in queue	t_Q	6:21 h/week
Time for loading/unloading	t_L	11:02 h/week
Time for a paperwork	t_W	2:12 h/week
Average number of orders	N_L	20 orders/week
Overall weekly distance	d	2475 km

8.3 Effects of Speed Increasing

Let $z \in \mathbb{N}$ be a number of additional orders for customers that a truck serves in a week due to an average speed increment. We provide a formula to calculate the average speed of a truck depending on additional number of customers. An average speed is a distance divided by the whole driving time. To get to the new average speed, one has to subtract the product of the number of additional orders and average time for an order

$$v(z) = \frac{d}{t_D - z \cdot \frac{t_L + t_Q + t_A + t_W}{N_L}}.$$

The inverse $v^{-1}(z)$ of this function gives a function for additional orders depending on the average speed

$$N_D(y) := v^{-1}(y) = \frac{(t_D \cdot y - d) \cdot N_L}{y \cdot (t_L + t_Q + t_A + t_W)}, \quad N_D^*(y) = \left\lfloor \frac{(t_D \cdot y - d) \cdot N_L}{y \cdot (t_L + t_Q + t_A + t_W)} \right\rfloor,$$

where $\lfloor y \rfloor$ is a function that returns the largest integer value less or equal to y. A plot of this dependence is presented on Fig. 8.1.

So every increment of the average speed by approximately ≈ 2.14 km/h gives an opportunity to serve one additional customer in a week-long run. The ways to increase average speed can be the following: increasing the maximum speed allowed for trucks (which is not too realistic because of safety reasons), route optimization, better connections between industrial parks and highways. Additionally, local sensor data and electronic toll collection data can be used for prediction and improvement of motorways traffic performance and average speed increment (Heilmann et al. 2011). On the other hand, saved driving time can extend the resting time of a truck driver with no additional revenue to the company. Our calculations show that the increasing of an average speed by 5 km/h leads to the increasing of resting time of the driver by 5 %. It will make the profession of truck driver more attractive and enlist more qualified and responsible workers to the industry.

Fig. 8.1 A plot of the function $N_D^*(y)$ starting at the current average speed of 61.875 km/h and ending with the maximum speed allowed for trucks in Germany. Realistically, the maximum possible improvement could be to generate around 4–5 more orders per week

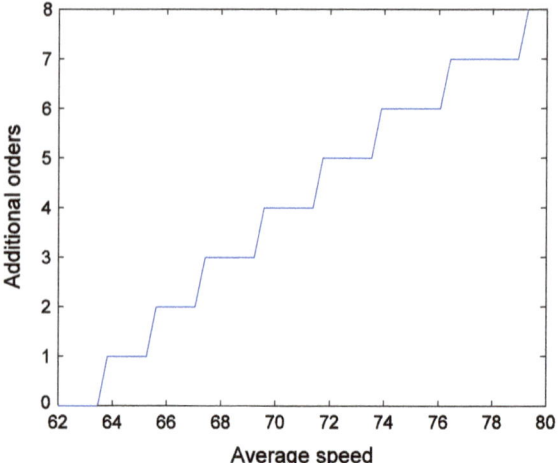

8.4 Pallet System Adjustments

An improvement of loading/unloading efficiency is one of the main ways to improve an overall performance of a logistic process. The Paletten-Tagging-Roboter (PaTRo 2015) for instance, is one of the techniques to make it. To calculate the benefits from the adjustments of the time spending on pallets, we need to modify the un/loading time with our parameter x which is the fraction of loading/unloading time that can be saved in percent. Then $t_L \cdot x$ is the time that can be saved, $t_L \cdot (1-x)$ is the remaining time within the process. So the new average time for an order is

$$t_{AL}(x) := \frac{t_L(1-x) + t_Q + t_A + t_W}{N_L}.$$

A division of the saved time through the new average time for an order leads to

$$N_P(x) = \frac{x \cdot t_L \cdot N_L}{t_L(1-x) + t_Q + t_A + t_W}.$$

Like the previous, it also has to be a natural number, so

$$N_P^*(x) = \left\lfloor \frac{x \cdot t_L \cdot N_L}{t_L(1-x) + t_Q + t_A + t_W} \right\rfloor.$$

This dependence is presented on Fig. 8.2.

Every 7.5 % of saved time enables the truck driver to serve an additional customer in a week-long run. Reducing this parameter by 50 % will benefit into 5.16 % additional leisure time for the driver. How such massive change in

Fig. 8.2 A plot of $N_P^*(x)$ beginning at the current point with 0 % saving and ending with a maximum of 50 %. As the figure shows, it is possible to generate up to 6 more orders per week

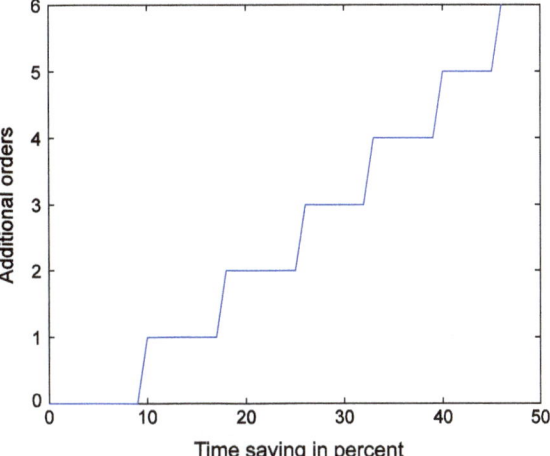

loading/unloading time can be reached? For example, by introducing one-way recyclable pallets. In this situation a truck driver will not care about loading empty pallets after unloading his trailer and will follow his next loading/unloading point. A detailed examination of reverse logistics practices can be found in (Rogers and Tibben-Lembke 2001).

8.5 Synergistic Effect

Now we try to apply these two approaches simultaneously. This chapter is divided into two parts. In the first one, one has to take a look to the summation of the speed increasing and pallet system formulas. Afterwards we can improve this process even more, as we can include the modification of the average time for an order to the speed increasing part.

We assume that the average speed has been increased as well as the time spending for pallets has been decreased. The combination of both formulas leads to the following:

$$N_{D+P}(x,y) = N_D(y) + N_P(x) = \frac{(t_D \cdot y - d) \cdot N_L}{y \cdot (t_L + t_Q + t_A + t_W)} + \frac{x \cdot t_L \cdot N_L}{t_L(1-x) + t_Q + t_A + t_W}$$

Like in the previous sections the number of additional orders that could be served, has to be a real number (Fig. 8.3).

$$N_{D+P}^*(x,y) = \left[\frac{(t_D \cdot y - d) \cdot N_L}{y \cdot (t_L + t_Q + t_A + t_W)} + \frac{x \cdot t_L \cdot N_L}{t_L(1-x) + t_Q + t_A + t_W} \right]$$

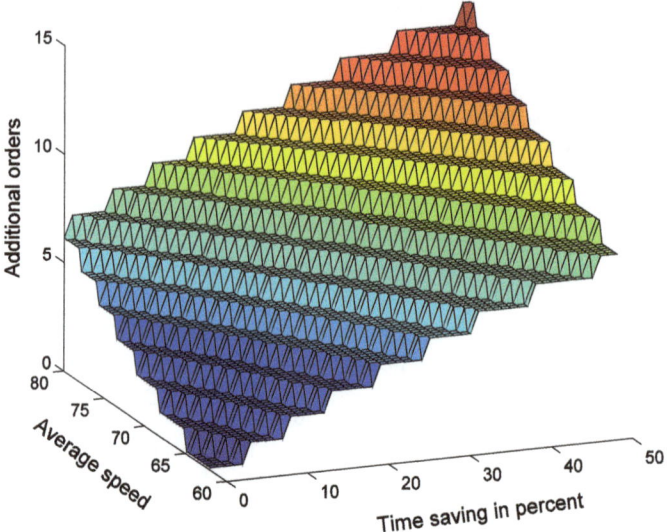

Fig. 8.3 A plot for the function $N^*_{D+P}(x,y)$. A synergy effect can be seen as the maximum amount of this combined function, the 15 orders cannot be reached with the results of the two other functions themselves

The synergy effect comes into play, when the results of both single functions with a specific improvement are slightly lower than it has to be for a further increase but it is enough for the combination of these functions.

Let us have a look at some examples to illustrate this:

Let the improvement for the speed $y = 62.72$ km/h and for the pallet system either (i) 6 %, (ii) 11 % and (iii) 15 %.

(i)	(ii)	(iii)
$N^*_D(62.72) = 0$	$N^*_D(62.72) = 0$	$N^*_D(62.72) = 0$
$N^*_P(6) = 0$	$N^*_P(11) = 1$	$N^*_P(15) = 1$
$N^*_{D+P}(6; 62.72) = 1$	$N^*_{D+P}(11; 62.72) = 1$	$N^*_{D+P}(15; 62.72) = 2$

One can see, that the synergy effect is effective for (i) and (iii) but has no impact for case (ii).

This calculations ensures that a slight increase of an average speed cannot benefit into an additional customer since $N^*_D(62.72) = 0$. The same situation can be seen with pallets—a 6 % of saved time does not lead to any new customers. So these two improvements do not benefit into an additional revenue of a freight company. However, these improvements make the working time of a truck driver less intensive. He can spend saved working time on the rest. On the other hand, a combination of these two improvements lead to an opportunity to serve a new customer since $N^*_{D+P}(6;62.72) = 1$.

Furthermore $N_D(y)$ can be updated with the implementation of the time saving x as from pallet system section to the average time for the orders to the following:

$$N^\dagger_D(x,y) = \frac{(t_D \cdot y - d) \cdot N_L}{y \cdot (t_L(1-x) + t_Q + t_A + t_W)}.$$

A combination of this with the previous leads to the formula:

$$N^\dagger_{D+P}(x,y) = N^\dagger_D(x,y) + N_P(x)$$
$$= \frac{(t_D \cdot y - d) \cdot N_L}{y \cdot (t_L(1-x) + t_Q + t_A + t_W)} + \frac{x \cdot t_L \cdot N_L}{t_L(1-x) + t_Q + t_A + t_W}$$
$$= \frac{(t_D \cdot y - d) \cdot N_L + y \cdot x \cdot t_L \cdot N_L}{y \cdot (t_L(1-x) + t_Q + t_A + t_W)} = \frac{((x \cdot t_L + t_D) \cdot y - d) \cdot N_L}{y \cdot (t_L(1-x) + t_Q + t_A + t_W)}$$

As this result must also be a natural number, we get (Fig. 8.4)

$$N^{\dagger*}_{D+P}(x,y) = \left[\frac{((x \cdot t_L + t_D) \cdot y - d) \cdot N_L}{y \cdot (t_L(1-x) + t_Q + t_A + t_W)} \right].$$

Let us have a look at two examples.

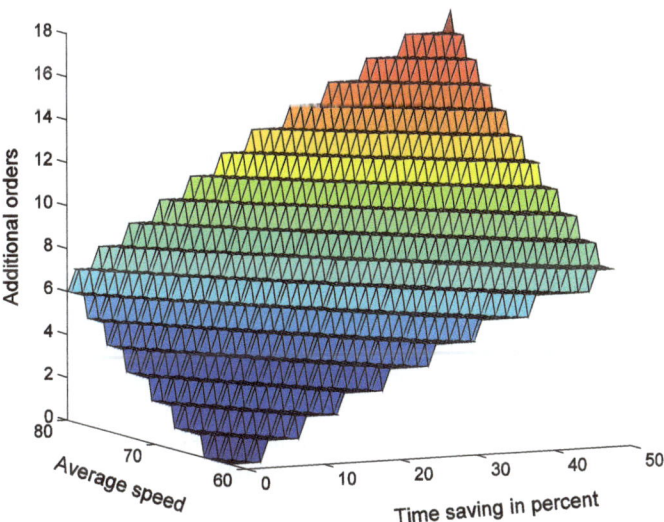

Fig. 8.4 A plot for the function $N^{\dagger*}_{D+P}(x,y)$. An additional effect is visible as the 18 possible orders are a further improvement in relation to the function $N^*_{D+P}(x,y)$

(i)	(ii)	(iii)
$N_D^*(65.96) = 2$	$N_D^*(65.96) = 2$	$N_D^*(65.96) = 2$
$N_P^*(18) = 1$	$N_P^*(23) = 2$	$N_P^*(25) = 2$
$N_{D+P}^*(18; 65.96) = 4$	$N_{D+P}^*(23; 65.96) = 4$	$N_{D+P}^*(25; 65.96) = 5$
$N_{D+P}^{\dagger *}(18; 65.96) = 4$	$N_{D+P}^{\dagger *}(23; 65.96) = 5$	$N_{D+P}^{\dagger *}(25; 65.96) = 5$

Here one can see, that the function $N_{D+P}^{\dagger *}(x, y)$ increases the advantage even more, as there is an overall synergy effect in (ii).

For a look at the upper limit, we calculate improvements with $x = 50\%$ and $y = 80$ km/h. The results are the following:

$$N_P^*(50) = 6 \quad N_{D+P}^*(50; 80) = 15$$
$$N_D^*(80) = 8 \quad N_{D+P}^{\dagger *}(50; 80) = 18$$

As one can see, the combination of two improvements lead to much better results than its separate applications. So $N_{D+P}^{\dagger *}(x, y) \geq N_{D+P}^*(x, y) \geq N_D^*(y) + N_P^*(x)$—this is a synergistic effect in logistics network. A synergy effect can be seen even at minor improvements of the parameters but it will stack even more for huge ones.

8.6 Influence of Saved Time on the Probability of Getting New Customers

In the previous sections, we considered the ways to save driving and working time of a truck driver and its influence on an overall performance of logistics network. As it was shown it benefits a lot into additional revenue of a freight company. The reason is that a truck can serve an additional number of customers. However challenging questions appear in this context. Is it easy to find new customers on the route? And how to estimate the change of probability of getting new customers depending on the saved driving time? With no additional driving time a driver should strictly follow the original route. However, an additional driving time enables to construct a route more flexible. The alternative routes between segments (even if these routes are not the shortest) may result in finding a new customer. The problem is how to distribute an additional driving time over the entire route. What segments should keep its original route and what segments should be replaced with an alternative way? The aim of this section is to propose a quantitative estimation of the influence of additional driving time onto a probability of finding new customers.

To answer aforementioned questions we create a mathematical model of this logistics process. Here we assume that customers are distributed uniformly over a route with the rate of δ potential customers per one kilometer of a route. Let $n \in \mathbb{N}$ be a number of segments, $x_i = (x_i^\alpha, x_i^\beta, \ldots) \in R^{k_i}$, $i = 1, \ldots, n$ be a vector of alternative distances between two points of i-th segment via different routes, x_{i_0} be a distance between two points of i-th segment via the initial route. Let T be a previously saved time, $v_i \in R_+$, $i = 1, \ldots, n$ be an average speed on i-th segment. For every value of saved time T our goal is to find vectors $\lambda_i \in R^{k_i}$, $i = 1, \ldots, n$ with elements from $\{0, 1\}$ such that

$$\delta \cdot \sum_{i=1}^{n} \frac{\langle \lambda_i, x_i \rangle}{x_{i_0}} \to \max, \quad \sum_{i=1}^{n} \frac{\langle \lambda_i, x_i \rangle}{v_i} \leq T, \quad \sum_{j=1}^{k_i} \lambda_i^j = 1, \quad i = 1, \ldots, n,$$

where $\langle \lambda_i, x_i \rangle = \sum_{j=1}^{k_i} \lambda_i^j \cdot x_i^{k_i}$ is a scalar product of two vectors, k_i is a dimension of vectors x_i and λ_i, $i = 1, \ldots, n$, e.g. k_i is a number of alternative routes between two points of i-th segment.

We have solved the following problem of optimization for a case scenario route of a truck with full load.

The results are presented on Fig. 8.5. The horizontal axis indicates an additional time T that a truck driver can use on driving in hours per day. The vertical one indicates the ratio of probabilities to find a potential customer between original and optimized route R. The probability of getting a new customer increases almost linearly. However at around $T = 35$ min of saved time it has a slight jump. A saving of 35 min of driving time increases the probability of finding new customers by 50 %.

As one may intuitively guess, the ratio of probability of getting new customers between original and optimized routes R behaves almost like a linear function of an additional driving time T. Indeed, a regression analysis of this dependence shows that it can be well approximated by a linear function $R(T) = 0.76 \cdot T + 0.99$.

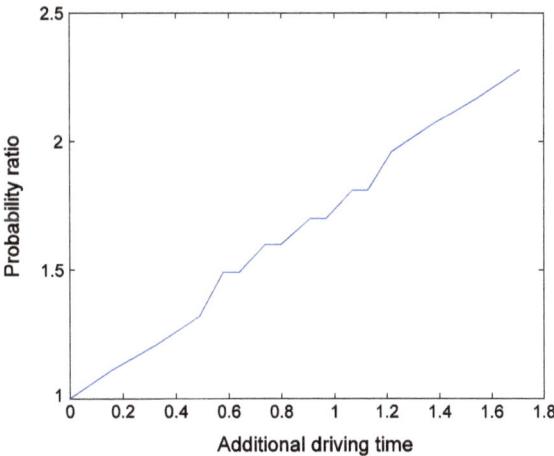

Fig. 8.5 A ratio of probabilities of finding new customers between the initial and optimized route depending on additional driving time

A residual sum of squares for this approximation is not more than 0.014. A correlation coefficient is $r \approx 0.997$.

This analysis ensures that there is a direct dependence between the ratio of probabilities R and an additional time T as a linear function with a slope of ≈ 0.76. So one may conclude: more additional time we have saved, more possibilities of finding new customers we get.

8.7 Conclusion

We have analyzed the influence of the average speed increasing and loading/unloading time decreasing on the overall performance of logistics network. A saved working time can be spent either on serving additional customers or on the resting time of a truck driver. Each of these options has its advantages and should be considered as a trade-off between freight company owner and a truck driver. It was shown that a simultaneous application of these improvements leads to a synergistic effect and benefits more than a separate usage of the proposed improvements. A change of probability of finding new customers depending on a saved driving time has also been estimated.

The results of the paper can be useful for justification of management decisions in logistics network functioning. The paper provides the implications of the proposed solutions and estimates them quantitative.

Acknowledgments This work was supported by the German Federal Ministry of Education and research (BMBF) as a part of the research project "LadeRamProdukt".

References

Bretzke W-R, Barkawi K (2010) Nachhaltige Logistik – Antworten auf eine globale Herausforderung. Berlin, Heidelberg

Dashkovskiy SN, Nieberding B (2014) Costs and travel times of cooperative networks in full truckload logistics. In: Dynamics in logistics—proceedings of the 4th international conference LDIC, 2014, Bremen, Germany

Hagenlocher S, Wilting F, Wittenbrink P (2013) Schnittstelle Rampe–Lösungen zur Vermeidung von Wartezeiten. Schlussbericht Anlagenband, hwh Gesellschaft für Transport-und Unternehmensberatung mbH (Hrsg.), Karlsruhe, http://www.bmvi.de/SharedDocs/DE/Anlage/VerkehrUndMobilitaet/laderampe-schlussbericht-schnittstelle-rampe.pdf. Accessed 29 Sept 2015

Heilmann B, El Faouzi N-E, de Mouzon O, Hainitz N, Koller H, Bauer D, Antoniou C (2011) Predicting motorway traffic performance by data fusion of local sensor data and electronic toll collection data. Comput-Aid Civil Infrastruct Eng 26:451–463

Lai K, Ngai EWT, Cheng TCE (2002) Measures for evaluating supply chain performance in transport logistics. Transp Res Part E 38:439–456

Meyer J (2006) Wirtschaftsprivatrecht: eine Einführung. Springer

PaTRo—Paletten-Tagging-Roboter (2015) BIBA – Bremer Institut für Produktion und Logistik GmbH, http://donar.messe.de/exhibitor/hannovermesse/2015/G392849/patro-paletten-tagging-roboter-projektbeschrei-ger-388662.pdf. Accessed 29 Sept 2015

Rogers DS, Tibben-Lembke R (2001) An examination of reverse logistics practices. J Bus Logist 22:129–148

Chapter 9
A Step Toward Automated Simulation in Industry

Himangshu Sarma, Robert Porzel and Rainer Malaka

Abstract Automation always plays an important role in industry. Today, it is a basic need for industry. To develop faster manufacturing or delivery, automation is an important need. Robots always play the main role for automation in the industry. Robots are mainly designed for specific task. But, the main problem is robots are too expensive for one task. Thats why, it is almost impossible to use robots for small industries. Therefore, we are aiming to develop a pipeline to design a multitasking robot, especially for different kinds of packaging tasks. Typical text-based instruction sheets are the main source of these automation robots, that means robots can pack different types of shapes using typical text-based packaging instructions. In robotics, learning by demonstration in robotics, could benefit from large body movement dataset extracted from textual instructions. The interpretations of instructions for the automatic generation of the corresponding movements thereof are difficult tasks. We examine methods for converting textual surface structures into the semantic representations and explore tools for analysis and automated simulation of activities in industrial and household settings. In our first step, we try to develop a pipeline from textual instructions to virtual actions that includes traditional language processing technologies as well as human computation approaches. Using the resulting virtual actions, we will train robots through imitation learning or learning by demonstration for multitasking packaging robots.

Keywords Industry · Robots · Textual · Instruction

H. Sarma (✉) · R. Porzel · R. Malaka
Digital Media Lab, TZI, University of Bremen, Bibliothekstr. 1, 28359 Bremen, Germany
e-mail: sarma@tzi.de

R. Porzel
e-mail: porzel@tzi.de

R. Malaka
e-mail: malaka@tzi.de

© Springer International Publishing Switzerland 2017
M. Freitag et al. (eds.), *Dynamics in Logistics*, Lecture Notes in Logistics,
DOI 10.1007/978-3-319-45117-6_9

9.1 Introduction

Automation is a very important need in industry. Already, automation took place in many areas in industry. But, again there are lot of areas which are difficult to automate until now in industries. For example if we look at the packaging in different industry, we can see that there are limitations during automation of packaging. The same robot cannot be used for different types of work. Robots are very expensive (Weinberg et al. 2001), and that is why researchers are trying to develop a low cost robotic vision system (Sandhu et al. 2015). Therefore, multitasking robots could be beneficial for the real world, which can perform different tasks in industry. Therefore, our main aim is to develop a pipeline by which we can develop a robot which can handle different kinds of packaging shapes and also different kinds of other industrial tasks too based on typical text-based instruction sheets. Where, input users will provide them with text-based instructions for the actions (e.g., packaging). To perform instructions-based task, we need to teach robots to do so. For that we need lot of data. Using those data, we can teach robots to perform different packaging works in industry. To collect those data we are going to use human computation approaches. So, as a first step we are going to generate 3D animations which will automatically be generated as an output from the provided instruction sheet. Using a human computation approach we determined which visualization serves best as an output of the video (Sarma et al. 2015). Assessing the quality of human body movement performances is an important task in many other application areas other than industry, ranging from sports to therapy, learning by demonstration in robotics, automated systems for generative animation, and many more. For example, the manual transformation of physical therapy exercises into computer-supported playful exercises in the form of so-called exergames requires a lot of time and effort, making it impractical for therapists or smaller practices to transform their preferred sets of therapeutic exercises into exergames to be used by their patients. Motivated by our use case of automatically generating movement patterns to be used in motion-based games for the support of physiotherapy, rehabilitation, and prevention, we thus set out to explore the potential of crowd-based quality of motion assessments, as a necessary intermediate step in the extraction and validation of motions (Fig. 9.1).

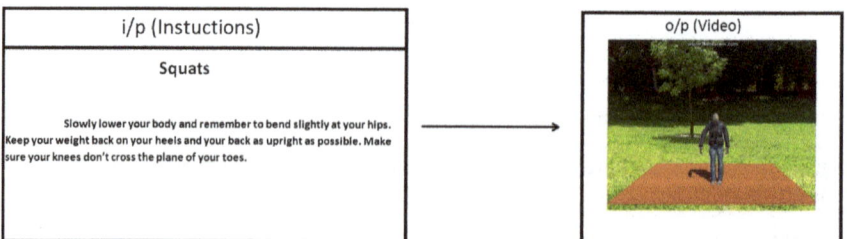

Fig. 9.1 Expected input and output

9.2 Background and Related Work

Based on the main idea to develop an instructions-based packaging robot, first we need a lot of human movement data. It is always better to collect this kind of data through some game or some applications rather than paying real humans. Data collections through a game are cost effective and it is also possible to collect lot of data in a small period of time. If we look at the past work we can find some many efforts, where lot of researchers tried to collect data through computer games. Below, we mention one work where a group of researchers try to collect data to help cooking robots (Walther-Franks et al. 2015).

Tower defense game: There are large sets of data available for imitation learning to robots. It is always difficult to collect such amounts of data. Again, as we mention earlier, it is also very expensive and time consuming to collect this amount of data. In this project researchers mainly tried to collect a large set of data which can be used for imitation learning mainly for cooking robots. Here they mainly designed a tower defense game using a human computation approach to gather data on human manual behavior (Fig. 9.2).

Based on the current state of the art, we set out to establish a human computation-based pipeline for extracting validated movements from instruction sheets, with the goal to then explore the potential of further automating the different steps involved in that pipeline, starting with a focus on the step of quality of motion assessment. In our previous work we have found the best visualization category to use in automated 3D animation from typical text-based instructions (Sarma et al. 2015). Below we show how we have found the best visualization, first step toward the autonomous packaging robots.

Data Collection and Results
At the early stage of the work we have developed a Physical Exercise Instruction Sheet Corpus(PEISC) of around 1000 physical exercise instructions. On the basis of different actions we categorize PEISC in different categories, e.g., standing, seating,

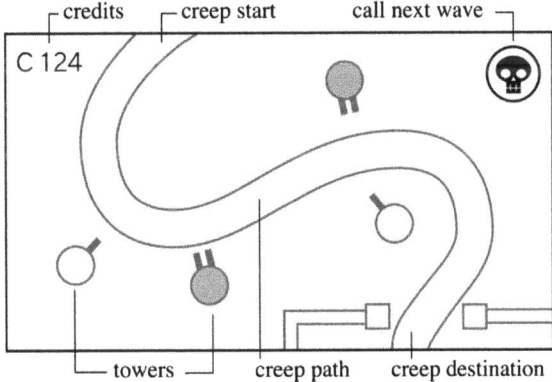

Fig. 9.2 Kitchen tower defense game

leaning, and lying. We have chosen five exercises from standing where there is no any external equipment used during exercise for our experimental work. The five exercises we have recorded are listed below

- Squats
- Lateral Lungs
- Standing IT
- Forward Lungs
- Reverse Lunges

With the help of a Kinect[1] we have recorded the performance of those five exercises from seven participants, three male and four female; four are from 15–25 and three from the 25–35 age group.

We have developed four different categories of video from the collected recording data. Below we list all the four categories with specific reason of why we selected those

- *RGB*: it is a simple color video, which can easily generated by anyone, people usually watch this kind of video in daily life
- *Depth*: the visualization is likely same as RGB videos, but the best part is that one cannot see the real person, so best when it comes to private matter
- *Skeleton*: this category is fully generated by joint tracking sensor of the human body, with this category we can easily find which joint is moving how much
- *Virtual Reality*: this category is also generated from the skeleton or joint tracking, we can add any virtual character to the skeleton, in case of privacy it is one of the best options for game-related area.

Using the above four categories we have developed a survey application to find the best visualization (Figs. 9.3 and 9.4).

Results

Rather than the above works, there are also certain works which somehow are related to our approach. A group of researchers (Chang et al. 2014) introduced a text to scene engine where user can also change the scene if it is not perfect. Also another multimodal text-to-animation system CONFUCIUS (Ma 2006) generates animations of virtual humans from single sentences containing an action verb.

9.3 Proposed Approach

Automated packaging simulations in industry are never an easy task. We are trying to achieve the target through different small tasks, which are shown in the pipeline below

The whole pipeline of the objectives is mainly divided into two main parts, as shown in Fig. 9.5. Below, we have analyzed the two parts

---

[1]https://dev.windows.com/en-us/kinect.

9 A Step Toward Automated Simulation in Industry

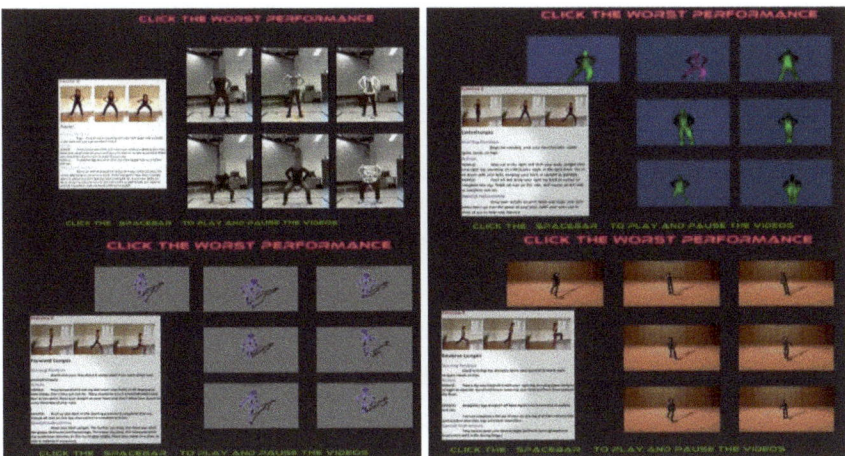

Fig. 9.3 Screenshots of the survey application showing different visualizations

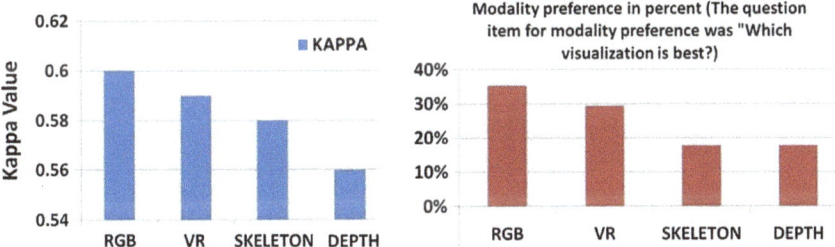

Fig. 9.4 Best visualization

- We try to divide this step in three small subtasks as below:
  1. In this step, our main aim is to extract the list of motions terms from a typical text-based instruction. First, we will try to generate the list of motions extracting the semantic information of the instruction using the Stanford parser (Marneffe et al. 2006) and Framenet (Baker et al. 1998). Using, human computation approach, crowdsourcing we will verify those motions to get a perfect list of motions.
  2. In second step, our main aim is mapping the motion terms to the database. Using, human computation we will label all motions that are stored in the database.
  3. This is the last step before we get to our main goal. Here, we try to set up some rules to generate new motion using human computation or crowdsourcing. In detail, if the required motion is not present in the database, then we need to generate new motions using the motion database. To generate new motions from the database, we need different rules. Then, we can use those rules to generate new motions.

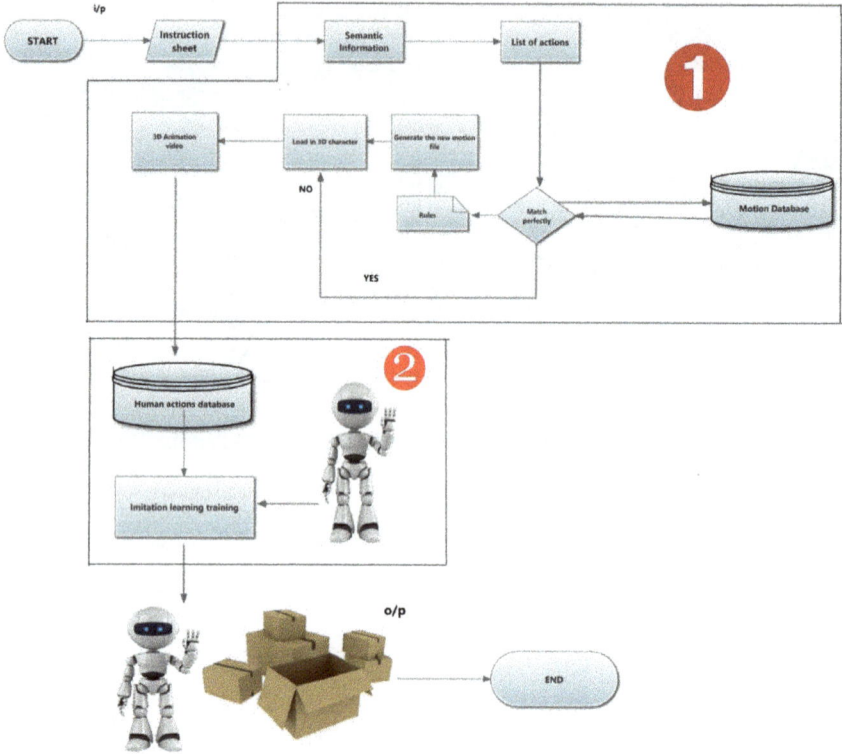

Fig. 9.5 Automatic packaging simulation pipeline

- This is the final step to achieve our desired target. After we get the animations from the typical instructions sheet, we will use those data as a database to teach robots. Using this database, we will train robots to perform automatic packaging simulations using imitation learning or learning by demonstration.

9.4 Challenges

To achieve the desired target as shown in Fig. 9.5, there are some challenges. Below we have described some of those challenges

- To extract the required motion action terms from the texts-based instructions.
- To generate new motions which are not yet available in the database.
- During generation of the virtual actions, it is an important task to ensure clean transitions between two different actions or poses, which is also a big challenge.
- Use this motion database as a training set for imitation training for robots.

9.5 Conclusion and Future Work

We have presented a pipeline for developing a multitasking robot, for automated simulation in industry, especially for packaging. Here we have divided the whole task into small subtasks, mainly two major tasks. We tried to show how we will solve these tasks and also we have shown the main challenges to complete these tasks.

References

Baker CF, Fillmore CJ, Lowe JB (1998) The Berkeley FrameNet project. In: Proceedings of the 17th international conference on computational linguistics, vol 1. Association for Computational Linguistics, pp 86–90

Chang AX, Savva M, Manning CD (2014) Semantic parsing for text to 3D scene generation. ACL 2014:17

De Marneffe MC, MacCartney B, Manning CD et al (2006) Generating typed dependency parses from phrase structure parses. Proc LREC 6:449–454

Ma M (2006) Automatic conversion of natural language to 3D animation. Ph.D. thesis, University of Ulster

Sandhu G, McGinn C, Kelly K (2015) Investigation into a low cost robotic vision system

Sarma H, Porzel R, Smeddnick J, Malaka R (2015) Towards generating virtual movement from textual instructions: a case study in quality assessment. In: Proceedings of the third AAAI conference on human computation and crowdsourcing (HCOMP-2015), AAAI

Walther-Franks B, Smeddinck J, Szmidt P, Haidu A, Beetz M, Malaka R (2015) Robots, pancakes, and computer games: designing serious games for robot imitation learning. In: Proceedings of the 33rd annual ACM conference on human factors in computing systems. ACM, pp 3623–3632

Weinberg JB, Engel GL, Gu K, Karacal CS, Smith SR, White WW, Yu X (2001) A multidisciplinary model for using robotics in engineering education. In: Proceedings of the 2001 ASEE annual conference and exposition

Part II
Predictive Analytics and Internet of Things and Services

Chapter 10
Application Potential of Multidimensional Scaling for the Design of DSS in Transport Insurance

Victor Vican, Ciprian Blindu, Alexey Fofonov, Marta Ucinska, Julia Bendul and Lars Linsen

Abstract Transport risks in supply chains have increasingly lead to significant capital losses. Insurance claims against such losses have grown accordingly, while simultaneous advances in technology lead to continuously larger volumes of data recorded. Traditional risk evaluation methods in insurance struggle to account for rising supply chain complexity which is reflected by growing amount and dimensionality of supply chain data. Therefore decision-makers in the transport insurance industry need new ways of appropriate knowledge representation to support transport insurance providers with daily tasks such as premium tariffing. This paper presents a method based on multidimensional scaling (MDS) for the identification of groups of similar claims as a first step towards the improvement of supply chain risk evaluation and forecasting. We show the application potential of transforming and visualising transport damage claims data as the basis for developing decision support systems (DSS) to support transport insurance providers in tasks such as premium tariffing as well as transport and supply chain managers in risk mitigation and prevention activities.

Keywords Multidimensional scaling · DSS · Transport · Insurance

---

V. Vican (✉) · C. Blindu · A. Fofonov · J. Bendul · L. Linsen
Jacobs University, Campus Ring 1, 28759 Bremen, Germany
e-mail: v.vican@jacobs-university.de

C. Blindu
e-mail: c.blindu@jacobs-university.de

A. Fofonov
e-mail: a.fofonov@jacobs-university.de

J. Bendul
e-mail: j.bendul@jacobs-university.de

L. Linsen
e-mail: l.linsen@jacobs-university.de

M. Ucinska
The Warsaw University of Life Sciences, Nowoursynowska 166, 02-787
Warsaw, Poland
e-mail: m.ucinska@gmail.com

© Springer International Publishing Switzerland 2017
M. Freitag et al. (eds.), *Dynamics in Logistics*, Lecture Notes in Logistics,
DOI 10.1007/978-3-319-45117-6_10

10.1 Introduction

Supply chains are increasingly faced with exogenous risk factors such as gradual pollution, floods, earthquakes and other natural disasters affecting the transportation of goods (Vilko and Hallikas 2012; Skorna and Fleisch 2012). Endogenous risks emerging from increasingly global and complex supply chains pose further equally important problems in transport operations. In order to cope with the resulting rise in capital losses, companies require significant financial insurance for their transport operations (in particular the cargo itself), which in turn results in a growing number of claims against insurers. This is reflected by a nearly 70 % increase in global transport insurance premiums between 2007 and 2014 to $ 32.6 billion (c.f. IUMI —International Union of Marine Insurance 2014). At the same time, recent advances in technology—such as new measurement devices—and the influence of big data lead to ever increasing volumes of data recorded for each insurance claim (Guelman 2012; Maas and Milanova 2014; Guo 2003). While insurers take actions to prevent and reduce the filing of claims (Skorna and Fleisch 2012), core activities in insurance are nevertheless exposed to an increase in data volume and dimensionality. Classic *No Claim Discount Systems* for example need to make use of this valuable amount of historical claims data to calculate risk-oriented customer premiums (Yeo et al. 2001). However, these conventional statistical models, which many insurers rely on, buckle under the dimensionality and volume of data (Guelman 2012). Insurance companies lack models and DSS that make use of the hidden knowledge in the available data (Schaerer et al. 2011).

Algorithmic modelling approaches have proven to be more efficient compared to conventional data modelling in insurance (Guelman 2012). Such approaches make no assumptions of underlying models or variable relationships, but rather treat the underlying data generating mechanism as unknown (Breiman 2001; Guelman 2012). Guo (2003) further suggests, that, with reference to data dimensionality and hence complexity, new ways of appropriate knowledge visualisation for decision-makers need to be developed. This paper therefore presents an algorithmic modelling approach based on MDS and cluster analysis to visualise qualitative transport insurance claims data. Through the transformation, reduction and visualisation of multidimensional data, groups of related claims become visible to decision-makers. This kind of information can be used as the basis for designing intuitive DSS for the transport insurance sector as well as for transportation and supply chain design tasks. Application potentials of the presented method in many key insurance tasks may include for example claims cost estimation or policy determination.

The remainder of this paper is structured as follows: a representative review of current knowledge generation and visualisation methods in insurance is given in Sect. 10.2. The proposed approach for this paper is presented in Sect. 10.3. In Sect. 10.4, the results of the applied method are discussed. Conclusions drawn in Sect. 10.5.

10.2 Overview of Existing Methods to Evaluate and Forecast Transportation Risk in Insurance

The insurance sector has a tradition of making use of statistical models in the DSS process to improve decision-making tasks such as tariffing. Table 10.1 gives an overview of various representative studies in their approach towards supporting

Table 10.1 Overview of quantitative methods in the insurance sector. Visualisation approaches are highlighted in bold

Author	Insurance field	Aim	Method	Limitations
Balaji and Srivatsa (2012)	Life	Discover policy preferences	Decision tree	Locally optimal rules
Broeksema et al. (2013)	**Car**	**Decision support**	**Multiple correspondence analysis, value cell merging**	**Only one distance measure employed**
Cerchiara et al. (2008)	Life	Analyse influence factors on lapse risk	Generalised linear models	Parameter relation investigated, not similarities among claims
Devale and Kulkarni (2012)	Life	Decision support	K-means clustering, K-nearest neighbours, association rules	No visualisation
Guelman (2012)	Car	Loss prediction	Gradient boosting trees	Requires training
Guo (2003)	Property	Decision support	Various data mining methods	No visualisation
Roux (2008)	**Health**	**Discover comorbidity**	**MDS, k-means clustering**	**Strong relation to medical research, not designed for DSS**
Schäfer et al. (2014)	**Health**	**Discover comorbidity**	**Factor analysis, MDS**	**Strong relation to medical research, not designed for DSS**
Skorna (2013)	Transportation	Discover customer groups	K-modes clustering	No visualisation
Thakur and Sing (2013)	Car	Discover customer groups	K-means clustering	No visualisation
Vilko and Hallikas (2012)	Transportation	Risk assessment	Monte Carlo simulation	Data based on expert judgement
Yeo et al. (2001)	Car	Risk classes	K-means clustering, linear regression	No visualisation

decision-making tasks and highlights how claims data is processed by appropriate knowledge generation methods.

The variety of methods reviewed primarily revealed that (a) commercial transport insurance is underrepresented in the literature, (b) identifying related claims is a common aim in many studies and (c) meaningful, intuitive output to the decision-maker through data visualisation is provided sparsely. Keim (2002) discusses a variety of data visualisation techniques to cope with large amounts of data and stresses in particular the use of MDS when working with qualitative data. MDS is a visualisation technique that produces a spatial configuration, in which the objects are arranged in such a way, that their distances correspond to the similarities of the object (Wickelmaier 2003). Contrary to similar methods like multiple correspondence analysis (MCA), MDS can be used easily with different underlying distance measures for representing the similarities among qualitative data. Considering the rising claims and data volume in transport insurance, this will become an increasingly important aspect and it has not been addressed in the literature so far. Despite its proven power, the application potential of MDS is underrepresented in the reviewed literature.

Unsupervised learning approaches—clustering algorithms—appear to be a widely used technique in insurance in order to identify a priori unknown common subgroups. An application by Skorna (2013) for example showed the discovery of previously unknown groups of customers based on insurance claims. When applied to data of reduced dimensionality (such as the output from MDS) rather than the original data, clustering approaches work faster and can provide automated visual recommendations for decision-makers in the transport insurance industry.

While visualisation methods have been applied for example in health insurance, so far, the potential of visualisation-based techniques with multiple views on data in transport insurance was not analysed. The following section therefore outlines a method how transport insurance claims of primarily categorical nature can be transformed and visualised so that insurance experts can easily harness valuable information from the structures in the data.

10.3 Applying MDS to Transport Insurance Claims

The proposed method consists of four steps. In a preliminary step of data preparation, relevant parameters for the analysis are selected and the data are cleaned from invalid values. In a first step, the categorical claims data are then transformed using a range of dis-/similarity measures and in a second step projected using a standard MDS approach. In a third step, a genetic clustering approach for supporting the detection of groups of similar claims is outlined.

Step 0: Preparation. The insurance data set contains 13,711 claims from commercial, intermodal transport with up to 42 characterising variables. While one of the advantages of the proposed approach lies in reducing data of great dimensionality, Table 10.2 describes six variables selected for this analysis based on

Table 10.2 Description of the data set

Parameter	Type	Description	Expressions
Damage payments	Numeric	Compensation payments for damages incurred	(Continuous)
Damage cause	Categorical	Nature of the damage cause, for example theft	8
Transport medium	Categorical	The transport medium, for example lorries	7
Type of goods	Categorical	The type of goods transported, for example chemicals	11
Journey from	Categorical	The country of origin of the transport	123
Journey to	Categorical	The destination of the transport	156

previous studies by Skorna (2013). Only the categorical values are fed to the projections for the purpose of showing the application potential in the transport insurance domain.

Step 1: Transformation of categorical data. Visualising categorical data using a spatial layout requires the expression of the degree of similarity between any two data points in the parameter space. As mentioned previously in Sect. 10.2, this study investigates a range of alternative categorical variable transformation, thereby aiming to offer decision-makers in transport insurance more alternatives in the visualisation process. Boriah et al. (2008) have recently given a holistic account of various measures to numerically express similarities among categorical data. To capture different views on the transport insurance claims, five measures were chosen from Boriah et al. (2008) based their varying assumptions (see Table 10.3). All selected measures scale linearly in their computational effort with the number of variables and their performance is measured by Boriah et al. (2008) in their ability to identify outliers. The application of each similarity measure yields one symmetric proximity matrix used as input for the subsequent projection step.

Table 10.3 Selection of similarity measures for categorical data based on Boriah et al. (2008)

Measure	Description	Performance
Eskin	More weight to mismatches that occur on attributes taking many values	Poor for large data
Goodall	Higher similarity when attribute value is infrequent; takes into account dependencies among attributes	Good
Inverse Occurrence Frequency (IOF)	Lower similarity to mismatches on more frequent values	Average performance
Lin	Higher weight to matches on frequent values, lower weight to matches on rare values	Good
Overlap	Simply assigns distance of 1 to identical and 0 to differing attribute values	Poor

Step 2: Data projection. Using the five proximity matrices as input, a classical multidimensional scaling algorithm (c.f. Wickelmaier 2003) is applied for the sake of its efficiency in visualising large amounts of data. We analyse the main principle components of the distance matrix for dimensions with the biggest influence. To increase the interpretability of the data structure for a decision-maker, we project the data into a three-dimensional (as opposed to a two-dimensional) space.

Step 3: Determining visual clusters. Open source viewers can be used for visualising the projection coordinates. Borg and Groenen (2005) give a profound overview such software, which decision-makers can then use to easily detect conjugate data points. However, this mainly depends on the visually perceivable separation of data points, which in turn may be impacted by the proximity measure applied and the underlying data. To support this process of visual detection for decision-makers and to ensure that the visualised data is a close representation of the underlying data, partitioning methods like facets are commonly used, however these often require prior knowledge or a typology for all points (Borg and Groenen 2005). More intuitive methods from the field of unsupervised learning can aid in this detection more easily. Clustering algorithms like a classical k-means are used to determine a priori unknown class labels for data points and a given number of k clusters (Krishna and Murty 1999). They are fast but tend to produce locally optimal results especially less clearly separated data sets. Genetic k-means algorithms can overcome this problem, but they are slower and rely on a priori parameter tuning (Krishna and Murty 1999). Particularly for poorly separated projections, the GKA may produce a more intuitive clustering for a decision-maker when compared to an ordinary k-means. If both the GKA and the manual selections of test users yield meaningless results (i.e. no clusters could be visualised), the noise in the data would prevent any valuable inferences. However, as noise removal is not the scope of this paper, the authors point to methods in data mining for the improvement of data quality.

To apply the GKA as implemented by Hornik et al. (2015), the reduced dimensionality of the projected data's matrix of three-dimensional coordinates is used as input to (a) overcome problems of data points' sparsity in the high-dimensional space and (b) reduce calculation times for high-dimensional data. As the parameter tuning for the GKA is frequently suggested to be the result of a case-based explorative setting (Hong et al. 2003), a mutation rate of 0.9 for the GKA's mutation operator was set for 100 iterations for this application. The k number of clusters to be provided to most nonhierarchical clustering algorithms such as the one applied can be proxied from the visual projections.

10.4 Results

Figure 10.1 illustrates the respective two-dimensional representations of the three-dimensional projections (corresponding to Steps 0–2). Different proximity measures yielded projections of different structures. The Eskin proximity measure,

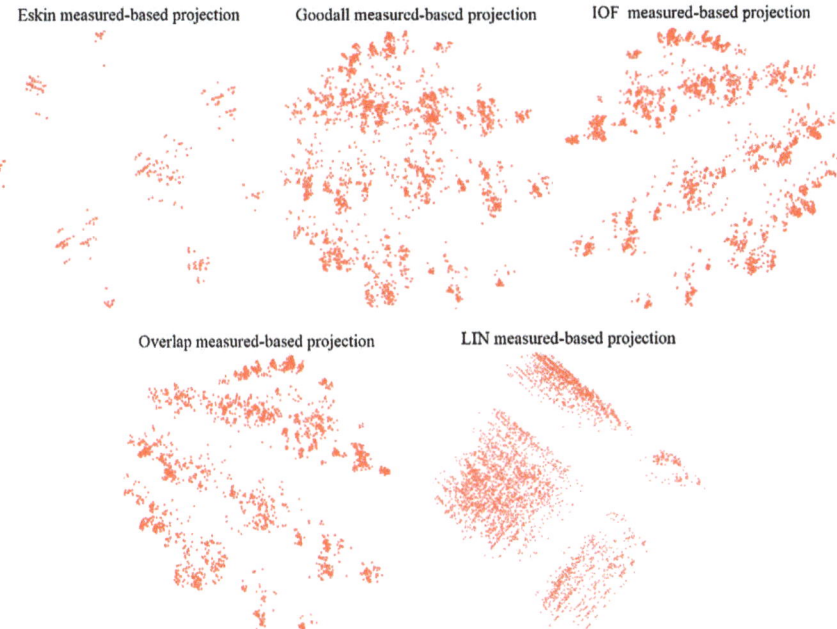

Fig. 10.1 Different views on the data are generated through the application of the various proximity measures

for example, resulted in a well-separated projection due to a very rough and undetailed measurement of data proximity. Here, decision-makers could easily identify groups of related claims. Projections, using a finer measurement of categorical proximity make this task difficult. To illustrate that the GKA can identify groups of related claims, which match a decision-maker's (visual) expectations, we compare the GKA clustering to the identification of related claims done manually by a test user in the well-separated Eskin projection (see Fig. 10.2).

The test user selected 9 groups of claims, and Fig. 10.3 shows to which degree (in %) these visually identified clusters (VC1-9) contain the same elements as the clusters identified by the GKA (with $k = 9$). Here, data points in visually perceived clusters are shown to be nearly identical to those identified by the GKA.

Using the GKA for less clearly separated projections, where decision-makers may struggle to find groups of similar claims, the GKA could be applied similarly to provide automated suggestions. The application of the proposed method enables decision-makers to harness subsequently the information about the projected data through the display of on-screen claims description as well as descriptive statistics of clusters. Figure 10.4 (left) for example illustrates the distribution of damage payments in one exemplary selection, using a classification into damage classes based on Skorna (2013). Figure 10.4 (right) describes in which proportion the most frequently occurring categorical expression per variable exists in one exemplary selection. Providing descriptive information of the identified structures in the data

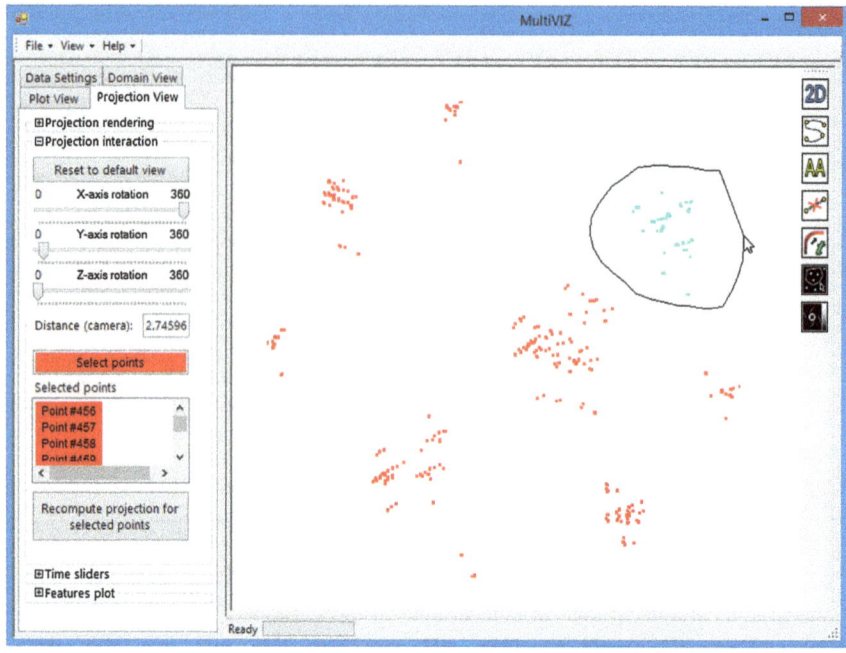

Fig. 10.2 A test user makes selections of perceived groups of related claims by using a manual selection feature

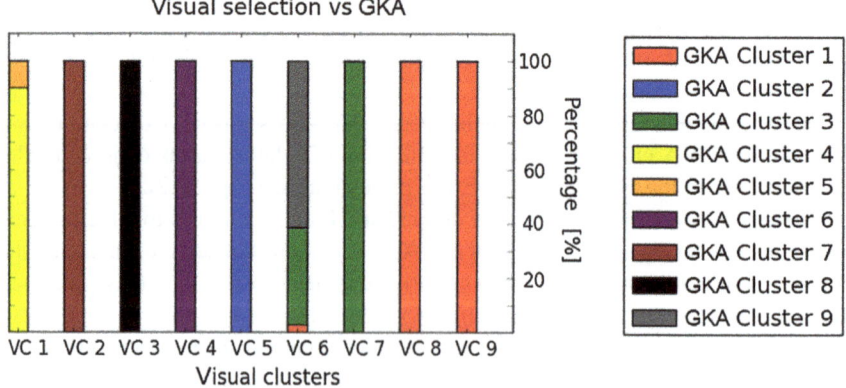

Fig. 10.3 Percent overlap of manual selections of visual clusters and the GKA's class label assignment in the Eskin projection

can for example support the validation of existing customer groups, the calculation of new premiums according to the transport damage distribution in manual selections, creating or validating country-specific risk classes (according to origin, destination, or transit routes), or many other tasks in transport insurance. The

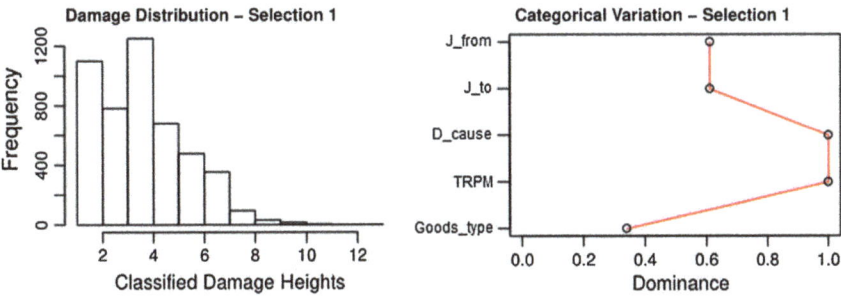

Fig. 10.4 On-screen information of one exemplary selection in the Eskin projection

investigation of causality or correlation for different practical problems in transport insurance, related to the presented data set, would exceed the scope of this paper but can provide an important outlook for future analyses in the practical transport insurance domain.

10.5 Discussion, Further Development and Conclusion

Traditional data modelling techniques in transport insurance may not produce meaningful output for a decision-maker when faced with increasing data volume and dimensionality. The presented algorithmic modelling approach provides multiple visualisations of claims data and hence forms the basis for the design of intuitive DSS tools in transport insurance. While for this application, the data dimensionality has been preliminarily restricted, the presented approach can handle data of greater dimensionality and volume, too. Extensions to the method can include linear combination of the distance measures with numerical distance measures and colour maps to visualise the damage payments in the projections. Limitations of the presented approach include the assumption of spherical clusters when applying the GKA to suggest groups of similar claims to the decision-maker. Joia et al. (2015) give a very good account for more elaborate ways of detecting clusters. Further work based on the findings presented should include the development of the presented methodology as a formal DSS tool. This should include further expert testing and subsequently an investigation into different practical problems in transport insurance to validate the presented approach in practice.

References

Balaji S, Srivatsa SK (2012) Decision Tree induction based classification for mining Life Insurance Data bases. Int J Comput Sci Inf Technol Secur (IJCSITS) 2:699–703

Breiman L (2001) Statistical modeling: the two cultures. Stat Sci 16:199–231

Broeksema B, Telea AC, Baudel T (2013) Visual analysis of multi-dimensional categorical data sets. Comput Graph Forum 32:158–169

Borg I, Groenen PJ (2005) Modern multidimensional scaling: theory and applications. Springer Science & Business Media

Boriah S, Chandola, V, Kumar V (2008) Similarity measures for categorical data: a comparative evaluation. red 30:3

Cerchiara RR, Edwards M, Gambini A (2008) Generalized linear models in life insurance: decrements and risk factor analysis under Solvency II. In: 18th international AFIR colloquium

Devale AB, Kulkarni RV (2012) Applications of data mining techniques in life insurance. Int J Data Min Knowl Manage Process (IJDKP) 2:31–40

Guelman L (2012) Gradient boosting trees for auto insurance loss cost modeling and prediction. Expert Syst Appl 39:3659–3667

Guo L (2003) Applying data mining techniques in property/casualty insurance. In: CAS 2003 winter forum, data management, quality, and technology call papers and ratemaking discussion papers, CAS

Hornik K, Feinerer I, Kober M (2015) skmeans: spherical k-Means clustering. R package version 0.2-7. https://CRAN.R-project.org/package=skmeans. Accessed 30 Oct 2015

Hong TP, Lin WY, Lee WY (2003) Adapting crossover and mutation rates in genetic algorithms. J Inf Sci Eng 19:889–903

IUMI—International Union of Marine Insurance (2014) Global marine insurance report

Joia P, Petronetto F, Nonato LG (2015) Uncovering representative groups in multidimensional projections. Comput Graph Forum 34:281–290

Keim D (2002) Information visualization and visual data mining. IEEE Trans Vis Comput Graph 8:1–8

Krishna K, Murty MN (1999) Genetic K-means algorithm. IEEE Trans Syst Man Cybern Part B: Cybernet 29:433–439

Maas P, Milanova V (2014) Zwischen Verheissung und Bedrohung – Big Data in der Versicherungswirtschaft. Die Volkswirtschaft 5:23–25

Roux I (2008) Application of cluster analysis and multidimensional scaling on medical schemes data. Doctoral dissertation, Stellenbosch University

Schaerer M, Wanner H, Grinyer CC (IBSG) (2011) Transforming the insurance industry to increase customer relevance. https://www.cisco.com/web/about/ac79/docs/fs/Insurance_Customer_Relevance_0301_IBSG.pdf. Accessed 30 Oct 2015

Schäfer I, Kaduszkiewicz H, Wagner HO, Schön G, Scherer M, van den Bussche H (2014) Reducing complexity: a visualisation of multimorbidity by combining disease clusters and triads. BMC Public Health 14:1285–1299

Skorna AC (2013) Empfehlungen für die Ausgestaltung eines Präventionskonzepts in der Transportversicherung. Doctoral Thesis, University St. Gallen

Skorna AC, Fleisch E (2012) Loss prevention in transportation to ensure product quality: insights from the cargo insurance sector. In: Advances in production management systems. Value networks: innovation, technologies, and management. Springer, Berlin, Heidelberg

Skorna AC (2013b) Empfehlungen für die Ausgestaltung eines Präventionskonzepts in der Transportversicherung. Verlag Versicherungswirtschaft GmbH, Karlsruhe

Thakur SS, Sing JK (2013) Mining customer's data for vehicle insurance prediction system using k-means clustering-an application. Int J Comput Appl Eng Sci 3:148–153

Vilko JP, Hallikas JM (2012) Risk assessment in multimodal supply chains. Int J Prod Econ 140:586–595

Wickelmaier F (2003) An introduction to MDS. Sound Quality Research Unit, Aalborg University, Denmark

Yeo AC, Smith KA, Robert JW, Brooks M (2001) Clustering technique for risk classification and prediction of claim costs in the automobile insurance industry. Intell Syst Account Finan Manage 10:39–50

Chapter 11
Methodological Demonstration of a Text Analytics Approach to Country Logistics System Assessments

Aseem Kinra, Raghava Rao Mukkamala and Ravi Vatrapu

Abstract The purpose of this study is to develop and demonstrate a semi-automated text analytics approach for the identification and categorization of information that can be used for country logistics assessments. In this paper, we develop the methodology on a set of documents for 21 countries using machine learning techniques while controlling both for 4 different time periods in the world FDI trends, and the different geographic and economic country affiliations. We report illustrative findings followed by a presentation of the separation of concerns/division of labor between the domain expert and the text analyst. Implications are discussed and future work is outlined.

Keywords Logistics · Transport system evaluation · Big data analytics · Data mining · Text mining · Machine learning

---

A. Kinra (✉)
Department of Operations Management, Copenhagen Business School,
Solbjerg Plads 3, 2000 Frederiksberg, Denmark
e-mail: aki.om@cbs.dk

R.R. Mukkamala · R. Vatrapu
Computational Social Science Laboratory, Department of IT Management,
Copenhagen Business School, Howitzvej 60, 2000 Frederiksberg, Denmark
e-mail: rrm.itm@cbs.dk

R. Vatrapu
e-mail: rv.itm@cbs.dk

R. Vatrapu
Mobile Technology Laboratory, Faculty of Technology,
Westerdals Oslo ACT, Oslo, Norway

11.1 Introduction

The assessment and appraisal of transportation systems is a central activity related to both national and regional policy making for passenger and freight movement. In this regard, there are a fixed set of widely accepted core methods for economic evaluation and appraisal for different project investments, e.g., related to transport infrastructure for the movement of people and goods. Most of these methods that are currently used for this purpose involve the calculation of benefit-cost ratios.

There is an inherent assumption in that such an evaluation and assessment process will aid economic goals, and even foster development in the case of developing countries by lowering the overall transportation and mobility costs of businesses and private users. However, the cross-border movement of goods involves international transportation, global supply chains, multinational businesses and users, where the costs and benefits of transportation systems are not only highly variable, but also tied to the broader global logistics and supply chain systems that embed these transportation systems. The Multi National Corporation (MNE), and its global supply chain then becomes a primary user and motor of economic growth in an era of globalization and global integration of businesses. Tracking the business considerations of MNEs, and how their assessments of the national and regional transportation, logistics and supply chain systems play a role in these considerations, becomes important for policy makers in order to increase international trade and to attract foreign direct investments in almost all sectors of the economy.

Toward this end, this paper explores and exploits text analytics methods and techniques from big data analytics to develop a methodology for information scanning, extraction, and retrieval of the logistics-related measures preferred by decision-makers of MNCs, and other firms involved in global supply chains. This information is relevant for the benchmarking and evaluation of transportation and logistics systems. We use text mining and text analytic approaches, and more specifically machine learning techniques in order to develop this methodology. We then test run this methodology on a global supply chain text corpus in order to illustrate and provide some feedback on the methodology. Our initial results illustrate that the methodology is rather useful, and it can be successfully applied to other logistics and transportation applications. However, we can also relate to the general observation that model specification is the Achilles' Heel of big data analytics as it is easy to under- or over-specify the model without deep involvement of domain experts and deep knowledge of domain-specific problems.

11.2 The Assessment of Country Logistics Systems

The assessment of country logistics systems is an emerging area and literature is sparse. Though most of the existing work on the assessment of country logistics systems can be related to the useful distinctions in existing location research (see

Beugelsdjik et al. 2010; Beugelsdjik and Mudambi 2013). Kinra (2015) traces the main literature streams in this area to the international economics, economic geography, and international business distinctions in existing location research, and these literature streams are now summarized.

The economics-oriented literature generally adopts the stance that spatial variation is generated by the structure of national resource endowments, and trading patterns. The main contributions in this stream come from Memedovic et al. (2008) and Bookbinder and Tan (2003), among others. For this group of literature spatial discontinuity generally occurs at national borders and is defined by national logistics systems. The geography-oriented literature tends to adopt the stance that connectivity and colocation effects at major trade interface points generate spatial variation. The contributions of Rodrigue and Notteboom (2010) and Rodrigue (2012) are prominent in this stream. For these contributors spatial discontinuity occurs at major nodes: transit points, hubs, and gateways, and is defined by regional logistics systems, and hence the focus on the ranking of regional logistics systems. Finally, the international business-oriented literature tends to adopt the stance that the managerial and organizational utility from locations generates spatial variation. Min (1994) is one of the earliest to demonstrate this stance. Similarly Carter et al. (1997), Kinra and Kotzab (2008) make important advances in this regard. For this group of contributors, spatial discontinuity generally occurs by country, and is defined by the country investment climate.

All these contributions seek to provide an evaluation of the national logistics systems. However some issues are common in the literature. First, the literature is not unanimous about the logistics-oriented determinants, or factors, of location. Second, it does not clarify the role and type of information, or measures, that are adopted in the assessments. The literature is therefore mute about the specific spatial impedance information and information categories that are relevant for decision-makers in country assessment, and it is indefinite about the relevance and the importance of the considered location determinants and the missing information. This is a problem because information about the foreign country is widely recognized as a major contributor in the complexity under conditions of globalization (see McCann and Mudambi 2004), and global supply chains.

We make a first attempt to the identification and categorization of the relevant, available information on country logistics systems in Kinra (2015). Apart from the identification of 187 different types of relevant and essential information measures that fall under 17 decision factors, the study establishes the main spatial transaction cost categories under which the information is made available, but also concludes on conditions of information-driven complexity for the ranking of country logistics systems.

However these findings are only based only complex manual procedures for discerning this information, and include observations from a limited dataset of countries for a limited time period (2006–2008). Nor any implications are presented for the ranking (assessment) of the different countries based on the spatial impedance measures that have been identified. This study will go beyond these limitations by developing more automated techniques that are capable of handling bigger information sets. This will take better into account factors such as longitudinality of data for determining the general conditions of information-driven complexity hypothesis, but also from the point of view of comparing different time periods. The next section now provides a detailed description of this methodology development, including the text mining and text analytic approaches that are employed to develop the methodology for automating the information scanning, extraction, and retrieval process.

11.3 Methodology

In this section, we describe the methodology that was developed to conduct big data analysis of global supply chain document corpus. The corpus primarily contains 21 text documents describing transportation and logistics systems pertaining to 20 countries spanning over period from 2006–2014 years. In order to analyze the text corpus, we have used both text mining and machine learning techniques as shown in Fig. 11.1.

As a first step (Text Extraction) toward applying text mining/machine learning algorithms on the text corpus, we have extracted texts from the global perspective documents which are in Portable Document Format (PDF) using an open source software component using Java Programming language. The size of a global

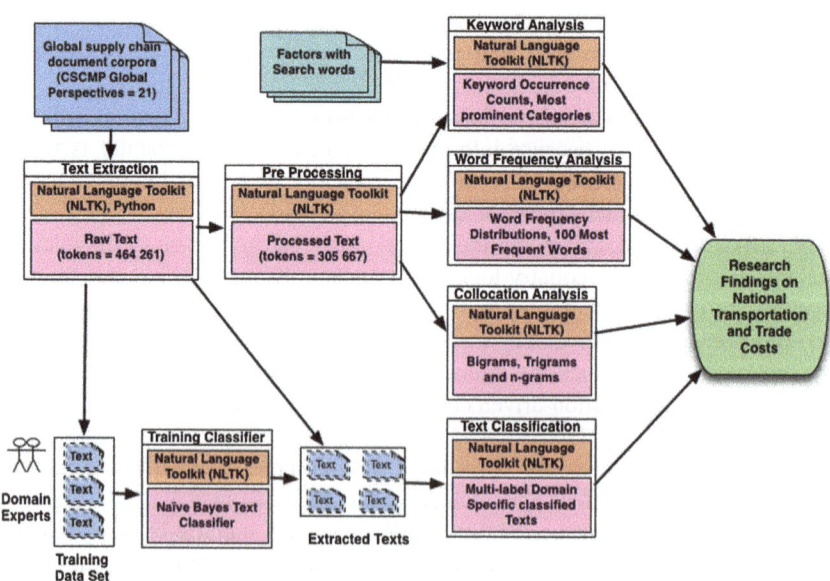

Fig. 11.1 Text analytics methodology of global supply chain document corpus

perspective document varies from 20 to 60 pages and text from each page has been extracted separately for further processing. Unfortunately, it was not possible for us to extract the PDF metadata such as links and bookmarks from the documents. The extracted texts are tokenized by breaking them sentences and words by using space as delimiting character between the words using Natural Language Toolkit (NLTK) (Bird 2006) and Python programming language. The final output of this step ended up in raw text containing approximately 464,000 tokens, which is used as an input for further processing using text mining and machine learning algorithms.

11.3.1 Text Mining

As part of text mining we have applied three different techniques: *Keyword Analysis*, *Word Frequency Analysis*, and *Collocation Analysis* as further described below. In order to prepare the raw text for applying these techniques, we had to clean and preprocess the raw text using Natural Language toolkit. As part of preprocessing step, we further processed the raw text tokens by removing nonalphabetical characters, numerical and stop words that are high-frequency words like *the, to, also, and* so on, which do not contribute much to the semantic meaning of the documents. Even though the raw text contains 464 000 tokens, after the preprocessing step, we ended up with approximately 300,000 tokens.

11.4 Keyword Analysis

As part of keyword analysis, we investigated which decision factors are more predominant in the country descriptions. In order to investigate that, each factor (e.g., Waterways) is supplemented with suitable search words (port, water, sea, shipping, etc.) that are representative of that factor. Even though 17 decision factors with suitable search words are adopted from Kinra (2015) initially, soon the list of factors got expanded to 21 factors to include more interesting factors. One of the challenges with the decision factors from Kinra (2015) is that the number of search terms varied from factor to factor which made the cross comparisons of decision factors among the countries difficult. Therefore, we have readjusted search terms by dropping few search terms for some factors and by adding new search terms for some other factors, thereby making the number of search terms equal to four for every decision factor. Soon after finalizing the search terms for decision factors, we used NLTK with Python programming language to conduct analysis to find out the number of occurrence of each search term (or keyword) in each perspective document for all the decision factors. The occurrence counts of search terms are summed up to find out the keyword occurrence count for a decision factor. Based on the keyword occurrence counts of decision factors, we were able to compare which decision factors are predominant in the entire corpus and as well as in each country descriptions. In addition to that, we could also be able to compare decision factors across country dimension such as which are the most prominent countries for any given decision factor.

11.5 Word Frequency Analysis

Word frequency analysis is a method of automatically identifying the frequent occurring words from a given text corpus, by using the term document matrix. In order to compute the word frequencies, raw tokens from the preprocessing step are further analyzed using NLTK to prepare term document matrix containing the frequency of words across perspective document collection. We have computed the term document matrix for the whole corpus as a single document and for each individual perspective document as well. The construction of term document matrices enables us to find most frequent words (e.g., 100 most frequent words) with the word frequencies, which is used to generate word cloud for a given document. Word clouds or tag clouds is a simple way of visualizing or highlighting the most frequently used words from a given text document. As part of the analysis, we have generated word clouds for each perspective document and also a word cloud for whole text corpus by combining all the perspective documents into one. Word frequency analysis and word clouds enabled us to get an overview of major concepts/topics discussed in the documents.

11.5.1 *Collocation Analysis*

Collocations are expressions of multiple words, which commonly co-occur in the documents and therefore a collocation is a sequence of words that occur together unusually often in the text documents. Finding collocation expressions involves standard statistics-based and linguistically rooted association measures against mere frequency of word occurrence counts. The collocation analysis provides insights about the documents by providing bigrams, trigram, and n-grams that contain words, which co-occur in the documents. The collocation analysis for each perspective document as well as for whole data corpus was conducted using Natural Language toolkit (Bird 2006) to find out bigrams and trigrams. It provided us with intuition about the topics and concepts that are discussed in the country descriptions of transportation and logistics. It also helped us notice the emerging trends (other than decision factors) in the perspective documents, which further helped us process the documents for emerging trends or factors.

11.5.2 Machine Learning and Text Classification

As part of the methodology, our plan is to apply machine learning algorithms on the text corpus to perform text classification tasks. Text classification approach comes under supervised machine learning and it can be defined as a process where assigning a predefined category of labels to new documents based on probabilistic measure of likely hood using a training set of labeled documents (Yang and Liu 1999). Out of several approaches available in text classification domain, we have chosen a simple text classification method (Zhang and Li 2007) based on Bayes rule that relies on a simple representation of documents using bag of words approach. In this project, we have used Naïve Bayes classification method to classify the extracted text pieces from the perspective documents.

First, we have extracted text portions from the perspective documents of country descriptions using NLTK based on occurrence of each search word that belong to a decision factor. In order to get proper understanding of the context of the occurrence of the search word, we have extracted few sentences of text on both sides of the occurrence location of search word. Altogether, we have extracted around 28,000 text portions from the total 21 perspectives documents. The accuracy of text classifier purely depends on the quality of manually encoded training set data. Therefore, in order to get a proper training set that is representative of the transportation and logistics domain, we are planning to get 10 % of the extracted text portions manually coded by the domain experts belong to Transportation and Logistics domain. As shown in Fig. 11.1, major part of the manually coded texts will be used for training the Naïve Bayes text classifier. After the classifier gets sufficiently trained we will use the rest of the training set to test the accuracy of the classifier. Finally, the classifier will be used to on the remaining 90 % of extracted texts to classify them automatically using the machine learning technique. Furthermore, for the text classification we would like to different domain-specific models such as sentiment, emotion, and any other models that are specific and suitable for the transportation and Logistics domain.

11.6 Illustrative Results

We present a selection of the results below to illustrate the methodological approach and to demonstrate the domain-specific utility of applying text analytics methods and techniques (Bird 2006; Yang and Liu 1999; Zhang and Li 2007). Table 11.1 presents the category analysis of the text corpus to extract the basic indicators of global supply chain perspectives.

Figure 11.2 shows the five most frequent categories and their respective cumulative keyword occurrences across the different time periods.

Table 11.1 Category analysis of global supply chain perspectives

Categories	2004–2007	2008–2009	2010–2013	2014
Waterways	1361	2232	2965	1117
Intermodal	668	701	1119	424
Economic structure	396	464	917	225
Railways	363	486	677	337
Economic policy	321	239	581	124
Airways	185	674	849	299
Geographical location	108	548	649	120

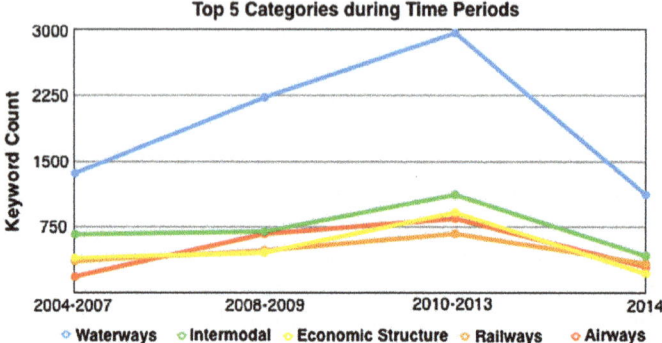

Fig. 11.2 Top 5 categories during time periods

Fig. 11.3 Top categories and measures for whole dataset

Figure 11.3 shows the most frequent categories and measures for the whole data corpus. Finally, Fig. 11.4 shows the emergent categories and measures for the whole data corpus.

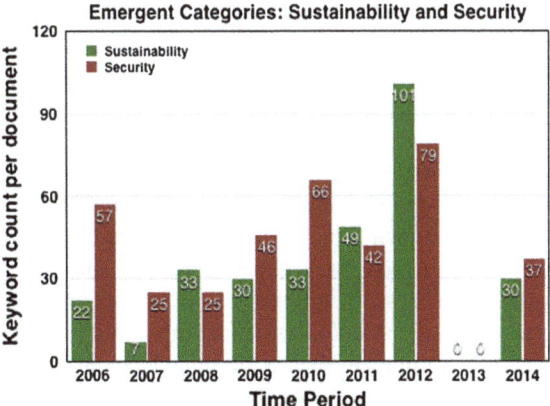

Fig. 11.4 Emergent categories of sustainability and security

11.7 Discussion and Conclusion

Figure 11.5 presents the separation of concerns/division of labor between the domain expert and the text analyst. As can be seen, the deep involvement of the domain expert during the model specification and training phases of the text analytics process is necessary but not sufficient. Subsequent involvement of the domain expert during the classification and model fine-tuning phases yields empirical results that are robust, reliable, and relevant. We will further explore the dynamics

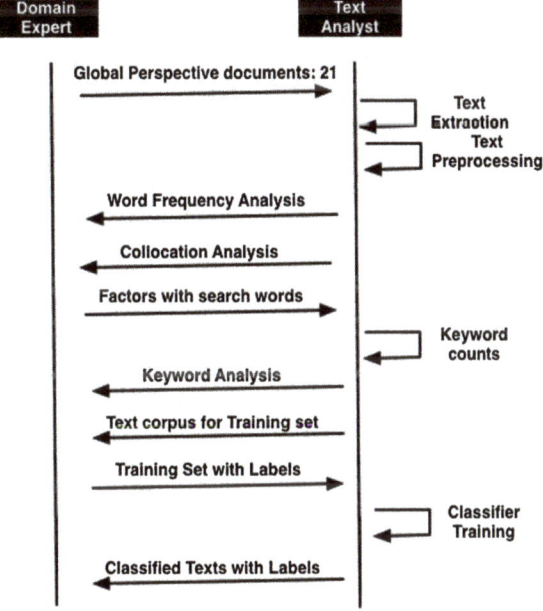

Fig. 11.5 Separation of concerns: domain expert versus big data analyst

of the relationship between specific domain expertise and generic big data analytics expertise in our future work.

This study has developed and demonstrated an approach for automating the information scanning, retrieval, and extraction process for logistics and transportation applications. The approach is especially useful because it will aid in the systematic study of big data while making sense of different types of logistics costs that are currently being incurred in a tumultuous market place. Companies are increasingly facing big data challenges in the global transportation of goods and services, and look for dedicated approaches in this regard in order to leverage their information system investments for optimizing their logistics and transportation costs. For example, Maersk Line, the world's largest container shipping company has recently implemented a "Remote Container Management" system that collects information using sensors and transceivers embedded in the containers. Such systems create a deluge of near real-time data that need to be modeled and analyzed using methods and techniques from the emerging field of data science in general and big data analytics in particular. However, we can also relate to the general observation that model specification is the Achilles' Heel of big data analytics as it is easy to under- or over-specify them without deep involvement of domain experts and deep knowledge of domain-specific problems.

In addition to traditional text mining and machine learning methods, the next steps involve employing the new method of "Social Set Analysis" to extract and model latent social information and the resulting social influence, if any, from the document corpus. Vatrapu et al. (2014a, b) have proposed a set theoretical approach to big social data analytics called Social Set Analysis (SSA). SSA is based on the sociology of associations and the mathematics of set theory and supports both interaction analytics in terms of actors involved, actions taken, artifacts engaged with as well as text analytics in terms of keywords employed, feelings expressed, pronouns used, and topics discussed (Mukkamala et al. 2014a, b; Vatrapu et al. 2014a, b). For the purposes of this paper and the domains of trade and transportation, we plan to employ SSA to uncover the temporal distribution of institutional actors' engagement as well and unique actor/keyword sets before, during, and after events of theoretical interest and the spatiotemporal dynamics of keywords and expressed feelings with regard to country assessments in the text corpus.

References

Beugelsdijk S, Mudambi R, McCann P (2010) Place, space and organization: economic geography and the multinational enterprise. J Econ Geogr 10(4):485–493

Beugelsdijk S, Mudambi R (2013) MNEs as border-crossing multi-location enterprises: The role of discontinuities in geographic space. J Int Bus Stud 44

Bird S (2006) NLTK: the natural language toolkit. Paper presented at the proceedings of the COLING/ACL on interactive presentation sessions

Bookbinder JH, Tan CS (2003) Comparison of Asian and European logistics systems. Int J Phys Distribut Logist Manag 33(1)

Carter JR, Pearson JN, Peng L (1997) Logistics barriers to international operations: the case of the people's republic of China. J Bus Logist 18(2):129–145

Kinra A, Kotzab H (2008) Understanding and measuring macro-institutional complexity of logistics systems environment. J Bus Logist 29(1):327–346

Kinra A (2015) Environmental complexity related information for the assessment of country logistics environments: implications for spatial transaction costs and foreign location attractiveness. J Transp Geogr 43:36–47

McCann P, Mudambi R (2004) The location behavior of the multinational enterprise: some analytical issues. Growth Change 35(4):491–524

Memedovic O, Ojala L, Rodrigue JP, Naula T (2008) Fuelling the global value chains: what role for logistics capabilities. Int J Technol Learn Innov Dev 1(3):353–374

Min H (1994) Location analysis of international consolidation terminals using the analytic hierarchy process. J Bus Logist 15(2):25–44

Mukkamala R, Hussain A, Vatrapu R (2014a) Fuzzy-set based sentiment analysis of big social data. In: Proceedings of IEEE 18th international enterprise distributed object computing conference (EDOC 2014), Ulm, Germany, pp 71–80. doi:1510.1109/EDOC.2014.1519. ISBN: 1541-7719/1514

Mukkamala R, Hussain A, Vatrapu R (2014b) Towards a set theoretical approach to big data analytics. In: Proceedings of the 3rd IEEE international congress on big data 2014, Anchorage, United States

Rodrigue J-P (2012) The geography of global supply chains: evidence from third-party logistics. J Supply Chain Manag 48(3):15–23

Rodrigue J-P, Notteboom T (2010) Comparative North American and European gateway logistics: the regionalism of freight distribution. J Transp Geogr 18(4):497–507

Vatrapu R, Mukkamala R, Hussain A (2014a) A set theoretical approach to big social data analytics: concepts, methods, tools, and findings. Paper presented at the computational social science workshop at the European conference on complex systems 2014

Vatrapu R, Mukkamala R, Hussain A (2014b) Towards a set theoretical approach to big social data analytics: concepts, methods, tools, and empirical findings. Paper presented at the 5th annual social media & society international conference 2014

Yang Y, Liu X (1999) A re-examination of text categorization methods. Paper presented at the proceedings of the 22nd annual international ACM SIGIR conference on research and development in information retrieval

Zhang H, Li D (2007) Naïve Bayes text classifier. Paper presented at the granular computing, 2007. IEEE international conference on GRC 2007

Chapter 12
Big Data Analytics in the Maintenance of Off-Shore Wind Turbines: A Study on Data Characteristics

Elaheh Gholamzadeh Nabati and Klaus Dieter Thoben

Abstract The aim of this study is to discuss the characteristics of the input data to the data analytical algorithms of a predictive maintenance system, from the viewpoint of big data technology. The discussed application is for the maintenance of off-shore wind turbines. The maintenance of off-shore wind turbines is an expensive and sensitive task. Therefore, making decision for planning and scheduling of maintenance in a wind farm (which is made by the operating company of a wind farm) is important and plays a critical role in the cost of maintenance. In this paper, the current state of the art for big data technology in the maintenance of off-shore wind turbines is presented. The dimensions of big data analytics and the technical requirements of data for the use of this technology in the maintenance of off-shore wind turbines are described. A contribution of this paper is to study the technical requirements of suitable data for decision-making. The outcomes of presented study are identifying the characteristics of input data to predictive maintenance in the era of big data and the discussion of these characteristics in the condition monitoring for off-shore wind energy.

Keywords Predictive maintenance · Big data · Off-shore wind energy · Data analytics

---

E.G. Nabati (✉)
International Graduate School of Dynamics in Logistics, University of Bremen, Bremen, Germany
e-mail: nab@biba.uni-bremen.de

K.D. Thoben
BIBA- Bremer Institute für Produktion und Logistik GmbH, Bremen, Germany
e-mail: tho@biba.uni-bremen.de

© Springer International Publishing Switzerland 2017
M. Freitag et al. (eds.), *Dynamics in Logistics*, Lecture Notes in Logistics,
DOI 10.1007/978-3-319-45117-6_12

12.1 Introduction

Maintenance process is a critical element of system's operation. The maintenance of off-shore wind turbines is usually done by the operating companies. The operating company is responsible for operating a wind farm and making the decision on the service in order to prevent the equipment stopping from producing the electricity. In the first five years of the operation, the maintenance and service tasks are done in cooperation with the wind turbine manufacturer.

The maintenance strategies of the off-shore wind mills which the operating companies perform have three categories: the corrective maintenance, preventive maintenance and the predictive maintenance. In the corrective maintenance the turbine runs until the failure. Failure causes the system to stop. At this time, the fault diagnosis is done and then it is decided whether to repair or replace (Puig 2011). The preventive maintenance is the interval visits of the wind turbines and service all the parts based on fixed time intervals. This type of maintenance usually requires the part to be changed before they reach the end of their life time. Therefore, this maintenance strategy is costly. The third strategy for maintenance is the predictive maintenance where the status of system is monitored with the condition monitoring system and the parts are replaced when the monitoring system shows, there is need for replacement. In other words as stated by (Karyotakis 2011) the equipment are *monitored for their condition and statistical reliability tools are used to preventively maintain or exchange critical items before any failure occurs.*

The off-shore wind equipment has several characteristics which calls for more control and monitoring of the equipment from remote. First, the wind farms are normally located far from the land and the accessibility to these equipments is only with ship and helicopter possible. The dispatching of ship or helicopter depends on the weather condition, whether if the sea is calm and safe enough to travel. Second, the wind turbines are exposed to moisture, salt and sever marine condition as well as constant vibration because of the sea waves. These factors cause the early erosion and consequently need for more service in comparison of on the land turbines.

Therefore the maintenance strategies such as predictive maintenance are the most preferable solution among the above maintenance strategies for the off-shore. It enables the control from the remote and if properly implemented causes the least expenses during the services.

From one hand, the predictive maintenance uses the data sources such as sensor measurements and health and environmental data for performing the predictions. From the other hand, currently the advances in the technology, caused a big data to appear, new types of sensors, new measurement methods are being generated. These new data sources cause challenges and need for modifications in the current predictive maintenance and reliability estimation approaches. It is necessary to clarify the characteristics of new data to enable faster adoption of maintenance system to the new situation. Moreover, to enable innovation in the data modelling approaches for the maintenance. Although recently several papers are published regarding the issue of big data dimensions in maintenance (Meeker and Hong 2014;

Göb 2014), still more work and applications are needed. In this paper the characteristics of input data to the data analytical algorithms of a predictive maintenance system is discussed from the viewpoint of big data technology. The application is for the maintenance of off-shore wind turbines.

This paper is organized as follows. Section 12.2 introduces the big data characteristics and the big data technology. Section 12.3 describes the relation of predictive maintenance with the big data analytics. Section 12.4 provides a review of literature on big data technology in the maintenance of wind energy. Section 12.5 presents the research methodology. Section 12.6 discusses the technical requirements of maintenance data to be used in the big data system. Section 12.6 is dedicated to the conclusion.

12.2 Big Data Technology and the Place of Data Analytics

Big data is a term which refers to the data which has a huge size, high speed of production and comes from several sources with different formats. In the context of decision-making, big data is considered as a new source of competitive advantage which has the potential to derive new insight from it, and can help in making more realistic decisions. Form the technical point of view, the definition of big data is the data which is more than the size of current relational data bases and so complex in format and dimension, which is hard or impossible to use and apply traditional analytical techniques on it. (Fosso Wamba et al. 2015) provided the different definitions of big data in the literature. The basic definition contains three characteristics (3 V).Volume, which refers to the size of data. Variety, that refers to the several data sources and data formats and Velocity, which is the increasing speed of data generation and transmission. Some authors add more dimensions to the 3 V. For example (Opresnik and Taisch 2015; White 2012), added the two dimensions of Veracity and Value to the big data. Veracity means the accuracy of the data and Value describes the insight which these information can provide for us. These authors applied the concept of big data on a certain business oriented domain such as IT system, product service system, business and organizational data sources. Therefore, taking into account the business value of the big data. From another perspective, (Göb 2014) named the fourth dimension of big data as the complexity. Indeed he looked at the big data problem from a more technical view. Complexity describes the data management difficulties, statistical analysis and algorithmic complexity aspects of big data (Göb 2014).

Big data analytics is a part of big data technology, where the analysis of data is performed. Data analysis basically is an interdisciplinary field between the sciences of probability theory, statistics, machine learning and data mining. From the application purpose, data analytics can be used for describing the characteristics of a certain entity such as the mean value of a measurement, exploring the correlation

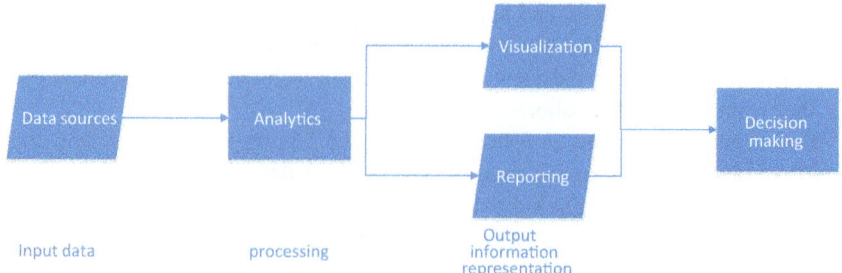

Fig. 12.1 Simple systemic view of data analytics

and relations between different variables, e.g. the relationship between the increase in the speed and increase of temperature, monitoring the current situation (diagnosis), e.g. fault detection, predicting the future condition (prognosis), e.g. predicting the life of a part. A good representation of different approaches of data analysis is illustrated in the research done by (Freitag et al. 2015).

In the next section the location of data analytics is discussed in the predictive maintenance.

Figure 12.1 shows a simple systematic view to the data analytics. First, different data from the different data sources are being preprocessed. Importantly, the missing values and errors are handled. Afterwards, the data analytical process is being done. It usually contains the feature selection, model fitting or the algorithm application and assessing the appropriateness of the output information. In the next step, the gained information are represented to the decision-maker for further interpretation and use in the decision-making process.

12.3 Data Analytics and Predictive Maintenance

Predictive maintenance uses a part of data analytical techniques. These techniques are diagnosis and prognosis models. With them, it is possible to monitor the condition of the products (Puig 2011). Therefore, the analytical module of the predictive maintenance is called Condition Monitoring System (CMS). It is always beneficial to distinguish between prognosis and diagnosis. While diagnosis is used to monitor and finding the faults, abnormalities and control if the process is functioning within certain standards, prognosis is done to understand the problem (fault) before it has occurred. The example of prognosing is estimating the remaining life of the parts in a wind turbine. Several algorithms exist for prognosing and diagnosing. Good references about the classification of these algorithms in the predictive maintenance and wind industry are available at (Hameed et al. 2009; García Márquez et al. 2012; Takoutsing et al. 2014; Sheng 2015; Lau et al. 2012).

12.4 Big Data Analytics in Wind Maintenance: Review of the Literature

Wind energy has used the data analytics from long ago in the maintenance practices for analyzing the system health monitoring data (SHM) and the environmental data (Meeker and Hong 2014). Nevertheless, still no comprehensive work about the application of big data technology in the off-shore wind energy maintenance exists. Only, some authors addressed different parts of this technology. In this section an overview of existing literature is presented.

By reviewing the literature on the applications of big data for the maintenance of wind energy, four main categories are recognized:

1. **SHM and environmental data Storage**: (Viharos et al. 2013) studied collecting and storing large volumes of data from the wind park site and store it for further fault prognosis. The study aims at providing guidelines for practitioners. This research groups the challenges of big data storage and processing as the following: "

   - *The volume of the data puts very high load on both the local industrial computers and the data centers.*
   - *End user applications must balance between flexibility of querying and quick query response times as distributed systems pose limitations to certain elements of SQL including the join operation".*

   Markovic et al. (2013) studied the cloud computing for real-time high volume data capturing and analysis. Nguyen et al. (2014) discussed the data management and metadata generation aspects.

2. **Data processing and analysis**: The current literature can be divided to two subgroups, first prediction and reliability analysis and second system health monitoring and controlling. For the first subgroup, (Xiang et al. 2009) studied the utilizing the sensor data for fault prediction. They applied fusion technique which is a data-driven approach to take advantage of the available measurements without installing extra sensors explicitly for fault diagnosis. Xiang et al. (2009 and Spahić et al. (2009) presented a reliability model for several wind mills in a large off-shore wind farm. They computed the reliability model with high volume of data and also they included new data characteristic such as wind farm location, single generator power and power of the entire wind farm, wind farm grid type (radial/meshed), switchgear and protection type, and automation technology type. They applied statistical methods on the data.

   For the second subgroup which is monitoring and controlling applications, Lau et al. (2012) provided a review of failure diagnosis types. Wan (2004) analyzed the long time data to see the behaviour of wind mills at the farm.

3. **Smart grinds**: The management and efficiency of produced electricity in the era of big data has been addressed by some authors. SunGard solutions (2013) named the advantages of big data for energy trading. IBM (2012) discussed the

characteristics and relevant system architectures for big data in smart grinds (IBM 2012). Diamantoulakis et al. (2015) addressed the energy management in smart grids with the big data analytics.
4. **Social media and unstructured data analysis**: Anninni (2014) reported the use of new sources of unstructured data in the wind energy. For example, the use of social media about the opinions of public towards the wind energy. He tested the sentiment analysis on the opinions of Vastas employee towards the finical benefits of big data business intelligent platform in this wind energy company (Anninni 2014).

12.5 Research Methodology

The models and algorithms described in the previous section need the input data to produce the output information, upon which the decision-makers can make decisions. In this section with a systemic view, we look at the characteristics of input data to the data analysis module of predictive maintenance.

Through a study on different characteristics of input data for the big data analytics, the following main categories were found.

Taking a closer look at these dimensions, several common characteristics are recognized. Considering their relevancy to the predictive maintenance these characteristics are to be seen: Volume, Variety, Veracity (quality structure), Complexity and Velocity.

In the next section, these dimensions of data will be discussed in the context of off-shore wind energy maintenance.

Some of the dimensioned mentioned by the authors in Table 12.1 are not relevant to the input data to a data analysis system. Two of them are "Sensitivity" and "Value" (Table 12.1). Géczy (2014) described sensitivity as one of the characteristics of big data. When we distinguish between the input data to and the output information from an analytical system, these characteristics belong to the output information and they are not relevant for the input data. As stated by Geczy,

Table 12.1 Different characteristics of big data

Big data characteristics	Reference
3 V: Volume, Variety, Velocity	(Wamba et al. 2015; Gandomi and Haider 2015)
4 V: Volume, Variety, Velocity, Value	(Wamba et al. 2015)
5 V: Volume, Variety, Velocity, Value, Veracity	(Opresnik and Taisch 2015; White 2012; Wamba et al. 2015)
Interdisciplinary characteristics framework: Sensitivity, Diversity (Variety), Quality, Structure, Volume, Speed (Velocity)	(Géczy 2014)

sensitivity means the information which contains know-how and the personal or confidential information. The raw data such as sensors and machine logs does not reveal any of such qualities before they are processed.

The same is true for the "Value". Value is the knowledge or insight which is being extracted. From the raw data and before analyzing the data, it is not possible to get the value.

12.6 Technical Characteristics of Input Data for Predictive Maintenance in the Era of Big Data

This section describes the technical characteristics/requirements of input data to the predictive maintenance in the era of big data. These characteristics as stated in Sect. 12.2 are volume, variety and veracity and complexity. We discuss these four main characteristics in the field of off-shore maintenance.

Volumn

The data are growing in the wind energy. The sensor data are growing in number and frequency of information broad casting. For a wind farm, we have the data of around 100 turbines, which each turbine has around 20–30 sensors installed on it (Viharos et al. 2013). So, the amount of information is huge when all needs to be processed. During the maintenance process usually not one but several turbines are being serviced together. Therefore, having the data of the status of all those turbines and making the maintenance in a coordinated way based on the real condition of turbines would be more effective. This needs processing the data of several turbines together. In this case, if three turbines are being maintained together and the results from the analysis shows the probability of failure in one turbine is higher than the rest two it is more practical to start the maintenance process not randomly but from that turbine which has the higher probability of failure.

The other source of big data (increase in data volume) comes from digitalization of the business processes. Traditionally the data on the paper such as maintenance reports from the practitioners, fault descriptions, were not included in the condition monitoring system. Currently by the advances in automation; emergence of mobile devices and smart tags, it is possible to integrate more of these data sources. For example, now the maintenance check lists can be filled by the technicians on the web-based applications installed on the tablets. Therefore, it facilitates the use of them in the analytics (Liebstückel 2012).

Velocity

Velocity of data generation shows the speed in which the data are produced and transmitted. The sensor measurements from the wind farms are broad casted every few minutes. The new sensor technologies such as wireless sensor and acoustic sensors are increasing this pace of data exchange event more. Improved speed of data availability calls for real-time analytics and data aggregation which in a large scale is a challenge. That is because, the transmitted data should be checked for the data quality quicker than before and the feature selection and the analytics should be done near real time. It means the predictive maintenance system should respond faster to the input data. As an example, feature selection for fault prognostic in gearbox uses the vibration sensor data. The axel gear vibration is monitored for this aim. The selection of important features which signal the fault can be recognized by the neural network algorithm (Lau et al. 2012). Incorporating the techniques such as sliding window to deal with the input data pace can help to convert this application to a real-time monitoring.

Variety

The traditional data for the predictive maintenance and reliability engineering are in the form of failure data (status code logs), service and maintenance activity list, and system health management logs. The latter is a combination of sensor measurement and the records of wind parameters such as stress and acceleration (Sheng 2015). Currently the SHM sources are expanding. For example in the wind energy not only the speed of the wind and temperature of environment are being measured and used in the analytics but also the weather condition such as the see state, the wave heights based on the season of the year, the salt in the water are being taken into account.

Audio, video multimedia data and unstructured text data are other new sources of data. For example the non-structured data of critical failures which is provided by the wind turbine technicians, in form of event annotations or the controllers at the control room are also very promising sources of information. They can provide many good insights to the real conditions which caused the failure to happen.

Veracity

Veracity refers to the correct structure and good quality of input data. The measured data should be of high quality. It means the missing values, the errors in the measures should be as less as possible. The format of input data to the algorithm should be correct. For example if the algorithm models the integer values the use of numeric values (with fraction) causes the error in the prediction results. One good advantage of big data sources as stated by (Meeker and Hong 2014) is, classically the estimation of life time of a part usually is done based on the data from laboratory tests during the product development. Those tests used the historical data of similar products (the failures of previous products). But today with the installation of sensors and chips on the products it is possible to increase the veracity.

Complexity

The modern filed data of wind energy, such as the sensor measures and health and environmental data, are captured more in form of vector (a batch of data at every measurement) rather than a single value. This property makes it possible to find more relations between the different variable and learn how a change in one parameter affects the other, also how change in both parameter together signal the degradation and abnormality in the function of a part.

An example of complex aspect of big data is *"sensor data collected in short time intervals are inevitably serially correlated"* (Göb 2014). So, this data cannot be used directly in the algorithms which use the assumption that the input data are normally distributed. As most of analytical algorithms has the assumption of normality, dealing with the autocorrelated data increases the complexity of analysis.

The other opportunity is we can monitor the degradation of the part over time rather than just wait/work with the failure data. The degradation shows the decrease in the performance of the part and with that we can recognize a failure before it happens. For example, performance degradation of bearing, or the colour coating (Meeker and Hong 2014).

12.7 Conclusion

This paper reviewed the current literature on the big data analytics technology in the predictive maintenance, with the focus on the off-shore wind energy maintenance. It is possible to control a few wind turbines from remote with the current data analytical techniques (currently available prognosis and diagnosis models) but for a wind farm with several turbines and sensors, it is hardly possible to do the efficient analysis of the data, either the offline analysis for prognosis of next failures or the online analysis for monitoring the health condition of the equipment without considering the big data technology. However, there is a need for more research which clarifies the characteristics of big data in the maintenance of wind power. This can enhance the development of new maintenance strategies. Further research in this area can be discussing the requirements of data analytics algorithms as well as the output information, also the other aspects of big data technology such as storage, parallel processing in the condition monitoring off-shore wind turbines.

References

Anninni A (2014) Big data in the wind power industry. Master thesis, Grenoble Graduate School of Business
Diamantoulakis PD, Kapinas VM, Karagiannidis GK (2015) Big data analytics for dynamic energy management in smart grids. Big Data Anal High-Perform Comput 2(3):94–101
Wamba FS, Akter S, Edwards A, Chopin G, Gnanzou D (2015) How 'big data' can make big impact: findings from a systematic review and a longitudinal case study. Int J Prod Econ (in press)
Freitag M, Kück M, Ait Alla A, Lütjen M (2015) Potentiazale von Data Science in Produktion und Logistik. Industrie 4.0:22–26
Gandomi A, Haider M (2015) Beyond the hype: big data concepts, methods, and analytics. Int J Inf Manag 35:137–144
García Márquez FP, Tobias AM, Pinar Pérez J, Papaelias M (2012) Condition monitoring of wind turbines: techniques and methods. Renew Energy 46:169–178
Géczy P (2014) Big data characteristics. Macrotheme Rev 3(6):94–105

Göb R (2014) Discussion of reliability meets big data: opportunities. Qual Eng 26(1):121–126

Hameed Z, Hong YS, Cho YM, Ahn SH, Song CK (2009) Condition monitoring and fault detection of wind turbines and related algorithms: a review. Renew Sustain Energy Rev 13(1):1–39

IBM (2012) Managing big data for smart grids and smart meters

Karyotakis A (2011) On the optimisation of operation and maintenance strategies for offshore wind farms. Ph.D. dissertation. University Collage London

Lau BCP, Ma EWM, Pecht M (2012) Review of offshore wind turbine failures and fault prognostic methods. In: IEEE prognostics & system health management conference, pp 1–5

Liebstückel K (2012) Plant maintenance with SAP contents at a glance. Galileo Press, Bonn, Boston

Markovic D, Zivkovic D, Branovic I, Popovic R, Cvetkovic D (2013) Smart power grid and cloud computing. Renew Sustain Energy Rev 24:566–577

Meeker W, Hong Y (2014) Reliability meets big data: opportunities and challenges. Qual Eng 26:102–116

Nguyen TH, Nunavath V, Prinz A (2014) big data metadata management in smart grids. Stud Comput Intell 546:189–214

Opresnik D, Taisch M (2015) The value of big data in servitization. Int J Prod Econ 165:174–184

Puig J (2011) Asset optimization and predictive maintenance in discrete manufacturing industry. Master thesis, École Polytechnique Fédérale De Lausanne, Switzerland

Sheng S (2015) Improving component reliability through performance and condition monitoring data analysis. National Revenewable Energy Labratory (NREL), Houston, Texas

Spahić E, Underbrink A, Buchert V, Hanson J, Jeromin I, Balzer G (2009) Reliability model of large offshore wind farms. In: Proceedings of the IEEE Power Tech conference, pp 2–7

SunGard Solutions (2013) White paper big data challenges and opportunities for the energy industry

Takoutsing P, Wamkeue R, Ouhrouche M, Slaoui-Hasnaoui F, Tameghe T, Ekemb G (2014) Wind turbine condition monitoring: state-of-the-art review, new trends, and future challenges. Energies 7(4):2595–2630

Viharos ZJ, Sidló CI, Benczúr A et al (2013) "Big data" initiative as an IT solution for improved operation and maintenance of wind turbines. In: EWEA, Austria, pp 184–188

Wan Y (2004) Wind power plant behaviors: analyses of long-term wind power data. NREL

White M (2012) Digital workplaces, vision and reality. Int J Bus Rev 29(4):205–214

Xiang Y, Veeramachaneni K, Yanjun Y, Osadciw L (2009) Unsupervised learning and fusion for failure detection in wind turbines. In: 12th international conference on information fusion, Seatel, USA, pp 1497–1503

Chapter 13
Mitigating Supply Chain Tardiness Risks in OEM Milk-Run Operations

Antonio G.N. Novaes, Orlando F. Lima Jr., Monica M.M. Luna and Edson T. Bez

Abstract The objective of this study is to investigate an optimal vehicle routing scheme to perform an OEM milk-run pickup service over a regional road network. The manufacturing of components is subject to varying tardiness among suppliers when fulfilling OEM orders. This often leads to non-accomplished orders at the end of the vehicle cycle time since the transport operation, predominantly composed by random variables, must comply with a strict delivery time limit set up by the OEM company. The mathematical model searches for the optimal vehicle routing sequence, together with searching for the best tardiness tolerance level that minimizes the sum of penalty costs levied against faulty suppliers, and transport expenses.

Keywords OEM · Milk run · Routing optimization · Tardiness risk

13.1 Introduction

A key issue in supply chain management is to develop mechanisms that will align member objectives and coordinate their activities as to optimize integrated performance (Li and Wang 2007). In this context, just-in-time (JIT) concepts are

---

A.G.N. Novaes (✉) · M.M.M. Luna
Department of Industrial Engineering, Federal University of Santa Catarina,
Florianópolis, SC, Brazil
e-mail: antonio.novaes@ufsc.br

M.M.M. Luna
e-mail: monica.luna@ufsc.br

O.F. Lima Jr.
Faculty of Civil Engineering, Unicamp - University of Campinas,
Campinas, SP, Brazil
e-mail: oflimaj.fec@gmail.com

E.T. Bez
Itajaí Valley University, São José, SC, Brazil
e-mail: edsonbez@gmail.com

widely used in original equipment manufacturer (OEM) outsourcing operations in order to reduce logistics expenditures (Ullrich 2013). The term OEM is frequently used to describe those companies that acquire components from manufacturing suppliers and assemble them into final products to be sold to end customers. The essence of JIT in OEM material procurement is the elimination of all possible waste when purchasing components from suppliers, replacing the traditional receiving, inspection, and storage pattern by commit-to-delivery business agreements. To deal with this issue, researchers have attempted to find integrated production and transport scheduling strategies in a dynamic way. At the tactical planning level, supply chain stakeholders, acting in a collaborative way, strive to align their objectives by specifying order delivery moments of the required components at each supplier, in order to coordinate their operations with the OEM production line. However, cost and lead-time savings possibly obtained with such integrated manufacturing and logistics strategies are often impaired, in practice, by unbalanced, unsynchronized, and unstable integration of production and logistics flows. And, in order to resolve specific poor schedules, the planning participants have to frequently adjust the basic plan at the operational level with the help of special algorithms (Scholz-Reiter et al. 2013).

Tardiness occurrences at supplier's premises along the milk-run collecting route can generate delivery time uncertainty at the OEM's production line. Safaei et al. (2013) have analyzed this phenomenon from a more general point of view, estimating the cumulative delivery time uncertainty for three general types of supply networks. In this paper, a similar problem is analyzed with a more detailed operational approach, in which the ordering of visits to collect components from a group of suppliers generates different tardiness cost values, represented by the sum of penalties assigned to the suppliers responsible for fault deliveries.

13.2 Literature Background

JIT philosophy calls for continuous work in reducing disruptions and losses along the production and distribution flow, such as shortening setup times, reducing breakdown frequencies, and avoiding scrap losses. Researchers have analyzed different settings of integrated production and transport commit-to-delivery models (Ullrich 2013; Melo and Wolsey 2010; Yang et al. 2009; Craighead et al. 2007; Tomlin 2006; Tang 2006; Zimmer 2002; Hajji et al. 2011; Ericsson and Dahlén 1997; Grout 1996). Tang (2006) and Tomlin (2006) compared different strategies for mitigating major supply chain disruptions which, when they occur, tend to break down the system and take a long time to recover. Ullrich (2013) treats the problem of integrating machine scheduling and vehicle routing by defining sets of jobs on parallel machines, together with a set of vehicle delivery jobs, with varying loading capacities and ready times. In a minor scale, when a milk-run collecting scheme has been previously defined, the optimizing objective can simply be the minimization of tardiness and transport costs associated with accumulated delivery time uncertainty observed in the OEM production line.

Milk run is a generic name of a logistics procurement method that uses planned routing to consolidate and collect goods to be delivered to a specific OEM company. It is a pickup method in which trucks are dispatched at specified time periods to visit various suppliers in order to collect components or subproducts, and delivering them to a central point (Brar and Saini 2011).

The objectives of this work are twofold: to search for an optimal milk-run routing scheme that reduces to a minimum the sum of penalties and transport costs due to delivery time uncertainties, and to define other management measures to dynamically improve the supply chain performance. Specifically, a dynamic assignment of vehicles to perform an OEM pickup service on a regional road network is investigated. Order fulfillment is frequently subject to varying delays among suppliers. This often leads to non-accomplished orders at the end of the delivery milk-run cycle time, impairing the commit-to-order arrangements with the OEM company. With regard to vehicle routing, the traditional way that transport operators optimize vehicle operation is to apply a Traveling Salesman Algorithm (TSP) to get the minimum distance (or minimum time) route. This approach minimizes vehicle operating costs, but when one considers integrated production and transport costs, such solution is often unsatisfactory and the problem must be dealt differently.

13.3 The Mathematical Model

An OEM company establishes periodic milk runs in order to collect components from n suppliers located at fixed points spread over the industrialized region of Campinas, SP, Brazil (Fig. 13.1). The vehicle starts its run by picking up special empty containers at the assembling company, to be later loaded with components at the suppliers' premises. The vehicle visits suppliers in a predefined sequence, leaving empty containers to be further loaded and collecting full containers already loaded with components. The component load is carried back to the central OEM manufacturing facility. The collecting visits occur in sequence until the milk-run process reaches an end. The logistics supplying operation must comply with a strict delivering time limit setup by the OEM company. Otherwise, substantial penalties are levied to the faulting suppliers. It is admitted that the logistics operator is reliable enough to be not fined for late deliveries, but the model can be easily modified to also incorporate this kind of penalty.

A *candidate solution* is represented by: (1) a vector $R\{E_0, E_1, \ldots, E_n, E_0\}$ containing a vehicle routing sequence in which E_0 stands for the OEM facility and E_i ($i = 1, \ldots, n$). represents the suppliers to be visited sequentially in the run; (2) C_{MR} representing the total cost of a milk-run cycle, composed by the transport cost of components (C_{TRP}), plus the total amount of penalties incurred by the suppliers for unperformed collecting tasks (C_{PEN}), and the total cost of special

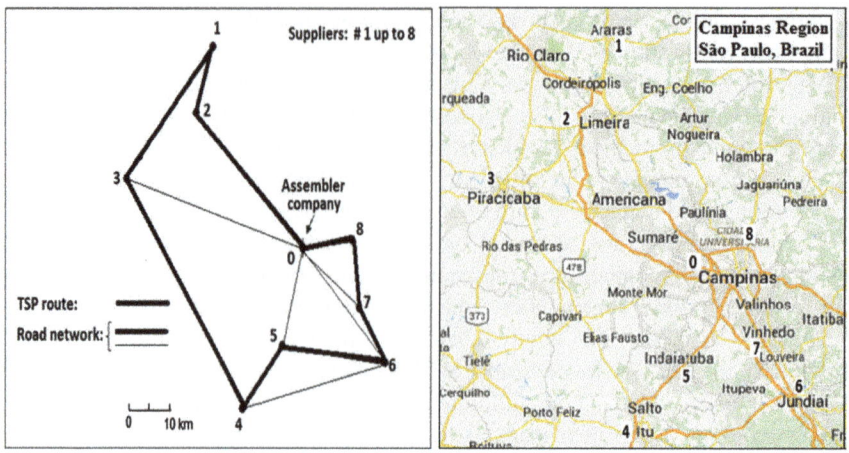

Fig. 13.1 Road network, the TSP route, and the served region

picking-up trips (C_{SPC}); (3) D_T expressing the total distance traveled by the vehicle in cycle R. The objective of the problem is to get the sequence $R\{E_0, E_1, \ldots, E_n, E_0\}$ that minimizes C_{CMR}. A matrix $d_{i,j}$ ($i, j = 0, 1, \ldots, n$), containing the shortest road distances between the relevant network points is known, in which $i = 0$ represents the OEM location. Since road traffic characteristics along the road network vary, a matrix $\bar{s}_{i,j}$ ($i, j = 0, 1, \ldots, n$) of the average speed in each road link is also available. Supplier i produces an average volume of $\bar{V}_i$ cubic meters of components per milk-run cycle, with a standard deviation σ_{V_i}. Whenever the components of a supplier E_i are not delivered in time at the OEM premises, a fine of ω_i monetary units per m^3 will be imposed by the leading organization, i.e., the OEM company.

In the model, a clock time is set to zero when the vehicle starts to be loaded with empty containers at the OEM premises. Along the routing cycle, the instant when a pickup has just been completed at supplier i, with the truck ready to depart to the next collecting point, is defined as a *stochastic regeneration point* in the model. At that point, the on-board system estimates the time τ_i^* that the vehicle will take to perform the remaining collecting visits, plus the traveling time from the last supplier visit to the OEM delivering point. Let T_{OEM} be the latest time, the pickup jobs must be performed as to have the components dully delivered at the OEM facility. Two situations may occur at the regenerating point i: (a) $\tau_i^* \leq T_{OEM}$, meaning the vehicle is free to continue its visits further than point i and according to the predefined routing plan; (b) $\tau_i^* > T_{OEM}$, indicating that one or more collecting tasks will probably not been performed according to the established routing plan, and the respective components will not be delivered in time. At this point, the probability $p_i^{(b)}$ of occurrence of situation (b) is also estimated.

13.3.1 Tardiness Tolerance Strategy

It is admitted that the probability of tardiness occurrence at the supplier E_i production line is $p_{E_i}^{(td)}$. Furthermore, the manufacturing tardiness duration at supplier E_i, when it occurs, is represented by a negative exponential distribution with expected value $Tard_{E_i}$. In the model, it is assumed that the vehicle, upon its arrival to supplier E_i, will wait up to a time T_{up} for a delayed order delivery. When this limit is surpassed, the truck continues its milk-run sequence without the components of supplier E_i, and a special vehicle is sent by the OEM organization to do the unperformed task. This special vehicle departs from the OEM facility, goes to the faulty supplier, collects the components, and takes them to the OEM Company. It is assumed that, for this special collecting trip, the unit transportation cost is $\gamma \times C_{km}$, with $\gamma \gg 1$, where C_{km} is the unit transport cost per km for the normal collecting run. The probability $p_{E_i}^{(Tup)}$ that the vehicle will depart without the subproducts from supplier E_i is

$$p_{E_i}^{(Tup)} = p_{E_i}^{(td)} \exp\left[-T_{up}/Tard_{E_i}\right], \quad (13.1)$$

generating a special collecting trip cost $C_{SPC}^{(E_i)}$ given by

$$C_{SPC}^{(E_i)} = p_{E_i}^{(Tup)} \gamma C_{km} D, \quad (13.2)$$

where D is the distance from the OEM facility to E_i, and back.

Since a negative exponential distribution has the memoryless property, the tardiness expectation at time T_{up} is still $Tard_{E_i}$, when the vehicle departs from supplier E_i without the cargo. One special operating situation occurs when full tardiness tolerance is accepted. In such strategic possible scenario, the vehicle is allowed to wait for the order delivery accomplishment independently of the manufacturing tardiness magnitude. In the model, this decision option is handled by making $T_{up} \to \infty$.

13.3.2 Cumulative Delays Along the Milk-Run Route

Upon getting the cargo from supplier E_i, the vehicle proceeds along the milk-run route to the next supplier until the on-board system detects it will not be possible to proceed further. When this happens, the truck is directed straight ahead to the OEM facility in order to comply with the limiting delivery time T_{OEM}. Suppose the vehicle performs m collecting visits ($m \leq n$) in a given route R, leading to the OEM facility as soon as the mth visit is completed, since otherwise it will not comply with the limiting time T_{OEM}. The total vehicle cycle time $T_R^{(m)}$ along such a route, up to the mth regeneration point and back to the OEM facility, is given by the following random variable

$$T_R^{(m)} = T_{cont} + Tlink_{o,E_1} + \sum_{i=1}^{m}\left(p_{E_i}^{(td)} * Tard_{E_i} + Tload_{E_i}\right) + \sum_{i=1}^{m-1} Tlink_{E_i,E_{i+1}} + Tlink_{E_{m,0}} + T_{unld},$$
(13.3)

where T_{cont} is the loading time of empty containers at OEM premises plus dispatching procedures, $Tlink_{j,k}$ is the vehicle traveling time from point j to point k in the road network, $Tard_{E_i}$ is the expected supplier tardiness in delivering OEM orders; (d) $Tload_{E_i}$—loading time of components at supplier j, $Tlink_{k,0}$ is the traveling time from the last supplier to the OEM facility, and $Tunld$ is the unloading time of all components at the assembling company. The OEM facility is represented by 0, i indicates the routing sequence order, and E_i is the supplier visited at the ith order position. For m sufficiently large, one can admit according to the central limit theorem of statistics that $T_R^{(m)}$ can approximately be represented by a normal distribution. Thus, one can determine $\overline{T}_R^{(m)}$, the expected value of (13.3), and the corresponding standard deviation $\sigma_R^{(m)} = \sqrt{vc_R^{(m)}}$, where $vc_R^{(m)}$ is the variance of $T_R^{(m)}$, the latter obtained by adding up the variances of every component in the right-hand side of expression (13.3). Since $T_R^{(m)}$ is approximately normal, the probability p_m that $T_R^{(m)} > T_{OEM}$ can be estimated via numerical calculus. For a given route $R\{E_0, E_1, \ldots, E_n, E_0\}$, p_m is computed for $m = 1, 2 \ldots n$. For a given m value, whenever the restriction $T_R^{(m)} > T_{OEM}$ is violated, all collecting tasks $m+1, m+2, \ldots, n$ will not be performed, and the corresponding suppliers will be subject to the penalties established by the OEM company. Thus, for a given route $R\{E_0, E_1, \ldots, E_n, E_0\}$, the estimated penalty cost is

$$C_{PEN} = \sum_{m=1}^{n} p_m \times \sum_{i=m+1}^{n} \omega_{E_i}.$$
(13.4)

The total milk-run cost per vehicle cycle is

$$C_{MR} = C_{PEN} + C_{TRP} + \sum_{i=1}^{n} C_{SPC}^{(E_i)},$$
(13.5)

where C_{PEN} is computed with (13.4), the vehicle milk-run traveling cost C_{TPR} is obtained by multiplying the distance D_T by the unit transportation cost C_{km}, and $C_{SPC}^{(E_i)}$ is given by (13.2). The objective of the analysis is to search for the route sequence $R\{E_0, E_1, \ldots, E_n, E_0\}$ and the best value of T_{up} that minimizes C_{MR}. A simulated annealing model (SA) was specifically developed to solve the corresponding optimization problem (Novaes et al. 2015; Breedam 1995). First, the SA model was applied setting $T_{up} \to \infty$ with the objective of getting a preliminary suboptimal route R. Then, the model was applied parametrically to get the optimal value of T_{up}.

13.4 Numerical Application

The milk-run operation includes eight suppliers located on a road network schematically represented in Fig. 13.1. The OEM Company is located at point 0, in Fig. 13.1. The minimum travel distance TSP route $R\{0, 8, 7, 6, 5, 4, 3, 1, 2, 0\}$ is also shown in Fig. 13.1. As mentioned before, the vehicle routing time is composed by the variables shown in Eq. (13.3). The loading time of empty containers at the OEM facility, plus dispatching procedures (T_{cont}), has an expected value of 8 min, and standard variation of 0.6 min. The average vehicle traveling times $Tlink_{i,j}$, in link $\{i, j\}$, with $i, j = 1, 2, \ldots, n$, are given, and its standard deviation is computed assuming a coefficient of variation equal to 0.05. The vehicle loading time at supplier E_i is composed by a constant segment of 3 min, plus a variable part of 0.6 min per m^3. The standard deviation of the loading times is computed assuming a coefficient of variation equal to 0.1. The same expression is used to compute the time to unload the vehicle at the OEM facility, considering now the total quantity of cargo in the vehicle. Table 13.1 presents the relevant data related to the suppliers' performance, including volume of produced components, tardiness characteristics and penalties incurred for undelivered OEM orders. Transport cost for the normal milk run is computed on a basis of $C_{km} = \$1.30/km$. On the other hand, for the special collecting trips cost, Eq. (13.2) is employed with $\gamma = 3$.

Table 13.1 Supplier's data

Supplier E_i	(a) $\overline{V}_i(m^3)$	(b) $\sigma_{Vi}(m^3)$	(c) ω_i ($/m^3$)	(d) p_{td}	(e) $Tard_{E_i}$ (min)
1	4.7	3.5	2,800	0.11	15.0
2	5.2	2.5	2,900	0.16	25.0
3	5.6	0.2	2,300	0.08	13.0
4	0.4	0.1	700	0.02	7.00
5	1.2	0.5	1,200	0.06	7.00
6	2.5	0.9	1,800	0.10	4.00
7	1.8	0.7	800	0.03	3.50
8	4.8	0.5	3,200	0.13	20.0
Total	26.2	–	–	–	–

(a) Average quantity of components per milk-run cycle; (b) standard deviation of (a); (c) penalty incurred by supplier when the manufacturing order is not accomplished; (d) probability that a manufacturing delay occurs in a collecting cycle; (e) average tardiness when a manufacturing delay occurs

Table 13.2 Total cycle cost as a function of tardiness tolerance

T_{up} (min)	Cost per cycle ($)	T_{up} (min)	Cost per cycle ($)
0	222.07	70	294.85
10	182.71	80	300.62
20	203.65	90	304.42
30	229.88	100	306.91
40	254.53	120	309.62
50	273.16	240	311.67
60	286.13	∞	311.69

Assuming $T_{up} \to \infty$ for the first SA run, the algorithm starts up by taking the TSP route as a candidate initial solution, namely $R_1\{0, 8, 7, 6, 5, 4, 3, 1, 2, 0\}$, or else its inverse $R_2\{0, 2, 1, 3, 4, 5, 6, 7, 8, 0\}$. The TSP milk-run trip distance is $D_T = 107.2$ km and the transportation cost $C_{TRP} = \$133.39$. The total cost for TSP route R_1 is \$15,074.39.46, and \$13,950.85 for TSP route R_2. The TSP route R_2 with the least cost was selected as the initial candidate solution to the SA problem. Since $T_{up} \to \infty$ in the first run, there is no special pickup trips, and consequently the cost C_{SPC} is nil. A large initial temperature $GT_0 = 100,000$ was assigned for the SA routine, with a minimum temperature $GT_{min} = 0.01$, and a decreasing temperature rate of 0.999 (Novaes et al. 2015). It took a total of 12,684 replications to get the optimal SA solution $R_{SA}\{0, 1, 2, 3, 4, 5, 8, 6, 7, 0\}$, with a total cost of \$311.69 and $D_T = 113.4$ km, a distance only 5.8 % greater than the obtained for the TSP solution.

Next, the SA procedure was parametrically applied to diverse values of T_{up}. The SA algorithm produced the same optimal R_{SA} route as before for all T_{up} values, of course with different total cost levels as shown in Table 13.2 for a selection of cases. This result indicates that, in this specific application, the vehicle routing prevails in determining total cost. Of course, this is not a general finding, and the optimal route R_{SA} may vary for different T_{up} values, when considering other configurations.

Figure 13.2 depicts the milk-run cost variation as a function of T_{up}. The curve shows a singularity at point $T_{up} \cong 20$ min (more precisely, at $T_{up} \cong 19.5$ min. The reason for this singularity is a discrete probability transition related to an additional component delivery failure. Cost wise, the model shows that the vehicle should wait a time $T_{up} < 20$ min for a delayed order accomplishment at a supplier facility. For $T_{up} \geq 20$ min the cost increases asymptotically from a minimum of \$167.49 at $T_{up} = 19.5$ min, to a maximum of \$311.69 when $T_{up} \to \infty$, a 86 % increase.

13.5 Dynamic Aspects and Research Prospects

When applying the optimal route R_A to a real milk-run problem, since most variables are random, the on-board system can analyze the prospective stochastic conditions ahead of any of the n regeneration points, as the pickup process goes on. This dynamic revision allows to look for advantages in terms of cost reduction and

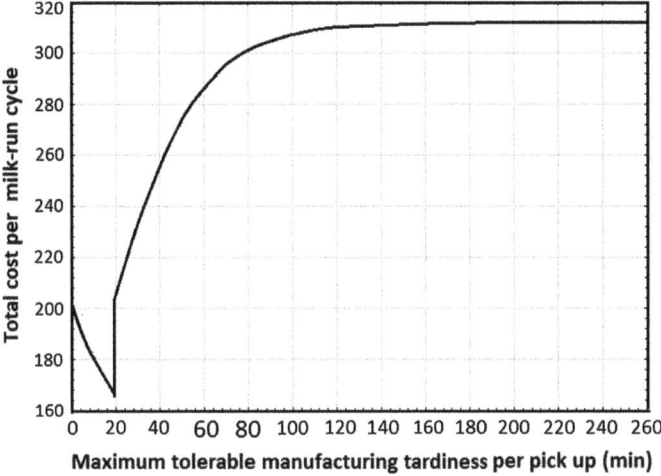

Fig. 13.2 Total C_{MR} cost as a function of T_{up}

possible improvements on the OEM service level, due to potential dynamic changes in the previous established plan. With this, changes in the incoming sequence of tasks could be beneficial. On the other hand, instead of choosing a static figure for T_{up}, the on-board system can initially select individual T_{up} values for the different suppliers based on their fault history, and review them at each regeneration point of the process as a function of dynamic variations observed along the routing process. In addition to the mentioned improvements, another research possibility is to leave the pickup of faulty components to the next milk-run round (Yang et al. 2009). This procedure would eliminate the special transport cost C_{PEN}, but it would imply in additional safety inventory cost for either the OEM company or its suppliers. Thus, a comparing investigation would be of interest. This research will proceed further focusing on other decision and procedure alternatives, and additionally comparing the present method with other strategies.

13.6 Conclusions

This paper has investigated a dynamic assignment of vehicles to perform an OEM pickup service over a regional road network. Manufacturing order fulfillment is subject to varying delays among suppliers. This often leads to non-accomplished orders at the end of the delivery cycle time, impairing the commit-to-order arrangements with the OEM company. In the model, it is assumed a varying tolerance tardiness level for a late order delivery. When the limit is reached, the truck continues its milk-run sequence without the supplier's components, and a special vehicle is sent by the OEM organization to collect the left-over material. The mathematical model searches for the optimal routing sequence, together with the best tardiness tolerance level that minimizes the sum of penalty costs assigned to the faulty suppliers, plus the transport cost. A simulate annealing model was developed to search for the optimal vehicle route. In addition, the model analyzed the maximum tolerance tardiness level that would lead to a further cost reduction. Finally, numerical results were provided and further research prospects were discussed involving dynamic improvements of the model.

Acknowledgments This research has been supported by the Brazilian CNPq Foundation, Projects 302527/2011-7 and 470899/2013-1.

References

Brar GS, Saini G (2011) Milk run logistics: literature review and directions. In: Proceedings of the world congress of engineering, July 6–8, London, vol I
Breedam AV (1995) Improvement heuristics for the vehicle routing problem based on simulated annealing. Eur J Oper Res 86:480–490
Craighead CW, Blackhurst J, Rungtusanatham MJ, Handfield RB (2007) The severity of supply chain disruptions: design characteristics and mitigation capabilities. Decis Sci 38(1):131–156
Ericsson J, Dahlén P (1997) A conceptual model for disruption causes: A personnel and organization perspective. Int J Prod Econ 52:47–53
Grout JR (1996) A model of incentive contracts for just-in-time delivery. Eur J Oper Res 96:139–147
Hajji A, Gharbi A, Kenne JP, Pellerin R (2011) Production control and replenishment strategy with multiple suppliers. Eur J Oper Res 208:67–74
Li X, Wang Q (2007) Coordination mechanisms of supply chain systems. Eur J Oper Res 179:1–16
Melo RA, Wolsey LA (2010) Optimizing production and transportation in a commit-to-delivery business mode. Eur J Oper Res 203:614–618
Novaes AG, Lima Jr. OF, Carvalho CC, Bez ET (2015) Thermal performance of refrigerated vehicles in the distribution of perishable food. Pesquisa Operacional 35(2):251–284. ISSN: 1678-5142
Safaei M, Issa S, Seifert M, Thoben KD, Lang W (2013) A method to estimate the accumulated delivery time uncertainty in supply networks. In: Kreowski HJ, Scholz-Reiter B, Thoben KD (eds) Dynamics in logistics (proceedings: third international conference LDIC 2012, Bremen, Germany). Springer, Berlin, Heidelberg
Scholz-Reiter B, Meinecke C (2013) Towards an integrated production and outbound distribution planning method. In: Kreowski HJ, Scholz-Reiter B, Thoben KD (eds) Dynamics in logistics (proceedings: third international conference LDIC 2012, Bremen, Germany). Springer, Berlin, Heidelberg
Tang CS (2006) Robust strategies for mitigating supply chain disruptions. Int J Logist Res Appl 9 (1):33–45
Tomlin B (2006) On the value of mitigation and contingency strategies for managing supply chain disruption risks. Manag Sci 52:639–657
Ullrich CA (2013) Integrated machine scheduling and vehicle routing with time windows. Eur J Oper Res 227:152–165
Yang Z, Aydin G, Babich V, Beil DR (2009) Supply disruptions, asymmetric information, and a backup production option. Manag Sci 55:192–209
Zimmer K (2002) Supply chain coordination with uncertain just-in-time delivery. Int J Prod Econ 77:1–15

Chapter 14
What Hinders the Implementation of the Supply Chain Risk Management Process into Practice Organizations?

Pauline Gredal, Zsófia Panyi, Aseem Kinra and Herbert Kotzab

Abstract Supply chain risk management process (SCRMP) is being advanced as a systematic and structured approach for identifying, assessing, mitigating, and monitoring all risks arising from complex supply chains. However, while the literature deems it necessary to implement such a process as the solution to the increasing vulnerability companies face, there is a lack of empirical evidence on whether the process model can be implemented. This paper shows possible hindrances in the implementation of SCRMP for companies with global supply chains based on the findings of an in-depth case study. Our empirical findings indicate that the unavailability of information and lack of proper data management hinders the implementation of SCRMP in the context global supply chains.

Keywords Risk management process · Global supply chains · Implementation · Barriers

14.1 Introduction

The rapid expansion processes of global supply chains has carried the weight of having to account for more types and forms of vulnerabilities and risks than previously needed (Peck 2005). Supply chain managers thus increasingly face an emerging concern: supply chain *risk* management (SCRM). The academic community has reacted by generating an increasing body of literature on supply chain risk typologies, mitigation strategies, and other theoretical developments. According to Sodhi and Tang (2012), there are three distinctive reasons for the emergence

---

P. Gredal · Z. Panyi · A. Kinra (✉)
Department of Operations Management, Copenhagen Business School,
Copenhagen, Denmark
e-mail: aki.om@cbs.dk

H. Kotzab
Department of Logistic Management, University Bremen, Bremen, Germany
e-mail: kotzab@uni-bremen.de

of this need: first, more complex supply chains have more points of possible unforeseen disruption; second, local problems may create complications in other parts of the supply chain; and third, longer supply chains imply less visibility, which suggests that in case of a disruption firms will be slower in responding and decision-making.

Moreover, several authors have proposed the concept of a supply chain risk management process (SCRMP), that is, a systematic and structured approach for identifying, assessing, mitigating, and monitoring all risks arising from complex supply chains (e.g., Chopra and Sodhi 2004).

However, existing research (e.g., Manuj and Mentzer 2008; Harland et al. 2003; Tummala and Schoenherr 2011; Grötsch et al. 2013; Norman et al. 2004) only highlights the different supply chain risk management process steps without providing any empirical insight into the hindrances for implementation of the process. So it remains an open question if the generic model does in fact provides a viable solution to the risk management of supply chains..

The purpose of this paper is to uncover the possible hindrances in the implementation of SCRMP for companies with extended supply chains. That is, it explores the aspects inherent in the process model or inherent in the company trying to apply the process model that might hinder the implementation of the supply chain risk management process. The paper thus aims to answer the following research question: What could be potential hindrances in the implementation of SCRMP for companies with extended supply chains?

In order to do answer the question, this paper reviews the current literature on SCRMP, and builds an analytical framework of notions of such hindrances to SCRMP implementation, derived from different lines of theory and enterprise risk management (ERM)-related empirical evidence. An in-depth case study of a company provides additional empirical insights into the analytical framework. This investigation into theory and practice leads us to present 10 final propositions. We find that procedural uncertainty, and a lack of data and data management can hinder the implementation of SCRMP. Further, we find that a lack of internal vision, significant historical discrepancies, external pressure, and the lack of management commitment toward, and communication of SCRMP might pose as obstacles to implementation.

Finally, we also found that insufficient allocation of financial resources, human resources, and capabilities as well as a lack of inbound SCRMP-related tacit information and codified resource transfers from network linkages seem to be an obstacle toward the implementation of SCRMP. Due to the fact that our findings are based on a single case study, the generalizability of these results can be questioned. However, our findings from this exploratory study presented as propositions provide a sound, empirically supported base for future investigation, while also providing practical implications for managers who are considering the implementation of SCRMP.

14.2 Literature Review

Recently there has been an increased momentum in the examination of SCRM with an urge to develop processes and frameworks for supply chain managers to better manage risks. A number of researchers have attempted to define SCRM and have provided frameworks (Chopra and Sodhi 2004; Manuj and Mentzer 2008; Tang 2006; Finch 2004; Zsidisin 2003; Harland et al. 2003; Jüttner et al. 2003; Norrman et al. 2004), without a reaching a general consensus. Though the definitions vary, these scholars do agree on central components such as management, identification, assessment/evaluation and mitigation, and seem to embrace a process perspective.

Several authors have elaborated on this process perspective under the umbrella of SCRMP, considering the actual execution of SCRM (e.g., Harland et al. 2003; Manuj and Mentzer 2008 or Tummala and Schoenherr 2011; see also Fig. 14.1).

All these three models consider a stage process of risk identification, risk assessment, risk mitigation, and monitoring—a model that has been outlined in various forms by various authors. The SCRMP is intended as a process for risk management of the entire global supply chain, and is therefore a holistic process where all aspects of supply chain risks are identified, evaluated, and addressed. There is a general agreement on what process steps SCRMP entail, however, as noted by several authors there is a lack of empirical testing of the concepts.

However, this academic literature generally takes for granted that SCRMP can be executed in the real world, leaving analyses of the prerequisites for

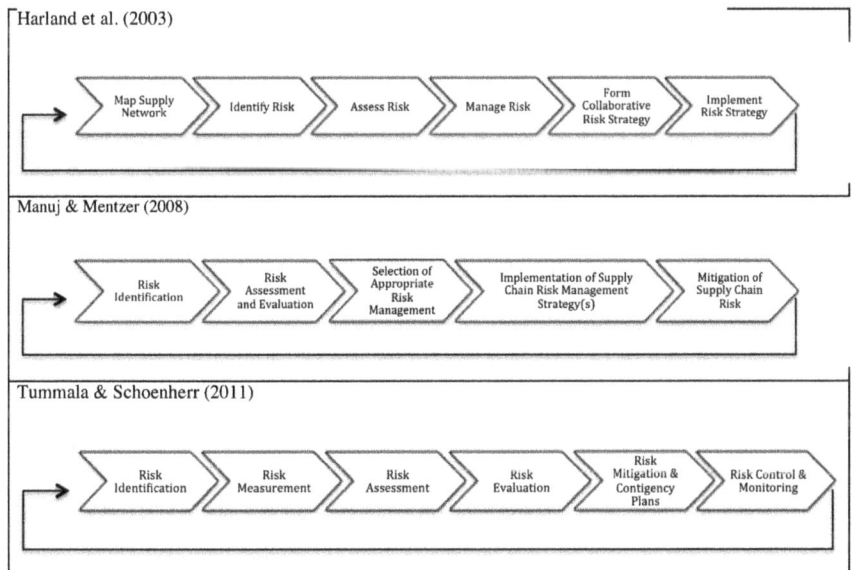

Fig. 14.1 Three academic frameworks of SCRMP

implementing SCRMP untouched. From these three approaches, a picture emerges that implementation of SCRMP as a whole is not treated as a factor, but is taken for granted.

Thus, there is a sizeable gap in the existing literature on SCRMP, as the causalities between the models and the organization implementing them have not been studied. Thus we were interested to see how SCRMP is put into business practice and which barriers hinder its execution. In order to do so, we developed an analytical framework which will be discussed in the next section.

14.3 Analytical Framework for Implementing Supply Chain Risk Management in Organizations

Our model is based on five building blocks representing Enterprise Risk Management (ERM), complexity and uncertainty, data and uncertainty and organizational change. The final building block is represented by the resource-based view of the firm which is required for the allocation of resources within the organization.

We argue that SCRM is a part of the total risk management of a company, where all kinds of processes can be a source of risk (Andersen 2006). There we were able to see that the implementation of ERM within an organization is driven by uncertainty, organizational change and the resource-based view of the firm. Consequently we transfer this understanding to the field of SCRMP.

The increasing appearance of global supply chains over the past decade led to an increased complexity on the one hand and an increased vulnerability and uncertainty on the other hand (Talluri et al. 2013; Tang 2006; Manuj and Mentzer 2008). Supply chains face risks at the value stream level, the asset and infrastructure level, the inter-organizational network level and at the level of the surrounding environment of the chain (Peck 2005). With hundreds of nodes spread out geographically, suffering of vulnerability at multiple levels, supply chain risk managers face a severe challenge when trying to identify potential risks.

Additional complexity of the supply chain makes it increasingly difficult for managers to define and fully understand the causalities of the drivers and sources of risks as well as their impact on the supply chain. This complexity makes it increasingly difficult to identify and analyze risks, and as such it might pose a significant hindrance to risk management. This hindrance, that is, the gap between the complexity of the supply chain and the risk managers' competence in processing the information is reflected in the concept of procedural uncertainty (Dequech 2001).

Over the past few decades, firms have experienced supply chain disruptions due to natural disasters, terrorist attacks, etc., that could not have been anticipated. On a smaller scale, a fire in a plant or a labor strike of employees can also cause supply chain disruptions. For the manager, it is impossible to objectively calculate the probability and impact of such "unknowns," due to the inherent lack of knowledge of such events (Simchi-Levi et al. 2014). Rather, the manager must base his assessment on belief, intuition, and subjective evaluation. The lack of quantitative data in risk management has been recognized by several studies (RMA 2006; Economist Intelligence Unit 2005) and has been identified as one of the most significant challenges of risk management. But not only is there a lack of data—the survey has also revealed that many respondents openly admitted to the use of intuition in their risk management.

Implementing SCRMP can therefore be viewed as an organizational change, since the deliberate introduction of supply chain risk management thinking, acting, and operation causes a difference in form, quality and/or long term state of the company. However, research suggests that failed change initiatives range from one-third to as high as 80 % of attempted change efforts (Appelbaum et al. 2012).

Empirical investigations from the ERM field suggest that lack of risk awareness, confidence in existing risk management practices and lack of a business case for risk management are very common among organizations, and keep managers from implementing operational risk management practices (Muralidhar 2010; AON 2010; Zhao et al. 2014; Roth 2006). With a clear vision of a better state and SCRMP being the vehicle to get there, such negative perceptions can be overcome. The challenge is underpinned by the fact that risk management often deals with "non-occurrences." The significance of management commitment in a risk management setting has been demonstrated by several studies. In a survey by the global risk management service provider (AON 2010), 31 % of respondents pointed toward lack of senior management sponsorship as the main barrier to implementation of ERM. Insufficient resources have been also empirically emphasized by the findings of Zhao et al. (2014), who in their study of 35 construction firms in Singapore find that "insufficient resources" have been the most influential hindrance to ERM, while being a major source of resistance for organizational change. We thus argue that having the appropriate resources *and* allocating them to SCRMP is essential for the firm to be able to proficiently deal with risk and meeting its objectives.

Based on this argumentation, we can propose the following five notions, which will impact the implementation of SCRMP in an organization (see Fig. 14.2):

N1: Implementation of SCRMP can be hindered by procedural uncertainty.
N2: Implementation of SCRMP can be hindered by substantive uncertainty.
N3: Implementation of SCRMP can be hindered by lack of perceived necessity of SCRMP.

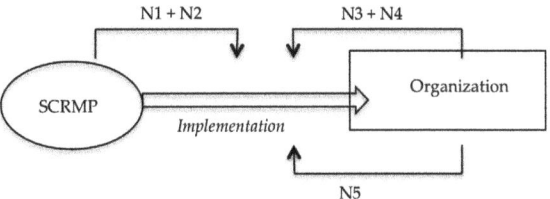

Fig. 14.2 Analytical framework for implementation of SCRMP

N4: Implementation of SCRMP can be hindered by lack of managerial commitment.
N5: Implementation of SCRMP can be hindered by insufficient allocation of financial resources, human resources, and capabilities.

14.4 Methodology

14.4.1 Case Study Approach

To assess the identified research gap, an instrumental exploratory case study methodology was applied, as this methodology allows for the researcher to explore a representative case to provide insight into SCRMP theory, without searching for a clear set of outcomes. The notions were applied to place limits on the scope of the study, provide focus and purpose, and guide the data collection, analysis, and discussion (Richey et al. 2007). The practical implementation of SCRMP has not been—to our knowledge—studied previously. The empirical findings for this paper are based on a single case study of a contract manufacturer of biopharmaceuticals with operations in Denmark and the US, which has some experience with SCRMP.

14.4.2 General Presentation of the Case Company

The case company has succeeded in becoming the largest independent biologics contract manufacturer in the world despite its young presence of 13 years in operation. The case company is a contract process development and manufacturer of biopharmaceuticals. Their primary daily operations include sourcing and processing raw materials that require extreme care and attention. Therefore, it is essential for all the employees to obtain the high level of expertise and dedication that is demanded. The operations of the company can be characterized as project-based, business-to-business manufacturing and the accompanying sourcing activities for each project. For pharmaceuticals, the U.S. Food and Drug Administration requires compliance with the "current good manufacturing practices (cGMP)" regulation, which implies that pharmaceutical-related products are handled by companies who have systems in place to assure the design, monitoring, and control of manufacturing processes and facilities (U.S. Food and Drug Administration 2015). With cGMP compliant facilities in both the United States and Europe, our case company is assuring that every project is performed with their absolute commitment to quality manufacturing and customer satisfaction.

14.4.3 Design and Objectives of the Case Study

The primary objective of the case study was not to provide empirical evidence with predictive value, but rather to add empirical observations from the SCRM context to our notions. Data was collected from the company in spring 2015, by interviews with relevant executives as well as through company visits and retrieval of relevant written resources. The case study involved both quantitative and qualitative as well as primary and secondary data collection, including confidential risk management practices and semi-structured interviews with top executives like the Senior Vice President and the Director of Supply Chain Management. The relevant data gathered through the company visits was analyzed and coded according to the theoretical notions that form our analytical framework, presented earlier in Fig. 14.2. This allowed us to examine our notions more carefully, and comment on the number and nature of the potential hindrances to the implementation of supply chain risk management process in organizations.

14.5 Hindrances for SCRMP: Findings and Discussion

Is the Implementation of the SCRMP Hindered by Procedural Uncertainty?
The case work revealed some nuances to this notion. It seems as though procedural uncertainty can incite two reactions: apathy or action. The overwhelming task of identifying and evaluating a complex and undiscovered portfolio of risks can leave the manager paralyzed, arguing that it is too big of a task, or that the existing risk management is sufficient. This response will most definitely hinder the implementation of SCRMP. However, the overwhelming task of dealing with complexity can also leave the competent manager longing for a framework to reduce the procedural uncertainty. SCRMP can function as a tool, enabling the manager to take on a systematic, structured approach toward risk management. We are, however, not suggesting that the reaction of managers induced by complexity needs to be either one or the other. As in the case of our case company, it can also be a combination—where the result might be the implementation of SCRMP to a certain extent, but not fully.

Our research also revealed that the concept of complexity has to be assessed with caution in a company, as there can be more than one level of complexity. Whether certain risk sources are perceived to be more or less complex by managers ultimately define the level of the associated perceived procedural uncertainty. As such, risk sources that are perceived to be less complex—such as supply risks in our case company—do not provoke a sense of procedural uncertainty, as the company feels competent in managing such risks. On the other hand, very complex risks—such as process risks in our case—do provoke a high sense of procedural uncertainty, as it is increasingly difficult to cognitively comprehend the risk. The ways in which the different levels of complexity and the accompanying perception of procedural uncertainty affect SCRMP implementation consequently differ as well. In the case of less complex risks, SCRMP implementation may be hindered by the fact that managers unconsciously do not *want* to implement it, to avoid an increased procedural uncertainty arising from the increased complexity. For risks with great complexity, on the other hand, holistic SCRMP implementation may be hindered by the incompleteness of the SCRMP model.

Is the Implementation of the SCRMP Hindered by Substantive Uncertainty?

This seemed to be the situation at the case company, as both interviewees argued for the use of experience as a good basis of risk management. Further, the case company had trouble obtaining data for their risk assessment, which made formal SCRMP seem unrealistic to implement. This was further enforced by a low commitment to data management, which highlighted a nuance to the concept of substantive uncertainty. Discussing data availability on "abnormal" risks, that is, events with low probability and high impact, is only one side of the coin. In such cases, historical data does not exist, and SCRMP can thus be difficult to implement as no probability or impact can be assigned due to the fundamental lack of data based on prior experience. On the other hand, in cases of "normal" risks, the company faces probabilistic certainty where probability, structure, and impact of future events are known, and risks can therefore be objectively evaluated. However, a lack of data management of historical data may transform this probabilistic certainty into structural uncertainty, where risk assessments are based on subjective beliefs, and formal SCRMP is unlikely to be conducted.

Is the Implementation of the SCRMP Hindered by a Lack of Its Perceived Necessity?

Our third notion argued that discrepancies from the current state, and a vision of a future better state lay the foundation of the perception that SCRMP can add value to the company and therefore should be implemented. Our case study, however, revealed a more distinct picture of what drives this sense of urgency. It was clear that the case company did not have a clear vision of what a fully implemented SCRMP would add to the firm. The concurrent lack of vision could be a complementing factor in why no full SCRMP was implemented. This might be explained by a notion of path dependency: once a practice has been implemented, the likelihood that alternatives will be considered in the future diminishes. This view was also evident in considering discrepancies within the company. The case company had experienced no significant losses from the lack of SCRMP. This caused managers to feel that the current level of SCRM was sufficient. Thus, it seems that "history matters," and that an absence of significant losses due to insufficient risk management can impede SCRMP.

The study also revealed a notion of external influence on the perception of necessity within the company. Regulative, normative, and cognitive pressures of the institutions within which the company is embedded highly influenced their SCRM decisions. To achieve legitimacy within its industry the case company conformed to regulations, adapted to expectations and mimicked practices from other companies. This was evident when exploring the origin of their SCRM practices, as the spreadsheet was implemented upon a demand from a customer, while the policy responded to a general expectation from the industry. This isomorphic aspect, revealed by the case study, certainly warrants exploration.

Is the Implementation of the SCRMP Hindered by a Lack Managerial Commitment?

From our empirical findings we found that the most observable way of measuring the degree to which top management is committed to formal risk management practices is the way in which they communicate it. If there is a lack of "push" from the top management to apply tools, frameworks, and practices, it will most definitely affect the way in which the organization functions, and ultimately how successful it is. In our case company, if the top management had enforced the formal practices of SC risk management more, it would have prevented the company from being in its current situation, where the way some managers perceives risk management is fundamentally different from the way it is applied throughout the organization—with most employees relying on their own experiences and gut feelings, rather than on the formal practice. Thus, as SCRMP is a complex process that requires widespread understanding and knowledge of risk management tools and techniques, it is essential for top management to align the perceptions, awareness and need with regard to such a practice, on all levels of the organization. Another sign of management commitment that seems essential for efficient supply chain risk management is the development of a company-specific SCRMP, which takes into consideration the operations and industry demands regarding that company, rather than the adaptation of an already existing tool. Therefore, we find that our initial notion can be a viable point in future research.

Is the Implementation of the SCRMP Hindered by the Insufficient Allocation of Resources?

The empirical findings regarding the allocation of internal resources are twofold. Financial resources within the case company at the certain point in time of our study seem to be allocated to SCRM to a large extent, as pointed out by both interviewees. However, as pointed out in the analysis on multiple occasions, the sufficient allocation of human resources and capabilities can be questioned. It seems as though it is not only satisfactory to appoint a CRO or risk manager, but it is also necessary to investigate the degree to which that assigned person is aware of the responsibilities and amount of knowledge about the operations of the company required to carry out a sound judgment regarding risks. Even though the case company do have a functional risk manager in place, they are considering the need for a person who is solely responsible for managing risks across different locations. Furthermore, it seems fundamental in the implementation of SCRMP for the employees to be formally trained and aware of the process. In our case company, employees do not receive workshops or seminars, but most of them are expected to perform risk management as part of their daily activities. It can be regarded as a prerequisite that is inherent in the job. However, if the employees had received a company-wide formal training, mistakes might have been avoided. Therefore, it seems reasonable to argue that in order to implement a holistic SCRMP, the sufficient financial and human resources need to be allocated, and the capabilities of the employees to perform risk management have to be ensured.

14.6 Conclusion and Future Directions

The objective of the study was to investigate how the implementation the SCRMP is practically realized and what may hinder this implementation. From these two aspects, we examined the literature as well as empirically investigated the phenomenon in a company setting.

The analysis of our case study identified that (1) procedural uncertainty, (2) lack of data and data management, (3) lack of vision, historic discrepancies and external pressures for implementation, (4) lack of management commitment toward and communication of SCRMP, (5) insufficient allocation of resources, and (6) lack of inbound transfer of information and resources from network linkages could all constitute hindrances to SCRMP implementation for companies with extended supply chains.

However, there are some limitations and related future research implications arising out of the present study. The main limitation is that the adoption of a single case study has amongst other things limited the observations to the industrial context of the case company. Nevertheless it will help in formulating some research propositions for further exploration, and provide a sound, empirically supported base for future investigation, while also providing practical implications for managers who are considering the implementation of SCRMP. There is currently a lacks of a clear vision of the challenges involved while implementing risk management programs across the company supply chain, both internal and external. The hindrances identified in this paper will provide some insight into what might hinder and/or slow down the development of SCRMP within the company. Besides developing the notions presented in this paper better, the next step will also be to be more aware of the simplicities of the model in order that the SCRMP does not blindly follow the same path as the mainstream ERM field.

References

Andersen (2006) Perspectives on strategic risk management. Copenhagen Business School Press
AON (2010) Global ERM survey. Chicago. http://doi.wiley.com/10.1002/jhrm.5600210106
Appelbaum et al (2012) Back to the future: revisiting Kotter's 1996 change model. J Manag Dev 31(8):764–782
Chopra and Sodhi (2004) Managing risk to avoid supply-chain breakdown. MIT Sloan Manag Rev 46:53–61
Dequech D (2001) Bounded rationality, institutions, and uncertainty. J Econ Issues XXXV(4)
Economist Intelligence Unit (2005) The evolving role of the CRO

Finch (2004) Supply chain risk management. Supply Chain Manag Int J 9(2):183–196
Grötsch et al (2013) Antecedents of proactive supply chain risk management—a contingency theory perspective. Int J Prod Res 51(10):2842–2867
Harland et al (2003) Risk in supply networks. J Purch Supply Manag 9(2):51–62
Jüttner et al (2003) Supply chain risk management: outlining an agenda for future research. Int J Logist Res Appl 6(4):197–210
Manuj, Mentzer (2008) Global supply chain risk management strategies. Int J Phys Distrib Logist Manag 38(3):192–223
Muralidhar (2010) Enterprise risk management in the Middle East oil industry: an empirical investigation across GCC countries. Int J Energy Sect Manag 4(1):59–86
Norrman et al (2004) Ericsson's proactive supply chain risk management approach after a serious sub-supplier accident. Int J Phys Distrib Logist Manag 34(5):434–456
Peck (2005) Drivers of supply chain vulnerability: an integrated framework. Int J Phys Distrib Logist Manag 35(4):210–232
Richey et al (2007) Design and development research: methods, strategies, and issues. Lawrence Erlbaum Associates
RMA (2006) Enterprise risk management survey 2006. Philadelphia
Roth (2006) An enterprise risk catalyst. Internal Audit, February
Simchi-Levi et al (2014) From superstorms to factory fires: managing unpredictable supply-chain disruptions. Harvard Bus Rev, (Jan–Feb). https://hbr.org/2014/01/from-superstorms-to-factory-fires-managing-unpredictable-supply-chain-disruptions. Accessed 28 Apr 2015
Sodhi and Tang (2012) Managing supply chain risk. Springer, London
Tang (2006) Perspectives in supply chain risk management. Int J Prod Econ 103(2):451–488
Talluri (2013) Assessing the efficiency of risk mitigation strategies in supply chains 344:253–269
Tummala, Schoenherr (2011) Assessing and managing risks using the supply chain risk management process (SCRMP). Supply Chain Manag Int J 16(6):474–483
U.S. Food and Drug Administration (2015) cGMP. http://www.fda.gov/Drugs/DevelopmentApprovalProcess/Manufacturing/ucm169105.htm. Accessed 13 May 2015
Zhao et al (2014) Enterprise risk management implementation in construction firms: An organizational change perspective. Manag Decis 52(5):814–833
Zsidisin (2003) A grounded definition of supply risk. J Purch Supply Manag 9:217–224

Chapter 15
The Fashion Trend Concept and Its Applicability to Fashion Markets and Supply Chains

Samaneh Beheshti-Kashi

Abstract This paper focuses on different aspects of fashion markets. In a first step, fashion levels will be classified; followed by definitions on fashion trends, and the suggestions on a fashion trend concept. In order to fill this concept and support decision-making processes along the supply chain, such as the catching of actual fashion trends, it is required to fill this concept with relevant information on different product features. Social media text data is considered as one relevant source. Showing previous researches, we assume that for instance fashion weblogs can be used for extracting this information. In a further step, we describe different fashion markets, namely fast fashion and luxury, in order to examine the applicability of the approach to real-life markets and their supply chain processes. The paper concludes by formulating hypotheses on a potential application of the approach.

Keywords Fashion trend concept · Fashion markets · Fashion supply chains · Social media · Fast fashion · Luxury markets

15.1 Introduction

The fashion industry faces challenges in meeting the demand of the customers. Often retailers are confronted with losses due to stock-outs or overstocked inventories (Fisher and Raman 2010). In addition, the demand for fashion is impacted by various factors such as changing weather conditions, celebrities, events, holidays, or the general economic situation (Thomassey 2010). A crucial challenge is that most fashion products are highly short-lived compared to their long time to market. Most production plants are situated in countries such as China, Taiwan, India, or Bangladesh. Lately, even these countries are turning too expensive for the companies due to an increased standard of living and higher salaries. Consequently, the

---

S. Beheshti-Kashi (✉)
International Graduate School for Dynamics in Logistics (IGS), Universität Bremen, Hochschulring 20, 28359 Bremen, Germany
e-mail: bek@biba.uni-bremen.de

© Springer International Publishing Switzerland 2017
M. Freitag et al. (eds.), *Dynamics in Logistics*, Lecture Notes in Logistics,
DOI 10.1007/978-3-319-45117-6_15

first firms have started to locate some production plants to African countries such as Ethiopia or South Africa (NTV.de 2015). Though, the target regions are often European countries or the United States. Therefore, retailers are confronted with long transportation and shipping routes. In traditional retailing, reproducing good selling items during a current season is hardly possible or rather solely with additional costs. Some companies, for instance, place productions in Northern African countries or in Turkey in order to meet the demand during the season. Alternatively, companies use air shipping to replenish stocks in time.

This paper looks at different aspects. Figure 15.1 illustrates the aspects and the relevant stakeholders: consumers, social media, demand uncertainty, fashion levels, fashion trend concept, fashion markets, and supply chain processes. The paper is organized as followed: First, fashion levels, the classification of them and the relation to fashion markets is described. Then, we focus on defining a fashion trends and suggest the building of a fashion trend concept. In a further step, we propose on filling the concept with information extracted from social media text data, since consumers publish content on these applications. Assuming that the extraction of valuable information is possible through for instance fashion weblogs, the next question is, if the approach is applicable for fashion companies to. Therefore, different real-life markets and companies are examined. In particular, fast fashion and luxury markets are focused, due to their huge differences. After introducing both markets, we will conduct a comparison on their processes along the supply chain. We will conclude the paper by formulating two hypotheses regarding the applicability of the presented approach by fast fashion and luxury markets. Supporting arguments will be presented along with the hypotheses. Though, the actual testing of them is considered as future work.

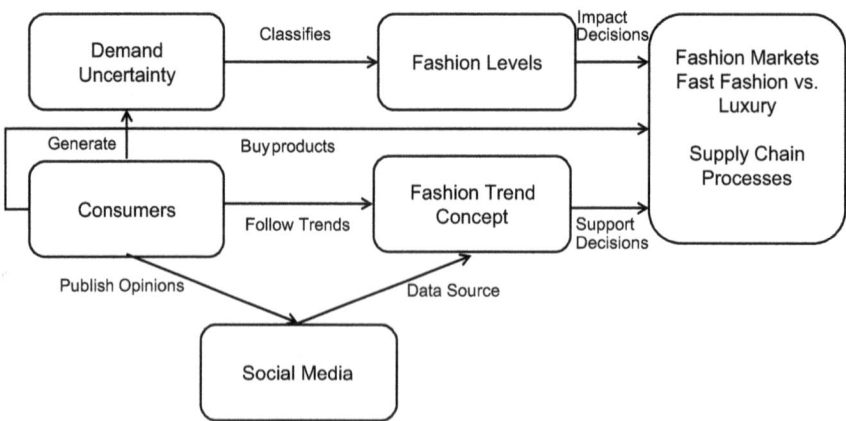

Fig. 15.1 Relation between the different mentioned aspects within the paper

15.2 Fashion Levels

The discussion of fashion levels is crucial in the context of this paper, since most decisions along the supply chains are based on the fashion level of the product. Literature suggests different ways for classifying the different levels of products. For instance, Wojaczek (1996) distinguishes different fashion levels and classifies them according to the criteria: novelty, life-time, sales risk, purpose, and replenishment options. He named the levels as high fashion, fashion, and basics articles. High fashion articles need continuous development and are highly short-lived. Furthermore, the fashion degree is the focus and replenishment of the products is hardly possible. Whereas, fashion articles are developed on a regular basis, with a short life-time and high sales risk. Similarly, the fashion degree is focused. Though, limited replenishment is possible. In contrast, basic articles are hardly further developed, have long life-times and the sales risk is low. The focus lies on the functionality of the product, and not on the fashion degree. Moreover, the replenishment of the articles is often possible. Hammond (1999) suggested a fashion triangle in which the relation of fashion grade and demand uncertainty is illustrated. He distinguished between basic products, fashion basic products, and fashion products and argued that the demand uncertainty increases with the fashion grade. This means that the demand uncertainty for basic products is lower than the fashion basic products, and for the fashion basic products is lower than fashion products. Lowson (2003) differs between basic products, seasonal (or fashion basic), and short-season (fast fashion) products. In all three classifications, the demand plays a crucial role with regard to the product type. This work considers fashion basic and fashion products, and ignores basic products due to low the demand uncertainty.

15.3 Fashion Trend Concept

Fashion basic products and in particular, fashion products often follow a certain trend. Since the term trend can be misleading, in the following, we look deeper at it. Starting from the word definition trend, we find on Merriam Webster (2015) the definition of a *current style or preference*. From an economic perspective, often deterministic trends or time series are mentioned. In the fashion literature, the term trend is stated as a *direction, movement or flow*. Combined with the term fashion, Cho and Lee (2005) see the meaning in trend as a *general tendency of the next fashion*. Looking at the literature on fashion trends, we can find works describing influences on a fashion trend such as social, political, economic, and cultural changes (Cho and Lee 2005). In addition, the authors assign the consumers a relevant role in setting a fashion trend (Perna 1987). Cho and Lee (2005) consider a fashion trend as something similar to a concept such as futurism, minimalism, or fun. Each trend is characterized by several different features. For instance,

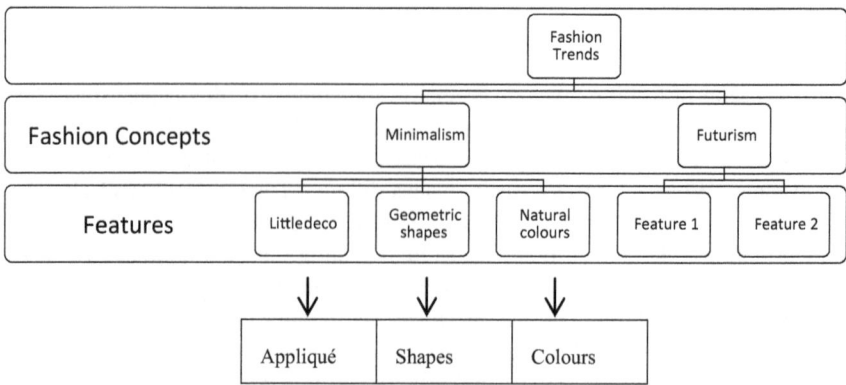

Fig. 15.2 Identifying relevant fashion product features out of past fashion trends

minimalism is characterized by *little decoration, frequent use of geometric shapes, neutral colors, and minimization of volume* (Cho and Lee 2005).

This research applies among others these fashion concepts in order to identify different features that a product can contain. In a further step, these features are used in generating the trend fashion concept. Figure 15.2 gives an overview on the process of building the concept. Taking a garment as an example, usually it has a particular color, fabric, material, cut, style, or pattern. Often a fashion trend is a combination of these features. Moreover, an emerging trend can also be manifested in each of the features. For instance, light blue is the leading color, silk the current preferred material, and vintage the leading style. Identified such an abstract concept, it is required to fill it with information on the different features.

As aforementioned, this research focuses on fashion products which are often consumer-driven. These consumers usually access social media applications in order to publish information on different topics. For this reason, we suggest considering the approach of Beheshti-Kashi and Thoben (2014) in which the authors propose to integrate social media text data into sales forecasting processes. Even though, the current paper is not dealing with sales forecasting, but more with the integration of potential relevant information extracted from social media, it is suggested for instance, to apply fashion weblogs as sources for filling the fashion trend concept.

15.4 Social Media and Fashion

With the emergence of the Web 2.0 and the development of a large number of social media platforms, every user with access to the internet has been empowered to publish his opinion and experiences on different platforms. Among the different applications, this research has considered weblogs and the microblogging service

Twitter so far. In Beheshti-Kashi (2015), so called hashtags posted along with Tweets (messages in Twitter) were focused in order to examine the posting behavior regarding fashion related content. It is reported, that indeed it is possible to extract information on product features such as the color, material, or style from the Tweets. In the case of the fashion weblogs, the primary question was not on the availability of fashion-related content, since their focus is reporting on fashion topics. In this case, the extraction of deeper information is targeted. Beheshti-Kashi et al. (2015) examined German fashion weblogs and showed that it is possible to extract more detailed information on different features. For the collection and processing of the text data, Text Mining and Natural Language Processing methods have been applied.

15.5 Fast Fashion Retailing

Fast fashion retailers strategy is oriented toward customers' demand and targets at responding to market changes in a timely manner. This demand is often tracked through sales data captured in the stores. In addition, mostly small and medium volumes are produced. In contrast, traditional apparel retailers place large volumes. Often fast fashion retailers follow a backward vertical integration. This strategy enables the companies to manage the different steps from design to distribution. Often a large portion of the supply chain is owned by the retailer, and the rest is often controlled due to close distances. One large difference between traditional and fast fashion retailing is the difference in lead times. Traditional retailing is characterized through long lead times, taking up to 29 weeks from the first design to the delivery to the stores. In contrast, fast fashion retailers take 15 days including the shipping to the stores. Also with regard to forecasts traditional retailers act differently: they do generate forecasts prior to the selling season. Whereas fast fashion is more closely to the season and also base often real time data from their stores in order to meet the demand of the customers. Though, often products are not replenished like in traditional retailing, where often replenishment is activated on an automatically based on the remaining level of supplies. Similar to the forecasts, designs are also forecasted 18–24 months prior to the selling season (Hines and Bruce 2014). For fast fashion, designs are often based on current runway fashion shows. Due to this strategy it is obvious that the fabrics have to be available in "gray" and not already colored, to be able to flexibly react to the catwalk designs during the season (Ghemawat and Nueno 2006). At this point mostly fast fashion companies follow a postponement strategy. Moreover, with regard to target groups differences occur: often female consumers between 16–24 years are the core target customers in the case of fast fashion. Reasons for that might be the fact that catwalk designs are offered by affordable prices. Therefore, young women who cannot afford the higher prices are able to purchase these articles and also be up to date. In regard of the quality, fast fashion companies do not claim highly qualitative garments; since they are *clothes to be worn 10 times* (Ghemawat and Nueno 2006).

15.5.1 The Case of Zara

Zara is considered as the innovator of fast fashion, and therefore, this section will describe their strategies more detailed (Hines and Bruce 2014). Zara, a Spanish retailer, is owned by Inditex and opened its first store in La Coruna in 1975. For each line, design teams consisting of designers, sourcing specialists, and product developer, are established. These teams work on designs for current seasons as well as for the following year at the same time. One crucial focus is on interpreting catwalk designs and makes them accessible for the mass market. In addition, Zara also employs trend spotters, who try also to catch the trends from the people for instance on university campuses. Beside the IT system in which all the sales data are captured, further information sources such as Internet, movies, TV, and industry publications are monitored. In order to reduce the risk of overstocked inventory, they often test products in key stores and only place larger productions if the customers react positively. With the flexibility of these systems, and the design team being able to modify current items, they are able to produce products with a slight modification and bring them to the stores within 2 weeks. All products independent from their production place have to be passed through the central distribution center in Arteixo, in north of Spain. 75 % of the products are transported via trucks to the stores in Spain, Portugal, France, Belgium, and Germany. For the rest air shipping is used. Usually, the merchandises are delivered to the stores within 24–48 h to European and outside European markets. They also do not participate in so-called ready-to-wear fashion shows. The products are first shown in the Zara stores (Ghemawat and Nueno 2006).

15.6 Luxury Markets

The luxury market is often considered as the counterpart to fast fashion retailing. Though, luxury markets are not homogenous. In order to compare fast fashion strategies with luxury markets, we need a deeper insight into this market. Table 15.1 gives an overview on variables and characteristics defined by different authors in order to classify luxury markets and products. It shows the variables positioning on the market, products, and duration of the shelf time which is measured in seasons or weeks.

As a general aspect, Fernie et al. (1997) state that most luxury companies provide besides to their "haute couture" also "diffusion lines" to reach a wider range of customers. These lines often differ in price, quality, and volumes. In addition, diffusion lines are mostly outsourced to Asian countries, while luxury lines are often manufactured in-house (Brun and Castelli 2015). However, values such as craftsmanship, high quality products, originality of design and exclusivity of the products are common values (Fionda and Moore 2008).

Table 15.1 Classification of luxury markets and products

Classification variables	References	Characteristics
Positioning on the market	Catry (2003)	Exclusive goods, limited editions
	Silverstein/Fiske (2003)	Addition of new luxury'
	Beverland (2004)	Mass level, premium, super-premium Icon level
Products	Altagamma (2008)	Absolute luxury products, aspirational luxury products, accessible luxury
	Brun/Castelli (2015)	Addition of mass market goods
Shelf time (weeks/seasons)	Brun/Castelli (2015)	Continuative, seasonal, fashion items
	Jacobs (2006)	Fashion products, continuative products

15.7 Fast Fashion Versus Absolute Luxury

This section shows a comparison between luxury and fast fashion companies along the supply chain: design, sourcing, manufacturing, distribution, and replenishment. Table 15.2 highlights the most relevant aspects. As shown in Table 15.1, luxury markets are not homogenous. In this comparison absolute fashion is considered as the counterpart for the fast fashion strategies. The design process includes also the stage of identifying or setting a fashion trend. This initial step differs largely in both cases. In most luxury company's designer's work independently on their designs, while in the case of fast fashion, usually designers are organized in teams. Besides to the different organizational structure, also the process differs.

Luxury designers focus on the originality of their work, whereas one purpose of fast fashion retailing is to interpret the catwalk's designs and make them accessible for the masses. Therefore, the originality of the design is considered as one

Table 15.2 Luxury versus fast fashion

Processes	Luxury	Fast fashion
Design	Single designers Originality of designs Setting a trend	Design teams Interpretation of catwalk design Tracking and following customer demands
Sourcing	Established relationships High quality	From external suppliers Quality of fabrics not the focus
Manufacturing	Often outsourcing to national companies Importance of craftsmanship	Often off-shored Product related strategies
Distribution	Established relationships, mono branded stores, official resellers	Central distribution centers, selling in own stores
Replenishment	Difficult due to long lead times	Not always desired

differentiating aspect. An additional relevant factor is the fact on what the demand is based on. Often haute couture and luxury designers' intention is to set a trend and to create the demand for their designs. In contrast, fast fashion companies follow a different strategy. These firms often consider their sales data which is tracked on a daily basis from the stores, as one crucial aspect for identifying the demand. Therefore, fast fashion usually follows a buyer-driven approach in the case of demand identification (Ghemawat and Nueno 2006). Luxury companies often have established relationships to their suppliers. One crucial factor is the quality of the fabrics and materials. In order to ensure the quality, investments in the infrastructure of the suppliers are not a rare strategy of luxury companies (Brun and Castelli 2015). Fast fashion companies often have larger numbers of external suppliers. One crucial difference is the dealing with the dyeing process. In the case of fast fashion, usually, a large portion of the dyeing is conducted during ongoing seasons in order to act responsively to market changes (Ghemawat and Nueno 2006). In contrast, luxury companies often finish this process prior to the season. Manufacturing in fast fashion mostly is off-shored. Though, it depends also on the companies. For instance, Zara follows a twofold production strategy depending on the product itself. Time-sensitive products are often produced in Spain/Portugal, while price-sensitive products are manufactured in Asian countries. In the case of absolute luxury, outsourcing is followed, but often to national high qualified companies since the craftsmanship still plays a crucial role (Ghemawat and Nueno 2006). Zara, for instance, owns a central distribution center in the North of Spain where all products weather they are produced in-house or outsourced/off-shored have to be passed through. Often the garments are sold in their own stores. Luxury companies often sell their products in mono-brand stores either owned or franchised stores. Often also they work with official resellers (Brun and Castelli 2015). In regard of replenishment strategies, most fast fashion companies follow the principle that there is no need for replenishment. With this strategy they want to give the feeling of limitedness to their customers, and want to attract them more often to their stores. Often the sold out products, are entered to the market and stores, in a variations within the same season (Hines and Bruce 2014). In contrast, replenishment in luxury companies is in some cases desired, but difficult to realize due to the long lead times.

15.8 Hypothesis

Describing the different segments and markets, we can formulate the two hypotheses shown in Table 15.3. At the right column some supporting arguments are presented. The testing of the hypothesis is considered as future work.

Table 15.3 Hypothesis on the applicability of the approach

Hypotheses	Supporting Arguments
Fast Fashion companies will have a higher interest in adapting or including information published on social media applications in their processes of identifying and evaluating a fashion trend.	Fast fashion companies: Focus on the right interpretation of the catwalk designs Chance to monitor the customers reactions of these design in real time Bloggers as tool for reaching the masses directly from the catwalks Luxury companies: Rely on the originality of their designs Reaching the masses not the focus
Fast fashion companies are more able to implement social media text data into their processes, and will have an additional value of it.	Fast fashion companies: Short lead times, often in 15 days shipped to the stores Luxury companies: Long lead times, often up to 6 months, Focus on craftsmanship (time intensive)

15.9 Conclusion and Future Work

This paper looks at different aspects related to fashion markets. Starting with the classification of different fashion levels and fashion trends, we suggest the process on building an abstract fashion trend concept which is required to be filled with information, in order to catch actual fashion trends. Showing results from two previous research works, we assume that social media text data is an adequate source for matching the information on product features and filling the trend concept. As a further aspect, we have focused on different fashion markets, namely fast fashion and luxury, in order to examine the potential applicability of the approach to real-life markets and companies. A comparison of both segments has resulted in the formulation of two hypotheses. The testing of these will remain as future research.

References

Altagamma (2008) Fashion and luxury insight. www.altagamma.it. Accessed 20 Oct 2015
Beheshti-Kashi S, Thoben K.-D (2014) The usage of social media text data for the demand forecasting in the fashion industry fashion sales forecasting and the predictive power of online. In: Fourth international conference on dynamics in logistics, LDIC 2014 Bremen, Germany, February 2014 proceedings (in press)
Beheshti-Kashi (2015) Twitter and fashion forecasting: an exploration of tweets regarding trend. In: Proceedings of the 35th international symposium on forecasting, Riverside, California, June 2015 (in press)

Beheshti-Kashi S, Lütjen M, Stoever L, Thoben KD (2015) TrendFashion—a framework for the identification of fashion trends. In: Proceedings of Informatik 2015, Cottbus, Germany

Beverland M (2004) Uncovering, theories in use: building luxury wine brands. Eur J Mark 38:446–466

Brun A, Castelli C (2015) Supply chain strategy in the fashion and luxury industry. In: Fernie J, Sparks L (eds) Logistics and retail management, 4th edn

Catry B (2003) The great pretenders. Bus Strategy Rev 14(3):10–17

Cho H-S, Lee J (2005) Development of a macroscopic model on recent fashion trends on the basis of consumer emotion. Int J Consum Stud 29(1):17–33

Fernie J, Moore C, Lawrie A, Hallsworth A (1997) The internationalization of the high fashion brand: the case of central London. J Prod Brand Manag 6(3):151–162

Fionda A, Moore C (2008) The anatomy of the luxury fashion brand. Brand Manag 16 (5/6):347–363

Fisher M, Raman A (2010) The new science of retailing: how analytics are transforming supply chains and improving performance

Ghemawat P, Nueno JL (2006) Zara: fast fashion. Harv Bus Rev

Hammond J (1999) Managing the apparel supply chain in the digital economy

Hines T, Bruce M (2014) Supply chain strategies, structures and relationships. In: Fashion marketing: contemporary issues. Butterworth-Heinemann, Oxford, 3rd edn, pp 27–53

Jacobs D (2006) The promise of demand chain management in fashion. J Fash Mark Manag 10 (1):84–96

Lowson RH (2003) Apparel sourcing: assessing the true operational cost. Int J Cloth Sci Technol 15(5):335–345

Merriam Webster (2015). http://www.merriam-webster.com/inter?dest=/dictionary/trend. Accessed 20 Oct 2015

NTV.de Asien wird zu teuer. Textilindustrie sucht. Made in Africa. http://www.n-tv.de/wirtschaft/Textilindustrie-sucht-Made-in-Africa-article15535926.html. Accessed 20 Oct 2015

Perna R (1987) Fashion forecasting. Fairchild Publications, New York, pp 243–245

Silverstein MJ, Fiske N (2003) Trading up: the new American luxury. Portfolio/Penguin Group, New York

Thomassey S (2010) Sales forecasts in clothing industry: the key success factor of the supply chain management. Int J Prod Econ 128(2):470–483

Wojaczek B (1996) Koordinationsorientiertes Logistik-Management in der Textilwirtschaft Ein Beitrag zur ganzheitlichen Optimierung der Logistik aus Sicht der Bekleidungsindustrie, Frankfurt am Main

Chapter 16
Intelligent Packaging and the Internet of Things in Brazilian Food Supply Chains: The Current State and Challenges

Ana Paula Reis Noletto, Sérgio Adriano Loureiro,
Rodrigo Barros Castro and Orlando Fontes Lima Júnior

Abstract In developing countries, significant food losses occur during distribution. The use of technologies such as Intelligent Packaging (IP) and the Internet of Things (IoT) may provide improvements in controlling the distribution of food products, minimizing losses. This paper identifies the Brazilian food supply chains current technological state and their receptivity to the IP and IoT technologies adoption. The analysis was based on a survey with logistics professionals from the country's largest food companies, which represent 75.3 % of this market (in sales). The results show that these companies do not currently use IP and that a few use what they consider to be IoT systems. Cost is the greatest barrier to the use of these technologies; however, the lack of knowledge about these technologies also represents a strong barrier to their use.

Keywords Food supply chains · Intelligent packaging · Internet of things · Logistics

16.1 Introduction

Food losses during the processing, distribution, and consumption stages represent 1/3 of the total produced in the world, accounting for approximately 1.3 billion tons per year. In developing countries, the rates of food loss during distribution are also significant (FAO 2011). To illustrate this point, when we consider the average percentage loss of meat products during distribution in Latin America (5 %), and apply it to the annual Brazilian production (26 million tons), we find that the

---

A.P.R. Noletto (✉)
Institute of Food Technology, Campinas, SP, Brazil
e-mail: anapaula@ital.sp.gov.br

S.A. Loureiro · R.B. Castro · O.F.L. Júnior
Logistics and Transportation Learning Laboratory – LALT/UNICAMP, Campinas, SP, Brazil
e-mail: saloureiro@gmail.com

country has a total waste of 1.3 million tons of meat per year. This volume of meat is not distant to the annual production of certain countries, such as Belgium (1.8 million tons per year) and Denmark (1.9 million tons per year) (FAO 2011).

The use of technologies such as Intelligent Packaging (IP) and Internet of Things (IoT) systems may enable advances in the monitoring of the product conditions and in the control of logistics activities, thus allowing a reduction in these food losses. According to Whitmore et al. (2014), López et al. (2012), the dissemination of information and the omnipresence provided by these technologies enable to overcome organizational and geographic barriers, improving the logistics and efficiency of supply chains, among other considerations (Holler et al. 2014; Heising et al. 2012; European Commission 2009).

High value-added food Brazilian packages present international standards level. The greatest challenge is the use of new technologies, with a suitable cost, to minimize losses along the supply chain. However, in Brazil, no consolidated data exist regarding the knowledge and use of these technologies. This study aims to understand the current technological state of food supply chains in Brazil and the receptivity of these chains to the use of IP and IoT. In addition, this study seeks to identify the main motivations for and barriers to the use of these technologies.

To accomplish this goal, we surveyed logistics professionals from the 20 largest food companies in Brazil.

The article comprises an introduction to the collaboration of IP and IoT in food distribution. Next, we describe the research methodology and results, and finally, we present our conclusions.

16.2 Intelligent Packaging in Logistics

The studies that address the subjects of "active packaging," "smart packaging," or "intelligent packaging" do not always define these terms in the same manner, and in some studies, there is "a certain confusion" regarding the meaning of each term. For this reason, in this section, we present the results of the literature review (Table 16.1) for which we selected certain articles that define the concept of "Intelligent Packaging." Through the described definitions, some attributes that are recurrently cited by the authors can be identified, such as the following: self-identity, the monitoring of product and environmental conditions, communication, traceability, decision-making, and the recording of information.

All of these attributes provide improvements in controlling the distribution of products throughout the logistics activities of distribution.

These attributes are similar to those described by Wong et al. (2002) to define a Smart Product: (a) possession of a unique identity; (b) ability to communicate effectively with its environment; (c) ability to retain or store data about itself; (d) possession of a language to display its features, production requirements, etc.; and (e) capability of participating in or making decisions relevant to its own destiny.

In all the selected studies, IP are used for packaging food products.

Table 16.1 Characteristics of Intelligent packaging and its applications

Author	Definition
Schilthuizen (1999)	Packages with intelligent functions such as identification, sensing, tracing, tracking, and monitoring
Kruijfy et al. (2002)	Intelligent packaging systems that monitor the condition of packaged products to provide information on their quality during transportation and storage
Yam et al. (2005)	Packaging systems that can perform intelligent functions "such as the detection, sensing, recording, communication, and application of scientific logic to facilitate decision-making to extend the product's shelf life, enhance safety, improve quality, provide information, and warn about possible problems"
Dainelli et al. (2008)	Packages that monitor the product's quality, identify critical points, and provide more detailed information along the supply chain
European Commission (2009)	Intelligent packages have a component that allows monitoring the condition of the food or its surrounding environment during transportation and storage
Vanderroost et al. (2014)	Intelligent packages have the ability "to sense, detect, or record changes in the product or its environment"
Goddard et al. (1997), Kerry et al. (2006), Dobrucka et al. (2014)	For these authors, the terms "Intelligent Packaging" and "Smart Packaging" have the same meaning and refer to the combination of advanced materials and sensors capable of monitoring the product and the environment in real time and responding properly. The devices attached to these packages must be able to report the integrity, quality, and safety of the product and are used to ensure authenticity, protection against theft, and the traceability of the product

16.3 The Internet of Things—IoT in Logistics

The term "IoT" has several definitions, including definition presented by the Cluster of European Research Projects (IERC 2010): "a dynamic global network infrastructure with self-configuring capabilities based on standard and interoperable communication protocols where physical and virtual 'things' have identities, physical attributes, and virtual personalities and use intelligent interfaces, and are seamlessly integrated into the information network."

Regarding logistics activities, the use of technologies such as Radio-Frequency Identification (RFID) tags and sensors to track and monitor products along the supply chain is not new; however, with the advent of IoT systems, their applications, and contributions can be amplified.

In transportation logistics, IoT systems can make significant contributions because:

(a) people (or intelligent cargo) can report, in real time, the performance of the transportation infrastructure;
(b) this sector already uses a sensor structure (cargo) and information database (managers); and
(c) the presence of different stakeholders provides a situation of information sharing in the search for improvements (Holler et al. 2014).

These considerations show that intelligent objects are essential for building IoT systems. López et al. (2012) present a classic example of the application of intelligent objects for monitoring the transportation of perishable goods with a cooling system. In their study, the authors demonstrate that, because it is an activity affected by a large number of variables, such as traffic conditions, the constant cooling system, and variability in the delivered amount, the more comprehensive management provided by IoT systems enables the reduction of product losses and costs through a higher level of autonomy in dynamic situations.

16.4 Methodology

We conducted a survey with the aim of identifying the current degree of knowledge, the use, and the barriers to the use of IP and IoT systems in Brazil. According to Babbie (1990), surveys are applied to make assertive descriptions about a population based on a group of participants (sample). For Fink (1995), the sample size must adequately represent the population to ensure that the results are accurate and reliable. Among the main precautions to be taken in the design of survey questionnaires, Babbie (1990) cites the following: (a) the clarity and objectivity of the questions; (b) the proper competence of participants; and (c) the use of relevant questions. We sent questionnaires to logistics professionals from the 20 largest food companies in Brazil, in terms of sales (in USD), excluding companies operating in the commodities area (grains and alcohol producers). We selected food sector because this type of cargo is mostly present in urban areas (Holguin-Veras et al. 2008; Munuzuri et al. 2010; Ibeas et al. 2012).

We initially contacted the professionals to convey information on the study and invite them to answer the questionnaire. Once they agreed to participate, we sent the questionnaire. We conducted all communication through email, and the Institute of Food Technology (ITAL), one of the leading research institutes in the Brazilian food sector, facilitated obtaining contact information. The questionnaire consisted of multiple-choice questions and open-ended questions. Below, we present the questionnaire answers, compiled, and analyzed.

16.5 Results

Of the 20 companies invited to participate, 13 companies responded to the survey, corresponding to a response rate of 65 %. These 13 companies represent 75.3 % of the food industry in Brazil (sales in USD), and 4 of them are among the largest food companies in the world. The results showed that 7.7 % of the professionals who participated in the survey occupy a Director position in their company, while 38.4 % are managers, 46.2 % coordinators, and 7.7 % supervisors. Regarding experience in logistics, 15.4 % of the professionals have between 1 and 3 years, 15.4 % have between 3 and 5 years, and 69.2 % have over 7 years of experience in the area.

16.5.1 Intelligent Packaging

In the first part of the questionnaire, we asked participants to answer what degree of knowledge they had on the subject of Intelligent Packaging. They also defined the level of importance of the attributes of an IP according to the attributes identified in the previous section of this article: self-identity, the monitoring of product and environmental conditions, communication, traceability, decision-making, and the recording of information. Upon completion of this part of the research, participants answered questions about the use of IP and its contributions to logistics activities. Figures 16.1, 16.2, 16.3, 16.4, 16.5, 16.6, and 16.7 present the aforementioned results.

Among the most relevant IP attributes, the monitoring of product and environmental conditions and traceability showed a much higher relevance (above 84.6 %, i.e., the sum of the answers "Very important" and "Extremely important") compared to the attributes of self-identity, scientific logic, and information recording (maximum of 46.2 %). Participants considered that the activities related to the consumer interaction (customer inventory control, automatic product replacement, and control of products returned by the customer) will cause a greater impact, in

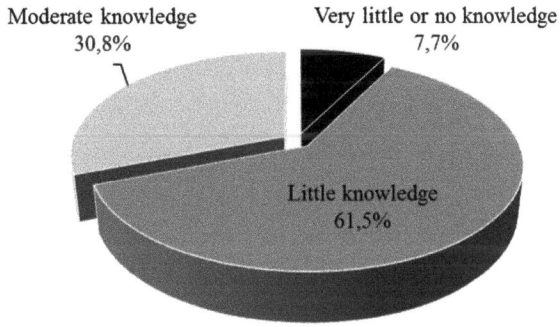

Fig. 16.1 Your degree of knowledge on the subject of IP

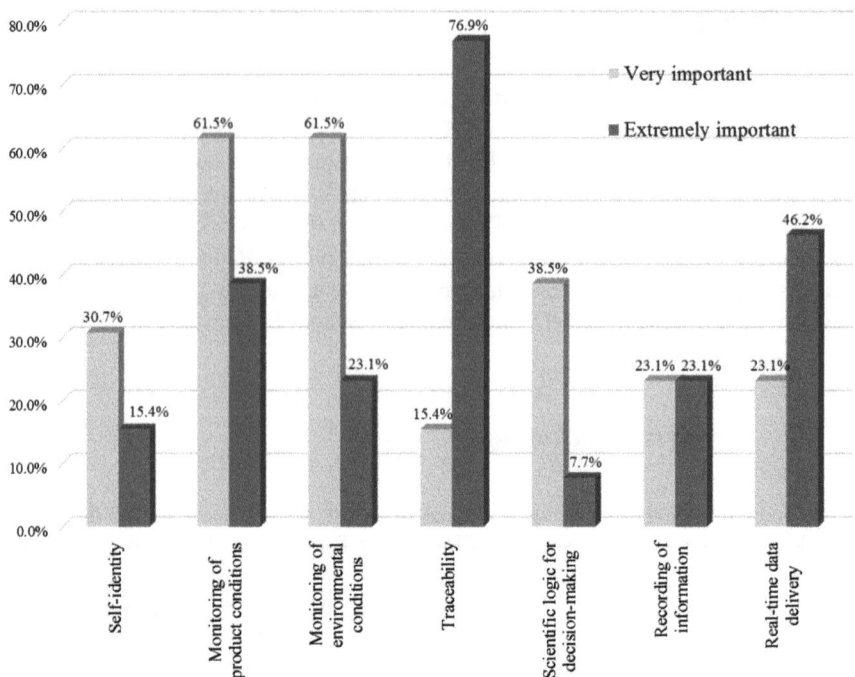

Fig. 16.2 Attributes level of importance

Fig. 16.3 Other attributes that an IP should have

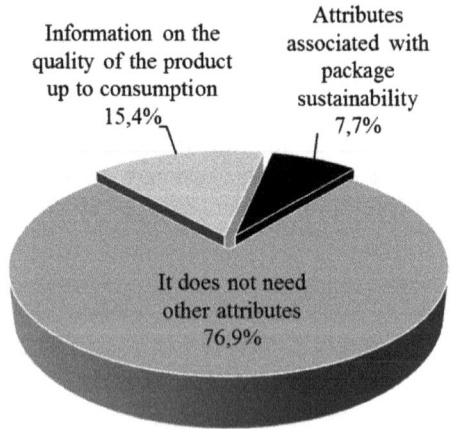

which the sum of "Large impact" and "Huge impact" responses represent around 53.8 % of the total. This result endorses the importance given to the real-time data delivery attribute, with 69.3 % of answers pointing to "Very important" and "Extremely important." All participants stated that they do not use IP, and therefore,

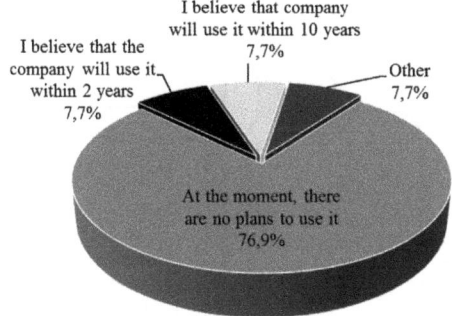

Fig. 16.4 Prediction of IP use in your company

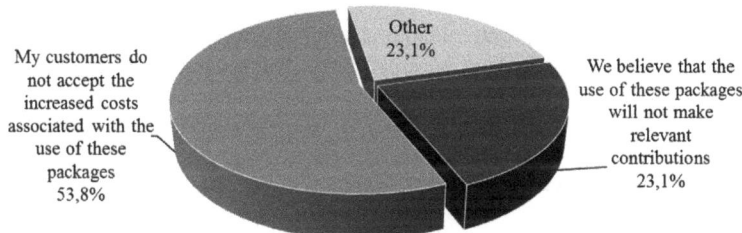

Fig. 16.5 Main barrier to IP use in my company

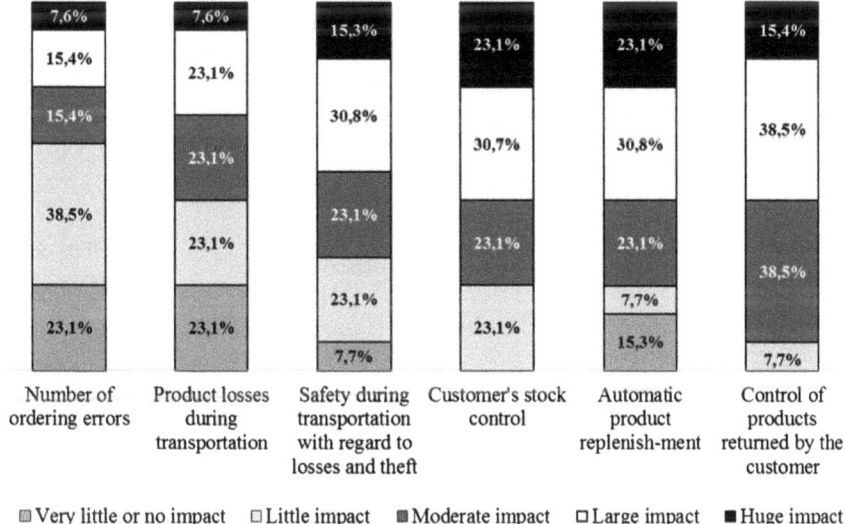

Fig. 16.6 If you use or will use IP, rate the degree of impact in the following activities

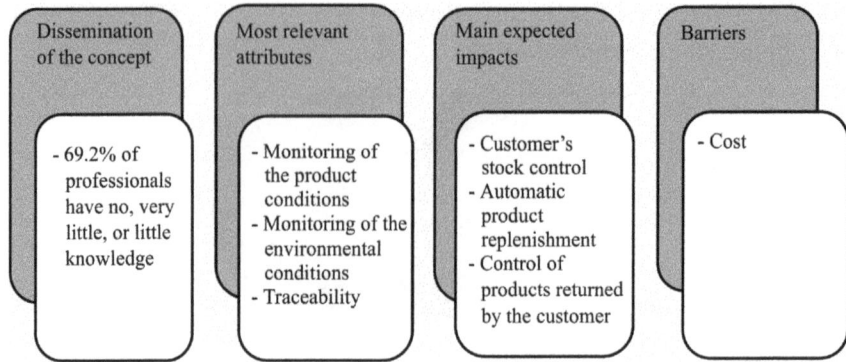

Fig. 16.7 Opportunities for IP in food industries in Brazil (*source* the author)

none of them replied to the open-ended questions regarding the description of these packages. Figure 16.7 presents a compilation of data collected on IP.

16.5.2 The Internet of Things (IoT)

The second part of the research focused on the subject of IoT. We asked the participants to answer what degree of knowledge they had on the subject of IoT and to predict the use of these systems in the logistics area of their companies. Figures 16.8, 16.9, and 16.10 show the results.

We asked participants to describe briefly their IoT systems, if their companies used them. Although four participants stated that they use IoT systems, only two participants reported the use of satellite and Global System for Mobile communication (GSM) management systems as their IoT systems but without presenting a description.

Fig. 16.8 Your knowledge on the subject of IoT

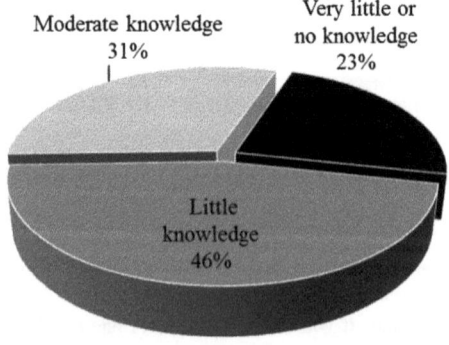

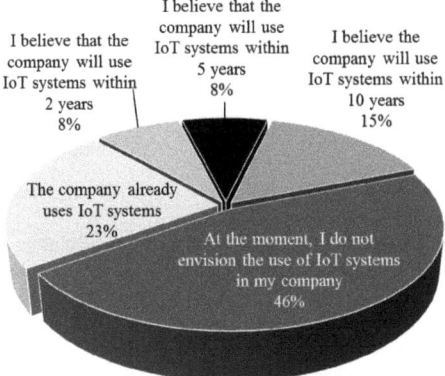

Fig. 16.9 Prediction of IoT use in the logistics area of your company

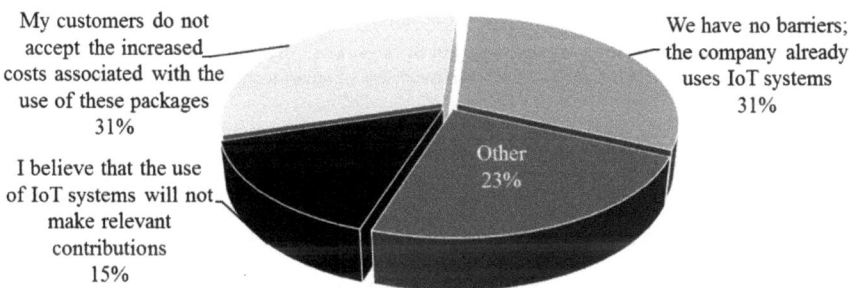

Fig. 16.10 Main barrier to IoT use in the logistics area of your company

16.6 Conclusion

Based on the results, we conclude that logistics professionals' lack of knowledge on these technologies represents a large barrier to their adoption because they are unable to envision the potential gains from their use. We observe that IP configurations with less attributes but with specific attributes that meet the users' needs at suitable prices may be adopted. Regarding IoT systems, the results show that, in addition to the lack of knowledge, some professionals reported inaccurate knowledge on this technology, considering technologies for vehicle tracking and satellite communications as IoT systems. In terms of future studies, we suggest the development of economic feasibility studies that consider, in addition to acquisition and implementation costs, the gains achieved in logistics operations and the reduction in food losses through the implementation of these technologies. Despite the focus of this work on the use of IP for packaging food, other areas, such as the pharmaceutical industry, are also studying its implementation.

Acknowledgments The authors thank the National Counsel of Technological and Scientific Development (CNPq), the Institute of Food Technology (ITAL), and the Logistics and Transportation Learning Laboratory (LALT) for the support provided to this study.

References

Babbie ER (1990) Survey research methods. CA, Wadsworth, Belmont
Dainelli D, Gontard N, Spyropoulos D, den Zondervan-van EB, Tobback P (2008) Active and intelligent food packaging: legal aspects and safety concerns. Trends Food Sci Technol 19: S103–S112
Dobrucka R, Cierpiszewski R (2014) Active and intelligent packaging food—research and development–a review. Pol J Food Nutr Sci 64(1):7–15
European Commission (2009) Commission regulation 450/2009 of 29 May 2009 on active and intelligent materials and articles intended to come into contact with food. http://ec.europa.eu/food/food/chemicalsafety/foodcontact/docs/guidance_active_and_intelligent_scofcah_231111_en.pdf. Accessed 24 June 2015
FAO (2011) Global food losses and food waste—extent, causes and prevention. Rome. http://fao.org/docrep/014/mb060e/mb060e00.pdf. Accessed 12 Aug 2015
Fink A (1995) How to analyze survey data, vol 8. Sage, Los Angeles
Goddard NDR, Kemp RMJ, Lane R (1997) An overview of smart technology. Packag Technol Sci 10:129–143
Heising JK, Dekker M, Bartels PV, van Boekel MAJS (2012) A non-destructive ammonium detection method as indicator for freshness for packed fish: application on cod. J Food Eng 110(2):254–261
Holguín-Veras J, Silas M, Polimeni J, Cruz B (2008) An investigation on the effectiveness of joint receiver–carrier policies to increase truck traffic in the off-peak hours. Netw Spat Econ 8(4):327–354
Holler J, Tsiatsis V, Mulligan C, Avesand S, Karnouskos S, Boyle D (2014) From machine-to-machine to the Internet of things: introduction to a new age of intelligence. Elsevier Academic Press, Amsterdam
Ibeas A, Moura JL, Nuzzolo A, Comi A (2012) Urban freight transport demand: transferability of survey results analysis and models. Procedia—Soc Behav Sci 54:1068–1079
IERC—European Research Cluster on the Internet of Things (2010). http://www.internet-of-things-research.eu/about_iot.htm. Accessed 20 July 2015
Kerry JP, O'grady MN, Hogan SA (2006) Past, current and potential utilisation of active and intelligent packaging systems for meat and muscle-based products: a review. Meat Sci 74(1):113–130
Kruijf NN, van Beest M, Rijk R, Sipiläinen-Malm T, Paseiro LP, De Meulenaer B (2002) Active and intelligent packaging: applications and regulatory aspects. Food Addit Contam 19(S1): 144–162
López TS, Ranasinghe DC, Harrison M, McFarlane D (2012) Adding sense to the Internet of Things. Pers Ubiquit Comput 16(3):291–308
Muñuzuri J, Cortés P, Onieva L, Guadix J (2010) Modelling peak-hour urban freight movements with limited data availability. Comput Ind Eng 59(1):34–44
Schilthuizen SF (1999) Communication with your packaging: possibilities for intelligent functions and identification methods in packaging. Packag Technol Sci 12(5):225–228
Vanderroost M, Ragaert P, Devlieghere F, De Meulenaer B (2014) Intelligent food packaging: the next generation. Trends Food Sci Technol 39(1):47–62
Whitmore A, Agarwal A, Xu LD (2014) The Internet of Things—a survey of topics and trends. Inf Syst Front 17(2):261–274

Wong CY, McFarlane D, Zaharudin AA, Agarwal V (2002) The intelligent product driven supply chain. In: Proceedings of the IEEE international conference on systems, man and cybernetics, vol 4, p 6

Yam KL, Takhistov PT, Miltz J (2005) Intelligent packaging: concepts and applications. J Food Sci 70:1–10

Chapter 17
Methodology for Development of Logistics Information and Safety System Using Vehicular Adhoc Networks

Kishwer Abdul Khaliq, Amir Qayyum and Jürgen Pannek

Abstract The Intelligent Transportation System (ITS) addresses issues regarding traffic management and road safety in the domain of Vehicular Ad hoc Networks (VANETs). With the evaluation of new applications, new goals regarding efficiency and security have been added for logistics and general user application, which demand time-bounded and reliable services. In this paper, we evaluate VANET with regard to its suitability in logistics scenarios. We simulated VANET by considering different application scenarios for logistics and transportation using varying parameters such as speed, number of nodes, traffic load and bit error rate etc. We observed that it performs well in most of the scenarios due to its highly suitability in vehicular environment.

Keywords VANET · Routing protocols · IEEE 802.11p · Logistics · Transportation

17.1 Introduction

In commercial applications, Vehicular Adhoc Networks (VANETs) (I. S. Association 2010) are used in traffic management applications to alert drivers of traffic jam, balance traffic flow and reduce traveling time (Olariu and Weigle 2009; Akbar et al. 2015), road safety (Yin et al. 2004) and efficiency. Furthermore, it helps in logistics by offering a number of applications like inbound/outbound traffic control for delivery of goods from ports and warehouses. Moreover, it also helps to provide commer-

---

K.A. Khaliq (✉) · J. Pannek
Department of Production, University of Bremen, Bremen, Germany
e-mail: kai@biba.uni-bremen.de

J. Pannek
e-mail: pan@biba.uni-bremen.de

A. Qayyum
CoReNeT, Capital University of Science and Technology (CUST), Islamabad, Pakistan
e-mail: aqayyum@ieee.org

© Springer International Publishing Switzerland 2017
M. Freitag et al. (eds.), *Dynamics in Logistics*, Lecture Notes in Logistics,
DOI 10.1007/978-3-319-45117-6_17

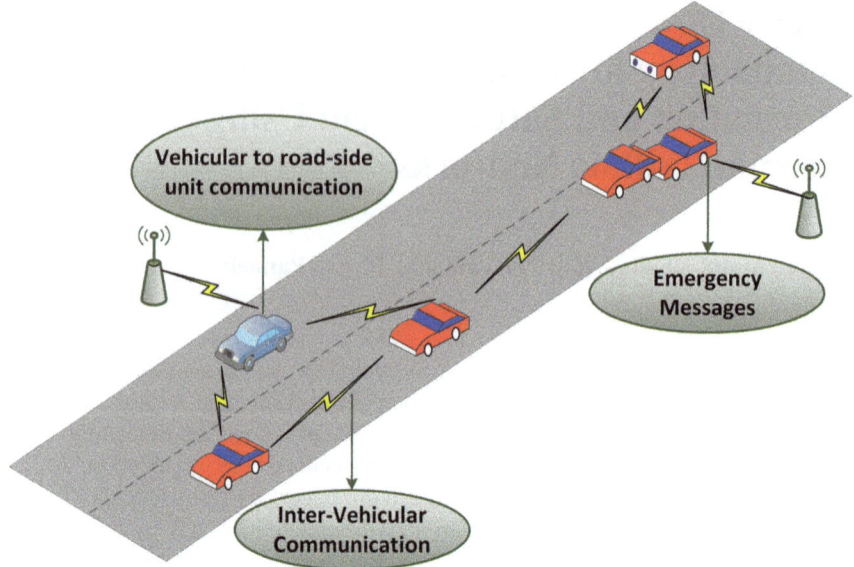

Fig. 17.1 Vehicular Adhoc Networks (VANET)

cial services on the road and offers attractive applications for infotainment (Amadeo et al. 2012), multimedia and online gaming applications (Akbar et al. 2014).

In the vehicular environment as sketched in Fig. 17.1, three types of communication may occur: Vehicle-to-vehicle (V2V), Vehicle-to-Infrastructure (V2I) and Infrastructure-to-Infrastructure (I2I). IEEE recommends 802.11p as MAC and PHY layer standard for VANET and the use of it has been frequently proposed for vehicular safety applications. The Crash Avoidance Metrics Partnership Consortium helped in the standardization of the SAE J2735 Basic Safety Message (BSM). These messages include information which is important for safety applications for vehicles state like position, speed, and acceleration (Ahmed-Zaid et al. 2011). By using such applications, a potential safety benefit is generated, but disseminated information is not always relevant for each participating node. For example, in the case of an emergency report, a distant vehicle is unlikely to be effected. Therefore, the estimation of relevance information is required to avoid incorrect alerts or to help ignore warnings. However, the determination of relevance can be uncertain as it may depend upon different factors like weather, road characteristics, positions, speeds, and accelerations of vehicles. Considering all these parameters, it can be difficult to develop a tool that estimates suitable conditions for new applications. For the channel access, IEEE 802.11p uses CSMA/CA at MAC layer. For safety applications, as safety messages are usually small in size and IEEE 802.11p uses 10 MHz channel which is enough for transmission of safety messages. Research shows that IEEE 802.11p performs well for safety application in vehicular environment (Katrin et al. 2008). The reason for its popularity is its claims about time bounded and high throughput services.

ITS was initially driven by the concept of safety applications, but the next generation ITS adds the use of bandwidth hungry applications like Video on Demand (VoD), Voice over IP (VoIP), video conferencing, online gaming and file transfer etc., which require less delay and high bandwidth (Akbar et al. 2015). The new trend in the current decade is to apply technologies in different areas to get the maximum benefits of them in real environment.

To get the competitive strength in market, logistics management is a key source and it has potential for cost reduction and the opportunity to increase market share, if it is used effectively (Christopher 1998). Here, in the urban environment, with its increasing population, its congestion of freight transport and impact on the environment is of particular interest. In this scenario, where the objective is to develop a more efficient and effective freight transport system with reduction of environmental issues such as pollution, noise, etc., a new concept of city logistics (Crainic and Feillet 2015; Taniguchi and Thompson 2014) has emerged. It has three main objectives; first, mobility of traffic to maintain smooth and reliable flow. Second, sustainability to make cities more environment friendly. Third, livability which consider risen residence of elderly people within city. Considering all these objective Intelligent Transport System (ITS) is one of the methods that used to solve this complicated problem. The scenarios in urban areas are considered a high priority because the half of the world's population lived in cities (Sands 2015) and it is estimated that this figure will increase to over 60 % by 2030 (Olsson and Head 2015). In this regard, VANET can help to design an efficient solution which will help to maintain information system and also provide road safety with reduction of the cost. As with the dense transport population, congestion will increase due to expansion in demand of home delivery of goods and services, specially to elderly people. With the help of a social concept, negotiation for services and quick (shortest path) delivery can lead to efficient and cost-effective solutions. The communication concept between vehicles not only helps to make vehicles able to communicate with other vehicles for road safety, but also they altogether form a cooperative network through which the services can be achieved in a more efficient way.

The rest of paper is organized as following; Sect. 17.2 gives an overview of the city logistics and Sect. 17.3 describes the evaluation of VANET for logistics applications, and also discusses the results in this scenario. Section 17.4 explains the methodology, which uses VANET as key technology to provide the solution for freight transport systems to provide efficiency, safety on road and information management for logistics and Sect. 17.6 concludes the paper.

17.2 State of the Art

VANET plays an important role for safety and traffic management applications. The authors in Martelli et al. (2012), Shakeel et al. (2015) performed field test for the real world scenarios for comfort and road safety. The paper (Anand et al. 2015) presented a review of urban freight transportation modeling efforts for analysis in particular environment and covered the fundamental aspects in selection process of modeling and review the trends of city logistics for relevance. The authors also attempted to identify gaps in modeling, specially for the city logistics domain.

In Taniguchi (2015), the authors presented concepts of city logistics for sustainable and liveable cities. City logistics can contribute to create more efficient and environmentally friendly urban freight transport systems. The application of innovative technologies of Information and Communication Technology (Tan et al. 2012; Perego et al. 2011) and ITS (Perego et al. 2011; Mondragon and Mondragon 2012), the change in mind-sets of logistics managers, and public-private partnerships can promote city logistics policy measures. The authors in Taniguchi et al. (2013) discussed the basic concept of logistics, and use of ICT and ITS in logistics. The authors also included methodologies to minimize incorporating risk in urban freight transport. The relationship between city logistics and quality of life is discussed in Witkowski and Kiba-Janiak (2012) using theory and a reference model. The authors discussed the survey results at the end that "moving around the city" does not improve the quality of life because of the heavy traffic movement in the city. In the current scenario, the need of proper methodology that not only consider economics factor, but also includes the environment, safety and management factors that can improve quality of life in city.

17.3 Evaluation of VANET for Safety Applications

For the simulation of VANET in logistics safety and infotainment applications, we selected open source network simulator NS-2. Table 17.1 describes parameters of our simulation setup. Simulations were performed Linux distribution Fedora and Constant-Bit Rate (CBR) traffic flows are used with UDP. The Table 17.2 shows

Table 17.1 General simulation parameters

Operating system	Linux distribution Fedora Core 17
NS-2 versions	NS-2.34 for IEEE 802.11p
Radio-propagation model	Two-Ray-Ground
Traffic environment	Urban area street traffic scenario
Traffic flows	Constant-bit rate (CBR)
CBR flow rates (Mb/s)	Varies from 1 to 40 Mbps
Transport layer protocol	UDP
Number of nodes	Varies from 1 to 600
Speed of nodes	Varies from 40 km/h to 100 km/h
Packet size	200–500 bytes

Table 17.2 Simulation parameters setting for MAC

Slot time	13 nano s
SIFS interval	32 nano s
Contention Window Min (CWMin)	15 slots
Contention Window Max (CWMax)	1023 slots
Transmission range	50–1000 m

the parameters settings for 802.11p. For the comparison, we kept the values of MAC layer parameters constant for each standard. For the comparative analysis, we selected Application traffic load, Node density, Inter-node distance, on Goodput simulation scenarios for urban areas with traffic load in single direction and for the comparison we calculated goodput and packet delay.

For evaluation, we examined the effect on the goodput and delay by varying traffic loads. Constant-bit Rate (CBR) traffic flows were used in the simulation with packet size of 256 bytes, which was kept constant. We imposed two other CBR flows of 500 Kbps, which acted as background traffic. 15 nodes were used in the scenario, which were moving at 100 km/h. CBR traffic rate was varied to check its effect on goodput and delay.

17.3.1 Goodput and Delay

Results are shown in Fig. 17.2a, b. We observe that the goodput decreases with increase in number of vehicles in the network. A constant goodput is observed in case of small sized packets, and when we increase the packet size, the decreases due to greater number of vehicles and number of hops involved. As we mentioned, the packet size in safety applications are small. Therefore, the graph shows a constant line for small packets with increased number of vehicles.

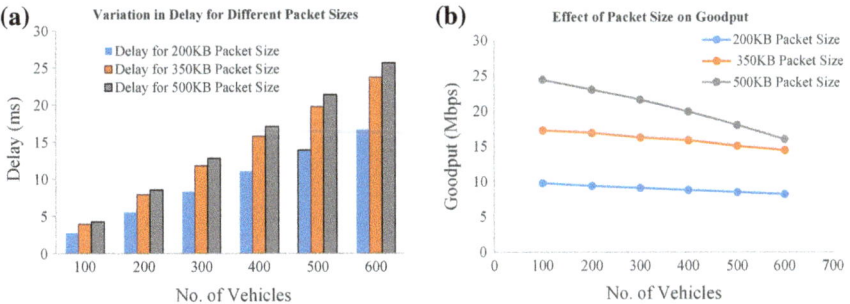

Fig. 17.2 **a** Variation of delay for different packet sizes; **b** Effect of packet size on goodput

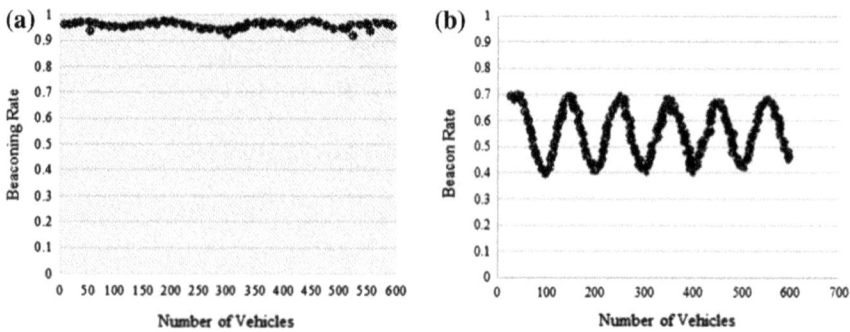

Fig. 17.3 a Beaconing rate when range is 50 m; b Beaconing rate when range is 250 m

17.3.2 Beacon Dissemination and Transmission Range

To provide safety in vehicular environment by deploying VANET safety applications, it is necessary to evaluate important parameter for success. In literature, the delivery rate and delay are common metrics used for the evaluation of performance. For safety applications, the latest information dissemination is important. For such applications, the individual vehicle is critical and should be monitored for message dissemination protocols. Both the message originating node to message receiving node are important for safety applications. Therefore, the communication range between these nodes or number of hops involved between these nodes are the relevant metrics for evaluation. As vehicles are also moving with high speed, though this movement is static with respect to each other (if they are in the same lane at the same speed) in highway scenarios, but in city these movement is more dynamic because the frequent change in neighboring vehicles gives dynamic topology. It may cause communication link breakage between sending and receiving vehicles. Within the vehicle range, the communicating vehicles can receive beacons for information sharing in safety applications i.e. alarm based and beacon based safety application. In this scenario, the success of safety application depends upon the fair beacon.

Figure 17.3 shows the relationship of beacon transmission and change in transmission range. It shows by increasing transmission range, there is an increasing trend in application's effective range until the delivery rate of its corresponding beacon dissemination protocol falls below the desired threshold. Beyond that threshold, increasing vehicle's transmission range leads to decreasing application's effective range.

17.4 Methodology

There are several steps involved from the problem statement to the implementation of proposed method. To get the best solution for a problem, it is necessary to review the previous facts. System requirements and strategies are important functions in any

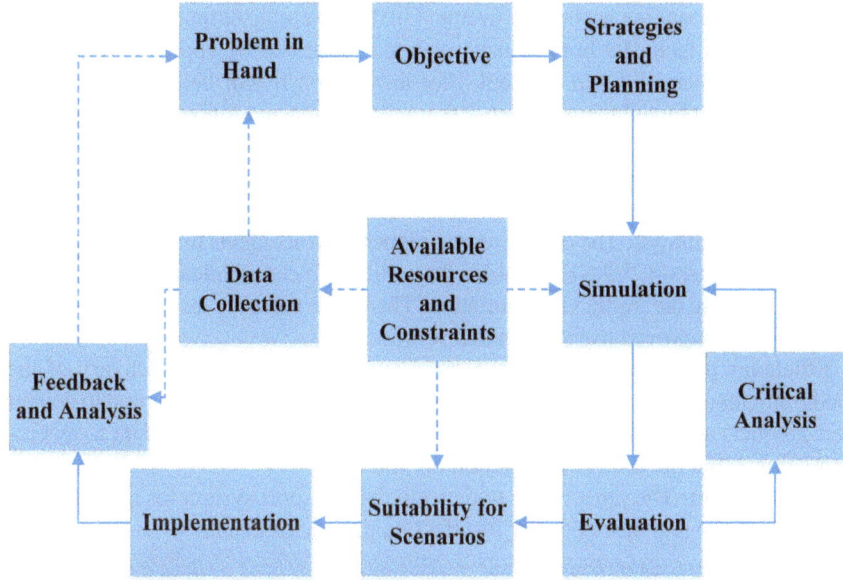

Fig. 17.4 Methodology to reach appropriate solution

kind of methodology. On the basis of objectives and requirements, we can model the system, and for required results simulation and evaluation are the next steps. These results help to implement the model in real world. Figure 17.4 sketches the flows of the city logistic system from data collection, problem statements to implementation. Each step has its own importance. For the logistic system, we identified objectives and requirements.

17.4.1 Objectives and Planning

In order to get the objective of safety on road and efficiency in city supply chain, planning plays a vital role in methodology. For planning several measures can be used. This may include following activities that play an important role:

1. Identify the objective and requirements
2. Identify the key performance indicator
3. Identify and define planning scenarios including simulation

apart from the objectives of the transportation (key performance indicator, scenarios), there are several other factors that can be part of the planning to achieve goal. There are different factors that are important for industries to achieve in long term; amongst them some are as following.

Economy: Economy as the improvement in the economic output of the study area is the important factor for both industry and end consumer. By applying VANET on city logistics, there is setup cost but after installation it will help to improve the economic output by reducing the network transportation and logistic costs (e.g., distributors, wholesalers, carriers, retailers, end consumers). Furthermore, it will also reduce selling prices of goods for end consumers.

Safety: For the transportation and traffic management, safety has high priority while finding solutions. The objective is to reduce road accidents linked to the freight and general transport. VANET enables the vehicles to communicate to others vehicles, and manage the traffic via negotiation. Therefore, there are less chances of road accidents and improved usage of shared resources.

Efficiency and Availability: From production of goods to the delivery of the goods, efficiency and availability are of importance. Industries are required to deliver their products at the right time to the right customer, and customers want to receive their shipment at the desired destination on time.

Environment: The objective is to develop the transport system for improvement of quality of life, but also consider its effect on the environment.

17.4.2 Strategies and Measures

In urban scenarios, citizens expect lively environment and easy access to the city. With the increase of population of the cities, the requirement for transportation also increases. In absence of good public transport, most of the people prefer to have individual transport. Hence, the population of vehicles in a city environment increases and effects city sustainability and livability. In addition to that, inventory management and shopping also produce negative impacts on the later. For city logistics, several measures can be implemented to reduce this impact; First, the requirement is to reduce the number of vehicles and the need to make it environmentally friendly. The freight transport should use optimized loading and unloading operations to reduce traffic congestion and interference with other vehicles.

17.4.3 Proposed Strategy

Considering strategies and measurement, we divided the system into three layers for understanding and implementation. Each layer has its own specific task. Figure 17.5 shows system layers, specific task of each layers and actors involves in the system. The first layer is called physical layer and involves the hardware requirement. For VANET implementation, each vehicle should have interface for communication and which type of vehicle mode to use (Which time to begin and end tour? which vehicle is suitable for delivery?). The second layer, i.e., management layer, contains tasks like location choices, planning (which type of retail type to buy and from where),

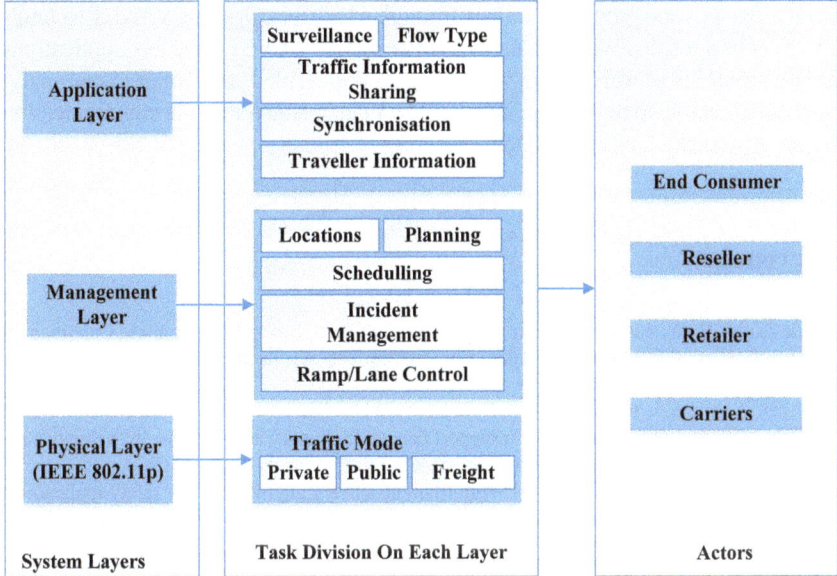

Fig. 17.5 Proposed system layers

scheduling (start and end time, total number of tours) for the transport. The incident management, lane control during traveling, and optimized path to follow are also the responsibility of this layer. The third layer is application layer, where applications are required to manage information of travelers, trips, which type of traffic flow is (How to share the traffic information? their synchronization and central control). These layers altogether lead to system for traffic.

17.5 Acknowledgment

This research was supported by the European Commission in the framework of Erasmus Mundus and within the project cLINK.

17.6 Conclusion

The integration of emerging technologies in different areas to get benefits is poplar since last many decades. Vehicular ad hoc network is one the wireless technology that is used in different traffic management applications, safety and multimedia applications. City logistics is the process of managing and optimizing the involved activities considering many factors. In this paper, we proposed a methodology for optimizing the city logistics and proposed strategy through which we can apply VANET technology to get best results. We evaluated the suitability of VANET for safety applications for city scenarios as well, where we achieved good results. According to our analysis, VANET can help to improve economy by manage the traffic via communication and improving the facility of shared resources.

References

Ahmed-Zaid F, Bai F, Bai S, Basnayake C, Bellur B, Brovold S, Brown G, Caminiti L, Cunningham D, Elzein H et al (2011) Vehicle Safety Communications–Applications (VSC-A) Final Report. Technical Report

Akbar MS, Khaliq KA, Qayyum A (2015) Vehicular MAC Protocol Data Unit (V-MPDU): IEEE 802.11 p MAC protocol extension to support bandwidth hungry applications. In: Vehicular ad-hoc networks for smart cities. Springer, pp 31–39

Akbar MS, Qayyum A, Khaliq KA (2015) Information delivery improvement for safety applications in VANET by minimizing Rayleigh and Rician fading effect. In: Vehicular ad-hoc networks for smart cities. Springer, pp 85–92

Akbar MS, Khan MS, Khaliq KA, Qayyum A, Yousaf M (2014) Evaluation of IEEE 802.11 n for multimedia application in VANET. Procedia Comput Sci 32:953–958

Amadeo M, Campolo C, Molinaro A (2012) Enhancing IEEE 802.11 p/WAVE to provide infotainment applications in VANETs. Ad hoc networks 10(2):253–269

Anand N, Van Duin R, Quak H, Tavasszy L (2015) Relevance of city logistics modelling efforts: a review. Transp Rev 35(6):701–719

I. S. Association et al 802.11 P-2010 IEEE Standard for Information Technologylocal and Metropolitan Area Networks specific Requirements part 11: Wireless LAN Medium Access Control (MAC) and Physical Layer (PHY) Specifications Amendment 6: Wireless Access in Vehicular Environments. http://standards.ieee.org/findstds/standard/802.11p-2010.html

Christopher M (1998) Logistics and supply chain management: strategies for reducing cost and improving service

Crainic TG, Feillet D (2015) Introduction to the special issue on city logistics. EURO J Transp Logistics 1–2

Katrin B, Uhlemann E, Store E, Bilstrup U (2008) On the Ability of the 802.11 p MAC Method and STDMA to support real-time vehicle-to-vehicle communication. EURASIP J Wirel Commun Networking 2009

Martelli F, Renda ME, Resta G, Santi P (2012) A Measurement-based study of beaconing performance in IEEE 802.11 p vehicular networks. In: INFOCOM, 2012 Proceedings IEEE. IEEE, pp 1503–1511

Mondragon AEC, Mondragon ESC (2012) Smart grid and wireless vehicular networks for seaport logistics operations. In: 19th ITS world congress

Olariu S, Weigle MC (2009) Vehicular networks: from theory to practice. CRC Press

Olsson L, Head BW (2015) Urban water governance in times of multiple stressors: an editorial. Ecol Soc 20(1):27

Perego A, Perotti S, Mangiaracina R (2011) ICT for logistics and freight transportation: a literature review and research agenda. Int J Phys Distrib Logistics Manage 41(5):457–483

Sands D (2015) An innovative scorecard for evaluating resiliency in our cities. In: Planet@ Risk, vol 3, no 1

Shakeel SM, Ould-Khaoua M, Rehman OMH, Al Maashri A, Bourdoucen H (2015) Experimental evaluation of safety beacons dissemination in VANETs. Procedia Comput Sci 56:618–623

Tan MII, Razali RN, Desa MI (2012) Factors influencing ICT adoption in Halal transportations: a case study of Malaysian Halal logistics service providers. Int J Comput Sci Issues 9(1):62–71

Taniguchi E (2015) City logistics for sustainable and liveable cities. In: Green logistics and transportation. Springer, pp 49–60
Taniguchi E, Thompson RG (2014) City logistics: mapping the future. CRC Press
Taniguchi E, Thompson RG, Yamada T (2013) Concepts and visions for urban transport and logistics relating to human security. In: Urban transportation and logistics: health, safety, and security concerns, p 1
Witkowski J, Kiba-Janiak M (2012) Correlation between city logistics and quality of life as an assumption for referential model. Procedia-Soc Behav Sci 39:568–581
Yin J, ElBatt T, Yeung G, Ryu B, Habermas S, Krishnan H, Talty T (2004) Performance evaluation of safety applications over dsrc vehicular ad hoc networks. In: Proceedings of the 1st ACM international workshop on vehicular ad hoc networks. ACM, pp 1–9

Chapter 18
Power Management of Smartphones Based on Device Usage Patterns

Lin-Tao Duan, Michael Lawo, Ingrid Rügge and Xi Yu

Abstract Smartphones provide rich applications and offer many crowdsensing services to end users. However, the power consumption of smartphones is still a primary issue in green computing. This paper presents a general power management framework including a data logger, an unsupervised learning algorithm of classifier, and a power-saving decision maker. The framework gathers and analyses the usage patterns of smartphones and separates end users into *non-active* and *active* ones, and their mobile devices into *low-power* ones and *high-power* ones using an unsupervised learning algorithm. If the smartphone of a *non-active* user belongs to the *high-power* group, we observe abnormal usage behaviour. The framework provides recommendations, e.g. as a power-saving notification to the user. We collected device usage and power consumption attributes on two kinds of Android–smartphones and evaluated the framework in experimental studies with 22 users. The results show that our framework can correctly identify *non-active* users' devices consuming much more power than *active* users by recognizing reasons of the abnormal usage behaviour of *non-active* users and providing recommendations for adjusting the device towards power saving.

L.-T. Duan (✉) · M. Lawo · I. Rügge
International Graduate School for Dynamics in Logistics, Bremen University, Bremen, Germany
e-mail: duanlintao@uni-bremen.de

M. Lawo
e-mail: mlawo@tzi.de

I. Rügge
e-mail: rue@biba.uni-bremen.de

L.-T. Duan · M. Lawo
Centre for Computing and Communication Technologies, Bremen University, Bremen, Germany

L.-T. Duan · X. Yu
School of Computer Science, Chengdu University, Chengdu, China
e-mail: oliveryx@cdu.edu.cn

© Springer International Publishing Switzerland 2017
M. Freitag et al. (eds.), *Dynamics in Logistics*, Lecture Notes in Logistics,
DOI 10.1007/978-3-319-45117-6_18

Keywords Power management · Unsupervised learning algorithm · Smartphones · Usage patterns

18.1 Introduction

In recent years, smart mobile devices became widely used because of their portability, interconnectivity and high performance. However, high-power consumption of applications seriously degrades the service quality of smart mobile devices. Green computing (Olsen and Narayanaswarni 2006; Anand et al. 2011; Kyu-Han et al. 2011) extensively investigated the power management of mobile devices for prolonging the battery life and preserving the quality of user experience.

Power management improves energy efficiency by adaptively adjusting the work states of hardware components based on system workload and user behaviour without affecting the quality of service (Han et al. 2015). However, improving energy efficiency of mobile devices using traditional power management technologies is challenging. The system workload is unique with random patterns, and the user behaviour is complex in nature. Most of the existing literature focuses on dynamic tuning the performance of hardware according to the current system workload and the behaviour of the individual end user resulting in a trade-off between the system performance and user experience. However, users often have similar usage patterns of attributes for a longer period. Therefore, we have the hypothesis that the energy consumption of mobile devices is similar for users of the same usage pattern group.

In this paper, we describe a power management framework to collect multi-user usage attributes. We use an unsupervised learning algorithm to split end users into just the group of *active* and *non-active* ones, and generate a power-saving decision if possible. Analysing usage patterns of smartphones and the related power consumption in a specific time interval, we had two findings: First, the device usage patterns of end users of the same group are similar although each user generally has a unique device usage pattern. Second, end users of the same group generally have similar battery-consuming characteristic. For example, most *non-active* users put their mobile device most of the time into *low-power* state, while *active* users run their devices in *high-power* state most of the time. However, sometimes devices of *non-active* users consume much more energy than those of *active* ones. Our aim is to recognize just this kind of users, identify the reasons and help them to improve the energy efficiency of their device.

The proposed power management framework uses multi-user usage patterns and separation of users into groups based on similar behaviours and habits. In addition, we have conducted a simulation study using real device usage patterns of groups of users. Our contributions mainly include three aspects of the following:

(1) Different to approaches from the literature, we use device usage patterns of groups of users. We propose a power management framework for smartphones based on these usage patterns.
(2) The designed and implemented data logger for Android–Smartphones collects device usage and power consumption patterns from users.
(3) The classifier splits users as *non-active* or *active* and their devices as *low-* or *high-power* based on the K-means clustering algorithm. *Non-active* users with *high-power* devices will receive a power-saving recommendation due to their abnormal usage when using the framework.

We organize the rest of the paper as follows: First, we discuss the related work. Next, we propose our power management framework. After that, we present our experimental results. Finally, we draw the conclusion and mention future work.

18.2 Background

Green computing has deeply studied power management methodologies and technologies for smart mobile devices. Most of the research work focuses on improving energy efficiency of mobile devices using dynamic power management (DPM) and dynamic voltage scaling (DVS) (Kahng et al. 2013; Khan and Rinner 2014). DPM policies select power manageable hardware components to switch them to low-power states for saving battery energy based on system workload or user behaviour. DVS techniques tune the CPU working frequency to save system dynamic power with little or negligible impact on user perception and the quality of user experience.

Min et al. (2012) proposed an energy efficient sleep-state selection (E2S3) framework, which places the system into the optimal low-power sleep state to minimize the energy consumption when the IDLE duration is longer than energy break-even time (EBT). An IDLE period longer than EBT gives an opportunity to save power because it can offset the energy cost for state transitions. That is one of the most important rules in DPM policies. Shih and Wang (2012) proposed to adjust the timeout using the average IDLE time to decide whether the system turns into a certain low-power state. Timeout policies have a low overhead and are easy to implement. However, the energy consumption of components staying in high-power state waiting for a timeout to expire waste energy when the IDLE period is longer than the EBT. Khan and Rinner (2014) proposed a DPM approach based on an online machine-learning algorithm. They utilize the RL-based DPM algorithm to learn the workload arrival patterns and estimate nonstationary workloads for power savings. Chen et al. (2012) designed the process-level power-profiling tool called pTopW, taking power-aware decisions based on real time process power information. Jung and Pedram (2010) proposed a power manager, which can dynamically scale the voltage of multi-core processors for the system-wide energy savings based on Bayesian classification algorithm under rapidly and widely varying workloads.

Rahmati et al. (2007) found that the end users often have inadequate knowledge on the power characteristics of mobile devices and their features. This was according to a large-scale international survey and a 1-month field data collection.

The work presented in this paper is inspired by the observations found by Rahmati et al. (2007), Kang et al. (2011), using unsupervised learning algorithms to separate the mobile users and their devices and to improve the energy efficiency for the end users who have inadequate knowledge on the system power characteristics.

18.3 Methodology

The overall framework of presented here consists of three main building blocks (see Fig. 18.1): (1) Data collector acquires every minute the usage pattern of mobile devices, system performance and battery information and records the data in the local file system for the data analyser. (2) Data analyser dividing end users and their devices into groups by using the K-means clustering algorithm. (3) Decision maker identifies abnormal device usage patterns and notifies the user to optimize the system settings.

The aim is to avoid using inefficient applications or update the HW/SW platform. It is important to realize that some applications and hardware platforms are not energy efficient. Generally, software developers are more concerned about performance and functionality but energy efficiency. Forerunner models have less

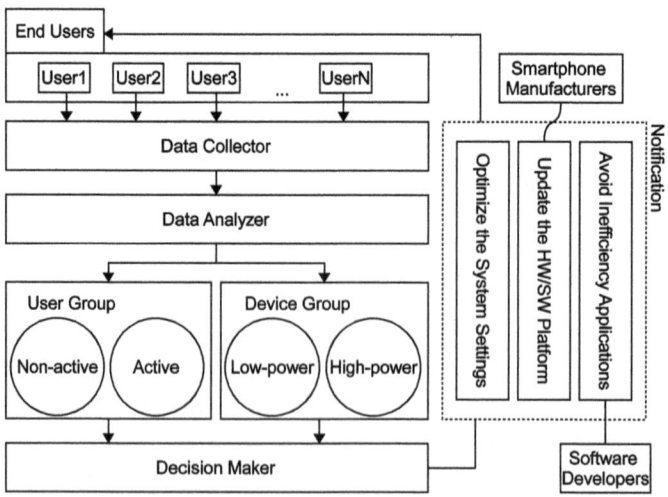

Fig. 18.1 The overall architecture of the framework

power efficiency than recent models. Our framework can help end users to recognize the high-power consumption of the device and of energy inefficient software.

18.3.1 Data Collection and Analysis

First, we look for the usage frequency on any device by the number of screen and foreground application switches. The interaction with mobile devices is stochastic and bursts. An *active* user in a specific period may change to non-active and vice versa. Therefore, we classify end users for predefined time intervals (set to 6 h in our experiments). We measure the battery energy drained executing applications for the duration of user interaction with the device. Hardware components executing instructions consume the system power. Normally, the energy consumption increases with the application run time. On average, the mobile device of the *active* user will consume much more energy than the one of the *non-active* user.

To classify the end users, we collected the following data as (Kang et al. 2011): (1) number of screen switches, (2) duration of Screen On in seconds (s), (3) number of foreground application switches, (4) volume of data transfer in kilobyte (kB) and (5) battery level n as percentage (0–100 %).

Out of the battery level n, one receives the energy consumption p measured in mW as $p = v*c*n/(100*t)$ with v [volt] as average voltage during the time t [s] of the measurement and c [mAh] as the total capacity of the battery.

We developed and deployed the data collector to two Android–smartphones and collected the usage pattern attributes mentioned above periodically. A smartphone has many status information, such as battery-plugged status, sensor status, phone call status, etc. Our application records only the previously defined data into a log file limiting the data logger application's power consumption itself.

Overall, 22 sets of usage patterns were collected. Half of them were collected from a Samsung Galaxy S II and the other half from a Samsung Galaxy Note III. We recorded each set of data for over 24 h.

We use the K-means clustering algorithm to classify end users according to the device usage pattern. The K-means clustering algorithm is an iterative, data-partitioning algorithm that assigns n observations to exactly one of K clusters defined by centroids and K chosen before the algorithm starts. In our experiments, we divide mobile users into two classes: *Active* users and *non-active* users. We classified two types of devices: *low-power* consumption and *high-power* consumption.

The obtained 22 samples of device usage patterns had three features each, i.e. the number of screen switches, the number of foreground application switches and the periods of screen on. We input these features into the K-means clustering algorithm to partition two kinds of mobile users: *active* and *non-active*. *Non-active* users tend to use their devices less than active users; their devices are into IDLE or SLEEP state most of the time.

Table 18.1 Attributes of two user groups

Attributes	Non-active users				Active users				
	User1	User5	User6	User11	User2	User7	User8	User9	User10
Data transfer volume (KB)	823	212	406	1942	6299	16404	7212	7211	4558
Period of screen on (s)	51	11	3	400	851	3227	1700	2183	901
Energy consumption (J)	7741	6877	6616	7043	9694	10381	8153	9111	5930

Table 18.1 shows the Data Transfer Volume (DTV) and the Period of Screen On (PSO) of different users. In general, *active* users have a much higher DTV than *non-active* users. In Table 18.1, User1, User5, User6 and User11 are *non-active* users, whose average data transfer volume is 846 kB. Whereas User2, User7, User8, User9 and User10 are *active* users with an average amount of data transfer of 8337 kB. This means that the network interface (Wi-Fi and Cellular) of active users consumes much more energy to transfer data. A detailed explanation of the power consumption of the network interface of mobile devices is available in Rodriguez Castillo et al. (2013), Bo et al. (2013).

From Table 18.1 we see *active* users' PSO is much higher than for *non-active* ones. Hardware components working at different states consume battery energy differently due to the fact that one operating mode corresponds to one work performance with exactly one power state. For example, mobile users find 5 working frequencies of a brand smartphone's processor are 300, 600, 800, 1000 and 1200 MHz from file /sys/devices/system/cpu/cpu0/cpufreq, and their power values are 203.5, 547.6, 921.3, 1509.6 and 2134.9 mW, respectively. Especially, when the processor is at IDLE state, its power consumption is only 7.4 mW (the data is of the power profile provided by the Android application framework on a target smartphone). Users interact with their devices according to different applications with different work states of hardware components. Extensive user interaction will drive components to high performance with high-power consumption. Vice versa, a small amount of interaction will allow most of the components to keep their low-power state, such as IDLE or SLEEP. As a result, one conserves battery energy by switching components into their low-power state and in the end to IDLE/SLEEP interval.

Table 18.1 shows that the energy consumption varies with the interaction frequency. The trend is similar with those of DTV and PSO.

From our experiments, we draw the conclusion that in most cases where users' activity corresponds to exactly working state of hardware components, the average energy consumption of the mobile device depends on the active usage time. In other words, *active* users will normally consume much more battery energy than *non-active* users in any specific interval. Therefore, device usage attributes may influence the power consumption characteristics of the devices. The *active* user group drives their devices at *active* state in most cases with *high-power* consumption. On the other hand, *non-active* users put their devices into IDLE/SLEEP state most of the time, which will result in less battery drain.

18.3.2 Decision Maker

However, two questions remain: (1) How much energy consumes a smartphone with *high-power* consumption. (2) Why use devices of *non-active* users more energy than expected? We used the K-means clustering algorithm to separate *low-power* devices and *high-power* devices to answer the first question. To answer the second one, we found mainly two reasons: abnormal user behaviour and energy inefficiency of old mobile devices. The abnormal behaviour mainly is due to inappropriate system settings, concerning brightness of background light, Bluetooth, Wi-Fi, GPS, etc. Beside this, background services as instant message applications, background compute-intensive tasks and energy inefficient applications cause energy drain. Old mobile devices have often installed complex software and batteries with ageing problems.

Based on the type of user and power consumption, our framework can make a decision to notify the *non-active* user with *high-power* consumption on the device. *Non-active* users receive the recommendation to shut down any unused network interface, tune the brightness of screen to an acceptable level and stop background services when the device stays in the sleep state. If an old device is the reason, we suggest changing it to an energy efficient device or installing a new battery.

With any new user, we gather both the device usage attributes and the energy consumption at a specific period, estimating the type of user and device for a respective notification. In case a user is an *active* one with a device consuming much less energy, we will recognize the respective device settings recommending them to others in the future.

18.4 Results

We evaluated our framework with 22 users and 2 mobile devices, i.e. Galaxy S II and Galaxy Note III. We deployed the data collection logger to the smartphones and gathered the power consumption and usage patterns of the devices periodically. We analysed the data as described above.

Figure 18.2 shows the classification of the users and their devices' power consumption in a 6 h interval. Twelve users have the Galaxy S II and ten the Galaxy Note III. Half of each group are *non-active* users with *lower power* consumption. Galaxy S II users (User2, User7, User8, User9 and User10) and Galaxy Note III users (User12, User14, User17, User20 and User21) are identified as *active* users.

Figure 18.3 shows the power consumption classification of mobile devices of all users. User13, User16, User17, User18, User19 and User22 belong to the *low-*

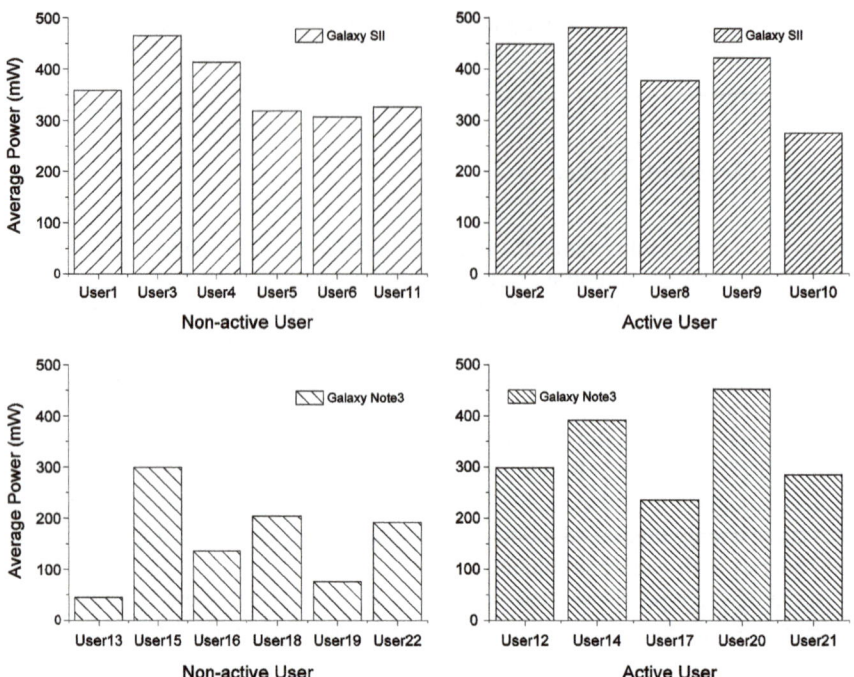

Fig. 18.2 Power consumption of *active* and *non-active* users

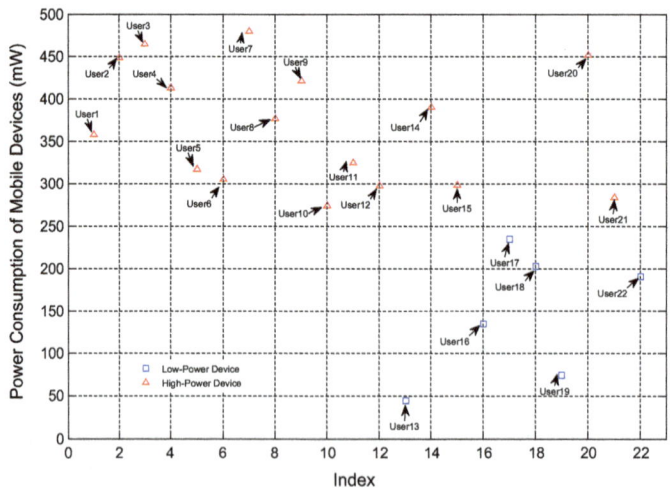

Fig. 18.3 Users classified by power consumption of their devices

Table 18.2 Impact of background service on power consumption in mW

Smartphones	Use Case1	User Case2	User Case3	Use Case4
Galaxy SII	307	375	461	483
Galaxy Note III	64	116	127	186

power consumption group while the rest of the users belong to the *high-power* consumption group.

However, there are two notable exceptions in Fig. 18.2. There is the *active* User10 with less energy consumption than all *non-active* users with the same device. And the *non-active* User15 has higher energy consumption than *active* User17 and User21 on the same device. There are two main causes for this: an objective and a subjective one. The objective cause is the energy efficiency of the mobile device itself determined by the smartphone manufacturer. The subjective cause is the user behaviour. Users with inadequate knowledge on the system power characteristics usually have abnormal usage behaviour, such as adopting high-power system settings and using energy inefficiency software. Therefore, User15 will receive a notification about the abnormal usage pattern from our framework whereas we use the settings of User10 as recommendation for others. The Galaxy S II users are all in the *high-power* group although some of them are *non-active* users. The Galaxy S II is in general less energy efficient than the Galaxy Note III. Our framework will thus send a notification to these users, too.

To validate our hypothesis on background services and old mobile devices as main factors for higher energy consumption, we compare the Samsung Galaxy S II and Galaxy Note III. The Samsung Galaxy S II is equipped with an Exynos 4210 dual-core processor, an 802.11 a/b/g/n Wi-Fi network card, OLED Plus capacitive touch screen, 1650 mAH Li-ion battery and Android 2.3 Operating System. The Galaxy Note III has a quad-core 1.4 GHz processor, an 802.11 a/b/g/n Wi-Fi network card, an AMOLED display, a 3200 mAH Li-ion battery and an Android 4.3 Operating System.

We gathered the power consumption of the two devices using battery indicator under four different use cases. Use Case1: All user-level background services shut down with buffer cleaned. Use Case2: As Use Case1 but running a test application switching the screen on for 30 s every 5 min. Use Case3 and Use Case4 install an internet instant message and an email client application under the previous use cases, respectively. We keep the same system settings without any user interaction in the experiments, such as network interface, screen, etc.

Table 18.2 shows the results: The systems with background service running consumed much more energy than without service execution. For example, on Galaxy SII, the power consumption of the overall system in Use Case4 is 483 mW, which means 5 % more than in Use Case3 but even 57 % more than in Use Case1 on the same device, but even 160 % more than in Use Case4 on the more recent Galaxy Note III. Thus, the type of smartphone has a great influence on the power consumption. The Samsung Galaxy Note III with Android 4.3 first came out in 2013; the Galaxy SII with Android 2.3 in 2011. Additionally, the larger battery capacity of the Galaxy Note III supports longer battery life. In our experiments the Galaxy Note III saved in Use Case1, 243 mW or nearly 80 % power compared to the Galaxy SII.

18.5 Conclusion

For this paper, we developed a power management framework for sensor data and data analytics purposes. We found reasons for *high-power* consumption of *non-active* users. We developed a notification service to motivate users to shut down background services or change old devices to new ones when appropriate. Our framework defines attributes under three aspects: (1) Multi-user usage and power consumption patterns gathered by a data collector. (2) Classification of users and their devices by an unsupervised learning algorithm. (3) Notification of non-active users with high-power characteristics. With our approach, one has less energy consumption, gains a longer battery life of devices and receives a better user experience. The results help smartphone manufactures as software developers to improve the platform and system settings as the efficiency of the applications. The power consumption of sensor nodes as of wearable computing devices is future research we plan based on the presented results here.

Acknowledgements The EU Erasmus Mundus project FUSION—featured Europe and south Asia mobility network (2013-2541/001-011) supported this research.

References

Anand B, Thirugnanam K, Sebastian J, Kannan PG, Ananda AL, Chan MC, Balan RK (2011) Adaptive display power management for mobile games. In: Proceedings of the 9th international conference on mobile systems, applications, and services

Bo Z, Qiang Z, Guohong C, Addepalli S (2013) Energy-aware web browsing in 3G based smartphones. In: IEEE 33rd international conference on distributed computing systems

Chen H, Li Y, Shi W (2012) Fine-grained power management using process-level profiling. Sustain Comput Inf Syst 2(1):33–42

Han S, Park M, Piao X, Park M (2015) A dual speed scheme for dynamic voltage scaling on real-time multiprocessor systems. J Supercomputing 71(2):574–590

Jung H, Pedram M (2010) Supervised learning based power management for multicore processors. IEEE Trans Comput Aided Des Integr Circuits Syst 29(9):1395–1408

Kahng AB, Kang S, Kumar R, Sartori J (2013) Enhancing the efficiency of energy-constrained DVFS designs. IEEE Trans Very Large Scale Integr Syst 21(10):1769–1782

Kang J-M, Seo S-S, Hong J-K (2011) Usage pattern analysis of smartphones. In: IEEE 13th Asia-Pacific network operations and management symposium

Khan UA, Rinner B (2014) Online learning of timeout policies for dynamic power management. ACM Trans Embedded Comput Syst 13(4):96

Kyu-Han K, Min AW, Gupta D, Mohapatra P, Singh JP (2011) Improving energy efficiency of Wi-Fi sensing on smartphones. In: INFOCOM

Min AW, Wang R, Tsai J, Ergin MA, Tai T-YC (2012) Improving energy efficiency for mobile platforms by exploiting low-power sleep states. In: Proceedings of the 9th conference on computing frontiers

Olsen CM, Narayanaswarni C (2006) PowerNap: an efficient power management scheme for mobile devices. IEEE Trans Mob Comput 5(7):816–828

Rahmati A, Qian A, Zhong L (2007) Understanding human-battery interaction on mobile phones. In: Proceedings of the 9th international conference on human computer interaction with mobile devices and services

Rodriguez Castillo JM, Lundqvist H, Qvarfordt C (2013) Energy consumption impact from Wi-Fi traffic offload. In: Proceedings of the 10th international symposium on wireless communication systems

Shih H-C, Wang K (2012) An adaptive hybrid dynamic power management algorithm for mobile devices. Comput Netw 56(2):548–565

Chapter 19
Integration of Wireless Sensor Networks into Industrial Control Systems

T. Raza, W. Lang and R. Jedermann

Abstract In this paper, a prototype is developed, which can easily integrate a wireless sensor with programmable logic controllers (PLC) using Profinet. The low-cost embedded system Raspberry Pi was used to connect a wireless gateway with 'ProfiNet'. A Wizziboard wireless sensor node, running the Dash7 communication protocol, periodically sends temperature and humidity data. This data is received at the UART of Raspberry Pi. A modified Snap7 library is used for communication between Raspberry Pi and SIMATIC Manager. The data is transferred using the Ethernet bus. The prototype is tested on SIMATIC Manager with a PLCSIM simulator installed on a PC. The Siemens HMI software Wincc flexible is used to monitor the state of PLC systems. Here, Wincc flexible is connected to the PLC over the TCP/IP protocol. SIMATIC Manager and Wincc flexible have been tested on a windows PC. Data logging is also implemented in the HMI software.

Keywords Wireless sensor · Profinet · Raspberry Pi gateway

19.1 Introduction

Logistic processes, e.g., production plants and warehouse/storage systems, need sensor information to improve their performance. Modern control systems have to react to dynamic changes of process parameters, e.g., different types or qualities of materials. Real-time feedback enables a fast reaction to faults and an optimization of control parameters.

T. Raza (✉) · W. Lang · R. Jedermann
Institute for Microsensors, -Actuators and –Systems (IMSAS),
University of Bremen, Bremen, Germany
e-mail: toqeer.iiee@gmail.com

W. Lang
e-mail: wlang@imsas.uni-bremen.de

R. Jedermann
e-mail: rjedermann@imsas.uni-bremen.de

Control systems are mostly implemented by programmable logic controllers (PLC). Commercial PLCs, such as STEP 7, provide various solutions for wired sensors by different bus protocols.

Industrial automation has many communication standards, which include Profibus, EtherCat, and Profinet. Profinet is the latest innovative protocol for the industrial Ethernet. Industrial automation is for production or electrical drives. Profinet is a much more efficient protocol used nowadays. In a harsh industrial environment, it is necessary that communication should be reliable and in real time. Due to the increase of the number of sensors and field devices, nowadays the cost for wiring is also increasing. There are also some industrial places where wireless sensors are rapidly being used. Wireless sensors or a wireless sensor network (WSN) can be placed far from the place where the actual phenomenon is occurring. In this type of sensor, the network position of the sensor and the communication topology are carefully implemented.

Wireless sensor networks are used to measure a variety of ambient conditions, such as temperature, humidity, pressure, noise level, and lightning. Nowadays, wireless sensors are widely used in industrial environments. They can be used for the control of robots in a manufacturing system. There are several other places where wireless sensor networks can be efficiently implemented, such as the process control of the automation industry and the instrumentation of any industry. There are several factors, which can be affected by the design of a wireless sensor network. These factors may include the level of fault tolerance, what the minimum cost will be, what the network topology involved will be and, last but not the least, the transmission media, which transmit data. Generally, WSN consists of a sensor node and the sensor node gateway. This gateway can be used to store data using different protocols (Cayirci et al. 2001).

Wireless sensor networks provide a promising solution for an increasing number of logistical systems in which control cannot be implemented by wired sensors alone. Although they have been in the focus of research for more than 10 years now, they have found little attention in industrial control. Also, hardly any researcher of wireless sensors knows about industrial control requirements and protocols.

Wireless sensors can be easily implemented by embedding them into a product itself. This can improve the accuracy of measurement. There are different places where wireless sensors can play an important role instead of wired sensors, for example, warehouses for the storage of vegetables and fruits. These types of refrigerated warehouses need a low temperature and ventilation to keep the vegetables and fruits fresh. Proper ventilation is done using ventilation units.

At the moment, there is no proper sensor information, which can give the actual airflow rate inside the boxes. Due to this, we cannot control the speed of the ventilator fans. They always run at the full speed, and a large amount of energy is wasted. To overcome this energy wastage, proper feedback control can be implemented using a wireless sensor. When we have proper information about the airflow, we can control the speed of the ventilator fans.

Wireless sensors can help to measure the temperature and other parameters at multiple locations as compared to wired sensors. In contrast to wired sensors, which are fixed and have a limited access, wireless sensors can be implemented in the boxes, which are temporarily placed in the warehouses.

Our paper shows how the number of applied sensors in industrial control can be increased by providing adequate protocol translators and interfaces to wireless technologies. A gateway for wireless sensor networks can be connected to industrial PLC systems. This prototype is currently compatible with Siemens PLC.

The paper is organized in the following way. The second section will explain about the basic block diagram and structure of the implementation. The third section will explain about the Raspberry Pi, which is used as a converter in this project. The fourth section will explain about the sensor module. The fifth section will explain the software implementation. The sixth section will be about the limitation of wireless sensors, and the last section will be the summary and conclusion.

19.2 Overall Structure

Due to the complexity of the Profinet protocol, it was not feasible to directly connect the wireless sensors with a PLC. A separate protocol translator was used instead, which also provided more flexibility. Different types of wireless sensors can be connected with only minor software changes. The complete chain of required hardware interfaces, units, and protocols is displayed in Fig. 19.1.

The temperature and humidity sensors are integrated on the WizziKit node. Wizzikit works on the Dash7 protocol. Dash7 stands for "Developer Alliance for standard harmonization of ISO 18000-7" and is used for low power consumption wireless devices. The Dash7 protocol uses a frequency range 433–434.97 MHz in Europe. The data wirelessly transmitted through the WizziMote is received at the WizziKit base, which can be called the Dash7 gateway because it works on the Dash7 protocol. Details about sensors and WizziKit will be discussed in the sensor module section of the paper.

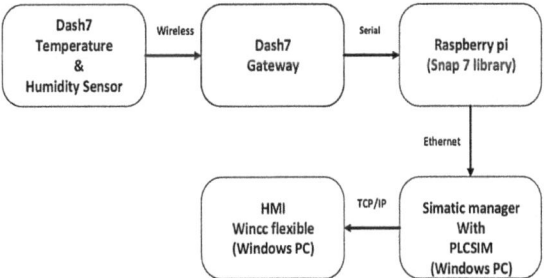

Fig. 19.1 Block diagram of the overall structure

The base station or gateway has a serial output. The serial output of the gateway is connected with the UART of the Raspberry Pi. The Raspberry Pi, which is used here, can be called a gateway, which translates one protocol to another protocol. The input protocol of the Raspberry Pi is a serial protocol, which can be seen from the block diagram, while the output protocol of the Raspberry Pi is Ethernet. This Ethernet is used as a Profinet to communicate with the Siemens PLC simulator.

The data, which is received through the serial port of the Raspberry Pi, is then transmitted to the PLC simulator using Ethernet. The communication between Raspberry Pi and the PLC software is done using an open source library named Snap7 library.

This data is then displayed on the HMI (Human Machine Interface) using Winccflexible. Winccflexible is the HMI software for Siemens PLC. This software is used to program the HMI panels. In this project, there is no real HMI panel used, so this software is used here to display the sensor values. The communication between the PLC simulator and Winccflexible is also done using a TCP/IP protocol.

19.3 Raspberry Pi as an Industrial Gateway

Raspberry Pi is a credit card size computer with lots of built-in functionalities. It is fully programmable and very flexible. There are different versions of Raspberry Pi available on the market. The Raspberry Pi, which is used here, belongs to the category of the 'B' model and has different properties. The processor of Raspberry Pi consists of 32-bit ARM. This CPU is based on the ARM11 architecture with 512 MB of RAM (Maksimovic and Vujovic 2014) for the model 'B' series.

One RJ45 connector is available, which is used to connect the Raspberry Pi with the Internet using an Ethernet cable; it can also be used for communication with other devices (Norair 2009).

19.3.1 Communication Protocol

Raspberry Pi has different communication protocols available, such as UART, I^2C and SPI. Every protocol has different aspects according to the requirements. UART was used to receive the data from wireless sensors. The detailed description about UART is given below.

19.3.2 UART

UART is also called universal an asynchronous receiver/transmitter for serial communication. Serial communication is a low level method of transferring data between Raspberry Pi and a computer. There are two pins, which are used for the UART communication; these pins are called TXD and RXD. The TXD pin is used for data transmission while the RXD pin is used for receiving data.

To connect the Raspberry Pi with other devices, it should be noticed that the TXD pin of the Raspberry Pi should be connected with the RXD pin of the other device, and, similarly, the RXD pin of the Raspberry Pi should be connected with the TXD pin. Raspberry Pi also has a voltage level difference, which makes it impossible to connect with RS232. Raspberry Pi has a voltage level of 0–3.3 V; it is necessary to use a voltage level converter to connect it with RS232 (Embedded Linux Forum 2015).

19.4 Sensor Module

A wireless sensor usually consists of a wireless sensor node and a base or gateway. A sensor node basically consists of four components that may include a processing unit, a sensing unit, a transceiver unit and a power unit. There are different applications for which additional components are included in the sensor node, such as a location finding system and power generator.

In the sensor node there is as processing unit, which is usually used to process data for communication with other nodes. The sensor node has a very important unit called a power unit. The power unit is used to provide the required power for the sensor node. A transceiver unit is used to connect with the network (WizziLab 2015).

The WizziKit, which is used, usually consists of two parts; these include

- WizziMote
- WizziBase

19.4.1 WizziMote

WizziMote has a built in microcontroller of Texas instrument TI CC430F5137. There is also an antenna fixed on WizziMote for the 433 MHz frequency (WizziLab 2015).

19.4.2 WizziBase

WizziBase is used to give a base board for WizziMote; there are different methods to connect WizziBase with other devices. JTAG and a serial connector are available on the WizziBase. The WizziBase provides power to WizziMote with the help of JTAG, FTDI or an external power plug (WizziLab 2015).

19.4.3 Sensors

There are three sensors mounted on the WizziMote node. The following three data types were received using Raspberry Pi.

- Tmp 100 temperature sensor
- SHT25 temperature sensor
- SHT25 humidity sensor

19.5 Software Design

The software design has three parts, which are explained in the following three sections (Fig. 19.2).

19.5.1 Raspberry Pi

The software module used for receiving data from the UART of the Raspberry Pi and transmitting it through the Ethernet port of Raspberry Pi can be explained by a flowchart. An open source library named as the Snap7 library was used to communicate Raspberry Pi with the Siemens PLC simulator using Profinet.

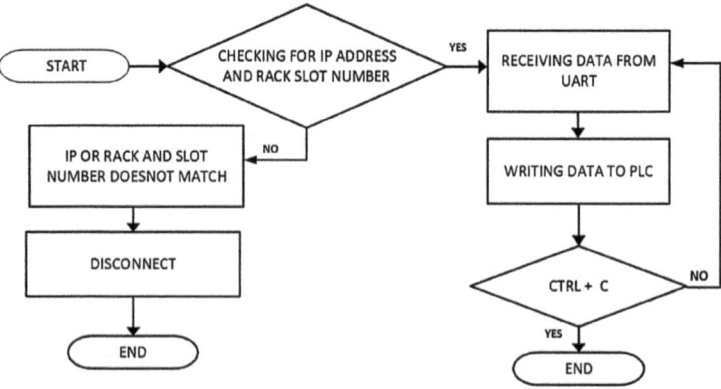

Fig. 19.2 Flowchart describing the software module

The software module first checks the IP address and the rack and slot of the PLC simulator. If it does not match the required IP address or the rack and slot of the PLC it will terminate the program. If the requirement is fulfilled it will start receiving data from the Raspberry Pi and transmitting it to PLC using Profinet. ctrl+C is used to quit the main program. This program continuously reads data from UART, so this command is used to quit the program at any time.

19.5.2 PLC (Programmable Logic Controller)

The Siemens software named SIMATIC Manager was used to program the PLC and PLCSIM simulator used instead of a real PLC. The data, which was transmitted through the Ethernet port of the Raspberry Pi, was received in the shared data block.

There are two different types of data blocks in the PLC. One data block is called the instant data block while the other data block is called the shared data block. Data blocks are the memory area, which can be easily defined for different types of memory, such as bits, bytes, words and even our data types. In SIMATIC Manager, OB1, also called the organization block, is the main block where all the function blocks and functions are. It is similar to the main function in other programming languages.

The instant data block is the data block, which is associated with the function block. The instant data block must be created before using the function block to make possible a call for the function block in the OB1. In this data block, different values are stored, which are used by the function block.

The second data block is called the shared data block. If the system does not have enough bits to store the data, then we can use the shared data block to store the data. The shared data block has different aspects, as compared to the instant data block. The instant data block can share its data only with the function block for which it is created while, on the other hand, the shared data block data can be shared anywhere in the system (Siemens AG 2006).

The shared data block has an ability to share data. In this project, the shared data block was used to receive data from the Raspberry Pi. The reason behind using the data block in this project was that the data, which was transmitted by Raspberry Pi, was in a byte format. The data blocks can have any data type depending on the requirement of the user.

19.5.3 HMI (Human Machine Interface)

The software design section's last part consists of HMI software programming named Winccflexible. Although this software is used to program the HMI panel, in this project it was used to display the sensor values (Fig. 19.3).

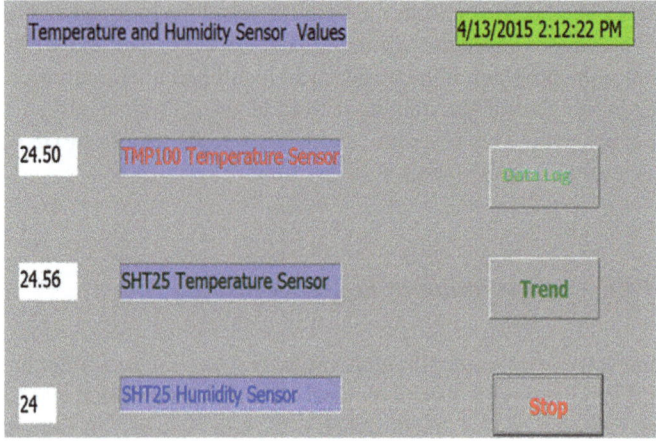

Fig. 19.3 HMI for sensor values

Basically, HMI has three parts, which can be classified as follows.

- Temperature and Humidity sensor values
- Data logging
- Sensor values trends

19.5.4 Temperature and Humidity Sensor Values

In this part of HMI, we can see that there are three boxes, which have some values. These three boxes correspond to each sensor's real-time value. These real-time values are calculated in the SIMATIC Manager and then transferred into Winccflexible.

19.5.5 Data Logging

Data logging was also done using this HMI. In this HMI, there is a special feature, which is defined by the box on the right-hand side. This box is named "Data Log." Due to this box, data logging can be started and controlled. Data logging was done in the Excel sheet format. All three sensors' values are connected with this single button. When we press this button, data logging will start automatically, and, when we press this button, again data logging will stop again. When the data logging starts, it will generate three Excel files at the back end as defined by the path of those files.

19.5.6 Sensor Value Trends

The third part of the HMI describes the trends among all three sensor values. There is a button on the right hand side of the HMI named trend. When we press this button, it will start generating the trend between the three sensor values. The trends give a better overview of any process. The trends, which are generated, are displayed as a bar graph.

19.6 Limitations of Wireless Control Applications

Industrial control systems have the highest requirements in terms of the reliability of the system components, whereas as wireless system always has to take communication failures into account. So, wireless communication should be made as reliable as possible, e.g., by using safe protocols with acknowledgements and automated retransmission of lost packets.

Depending on the application, process control consists of tasks with different reliability requirements. Whereas wireless sensors might not be suitable for some critical security tasks, they can still be applied to other tasks, such as the optimization of process parameters. In case of lost sensor data, optimization can be skipped for one circle, but process control can continue without interruption, only with slightly less optimized parameters.

Wireless reading of product item parameters is another safe application in industrial control systems. The transmission of sensory information, such as recorded statistics of use and aging effects is not time critical. If necessary, the start of processing can be delayed until all required data have been transmitted.

19.7 Summary and Conclusions

In this paper, an idea of integrating a wireless sensor into industrial control systems was successfully implemented. The idea was not only successfully implemented, but an economical solution for the industrial gateway was also provided in the shape of Raspberry Pi. The idea was to integrate the wireless sensor using Profinet and to give the wireless sensor a serial base station. So it was necessary to convert the UART protocol into the Profinet protocol. Industrial gateways, which translate one protocol into another industrial protocol, are very expensive as compared to the solution provide here, which is only a $35 Raspberry Pi. This solution is also very flexible. In future, it can also be used for wireless sensors having a base station with different protocols such as I2C or SPI. The software is designed in such manner that this can be done easily with simple modification.

During the implementation, certain considerations were taken such that there was no real PLC available, so it would behave as if it were implemented in a real control system. For this reason, it was tested on a simulator. Open source software named Snap7 was used to make this idea into reality. This library was designed and implemented for real Siemens systems. So this was implemented the first time for a simulator system.

HMI was designed not only for a better understanding of the real time values but also to keep the values saved for further processing. For this implementation, data logging was performed for the all three sensor values. The real time values were displayed in HMI, and trends were also shown using HMI.

In future, there are some additional features, which can be added. This idea can be used currently for Siemens PLC. But, in future, this can be enhanced and made vendor independent. On Raspberry Pi, we can make further enhancement such that Raspberry Pi can be used as a plug and play device. There is also a possibility to make a GUI on Raspberry Pi to make adjustments in the program as per requirements from the GUI itself.

Acknowledgments The project COOL is supported by the German Federal Ministry for Economic Affairs and Energy on the basis of a decision by the German Bundestag.

References

Cayirci E, Akyildiz IF et al (2001) Wireless sensor networks: a survey. Comput Netw Int J Comput Telecommun Networking. doi:10.1016/s1389-1286(01)00302-4
Embedded Linux forum (2015) RPi low-level peripherals. http://elinux.org/RPi_Low-level_peripherals. Accessed 9 Dec 2015
Maksimovic M, Vujovic V (2014) Raspberry Pi as a wireless sensor node: performances and constraints. In: 37th international convention on information and communication technology, electronics and microelectronics (MIPRO). doi:10.1109/MIPRO.2014.6859717
Norair JP (2009) Introduction to Dash7 technologies. RFID J
Siemens AG (2006) Automation and drives manual for STEP 7. http://www.fer.unizg.hr/_download/repository/STEP7.pdf. Accessed 9 Dec 2015
WizziLab (2015) WizziMote product page. http://www.wizzimote.com/. Accessed 9 Dec 2015

Chapter 20
Advantages of Sub-GHz Communication in Food Logistics and DASH7 Implementation

Chanaka Lloyd, Sang-Hwa Chung, Walter Lang and Reiner Jedermann

Abstract The benefits of remote monitoring of perishable food items in transport and storage of a cool chain have been established and much talked about in the past decade. In order to convey the measured parametric data over for processing, wireless sensor networks are used, mostly in 2.4 GHz communication range, but it suffers from high signal attenuation in environments with high water content. This paper analyzes the possibility of wireless communication in 433 MHz and its lower sensitivity to water containing environments by means of a case study in an apple storage warehouse. The experiment shows near conformity to a previously implemented theoretical model of signal attenuation, the only error being due to ca. 10 % more water in apples compared to bananas. Second, the papers focuses on practical experiments with the focus on the DASH7 protocol, which is dedicated to sub-GHz communication, the OSS-7 software stack, and a multi-sensor, small-footprint hardware platform.

Keywords DASH7 · 433 MHz · Signal attenuation · RFID · Apples

---

C. Lloyd (✉) · W. Lang · R. Jedermann
Institute for Microsensors, -Actuators and -Systems (IMSAS), University of Bremen, Bremen, Germany
e-mail: clloyd@imsas.uni-bremen.de

W. Lang
e-mail: wlang@imsas.uni-bremen.de

R. Jedermann
e-mail: rjedermann@imsas.uni-bremen.de

C. Lloyd
International Graduate School for Dynamics in Logistics (IGS) LogDynamics,
Bremen Institute of Production and Logistics (BIBA), Bremen, Germany

S.-H. Chung
Department of Computer Engineering,
Pusan National University (PNU), Busan, Republic of Korea
e-mail: shchung@pusan.ac.kr

20.1 Introduction

Agricultural products are accounted to as much as 25 % in EU transports (Verdouw et al. 2013). That alone illustrates the significance of investing on food waste reduction during transportation. The latter is substantiated by the food waste statistics by volume and monetary value presented in Gustavsson et al. (2011) and the alarming loss of 30–50 % (i.e., ca. 1–1.2 Billion tonnes) of food in the planet as indicated by Institute of Mechanical Engineers (2013). The common culprits for the degradation and ultimate destruction of perishable goods in transport are unsuitable temperature and humidity, atmospheric composition, lack of ventilation, nonoptimal packaging methods and mechanical injury during loading (Vigneault et al. 2009). Among the latter are certain aspects and conditions of transportation that can be monitored and controlled remotely, e.g., temperature, humidity and ventilation systems. A lot of research has been conducted on such remote monitoring schemes where the underlying software and hardware infrastructure is already in place, as indicated by Lang et al. (2011), Nascimento Nunes et al. (2014), Wang et al. (2011).

In order to convey the sensor data that is measured within the food items, wireless technology is the best suited for the job. In the recent past, some tests such as measurement of temperature, humidity, and air flow speed were done using 2.4 GHz communication (Jedermann et al. 2011). However, in order to overcome the challenges presented by high water content goods, such as fruits—that hinder wireless communication in containers and trucks that transport fresh agricultural products—authors in Jedermann et al. (2014) discuss how to circumvent the latter issue by using sub-GHz frequency communication instead.

The high water content of fruits being the major obstacle for wireless communication, and since most wireless sensors operate at 2.4 GHz, most experiments in perishables containing high water content are better served with lower frequency communication. The communication range of tests done in banana containers in The Intelligent Container (Lang et al. 2011) drops to 0.5 m inside densely packed bananas, and only 52 % of the data packets were received at this distance in a container filled with bananas. One-third of all links failed completely, resulting in 2 out of 20 sensors that could not be contacted all. According to the theoretical model presented in Jedermann et al. (2014), the signal attenuation by water should be much lower at frequencies below 1 GHz.

A new air management system could control the fan speed using multi-point air flow measurement between and inside the boxes with apples concerning the field test in this paper. Under practical conditions, the installation of such a multi-point measurement system is only feasible with wireless communication. Therefore, an experimental verification of this prediction with regard to the communication frequency is provided using the WizziMotes (WizziLab 2013) hardware operating at 433 MHz. With regard to the multi-point measurement case, Wireless Anemometers (WAMs) were used utilizing a distributed network protocol CITE. The results of the latter are impertinent here and not presented in this paper.

The field tests were carried out in an apple warehouse (Fig. 20.1a). However, the results are valid for a truck or a container in transportation as well. The main aim of this paper being the validation of the attenuation model and presentation of a new ubiquitous RFID tag that is usable in multiple test scenarios, the location where it is used has less importance. The main reason for that is that warehouses, trucks, and containers (used for transporting fruit and vegetable items) used in storage and transportation have very similar settings: metallic walls, floors, doors, and ceilings; cooling units mounted at one side; and, porous pallets used to pack the transported goods.

Hardware and software used for the test and well as the warehouse details are given in Sect. 20.2. The software platform OSS-7 of the DASH7 Alliance (International Organization for Standardization 2009) is used to facilitate 433 MHz communication (see Sect. 20.2.1). In addition, the paper also presents a small form-factor ubiquitous active RFID tag for 433 MHz communication that is proposed for future tests (see Sect. 20.4).

20.2 Field Tests

A field test was carried out on the October 20 and 21, 2015 at the Kompetenzzentrum Obstbau-Bodensee (KOB), Ravensburg, Germany. One experiment was conducted on 433 MHz communication in an apple warehouse. KOB stores the apples in their warehouses up to 6 months until they are ready for distribution to retail stores. Apples are stored at 1 °C and the typical humidity during storage is 93–96 %. The plastic containers that hold ca. 270 kg of apples have the dimensions of 118 cm (L) × 100 cm (W) × 75 cm (H). The evaporator fans used to cool the warehouse require about 40 % of the total energy consumption of the cold storage rooms. A set of wireless sensors (WizziMotes as shown in Fig. 20.2b) was placed in a storage room that contained 163 pallets of apples weighing ca. 40 tonnes.

20.2.1 Sensor Hard- and Software

The WizziMote module (WizziLab 2013) is based on the System-on-Chip (SoC) CC430F5137 processor from Texas Instruments (Texas Instruments Incorporated 2016). The module was equipped with housing and a lithium battery (Fig. 20.2b left). In earlier tests, the radio range of the module in free air was measured to about 60 m. If the antenna is in direct contact with a medium with higher dielectric constant, the antenna would be detuned. In order to avoid this effect, a distance holder with 5.5 cm diameter was mounted on the antenna of 5.5 cm length (Fig. 20.2b right). WizziMote is held in a durable and hard plastic enclosure that does not hinder its wireless communication. The field tests presented in this paper and previous tests attest to its robustness.

We used the OSS-7 open source implementation of the DASH7 Alliance protocol (International Organization for Standardization 2009) in the version from July 2014 as the basis for the communication software. The base-station sends a ping-query every 10 s. All sensor nodes send a response after a short delay depending on their identification number. Collisions are avoided by different delay periods. The radio chip measures the Received Signal Strength Indicator (RSSI). Both RSSI values of the query and the response were recorded and displayed on a laptop. The register value R_{Reg} was converted to the unit dB m by Eq. 20.1. The offset value R_{Off} has a typical value of 74 dB according to the processor's family manual (Texas Instruments Incorporated 2016). The transmission power was set at 10 dB m.

$$R_{dB} = \frac{1}{2}R_{Reg} - R_{Off} \qquad (20.1)$$

20.2.2 Stowage Schema and Sensor Positions

The warehouse was filled with 3 rows (R1 to R3) of 7 stacks (S1 to S7) of apple pallets. Each stack was 8 pallets (Levels of K1 to K8) high. Each box contained about 270 kg of apples. Between the pallets of the stacks, there was a horizontal gap of about 10 cm height. Each pallet had a 10 cm gap on either side of it along the *Rows* direction as indicated in Fig. 20.1b to improve the air flow. But the front and the back side of the pallets along the *Stacks* direction were in touch with each other, except for the pallet fronts of Stack 1 (S1) and backs of Stack 7 (S7). After all boxes were stowed in the room, cooling and ventilation fans were switched on. The control of the cooling unit was set to a target temperature between 0.7 and 1.4 °C.

The communication was tested for a range between 0.25 and 2.75 m in the first 3 of the 7 stacks in the middle row R2. Figure 20.1b highlights the pallets in and on which the WizziMotes are mounted. Figure 20.1c shows the first 3 Stacks of R2 and where the motes are placed on K2. This is a reduced perspective of the direction indicated by the arrow in Fig. 20.1b. Three motes per pallet were placed within the apples in K2 of the stacks (Figs. 20.1c and 20.2a) and approximately 0.25, 0.5, 0.75 m from the front side at half the height of the pallet, i.e., 1.2 m above the floor level. Three additional motes were placed in the horizontal gap between the boxes. The radio base-station was mounted at the front of the first box (Fig. 20.1c). The test was split into two because of the limited number of WizziMotes: During the first run, 102 intervals were recorded for the 1st and the 2nd pallets; during the second run, 402 intervals for the 2nd and the 3rd pallets.

20.2.3 Model for Signal Attenuation

The free-space path loss is proportional to squared distance. In a lossy medium with an imaginary part of dielectric constant greater than zero, the signal is additionally attenuated exponentially to the distance. The constant P_0 includes the transmission

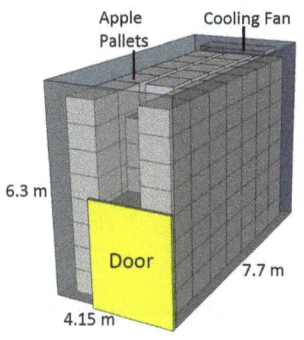
(a) KOB apple warehouse with its dimensions

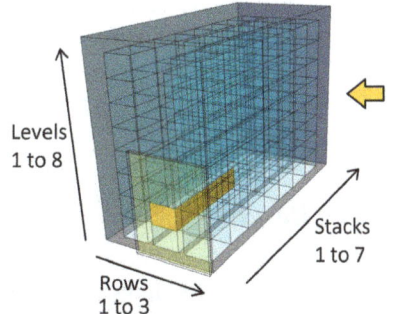
(b) Highlighted pallets where the WizziMotes are placed on or within

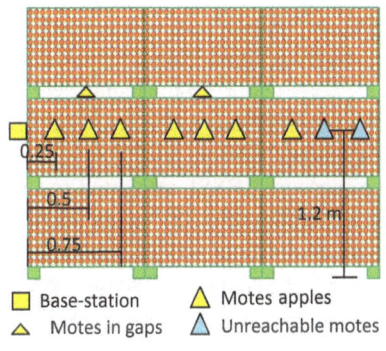
(c) WizziMote and Base-station positions

Fig. 20.1 KOB apple warehouse showing its loading scheme and WizziMote positions

(a) WizziMote being buried in an apple pallet

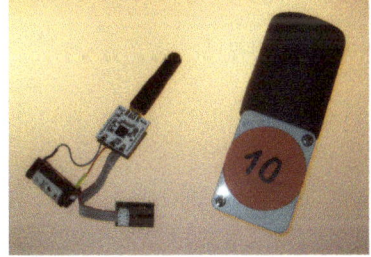
(b) WizziMote and its enclosure with the antenna protector

Fig. 20.2 WizziMote components and WizziMote installation for testing

power and the antenna gain. Conversion to logarithmic dB scale results in Eq. 20.2 with distance d and attenuation per meter a

$$P_{dB} = P_0 - 20\log_{10}(d) + ad \tag{20.2}$$

20.2.4 Results

The quality of the radio links to the sensors was evaluated based on the number of received responses. A response is received only if communication is possible over both the forward and the backward link. Responses were received up to a distance of 2.25 m. At this range the rate of received responses dropped to 81 % of the total 402 frames. For the motes at a distance of 1.75 m and below, all packets were received. At a distance of 2.5 and 2.75 m no communication was possible (i.e., blue color motes in Fig. 20.1c). The used non-rechargeable Lithium batteries turned out to be unsuitable for this temperature range. After about 3 h the battery voltage dropped below 2.5 V, but recovered to 3.3 V at room temperature. The sensor at 1.5 m distance failed during the second test due to low battery voltage.

Figure 20.3 shows the measured RSSI converted to dB m as a function of distance. Data packets were received with RSSI values down to −110.3 dB m, which is slightly better than the typical threshold −104 dB m at 38.4 kBaud data rate according to the data sheet. The RSSI values were stable over time, with fluctuation of ±0.5 dB maximum. The links showed good symmetry; the root mean square (RMS) difference between the query and the response RSSI was 0.78 dB, except one sensor with query 2.5 dB always better than response.

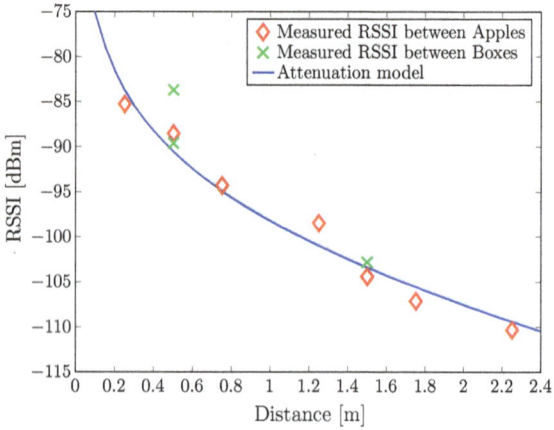

Fig. 20.3 Measured signal strength and attenuation model as a function of distance

Table 20.1 Signal attenuation of commonly used frequencies

Frequency (MHz)	433	868	915	2400
Attenuation (dB/m)	−2.2	−8.4	−9.6	−61.6

Fitting the measured RSSI values with the model of Eq. 20.2 resulted in an attenuation of 3.3 dB/m and an offset of $P_0 = -94.9$ dB m. The root mean square error of the fit was 1.59 dB.

For an overview of the attenuation values as per Jedermann et al. (2014), we present Table 20.1. The values are for bananas. Therefore, for apples, all values are a bit higher as can be seen from the attenuation in apples for 433 MHz.

The 3 sensors in horizontal gaps (the green cross markers in Fig. 20.3) showed only slightly different values. The gap height was only 14 % of the wavelength of 69.14 cm. Therefore, the gaps had no significant contribution to signal propagation. The experimentally estimated attenuation of −3.3 dB/m fitted very will the prediction of $a = -2.2$ dB/m at 433 MHz of the theoretical model for bananas in Jedermann et al. (2014). The difference can be explained by higher water content of apples (ca. 84 %) in boxes compared to bananas (ca. 74 %).

Motes shown in *blue* color in Fig. 20.1c were not reachable and never recorded any data.

20.3 433 MHz Band and Open Source Stack for DASH7 Alliance Protocol (OSS-7)

The OSS-7 (International Organization for Standardization 2009) is a software stack implementation of an active RFID standard for communication in 433 MHz in the DASH7 Alliance (D7A 2016). It is an implementation of the ISO/IEC18000-7 that defines the air interface for 433 MHz communication. In contrast to typical WSN standards such as ZigBee (built on top of IEEE 802.15.4), the DASH7 specification defines a full functional active RFID tag (Weyn et al. 2013).

ISM (Industrial, Scientific and Medical) band includes 433 MHz and is available worldwide without requiring a license for operation except a few restrictions based on the locality (Tuset-Peir et al. 2014). 433 MHz band spans from 433.05 to 434.79 MHz. Attenuation models such as indicated in Jedermann et al. (2014) and the empirical studies as described in Tuset-Peiró et al. (2014) suggests that 433 MHz is much better suited for long range than 2.4 GHz. In addition, path propagation studies done by Isnin (2016) indicates that the higher frequencies perform better in LoS (Line of Sight) scenarios but underperforms where the signal has to propagate through multiple obstacles. Therefore, it justifies the use and experimentation of using 433 MHz in perishables in transport or storage, where LoS is almost nonexistent.

DASH7 utilizes the so-called *BLAST* technology (Weyn et al. 2013), where *B* stands for *Bursty* meaning abrupt data transfer unlike audio/video data; *L* stands for *Light* which specifies the 256-byte packet limit; *A* stands for *Asynchronous* which indicates that communication takes place on a command-response basis, except periodic synchronization; *S* represents *Stealth* meaning that the communication nodes only talk to each other in trusted environments without broadcasting their addresses periodically; and, *T* stands for *Transitive* which refers to the mobile transitional behavior.

There are four devices classes defined in DASH7: Blinker, Endpoint, Subcontroller, and Gateway. The differentiation of these devices depends on transmission/reception rights/ability, feature set, wake-on scan cycle, and always on receiver status. The communication of all these devices are based on two basic models: pull and push. These models and the full 7-layered OSI protocol is comprehensively explained in DASH7 Version 1.0 that is found free to download at (D7A 2016).

20.4 Multi-sensor, Multipurpose DASH7 Tag

Figure 20.4 shows a prototype PCB of the current DASH7 tag design. The key features of this design are small form-factor chip antenna, commonly used microchip (SoC CC430F5137, the same as the WizziMote), 433 MHz balun RF circuit, 8 Mbit flash data logger, Sensirion temperature/humidity sensor, 3 debug LEDs, JTAG port with 0.5 mm FFC, dual power supply (battery and harvested power), and energy harvester circuit (with the charger/protector chip MAX 17710 (Maxim Integrated 2016)). The energy harvester is capable of harvesting energy from sources such as solar, vibration, and thermal and from energy levels as low as 4 μW.

In the case of deployment in containers carrying fresh produce and placement in warehouses, etc., the tag can be further miniaturized by removing the LEDs, the energy harvesting circuitry and the extra oscillator. Miniaturization is required to make sure that the tag is placed as discretely as possible within the produce when

Fig. 20.4 DASH7 prototype tag with energy harvesting

taking measurements. The current range of chip antenna is as low as 2 m at 0 dB m due to an impedance mismatch between the antenna and the balun chip. It can be improved and replaced with a whip antenna for better range.

This prototype RFID tag was designed to replace the wireless air-flow sensor used in Lloyd et al. (2015) for banana container experiments. Our goals was to integrate processor, radio, sensors and other electronic components on a small board. Due to the lack of sufficient number of boards for the field tests, the commercial WizziMotes were used. Furthermore, the prototype presented above had problems with antenna coupling (0 dBm communication range was limited to 2 m) which were not solved in time. Subsequent to correction of errors and mounting other electronics, it can be used in the apple warehouse for further testing and also in containers and trucks alike.

Another critical reason for the design of this prototype model (with energy harvesting) is that the authors of this paper recognize the importance of using 433 MHz in a variety of applications that suffer due to limitations in 2.4 GHz communication and lack of energy efficient standalone active RFID tags with low maintenance. DASH7 software solutions applied alongside the hardware solution of above nature, minimizes the need to change/charge batteries frequently and makes the deployment of a tag non-intrusively due to its small footprints.

20.5 Conclusion

The low signal attenuation of 3.3 dB/m makes the 433 MHz frequency range much more suitable for monitoring applications in food transportation and storage than the commonly used range of 2.4 GHz. The communication range of the latter one is limited to about 0.5 m by the very high attenuation of 61.6 dB/m.

The radio range can be improved by better circuit layout and careful selection of the circuit board material to optimize the coupling of the antenna. Most of the advantages can be achieved at lower frequencies. For example, an improvement of P_0, including the antenna gain, of 10 dB results in an extension of the range by about 3 m at 433 MHz, but only by 0.16 m at 2.4 GHz.

Since only low data rates are necessary for transmission of measured temperature and flow values in applications as the mentioned in this paper, 433 MHz a suitable solution for monitoring of cold storage rooms. The drawback of low bandwidth in 433 MHz is therefore non-significant in cold storage rooms and containers.

Due to lack of LoS between the base-station and the motes, the signal attenuation of 433 MHz was no better in gaps than within apples. Another reason could be that the wavelength 69.14 cm is much greater than the gap width of 10 cm.

Acknowledgments The project COOL "Flow sensor based air management in fruit and vegetable storage" is supported by the Federal Ministry for Economic Affairs and Energy on the basis of a decision by the German Bundestag (Ref. No. VP2869706CL4). We thank the support of Stiftung Kompetenzzentrum Obstbau-Bodensee (KOB), Ravensburg-Bavendorf, Germany for the provision of their warehouse and test facilities.

References

D7A. DASH7 Alliance. Oct 1. http://www.dash7-alliance.org/

do Nascimento Nunes MC et al (2014) Improvement in fresh fruit and vegetable logistics quality: berry logistics field studies. Philos Trans R Soc A Math Phys Eng Sci 372(2017):20130307–20130307. doi:10.1098/rsta.2013.0307

Gustavsson J et al Global food losses and food waste extent causes and prevention. In: Study conducted for Interpak2011 international congress. Food and Agriculture Organization of the United Nations. http://www.fao.org/docrep/014/mb060e/mb060e00.htm

Institute of Mechanical Engineers (2013) Global food. Waste not, want not. http://www.imeche.org/docs/default-source/news/Global_Food_Waste_Not_Want_Not.pdf?sfvrsn=0

International Organization for Standardization (ISO) (2009) ISO/IEC 18000-7—Information technology—Radio frequency identification for item management—Part 7 Parameters for active air interface communications at 433 MHz. http://www.iso.org/iso/iso_catalogue/catalogue_ics/catalogue_detail_ics.htm?csnumber=50368

Isnin I A study on wireless communication error performance and path loss prediction. Ph.D. dissertation. University of Plymouth, United Kingdom. https://pearl.plymouth.ac.uk/handle/10026.1/324

Jedermann R et al (2011) Testing network protocols and signal attenuation in packed food transports. Int J Sens Netw 9(3/4):170-181. ISSN: 1748-1279. doi:10.1504/IJSNET.2011.040238

Jedermann R, Pötsch T, Lloyd C (2014) Communication techniques and challenges for wireless food quality monitoring. Philos Trans R Soc Lond A Math Phys Eng Sci 372(2017):20130304. http://rsta.royalsocietypublishing.org/content/372/2017/20130304

Lang W et al (2011) The intelligent container a cognitive sensor network for transport management. IEEE Sens J 11(3):688–698. ISSN: 1530-437X. doi:10.1109/JSEN.2010.2060480

Lloyd C, Jedermann R, Lang W (2015) Airow behavior under different loading schemes and its correspondence to temperature in perishables transported in refrigerated containers. In: Kotzab H, Pannek J, Thoben K-D (eds) Dynamics in logistics. Lecture notes in logistics. Springer, Berlin, Heidelberg [in-press]. http://www.springer.com/us/book/9783319235110

Maxim Integrated. MAX17710. Energy-Harvesting Charger and Protector. Datasheet. https://www.maximintegrated.com/en/products/power/battery-management/MAX17710.html

Texas Instruments Incorporated. CC430 Family User's Guide. SLAU259E. http://www.ti.com/product/CC430F5137/technicaldocuments

Tuset-Peiró P et al (2014) On the suitability of the 433MHz band for M2M lowpower wireless communications: propagation aspects. Trans Emerg Telecommun Technol 25(12):1154–1168. ISSN: 2161-3915. doi:10.1002/ett.2672

Verdouw CN et al (2013) Smart agri-food logistics: requirements for the future internet. In: Kreowski H-J, Scholz-Reiter B, Thoben K-D (eds) Dynamics in logistics. Lecture Notes in logistics. Springer, Berlin, Heidelberg, pp 247-257. ISBN: 978-3-642-35965-1. doi:10.1007/978-3-642-35966-8_20

Vigneault C et al (2009) Transportation of fresh horticultural produce. In: Benkeblia N (ed) Postharvest technologies for horticultural crops. Transport vol 2, pp 1–24. http://ucanr.edu/datastoreFiles/234-1291.pdf

Wang R et al (2011) Data analysis and simulation of Auto-ID enabled food supply chains based on EPCIS standard. In: 2011 IEEE international conference on automation and logistics (ICAL). Institute of Electrical & Electronics Engineers (IEEE). doi:10.1109/ical.2011.6024684

Weyn M et al (2013) Survey of the DASH7 alliance protocol for 433 MHz wireless sensor communication. Int J Distrib Sens Netw 2013. http://www.hindawi.com/journals/ijdsn/2013/870430/

WizziLab (2013) WizziMote—The smallest and most versatile DASH7 development kit. http://www.wizzimote.com/

Part III
Transport, Maritime and Humanitarian Logistics

Chapter 21
Pre-selection Strategies for the Collaborative Vehicle Routing Problem with Time Windows

Kristian Schopka and Herbert Kopfer

Abstract In horizontal coalitions for auction-based exchange of transportation requests, (freight) carriers have to identify requests that are selected for offering to coalition partners. This paper contributes pre-selection strategies, where requests are selected based on their approximated potential to increase the carriers' profit. In a computational study, the strategies' ability to achieve savings is analyzed.

Keywords Collaborative transportation planning problem · Horizontal carrier collaboration · Request exchange · Pre-selection of own request pool

21.1 Introduction

Freight carriers are confronted with increasing customer claims, e.g., just in time pickups and deliveries or real-time transportation solutions. Moreover, rising petrol and labor prices, shorter life cycles of products, and a growing body of transport legislation reduce the profit margins of freight carriers (Cruijssen et al. 2007). Due to a limited portfolio of resources for disposal and a weak market position, particularly small and mid-sized carriers (SMCs) have difficulties in creating highly efficient transportation plans (tour plans). To enhance competitiveness against dominating freight forwarders, SMCs ally in horizontal coalitions for the auction-based exchange of (transportation) requests. The exchange of requests enables an improved request clustering and reduces the fulfillment costs for all members (Krajewska and Kopfer 2006). Despite the collaborative advantage (i.e., reduced fulfillment costs for all members), some barriers have to be overcome to ensure the viability and stability of long-term coalitions. In this context, the "carrier-fear" for abandoning autonomy can be observed; in particular potential

---

K. Schopka (✉) · H. Kopfer
Chair of Logistics, University of Bremen, Bremen, Germany
e-mail: schopka@uni-bremen.de

H. Kopfer
e-mail: kopfer@uni-bremen.de

© Springer International Publishing Switzerland 2017
M. Freitag et al. (eds.), *Dynamics in Logistics*, Lecture Notes in Logistics,
DOI 10.1007/978-3-319-45117-6_21

coalition members are not willing to share with their partners (former competitors) all information about their request structure. That is why SMCs intend to offer only those requests that are improper to their current tour plans. Resulting, one key question for SMCs within a horizontal coalition is which of their private (i.e., own) requests should be reserved for the private fleet (self-fulfillment requests) and which should be offered for request exchange (collaborative requests). The contribution of this paper is to present pre-selection strategies that divide the set of requests of collaborating SMCs in self-fulfillment requests and collaborative requests. Thereby, the pre-selection strategies estimate the requests' potential to increase profits. Based on these estimations the requests with the highest potential are combined to routes for self-fulfillment. The remaining requests are released for request exchange. The paper is structured as follows. Section 21.2 gives a literature review. In Sect. 21.3 the basic problem is introduced. Section 21.4 introduces a solution procedure, where the focus is on pre-selection strategies. The pre-selection strategies are compared in a computational study in Sect. 21.5. The paper closes with a conclusion in Sect. 21.6.

21.2 Literature Review

Some opportunities to solve the pre-selection of the own request pool are presented in the literature. One option to solve the problem is the adaption of existing procedures that use "cherry-picking" to select requests for self-fulfillment. The idea of those "cherry-picking" procedures is that the most profitable requests are included in the tour plan of the private fleet. Less profitable requests are excluded from the tour plans and for example, are forwarded to external carriers. In this context, the procedures for the integrated operational transportation planning problems (e.g., Krajewska and Kopfer 2009), the vehicle routing problems (VRPs) with private fleet and common carriers (e.g., Chu 2005) or the selective VRPs (e.g., Archetti et al. 2012) are established. For solving the pre-selection by the adaption of a "cherry-picking" procedure, the requests that are included in the tour plan are assigned to self-fulfillment and the remaining requests are assigned to collaborative requests.

Besides the adaption of established "cherry-picking" procedures some strategies are particularly developed for SMCs that use an auction-based request exchange. Berger and Bierwirth (2010) recommend that collaborating SMCs should offer all requests with the lowest marginal profits to the request exchange process until no further improvement can be realized. The remaining requests are reserved for self-fulfillment. Schwind et al. (2009) perform an advanced outsourcing phase for their auction-based request exchange, where requests that are located close to each other are combined to clusters. The clusters with the highest distance to the depot are released to the request exchange until a defined rate of collaborative requests is reached. Dai and Chen (2011) formulate an outsourcing requests selection problem, where the willingness to pay for the requests decides which requests are reserved.

Schopka and Kopfer (2015a) introduce tour potential valuation strategies (TPVs) for request selection. The basic idea is to generate a complete tour plan that includes all own requests by construction heuristics (e.g., sweep algorithm, savings algorithm). The requests included in tours that increase the own profits most are reserved for the private fleet until a defined percentage is achieved. The requests of the remaining tours are released to the request exchange. Furthermore, Schopka and Kopfer (2015a) introduce request potential valuation strategies (RPVs), where the idea is to evaluate the requests' potential to increase the own profits based on available information. A percentage of the requests with the highest calculated potential value is reserved for the private fleet. Similar to the RPVs, the request evaluation strategies of Gansterer and Hartl (2015) also estimate the requests' potential to increase profits.

21.3 Pre-selection of Own Request Pool

A horizontal carrier network that uses an auction-based mechanism to exchange requests becoming known during the execution of existing tour plans (unknown requests) is considered. All m independent SMCs ($P = \{1, 2, \ldots, m\}$) are established in the operational freight business and possess an individual tour plan for the forthcoming planning interval. The tour plan of each SMC results from solving a VRP with time windows (VRPTW) with the objective to find the cost minimal combination of vehicle routes that start and end at the SMC's depot (o_p) and serve an initial set of customer requests (I_p). For a mathematical formulation of the VRPTW, we refer to Solomon (1987). Let $\mu_p^k = \{o_p, i_1, i_2, \ldots, o_p\}$ the tour plan for the forthcoming planning interval of any vehicle k of the private vehicle fleet V_p for each SMC $p \in P$, under the observance of capacity and time restrictions as well as that each request $i \in I_p$ is served once by one vehicle $k \in V_p$. During the execution of those tour plans, each SMC $p \in P$ receives a set of unknown requests (R_p) that have also to be served during the forthcoming planning interval. To avoid uncertainty and to reduce fulfillment costs the SMCs have the ability to offer (some of) those unknown requests to their partners at stipulated common planning updates. Because of the "carrier-fear" to lose autonomy, the SMCs are not willing to offer all requests and the belonging business-relevant information to their partners. Hence, the SMCs want to reserve profitable requests for their private fleet and integrate them into their existing tour plan. To achieve this intention, all SMCs $p \in P$ have to divide their own request pool R_p in the two subsets collaborative requests R_p^c and self-fulfillment requests R_p^s. Each request $r \in R_p$ can either be assigned to R_p^c or R_p^s. Hence, let $R_p = R_p^c \cup R_p^s$ and $R_p^c \cap R_p^s = \{\}$ apply. To support the request exchange, for each SMC, it is not allowed to select more than σ requests for R_p^s. Resulting, each SMC $p \in P$ has to decide which requests in R_p have to be reserved for the private fleet and which are included in the collaborative request exchange. Therefore, the SMCs exhibit only their own information, i.e., the tour plan for the forthcoming planning interval and request specific information. The information of any request $r \in R_p$ includes the distance (g_{rl}) to any other location $l \in I_p \cup R_p \cup \{o_p\}$, the demand ($d_r$), the earnings ($e_r$), and time restrictions. Based on this information and a correspond strategy, each SMC $p \in P$ builds the subsets R_p^s and R_p^c. Afterward, the self-fulfillment requests R_p^s are integrated in the tour plan by resolving the VRPTW for each SMC $p \in P$. The collaborative requests are included in the request exchange that is organized by the execution of an auction. Notice that requests that generate no bid are reassigned to the SMC that offered the request for exchange. Based on the results, the tour plans of all SMCs are updated by a last resolving of the VRPTW and the execution of the updated tour plans is conducted.

21.4 Solution Procedure

In this section our solution procedure is presented. Therefore, in a first step a general framework for organizing the collaborative request exchange of the SMCs is introduced. In the second step we focus on new pre-selection strategies.

21.4.1 General Framework

To organize the general problem of the exchange of unknown requests of a horizontal carrier coalition, we use a stepwise solution framework. The sequence of this framework is visualized in Fig. 21.1 and introduced briefly. For further information on the individual solution techniques of the framework, we refer to the designated references. The considered problem includes that some requests become known during the planning interval. To avoid this feature, we use a periodic

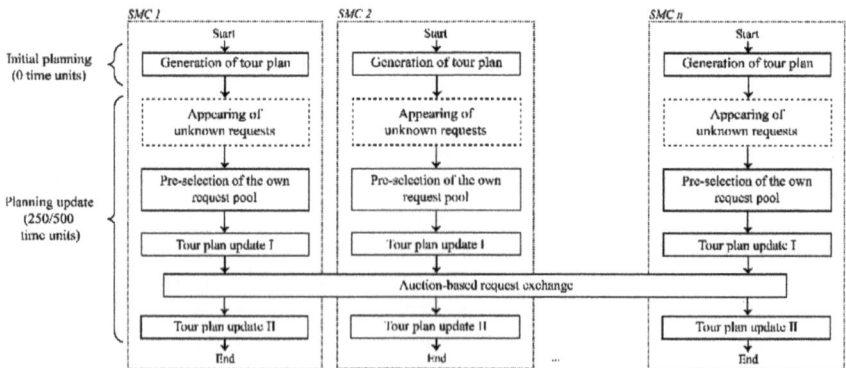

Fig. 21.1 Sequence of the stepwise solution framework

re-optimization strategy, where planning updates are performed after a predefined time interval (Pillac et al. 2013). Our periodic re-optimization strategy recommends only one planning update that is executed when all unknown requests become known. However, at the beginning of the forthcoming planning horizon (at 0 time units), each SMC ($p \in P$) generates an initial tour plan by solving a VRPTW with the initial requests (I_p). The initial tour plans of our stepwise solution framework result from an adaptive large neighborhood search (ALNS) (Pisinger and Ropke 2007). After the generation, the tour plan is executed by the vehicles of all SMCs until the common planning update (e.g., at 500 time units) is reached and the execution process is paused. At this point in the planning interval for each SMC ($p \in P$) the former unknown requests (R_p) become known and are able to be exchanged within the horizontal coalition. In the next step of the sequence each SMC builds the subsets self-fulfillment requests (R_p^s) and collaborative requests (R_p^c) by one of the pre-selection strategies. Based on this decision each SMC $p \in P$ supplements the existing tour plan by the self-fulfillment requests (by using the ALNS). After the first update of the tour plan, the actual request exchange is organized by an auction. To avoid a possible interdependency between the considered pre-selection strategy and the auction mechanism used in the computational study, we organize the request exchange by both, a first price sealed bid auction (FA) and a combinatorial auction (CA). Whereat the FA calculates bids by the marginal profit increase of the SMCs, the CA uses a greedy insertion operator in combination with a tabu search for the generation of cluster bids. For a detailed description of both auction-based request exchange mechanisms, we refer to Schopka and Kopfer (2015b). Afterward, all SMCs perform a second update of their tour plan and the execution of the transportation process is resumed.

21.4.2 Pre-selection Strategies

Based on request specific information our pre-selection strategies estimate the potential of each own request $j \in R_p$ to increase the profits for the associated SMC $p \in P$. Therefore, a potential value φ_j for each request $j \in R_p$ is calculated and the σ requests with highest potential values are reserved for the private fleet or rather build the subset self-fulfillment requests R_p^s. The remaining requests are included in the request exchange and stored as collaborative requests R_p^c. Figure 21.2 presents the general procedure of building the two subsets R_p^s and R_p^c for any SMC $p \in P$ by using a pre-selection strategy. In common, all pre-selection strategies calculate the potential value φ_j of preserving an own request $j \in R_p$ for self-fulfillment by one of the pre-selection strategies (lines 2–4). The requests with the highest potential values φ_j are combined to the subset self-fulfillment requests R_p^s until σ requests are reserved (lines 5–9). The remaining requests build the set collaborative requests R_p^c and are included in the request exchange (line 10).

```
Input: R_p; Output: R_p^s, R_p^c;
1   R_p^s ← {}; R_p^c ← {};
2   forall the j ∈ R_p do
3       φ_j ← calculate(φ_j);
4   end forall
5   while |R_p^s| < σ do
6       j' ← j ∈ R_p, with highest φ_j;
7       R_p^s ← R_p^s ∪ {j'};
8       R_p ← R_p \ {j'};
9   end while
10  R_p^s ← R_p
```

Fig. 21.2 Sequence of pre-selection strategies

Earnings proxy (EP): The EP is a greedy procedure and reserves the requests for self-fulfillment that exhibit the highest earnings, where no other characteristic is analyzed. The potential value φ_j^1 results from the earnings e_j, $j \in R_p$ (Eq. 21.1).

$$\varphi_j^1 = e_j \quad (21.1)$$

Earnings per demand proxy (DP): The DP extends the EP by the inclusion of the demand d_j per request $j \in R_p$. The demand of a request claims capacity of the private fleet and this capacity should be used effectively. Resulting the DP rates the earnings per obligated capacity unit for any request $j \in R_p$ (Eq. 21.2).

$$\varphi_j^2 = \frac{e_j}{d_j} \quad (21.2)$$

Cluster proxy (CP): The CP analyzes whether a clustering of the requests is sensible. Therefore, the CP calculates for any request $j \in R_p$ the average traveling distance to the ω_1 nearest requests $i \in R_p \setminus \{j\}$. Let $\Omega_1 := \{B_1 \subset R_p \setminus \{j\} : |B_1| = \omega_1\}$ represent all subsets of requests in R_p with a cardinal number of ω_1. The potential value φ_j^3 of a request $j \in R_p$ results from the earnings e_j reduced by the average traveling distance of the nearest requests (Eq. 21.3).

$$\varphi_j^3 = e_j - \frac{\min_{A \in \Omega_1} \sum_{i \in A} g_{ij}}{\omega_1} \quad (21.3)$$

Greedy insertion proxy (GP): The GP identifies for any request the insertion position that allows the highest increase of profits on the current tour plan. Therefore, Eq. 21.4 is used to identify the cheapest insertion position Δf_{jk} for each vehicle $k \in V_p$ on the existing tour plan μ_p^k. If a request $j \in R_p$ cannot be integrated (i.e., time or capacity restrictions are violated) on vehicle $k \in V_p$, let $\Delta f_{jk} = \infty$.

$$\Delta f_{jk} = \min_{i \in \mu_p^k \setminus \{0\}} \left(g_{i-1,j} + g_{ij} - g_{i-1,i} \right), \forall j \in R_p, k \in V_p \qquad (21.4)$$

To calculate the potential value φ_j^4 of a request $j \in R_p$, the earnings e_j are reduced by the costs of the cheapest insertion position of any vehicle $k \in V_p$ (Eq. 21.5).

$$\varphi_j^4 = e_j - \min_{k \in V_p} \Delta f_{ik} \qquad (21.5)$$

Regret proxy (RP): An issue of the GP is that only the cheapest insertion position is considered. Since several requests compete for the same insertion positions, the RP calculates the averaged insertion costs of the ω_2 cheapest insertion positions for all requests $j \in R_p$. Let $\Omega_2 := \{B_2 \subset R_p \setminus \{j\} : |B_2| = \omega_2\}$ represent all subsets of the vehicle pool V_p with the cardinal number ω_2. Resulting Eq. 21.6 calculates the scores φ_j^5 for all requests $j \in R_p$.

$$\varphi_j^5 = e_j - \min_{A \in \Omega_2} \frac{\sum_{k \in A} \Delta f_{jk}}{\omega_2} \qquad (21.6)$$

Function proxy (FP): The FP enables to consider all mixed-strategy of the previous presented basic pre-selection strategies with a different weighting. The potential values of the individual strategies are totalized (Eq. 21.7). The individual potential values are weighted by $\beta_1, \beta_2, \beta_3, \beta_4,$ and β_5.

$$\varphi_j^6 = \beta_1 \cdot \varphi_j^1 + \beta_2 \cdot \varphi_j^2 + \beta_3 \cdot \varphi_j^3 + \beta_4 \cdot \varphi_j^4 + \beta_5 \cdot \varphi_j^5 \qquad (21.7)$$

21.5 Computational Study

To analyze the pre-selection strategies, we use existing and previously published instances of the dynamic collaborative vehicle routing problem with time windows (http://www.logistik.uni-bremen.de/instances). The instances differ in terms of the request structure (R), the number of SMCs (P), and the size of the forthcoming planning interval (T). The experiments include instances with 4, 6, or 8 SMCs and the forthcoming planning interval consists either of 1,250 time units (T1) or 1,000 time units (T2). The planning update for T1 is performed at 500 time units and for T2 at 250 time units. Because of the earlier performed planning update, the initial tour plans of instances with T2 include a high rate of initial requests (I_p). All computational studies are performed on a Windows 7 PC (i7-2600 processor, 3.4 GHz, 16 GB RAM). For the solving of the instances, we implement the sequence of the stepwise solution framework in a C++-application, where the used pre-selection strategy and auction type are matched to the considered instance. The iteration number for generating/updating the tour plans by the ALNS is limited to 20,000 iterations. Since all requests have to be served by one member of the coalition and, consequently, the earnings are preexisting and invariable, in the computational experiments, we compare the ability of the pre-selection strategies to minimize the sum of the individual transportation costs of all members. In this context, we assume that within the horizontal coalition a fair cost allocation scheme exists and the savings result in financial advantages for all members. To increase the comparability among the pre-selection strategies, we generate for each instance only one initial tour plan that is stored and used for each pre-selection strategy. To avoid the heuristically deviation of the following sequence (i.e., by ALNS/request exchange), we solve each combination of instance and pre-selection strategy ten times. For the evaluation, we use the average values of ten runs. Besides the five basic pre-selection strategies, three versions of the FP are included in our tests. The chosen parameter settings for all strategies (see Table 21.1) are analyzed in advanced experiments.

The achieved transportation costs of the pre-selection strategies are presented in Table 21.1. As the results of the five basic pre-selection strategies indicate, especially the GP and the RP are able to generate solutions with low transportation costs. The strict greedy procedures EP and DP generate tour plans of inferior quality. Since the solutions generated by CP depend on the possibility to cluster the requests of the instances, the quality of the CP is between the ED and the GP. Several best results of the CP, GP and RP can be increased by the application of mixed strategies (i.e., FP1, FP2, FP3). Thereby, especially the FP3 achieves best solutions. The solution qualities are dependent on uncertain forthcoming factors (i.e., the auction-based request exchange) and it is impossible to identify the superior strategy. To identify pre-selection strategies that are preferable for specific scenarios, Table 21.2 presents the percentage of best results for the pre-selection strategies based on specific transportation settings. In total the FPs are able to generate for 62.50 % of all instances the best results. Since similar percentages are achieved by considering the FA- and CA-instances separately, any interdependency among pre-selection strategies and auction type is not observable. Only the GP is more appropriate for CA-instances. Considering the planning intervals separate, it seems notable that GP, RP and FP2 achieve 68.75 % of the best solutions for instances with a higher rate of initial re quests (i.e., T2). Hence, pre-selection strategies that consider a clustering (i.e., CP, FP1, FP3) are more proper for instances with a lower rate of initial requests (i.e., T1). Whereat the CP, the RP, the FP1, and the FP3 are preferable for instances with a low number of members (i.e., 4–6), the GP and the FP2 generate best solutions, especially for instances with numerous members (i.e., 6–8).

Table 21.1 Overall traveling costs on generated test instances achieved by the using of different pre-selection strategies and auction types (best results are marked bold); CP ($\omega_1 = 5$), RP ($\omega_2 = 5$), FP1 ($\beta_1 = 0, \beta_2 = 0, \beta_3 = 0.5, \beta_4 = 0, \beta_5 = 0.5$), FP2 (0, 0, 0.5, 0.5, 0), FP3 (0, 0, 0.4, 0.4, 0.2)

Instances	EP	DP	CP	GP	RP	FP1	FP2	FP3
R1P4T1 – FA	55392.3	55376.6	**54291.0**	54318.4	54473.8	54582.1	54380.0	54301.8
R2P4T2 – FA	59581.5	58264.0	57363.3	56766.8	57178.0	57046.9	56822.8	**56678.0**
R3P4T1 – FA	58211.3	57977.0	57072.9	56986.4	56723.8	**56621.5**	57042.6	57014.6
R4P4T2 – FA	59529.6	58217.3	57420.8	56776.2	57175.1	57276.9	56806.2	**56775.9**
R5P4T1 – FA	55852.2	55966.4	**54753.5**	54826.1	55147.6	55111.9	54935.2	54747.7
R6P4T2 – FA	59491.3	58237.1	57416.0	56702.4	57079.2	57139.3	**56502.9**	56672.2
R7P4T1 – FA	54878.9	54797.1	54362.7	54010.0	54171.3	54099.4	54011.0	**53889.2**
R8P4T2 – FA	59530.1	58298.8	57348.3	56970.0	57270.8	57131.4	56956.5	**56813.8**
R9P6T1 – FA	82415.5	82259.9	81626.8	81647.6	81607.2	81623.7	81579.2	**81429.1**
R10P6T2 – FA	68648.6	68566.8	58766.2	68040.7	68633.2	68650.9	**68098.6**	68129.5
R11P6T1 – FA	84187.2	84039.1	83928.4	83541.9	83878.3	83583.1	83638.7	83605.2
R12P6T2 – FA	65636.7	65264.1	64904.9	64723.1	64715.4	64860.6	**64632.2**	64642.0
R13P6T1 – FA	83378.9	83241.6	82054.5	81612.2	81914.1	81871.9	81730.6	**81556.1**
R14P6T2 – FA	66511.3	66463.2	67134.3	**65788.9**	66083.0	66043.6	65850.0	65925.8
R15P6T1 – FA	83351.3	83865.2	81638.8	**81163.4**	81820.7	82032.9	81351.1	81962.3
R16P6T2 – FA	67855.0	67039.4	67507.4	66416.0	**66247.4**	66307.7	66517.3	66467.4
R17P8T1 – FA	87390.6	87545.1	86906.6	86203.7	86172.1	86179.8	86146.5	**86110.8**
R18P8T1 – FA	88901.8	88445.3	88298.6	88383.2	88283.3	88342.5	88363.5	**88171.9**
R19P8T1 – FA	88085.3	88398.4	87044.1	86909.9	87201.2	87140.7	**86876.7**	87040.9
R20P8T1 – FA	89344.0	89285.8	88764.9	**88461.7**	88651.8	88674.1	88587.7	88651.8
R1P4T1 – CA	54637.2	54397.7	**53177.2**	53229.5	53490.0	53395.1	53426.6	53374.1
R2P4T2 – CA	58772.7	57587.9	56639.4	56252.0	56344.5	56321.6	**56114.7**	56221.1
R3P4T1 – CA	57466.3	57268.0	56316.1	56381.6	**56034.3**	56168.9	56341.2	56217.2

(continued)

Table 21.1 (continued)

Instances	EP	DP	CP	GP	RP	FP1	FP2	FP3
R4P4T2 – CA	58529.1	57425.0	56725.4	56148.6	56461.3	56552.3	56168.7	**56141.4**
R5P4T1 – CA	55578.9	55335.0	54485.2	54169.8	54495.7	54564.6	54210.7	**54138.8**
R6P4T2 – CA	58700.4	57506.8	56762.0	56174.3	56410.1	56547.8	**56020.0**	56128.6
R7P4T1 – CA	54308.4	53955.0	53710.4	53639.2	53428.5	**53300.2**	53500.9	53385.2
R8P4T2 – CA	58648.9	57590.4	56708.8	56333.5	56330.1	56278.2	56486.2	**56077.4**
R9P6T1 – CA	81565.8	81237.1	80036.8	79911.6	80168.5	80012.7	80007.9	**79810.9**
R10P6T2 – CA	67242.1	67409.7	67510.4	66684.0	67278.7	67216.6	**66664.6**	66687.4
R11P6T1 – CA	82976.1	83243.4	82257.5	**80939.4**	81501.9	81498.8	81034.4	81467.1
R12P6T2 – CA	64780.1	64176.4	63696.8	63394.9	63349.4	63430.1	**63284.2**	63309.6
R13P6T1 – CA	81888.3	81555.8	80741.2	79870.5	80099.7	80061.3	79899.3	**79761.3**
R14P6T2 – CA	65381.3	64582.5	65803.0	**64308.5**	64648.8	64356.1	64356.1	64685.7
R15P6T1 – CA	82049.8	82438.7	80285.3	79935.2	80220.8	80078.7	**79717.8**	79853.2
R16P6T2 – CA	66826.5	66062.0	66539.7	**65261.4**	65503.6	65516.7	65284.1	65326.3
R17P8T1 – CA	86731.3	86593.7	85503.9	84786.4	84708.9	84790.1	**84701.6**	84866.2
R18P8T1 – CA	88029.3	87727.0	87229.3	**87271.0**	87516.6	87515.4	87346.0	87457.0
R19P8T1 – CA	87625.0	87634.2	86059.0	**85417.6**	85610.6	85596.5	85569.2	85594.9
R20P8T1 – CA	88689.2	88655.5	87843.9	**87364.7**	87377.7	87386.4	87522.5	87423.8

Table 21.2 Frequency of achieved best results by the pre-selection strategies

	Overall (%)	Auction type		Planning horizon		Member		
		FA (%)	CA (%)	T1 (%)	T2 (%)	P4 (%)	P6 (%)	P8 (%)
CP	7.50	10.00	5.00	12.50	0.00	18.75	0.00	0.00
GP	22.50	15.00	30.00	25.00	18.75	0.00	31.25	50.00
RP	5.00	5.00	5.00	4.17	6.25	6.25	6.25	0.00
FP1	5.00	5.00	5.00	8.33	0.00	12.50	0.00	0.00
FP2	25.00	20.00	30.00	12.50	43.75	18.75	37.50	25.00
FP3	32.50	40.00	25.00	33.33	31.25	43.75	25.00	25.00

21.6 Conclusion

This paper introduces pre-selection strategies for identifying requests that should be reserved for the private fleet of collaborating SMCs. In common, the pre-selection strategies estimate the potential for requests to increase the own profits. Based on estimated values for request profits a percentage of the requests are reserved for the private fleet and the remaining requests are included in an auction-based request exchange with other SMCs. In experiments, the pre-selection strategies are analyzed on the ability to generate collaborative savings. Since the solution quality of pre-selection strategies dependent on uncertain forthcoming factors, it is impossible to identify the superior strategy. Nevertheless, the experiments give some advice which pre-selection strategies are preferable for specific transportation settings. Since dynamic collaborative transportation planning problems are characterized by immediate decisions, the pre-selection strategies and further approximation strategies may support reconcilements among collaborating SMCs in nearly real time.

Acknowledgments The research was supported by the German Research Foundation (DFG) as part of the project "Kooperierende Rundreiseplanung bei rollierender Planung."

References

Archetti C, Speranza MG, Vigo D (2012) Vehicle routing problems with profits. Technical report, Department of Economics and Management, University of Brescia, Italy
Berger S, Bierwirth C (2010) Solutions to the request reassignment problem in collaborative carrier networks. Transp Res Part E: Logist Transp Rev 46:627–638
Chu C (2005) A heuristic algorithm for the truckload and less than truckload problem. Eur J Oper Res 165:657–667
Cruijssen F, Cools M, Dullaert W (2007) Horizontal cooperation in logistics: opportunities and impediments. Transp Res Part E: Logist Transp Rev 43:129–142
Dai B, Chen H (2011) A multi-agent and auction-based framework and approach for carrier collaboration. Logist Res 3:101–120
Gansterer M, Hartl RF (2015) Request evaluation strategies for carriers in auction-based collaborations. OR Spectrum. doi:10.1007/s00291-015-0411-1

Krajewska MA, Kopfer H (2006) Collaborating freight forwarding enterprise. OR Spectrum 28:301–317

Krajewska MA, Kopfer H (2009) Transportation planning in freight forwarding companies: Tabu search algorithm for the integrated operational transportation planning problem. Eur J Oper Res 197:741–751

Pillac V, Gendreau M, Guret C, Medaglia AL (2013) A review of dynamic vehicle routing problems. Eur J Oper Res 225:1–11

Pisinger D, Ropke S (2007) A general heuristic for vehicle routing problems. Comput Oper Res 34:2403–2435

Schopka K, Kopfer H (2015a) Pre-selection strategies for dynamic collaborative transportation planning problems. Operations research proceedings 2014 to appear

Schopka K, Kopfer H (2015b) Proxies for the auction-based request exchange of dynamic and collaborative transportation scenarios. Working paper, University of Bremen

Schwind M, Gujo O, Vykoukal J (2009) A combinatorial intra-enterprise exchange for logistics services. Inf Syst E-Bus Manage 7:447–471

Solomon MM (1987) Algorithms for the vehicle routing and scheduling problems with time window constraints. Oper Res 35:254–265

Chapter 22
Transportation Planning with Forwarding Limitations

Mario Ziebuhr and Herbert Kopfer

Abstract Small- and medium-sized forwarders are confronted with thin margins and high demand fluctuations in competitive transportation markets. In these markets forwarders try to improve their planning situation by using external resources besides their own resources. These external resources might belong to closely related subcontractors, common carriers or cooperating forwarders in horizontal coalitions. In recent publications, it is assumed that some transportation requests are prohibited to be fulfilled by certain external resources due to contractual obligations. These requests are known as compulsory requests. In this paper, a transportation planning problem including external resources is extended by the mentioned compulsory requests. It is proposed to consider different types of compulsory requests depending on the applicable external resources for fulfilling these requests. As a solution approach a column generation-based heuristic is applied, which uses a strict composition procedure and a strict generation procedure for handling compulsory requests. In a detailed computational study, the increase of transportation costs is analyzed.

Keywords Transportation planning · Subcontracting · Collaboration · Compulsory requests · Column generation-based heuristic

22.1 Introduction

In transportation markets, forwarders are able to reduce their operational costs and to improve their flexibility by using external carriers (subcontracting) and cooperating forwarders in a horizontal coalition (collaborative planning) besides their own fleet (self-fulfillment) (Chu 2005). One option of subcontracting is the

---

M. Ziebuhr (✉) · H. Kopfer
Chair of Logistics, University of Bremen, Bremen, Germany
e-mail: ziebuhr@uni-bremen.de

H. Kopfer
e-mail: kopfer@uni-bremen.de

employment of a carrier on a spot market where a common carrier is employed for a request in exchange of a freight charge. A second option is the possibility of using long-term contractual agreements with subcontractors, where forwarders hire transportation capacities of carriers to an agreed limit and take over the planning for the hired capacities. These subcontractors can be paid on a tour basis (TB) or on a daily basis (DB). Simultaneously solving the combined problem of vehicle routing for the private fleet and the optimal employment of common carriers and subcontractors is known as integrated operational transportation planning (IOTP) (Krajewska and Kopfer 2009). In addition to subcontracting, a third fulfillment option is represented by collaborative planning. In collaborative transportation planning (CTP), independent forwarders try to improve their planning situation by reallocating their transportation requests or capacities in a horizontal coalition (Wang and Kopfer 2014). The difference between IOTP and CTP is the relationship among the forwarders and carriers which is either a hierarchical-related partnership in terms of IOTP or an equal partnership in terms of CTP. In this paper, the IOTP and CTP problem are combined to one transportation planning problem which is denoted as collaborative operational transportation planning (COTP). In COTP, a request can be fulfilled either by self-fulfillment or subcontracting or collaborative planning.

In transportation planning, a request is denoted as a compulsory request in case that the request has to be fulfilled by certain resources due to contractual obligations (Özener et al. 2011, Schönberger 2005). In previous publications, the topic of compulsory requests is analyzed for IOTP (Schönberger 2005, Ziebuhr and Kopfer 2014) and CTP (Ziebuhr and Kopfer 2016). This paper analyzes the COTP with respect to the increase of costs which is caused by request compulsiveness. Different types of compulsory requests are examined which differ in terms of the applicable external resources for fulfilling these requests. The idea behind this proposal is that forwarders are able to offer different services related to their requests. The following listing of transportation services is proposed.

- premium 1: self-fulfillment
- premium 2: self-fulfillment, long-term carrier
- premium 3: self-fulfillment, collaboration
- premium 4: self-fulfillment, long-term carrier, collaboration
- standard: self-fulfillment, subcontracting, collaboration

For example, a premium 3 service demands the application of a vehicle of the private fleet of any member in the horizontal coalition, while the application of subcontracting is prohibited. The corresponding requests of the mentioned services are denoted as standard (S1), premium 1 (P1), premium 2 (P2), premium 3 (P3), and premium 4 (P4) requests. All premium requests are compulsory requests and have in common that common carriers cannot be used for these requests.

This paper contributes by introducing and analyzing the impact of four types of compulsory requests for COTP. The corresponding COTP problem with forwarding limitations is denoted as COTPP-FL. To solve the COTPP-FL a column generation-based heuristic (CGB-heuristic) is applied which divides the COTPP-FL into a master problem and a subproblem. The master problem is solved by a commercial solver while the subproblem is solved by an adaptive large neighborhood search (ALNS). To handle compulsory requests two strategies are applied, which are known as strict generation and strict composition procedure. Based on a benchmark study the best strategy has to be identified and is used for an additional computational study which analyzes the increase of transportation costs.

The paper is structured as follows. In Sect. 22.2 the COTPP-FL is formulated and described. Section 22.3 describes the CGB-heuristic with the strict generation and strict composition procedure. Section 22.4 presents the computational studies and Sect. 22.5 concludes the paper and gives some ideas for future research.

22.2 Transportation Planning with Forwarding Limitations

In the COTPP-FL, m forwarders align their individual transportation plans by exchanging requests with each other. Besides the option of exchanging requests, each forwarder faces an IOTP problem, where a request can be fulfilled by self-fulfillment or subcontracting. First, the IOTP problem of a forwarder is presented and second the CTP problem is described.

The IOTP problem of a forwarder c can be defined on a graph $G_c = (V_c, A_c)$ where V_c represents the set of nodes and A_c represents the set of edges with $A_c = V_c \times V_c$. The set of nodes contains the set of pickup nodes $P_c = \{1, \ldots, n_c\}$, the set of delivery nodes $D_c = \{n_c + 1, \ldots, 2n_c\}$, and the depot $\{o_c\}$, i.e., $V_c = P_c \cup D_c \cup \{o_c\}$. The difference between the IOTP problem and the IOTP problem with forwarding limitations is that by considering compulsory requests the set of pickup nodes P_c is separated into five disjoint sets corresponding to the type of service that a request and its pickup node belong to. In conclusion, the set of pickup nodes contains the set of standard pickup nodes S_c, premium 1 pickup nodes P_c^1, premium 2 pickup nodes P_c^2, premium 3 pickup nodes P_c^3, and premium 4 pickup nodes P_c^4, i.e., $P_c = S_c \cup P_c^1 \cup P_c^2 \cup P_c^3 \cup P_c^4$. In the following, premium pickup nodes are denoted as premium requests. For the fulfillment of a request with load $l_i \geq 0$, goods have to be transported from their pickup location i to their delivery location j with $l_j = -l_i$. At node i a service with the duration s_i has to be started within a time window $[a_i, b_i]$. The corresponding travel time t_{ij} and distance d_{ij} are given for each edge $(i, j) \in A_c$. In IOTP three kinds of vehicles are applicable to the fulfillment of a pickup and delivery pair. The set of own and foreign vehicles $K_c = K_c^1 \cup K_c^2 \cup K_c^3$ is represented by the set of private vehicles (K_c^1), rented vehicles based on mode TB (K_c^2), and rented vehicles based on mode DB (K_c^3). Each vehicle $k \in K_c$ has the same capacity Q for which different fixed rates α_k and variable rates β_k are considered. Vehicles on mode DB have a maximal route length L^{DB}, which cannot be exceeded. The third possibility of subcontracting is the employment of a common carrier (CC), who charges a fee γ_i for fulfilling the request at node i.

Five types of decision variables are used. The binary variable x_{ijk} is one if vehicle k travels from i to j and zero otherwise. Moreover, y_k^{DB}, respectively, y_i^{CC} indicate whether a rented vehicle $k \in K_c^3$ on mode DB is used, respectively a common carrier is employed for the fulfillment of the request with pickup location i. The starting time of a service at node i by vehicle k is represented by w_{ik} while the variable L_{ik} defines the load of vehicle k after the service is completed at node i. The IOTP problem with forwarding limitations of forwarder c can be modeled as follows:

$$\min IP_c = \sum_{k \in K_c^1 \cup K_c^2} \sum_{(i,j) \in A_c} \beta_k d_{ij} x_{ijk} + \sum_{k \in K_c^3} \alpha_k y_k^{DB} + \sum_{i \in P_c} \gamma_i y_i^{CC} \tag{22.1}$$

$$\sum_{k \in K_c} \sum_{j \in V_c} x_{ijk} + y_i^{CC} = 1, \forall i \in S_c, \tag{22.2}$$

$$\sum_{k \in K_c} \sum_{j \in V_c} x_{ijk} = 1, \forall i \in P_c^2, \forall i \in P_c^4, \tag{22.3}$$

$$\sum_{k \in K_c^1} \sum_{j \in V_c} x_{ijk} = 1, \forall i \in P_c^1, \forall i \in P_c^3, \tag{22.4}$$

$$\sum_{j \in V_c} x_{ijk} - \sum_{j \in V_c} x_{j,n+i,k} = 0, \forall k \in K_c, \forall i \in P_c, \tag{22.5}$$

$$\sum_{j \in P_c \cup \{o_c\}} x_{o_c,j,k} = 1, \forall k \in K_c, \tag{22.6}$$

$$\sum_{i \in D_c \cup \{o_c\}} x_{i,o_c,k} = 1, \forall k \in K_c, \tag{22.7}$$

$$\sum_{i \in V_c} x_{ijk} - \sum_{i \in V_c} x_{jik} = 0, \forall k \in K_c, \forall j \in V_c, \tag{22.8}$$

$$\sum_{j \in P_c} x_{o_c,j,k} = y_k^{DB}, \forall k \in K_c^3, \tag{22.9}$$

$$\sum_{(i,j) \in A_c} d_{ij} x_{ijk} = L^{DB}, \forall k \in K_c^3, \tag{22.10}$$

$$x_{ijk} + s_i + t_{ij} - M(1 - x_{ijk}) \leq w_{jk}, \forall k \in K_c, \forall (i,j) \in A_c, \tag{22.11}$$

$$a_i \leq w_{ik} \leq b_i, \forall k \in K_c, \forall i \in V_c, \tag{22.12}$$

$$w_{ik} \leq w_{n+i,k}, \forall k \in K_c, \forall i \in P_c, \tag{22.13}$$

$$L_{ik} + l_j - M(1 - x_{ijk}) \leq L_{jk}, \forall k \in K_c, \forall (i,j) \in A_c, \qquad (22.14)$$

$$L_{ik} \leq Q, \forall k \in K_c, \forall i \in V_c, \qquad (22.15)$$

$$L_{o_c,k} = 0, \forall k \in K_c, \qquad (22.16)$$

$$x_{ijk} \in \{0,1\}, \forall k \in K_c, \forall (i,j) \in A_c, \qquad (22.17)$$

$$y_k^{DB} \in \{0,1\}, \forall k \in K_c^3, \qquad (22.18)$$

$$y_i^{CC} \in \{0,1\}, \forall i \in P_c, \qquad (22.19)$$

$$w_{ik} \geq 0, \forall k \in K_c, \forall i \in V_c, \qquad (22.20)$$

$$L_{ik} \geq 0, \forall k \in K_c, \forall i \in V_c. \qquad (22.21)$$

The goal of the IOTP problem with forwarding limitations is to minimize the total fulfillment costs (22.1), which are summarized by the fixed and variable costs for the different transportation modes and the sum of the freight charges paid to the common carriers. The fixed cost term for self-fulfillment is omitted in the objective function. By considering the problem formulation, it is obvious that most of the constraints are well-known from the pickup and delivery problem with time windows (PDPTW) and the IOTP problem. It is referred to Ropke and Pisinger (2006) and Wang et al. (2014) for a detailed explanation. The modified constraints of the IOTP problem with forwarding limitations are: (22.2), (22.3), and (22.4). Constraint (22.2) ensures that either a private vehicle or a rented vehicle or a common carrier is used for fulfilling a standard request. For premium 2 and premium 4 requests, the private fleet and a rented fleet can be used (constraint (22.3)), while premium 1 and premium 3 requests can only be served by the private fleet (constraint (22.4)).

For enabling the request exchange among independent freight forwarders in a horizontal coalition, the described IOTP problem with forwarding limitations has to be extended. Depending on the transportation plans of the coalition members each forwarder c offers a request portfolio P_c^- for exchange and receives a new request portfolio P_c^+ after the request exchange process is completed. The offered request portfolio P_c^- is defined by $P_c \backslash P_c^0$, where P_c^0 represents the set of not transferred requests. Thereby, it is worth mentioning that the set of not transferred requests contains at least all premium 1 and premium 2 requests, while P_c' contains standard, premium 3, and premium 4 requests. As soon as the request exchange process is completed, a forwarder c is responsible for producing the assigned request portfolio P_c' with $P_c' = P_c^0 \cup P_c^+$ by fulfillment costs defined by IP_c'. The COTPP-FL can be modeled as follows:

$$\min CTP_c = \sum_{c=1,\ldots,m} IP_c' \qquad (22.22)$$

$$P'_c \cap P'_h = \emptyset, \quad \forall c, \ h = 1, \ldots, m, \ c \neq h, \tag{22.23}$$

$$\cup_{c=1}^{m} P_c^+ = \cup_{c=1}^{m} P_c^-. \tag{22.24}$$

The goal of the COTPP-FL is to minimize the individual costs of each member of the coalition. Constraint (22.23) ensures that each exchanged request is fulfilled by exactly one coalition member, while constraint (22.24) ensures that all offered requests are assigned to the coalition members.

22.3 Solution Approach

To solve the COTPP-FL the CGB-heuristic introduced by Wang et al. (2014) is applied. For handling compulsory requests the CGB-heuristic is extended by two strategies: the strict generation procedure and the strict composition procedure. The strict generation strategy was introduced by Ziebuhr and Kopfer (2014). In the following, the CGB-heuristic as well as the strategies are briefly described.

The CGB-heuristic is a column generation approach. A column generation is devised for linear programs and is a popular solution approach for solving large-scale integer programming problems (Lübbecke and Desroisiers 2004). The CGB-heuristic proposes the reformulation of the transportation problem into two problems: the master problem (selection of vehicle routes) and the subproblem (generation of vehicle routes). In terms of the COTPP-FL, the master problem is formulated as a set partitioning problem (SPP) and is responsible for selecting routes which constitute the best feasible transportation plan among all plans which can be built based on the set of existing routes. In a column generation approach, the master problem is solved by the simplex method which is why the SPP formulation of the master problem is relaxed to a linear program (SPP-LP). Thereby, a binary variable, which is responsible that each request is served once, is relaxed to a continuous variable. By using the commercial solver ILOG CPLEX the SPP-LP formulated master problem is solved and the generated dual values are used for identifying new vehicle routes with negative reduced costs (minimization problem). These dual values are forwarded to the subproblem which is responsible for identifying vehicle routes with negative reduced costs. To solve the subproblem each coalition member uses the ALNS presented by Ropke and Pisinger (2006) (Fig. 22.1).

An ALNS is a local search heuristic with different removal and insertion heuristics and a simulated annealing acceptance criterion. The best vehicle routes are forwarded to the master problem. Thereby, it is ensured that the generated vehicle routes fulfill the constraints (22.5)–(22.21). Due to the limited set of vehicle routes the master problem is often denoted as a restricted master problem (RMP). At the end of the iterative procedure, the master problem is formulated as a set covering problem (SCP) allowing that a request can be served several times. It is

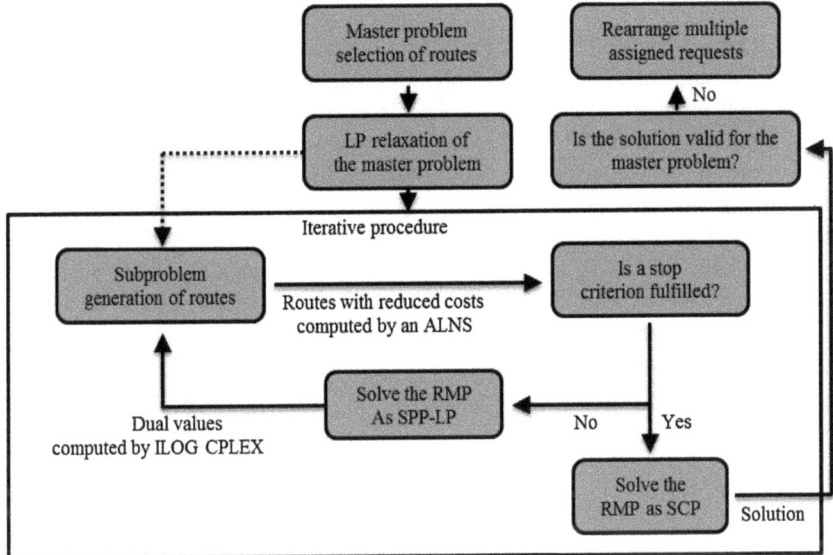

Fig. 22.1 Overview about the CGB-heuristic

identified that a solution of an SCP formulated master problem is easier to rebuild to a feasible COTTP solution compared to a solution of an SPP-LP formulated master problem. Furthermore, it is recommended to execute the CGB-heuristic two times. First, the approach is applied to the complete request portfolio P which means that each coalition member is able to bid on all requests of the coalition. Second, each coalition member uses this approach for the winning bids of the first application of the CGB-heuristic. For a detailed explanation it is referred to Wang and Kopfer (2014) and Ziebuhr and Kopfer (2014).

The proposed strict generation procedure observes the compulsiveness of requests within the subproblem. Thereby, only feasible vehicle routes are accepted by the simulated annealing acceptance criterion during the ALNS. Due to this procedure, a request is served by a transportation mode corresponding to the type of service attached to the request. To generate as many feasible vehicle routes as possible, three modifications are recommended for the ALNS. First, the external carrier option is penalized by using penalty costs for compulsory requests. Second, the fulfillment of premium 1 and premium 2 requests by coalition members as well as the fulfillment of premium 3 and premium 4 requests by long-term carriers are prohibited by infeasible insertion positions during the ALNS. Third, the request portfolio of the insertion heuristics is split into one for compulsory requests and one for standard requests in order that compulsory requests are preferred for reinsertion.

A second strategy for handling compulsory requests is the strict composition procedure where forwarding limitations are ignored by the ALNS and considered by the solution of the master problem. This means that many of the submitted routes may contain compulsory requests which are served by an improper fulfillment mode. To ensure that vehicle routes are selected, which are feasible for the considered CTP problem, the master problem is extended by new constraints which observe the applied fulfillment modes for compulsory requests. Therefore, the vehicles are numbered in an ascending order and for each compulsory request a certain range is determined. For example, in a scenario with two coalition members, where each forwarder is in charge of three vehicles per transportation mode (self-fulfillment, mode DB, and mode TB), a premium 3 request has to be served by a vehicle with the number 1–3 or 10–12. Based on the generated dual values, the ALNS is guided to generate feasible vehicle routes in terms of compulsory requests. To ensure feasible solutions by high ratios of compulsory requests in the first round of the column generation it is proposed that the ALNS only accepts feasible vehicle routes.

22.4 Computational Study

The impact of compulsory requests is determined by executing computational experiments. Therefore, existing COTP instances are modified by the definition of compulsory requests. In Wang et al. (2014) 24 COTP instances are presented where 2–5 IOTP instances with the same location structure (R1, C1, and RC1) are combined to one COTP instance with modified coordinates of the nodes. The IOTP instances are derived by extending the PDPTW instances from Li and Lim (2001) in terms of external resources. Thereby, the size of the available fleet size is set to the number of vehicles used in the best-known solutions for the PDPTW. This means that each instance could be solved without common carriers. The total vehicle fleet of each coalition member $|K_c|$ is composed of 40 % private vehicles $|K_c^1|$, 30 % vehicles on mode TB $|K_c^2|$, and 30 % vehicles on mode DB $|K_c^3|$. To consider the characteristic of the COTPP-FL, it is necessary to define different types of requests. In our experiments, instances with different ratios of compulsiveness are generated. Ratios of 5, 10, and 15 % are considered, which means, e.g., that 15 % of all requests are compulsory requests and 85 % are standard requests. For each ratio value and instance, 15 samples, which are solved once for premium 1, premium 2, premium 3, and premium 4 requests, are generated.

Each solution approach uses the same parameter setting as suggested by Ziebuhr and Kopfer (2014). The approaches are implemented in C++ (Visual Studio 2012) and the computational experiments are executed on a Windows 7 PC with Intel Core i7-2600 processor (3.4 GHz and 16 GB of main memory). To compute the dual values of the master problem, the commercial solver ILOG CPLEX (version 12.51) is applied.

In a benchmark study, the performance of the strict generation and the strict composition procedure are analyzed by solving all instances with four coalition members by considering a ratio of 10 % compulsory requests. In terms of the computational effort, both strategies perform similarly. However, in terms of the solution quality it is identified that the strict generation procedure is preferable for premium 1 and premium 3 requests while the strict composition procedure is preferable for the remaining types of compulsory requests. It is assumed that the strict generation procedure has problems identifying promising vehicle routes in case that the solution space is widely extended by additional fulfillment options.

In a second study, the impact of compulsory requests is analyzed. Thereby, the COTPP-FL is solved by the CGB-heuristic combined with the best strategy for each type of compulsory requests. The increase of transportation costs per compulsory request as well as the best COTP solutions are presented in Table 22.1.

Several findings can be derived that depend on the location structure, ratio of compulsory requests, type of compulsory request, and number of coalition members. First, it is observed that on average premium 1 requests have the highest additional costs, premium 3 requests the second highest, premium 2 requests the third highest, and premium 4 requests have the lowest additional costs. Concerning the premium 1 and premium 4 requests, the result is expected due to the different sets of applicable external resources, while the result is not obvious in case of premium 2 and premium 3 requests. Thereby, it is worth mentioning that in scenarios with low ratios and random clustered location structures premium 3 requests often have lower additional costs than premium 2 requests. In such scenarios the benefit of using the private fleet of partners is higher than the benefit of using long-term carriers. A third observation is that random location structures lead to the lowest additional costs for premium 1, premium 2, and premium 3 requests, while clustered structures are preferable for premium 4 requests. By increasing the number of coalition members the averaged impact of compulsory requests can be reduced. Another observation is that higher ratios lead to different results depending of the type of compulsory requests, while premium 1 and premium 3 requests get higher additional costs, the impact on premium 2 and premium 4 requests can be reduced by higher ratios. Thereby, it is worth mentioning that we analyzed higher ratios of compulsory requests than 15 %, but then it is observed that with higher ratios, it is difficult to identify feasible solutions for the COTPP with premium 1 and 3 requests. That is why higher ratios are skipped. Actually, in terms of a ratio of 15 % it is observed that on average with our CGB-heuristic 10 % of COTPP with premium 1 requests and 7 % of COTPP with premium 3 requests cannot be solved. Otherwise, all considered COTPP with premium 2 and premium 3 requests can be solved.

Table 22.1 Percentaged increase of transportation costs per compulsory request

Instance			Best	Premium 1 (%)			Premium 2 (%)			Premium 3 (%)			Premium 4 (%)		
id	m	N	CTP	5	10	15	5	10	15	5	10	15	5	10	15
C101	2	105	5348	0.97	1.00	1.15	0.37	0.30	0.29	0.83	0.86	1.08	0.03	0.02	0.06
C102	2	106	5340	1.08	0.86	0.99	0.31	0.22	0.18	0.93	0.75	0.84	0.00	0.00	0.00
C103	3	159	7910	0.61	0.60	0.63	0.23	0.26	0.23	0.50	0.46	0.57	0.07	0.06	0.06
C104	3	159	7335	0.60	0.67	0.72	0.27	0.24	0.20	0.44	0.55	0.68	0.02	0.01	0.02
C105	4	212	9912	0.37	0.38	0.48	0.19	0.14	0.20	0.28	0.31	0.45	0.03	0.03	0.03
C106	4	211	9721	0.56	0.56	0.58	0.32	0.30	0.29	0.47	0.43	0.56	0.05	0.04	0.03
C107	5	264	12,296	0.38	0.35	0.47	0.21	0.16	0.17	0.32	0.30	0.44	0.02	0.01	0.02
C108	5	264	12,386	0.33	0.39	0.47	0.16	0.17	0.19	0.31	0.32	0.47	0.00	0.00	0.01
R101	2	104	7052	0.34	0.41	0.48	0.25	0.23	0.21	0.27	0.31	0.39	0.05	0.08	0.06
R102	2	104	7504	0.53	0.60	0.80	0.10	0.13	0.12	0.51	0.59	0.75	0.03	0.03	0.03
R103	3	160	10,678	0.32	0.35	0.40	0.19	0.18	0.20	0.24	0.25	0.35	0.06	0.06	0.05
R104	3	154	9854	0.33	0.39	0.48	0.14	0.16	0.14	0.27	0.30	0.47	0.02	0.01	0.03
R105	4	208	13,333	0.31	0.31	0.41	0.18	0.17	0.16	0.21	0.24	0.43	0.06	0.05	0.04
R106	4	215	15,508	0.23	0.23	0.27	0.14	0.16	0.14	0.15	0.18	0.24	0.02	0.05	0.03
R107	5	265	16,222	0.20	0.24	0.27	0.16	0.15	0.18	0.14	0.19	0.26	0.00	0.04	0.03
R108	5	262	16,908	0.24	0.24	0.27	0.21	0.15	0.11	0.17	0.18	0.30	0.12	0.08	0.07
RC101	2	106	7415	0.52	0.46	0.61	0.35	0.27	0.25	0.37	0.39	0.55	0.19	0.18	0.13
RC102	2	107	8058	0.54	0.46	0.62	0.35	0.26	0.24	0.48	0.41	0.58	0.26	0.18	0.14
RC103	3	160	10,801	0.35	0.37	0.40	0.26	0.25	0.23	0.20	0.24	0.40	0.07	0.07	0.06
RC104	3	161	11,485	0.38	0.40	0.45	0.25	0.27	0.26	0.22	0.26	0.45	0.04	0.05	0.06
RC105	4	211	13,783	0.32	0.36	0.42	0.26	0.26	0.24	0.18	0.28	0.40	0.08	0.06	0.06
RC106	4	213	14,711	0.29	0.31	0.37	0.24	0.22	0.20	0.18	0.23	0.36	0.06	0.05	0.06
RC107	5	265	17,231	0.29	0.28	0.35	0.25	0.22	0.21	0.17	0.25	0.34	0.06	0.05	0.03
RC108	5	266	18,339	0.25	0.27	0.31	0.22	0.21	0.21	0.17	0.21	0.30	0.07	0.06	0.06
Mean			11,214	0.43	0.44	0.52	0.23	0.21	0.20	0.33	0.35	0.49	0.06	0.05	0.05

22.5 Conclusion and Outlook

This paper introduces a collaborative operational transportation planning problem with forwarding limitations denoted as COTPP-FL where four different types of compulsory requests are analyzed in terms of their increase of costs compared to a solution without forwarding limitations. In previous publications, the topic of compulsory requests is analyzed for transportation planning problems with either self-fulfillment and subcontracting or self-fulfillment and collaborative planning as fulfillment modes. Now, a comprehensive approach is presented where self-fulfillment, subcontracting, and collaboration are applicable as transportation modes. As a solution approach a CGB-heuristic with two different strategies for handling compulsory requests is presented.

In a benchmark study, we identified that the strict generation procedure is preferable for premium 1 and premium 3 requests and the strict composition procedure is preferable for premium 2 and premium 4 requests. In a second study, the impact of compulsory requests is analyzed. Several findings can be derived. For example, random location structures are preferable in terms of additional costs and premium 1 requests lead to the highest additional costs. For future research, it might be interesting to use our solution approach on real-life data sets.

Acknowledgments This research was supported by the German Research Foundation (DFG) as part of the project "Kooperative Rundreiseplanung bei rollierender Planung."

References

Chu C-W (2005) A heuristic algorithm for the truckload and less-than truckload problem. Eur J Oper Res 165(3):657–667

Krajewska M, Kopfer H (2009) Transportation planning in freight forwarding companies: Tabu search algorithm for the integrated operational transportation planning problem. Eur J Oper Res 197(2):741–751

Li H, Lim A (2001) A metaheuristic for solving the pickup and delivery problem with time windows. In: IEEE Computer Society (ed) Proceedings of the 13th IEEE International Conference on Tools with Artificial Intelligence (ICTAI). IEEE Computer Society, Los Alamitos, California: 160–167

Lübbecke M, Desrosiers J (2004) Selected topics in column generation. Oper Res 53(6):1007–1023

Özener OÖ, Ergun Ö, Savelsbergh M (2011) Lane-exchange mechanisms for truckload carrier collaboration. Transp Sci 45(1):1–17

Ropke S, Pisinger D (2006) An adaptive large neighborhood search heuristic for the pickup and delivery problem with time windows. Transp Sci 40(4):455–472

Schönberger J (2005) Operational freight carrier planning. Springer, Berlin, Heidelberg

Wang X, Kopfer H (2014) Collaborative transportation planning of less-than truckload freight. OR Spectrum 36(2):357–380

Wang X, Kopfer H, Gendreau M (2014) Operational transportation planning of freight forwarding companies in horizontal coalitions. Eur J Oper Res 237(3):1133–1141

Ziebuhr M, Kopfer H (2014) The integrated operational transportation planning problem with compulsory requests. In: Gonzlez-Ramrez RG et al (eds) Computational logistics. Lecture notes in computer science, vol 8760. Springer, Berlin, Heidelberg, pp 1–15

Ziebuhr M, Kopfer H (2016): Collaborative transportation planning with forwarding limitations. In: Accepted for publication in Operation research proceedings 2015. Springer

Chapter 23
Sustainable Urban Freight Transport: Analysis of Factors Affecting the Employment of Electric Commercial Vehicles

Molin Wang and Klaus-Dieter Thoben

Abstract Commercial vehicles are a common mode of transport employed in the urban freight transport. Normally, they are internal combustion engine (ICE) vehicles powered by burning fossil fuels. However, with deteriorating air quality and decreasing energy resources, ICE vehicles are experiencing a significant transition. Electric commercial vehicles (ECVs) as a feasible alternative to alleviate emissions and save energy resources are proposed in the field of urban freight transport. Nevertheless, with the increasing intention of adopting ECVs, the number of employed ECVs is lower than ICE vehicles and electric passenger cars. Therefore, this paper reviewed and analyzed 25 related articles to collect factors affecting the employment. Furthermore, we classified the factors into the three pillars of sustainability with integrating technological dimension. The results illustrate the influence of positive factors and negative factors on the employment. The future work will focus on ranking the priorities of factors and suggesting the logistics company to consider ECVs in their fleets thereby improving the adoption and the sustainable urban freight transport.

Keywords Electric commercial vehicle · Urban freight transport · Factors · Sustainability

---

M. Wang (✉) · K.-D. Thoben
Bremer Institut für Produktion und Logistik GmbH, University of Bremen, Bremen, Germany
e-mail: wag@biba.uni-bremen.de

K.-D. Thoben
e-mail: tho@biba.uni-bremen.de

23.1 Introduction

Fossil fuels as a primary fuel for powering the ICE vehicles are nonrenewable. It results in the reduction of global energy resources and increasing the energy imported dependency. On the other hand, ICE vehicles release harmful emissions such as the greenhouse gas (GHG) and the particulate matter (PM). They accelerate global warming and deteriorate air quality. In the early 1970s, electric vehicles (EVs), which are powered by electricity, reappeared because of the energy crisis. Nowadays, EVs are still a feasible alternative to solve above problems, because of low or zero emissions, quiet driving and independent on oil.

Urban freight transport (UFT) is a segment of freight transport to carry goods by or for commercial entities into, out of and within urban area. It plays an essential role in satisfying the needs of citizens, supporting efficient economic and social development. (Dablanc 2009; MDS Transmodal 2012; Comi et al. 2013; Taniguchi 2013). Currently, the rapid urbanization results in a large number of commercial vehicles running on the urban road. Thus, low emission vehicles have become one of the recommended measures proposed in the European Commission's Transport White Paper (European Commission 2011). However, because of high intention and low amount of ECVs, this paper collected, classified and analyzed the factors affecting the employment to explore the reasons.

23.2 State of the Art

Sustainable development as a concept to improve urban freight transport is attracting increasing attentions. There is no general accepted definition of sustainable development and sustainable transportation, but Brundtland proposed the most cited definition in the report Our Common Future. It defined the sustainable development is development that meets the needs of the present without compromising the ability of future generations to meet their own needs (Brundtland 1987). There are a number of researchers promoting the development of sustainable urban freight transport to satisfy the definition. Taniguchi proposed a concept of city logistics to optimize the logistics and transport activities with the support of advanced information systems in urban areas (Taniguchi et al. 2001). Moreover, Maden analyzed using traffic information with time-varying speeds to plan routes can reduce CO_2 emissions when compared with plans based on constant speeds and a general contingency allowance (Maden et al. 2010). In addition, Allen reviewed 114 urban consolidation centers (UCC) schemes, which are able to reduce goods vehicle traffic, vehicle-related greenhouse gas emissions and local air pollution in 17 countries (Allen et al. 2012).

The electric vehicle is one of technological measures to achieve sustainable urban freight transport. There are three categories in electric vehicles. They are Battery Electric Vehicles (BEVs), Hybrid Electric Vehicles (HEVs), and Fuel Cell

Vehicles (FCVs). For the HEV, depending on the type of refueling, there is another category called Plug-in Hybrid Electric Vehicles (PHEV), whose batteries are able to be charged by electricity grid. Due to comprehensive studies of EVs about performance, attitude, emissions and costs (Lu et al. 2013; Chau et al. 2008; Moreno et al. 2006; Hackbarth and Madlener 2013; Zhang et al. 2011; Hao et al. 2012; Granovskii et al. 2006; Van Vliet et al. 2011), EVs have matured gradually and been sold or leased by major automobile manufacturers in recent years. Nevertheless, the amount of adoption of EVs is still low. In 2013, alternative fuel vehicles, which include electric vehicles, hybrid electric vehicles and natural gas vehicles, accounted for only 4 % of total sales of passenger cars and only 1 % of light commercial vehicles in EU-28 (The international council on clean transport 2014). Based on statistical data from Federal Motor Transport Authority of Germany, the total electric and hybrid commercial vehicles are 3.069 on January 1, 2014. It accounts for only 0.12 % of total new registration commercial vehicles in 2013 (Statista 2014). The data implies that the employment of electric commercial vehicles is still a challenge.

23.3 Method

This paper reviewed the literatures, which combined the research field of electric commercial vehicles to the urban freight transport from 2005 to 2015. Web of Science, Science Direct, IEEE Explore and Google Scholar are selected database. Since there are a number of terms expressing same meanings, the research selected a set of keywords for searching papers in the database. They are electric vehicles, electric trucks, electric vans, electric commercial vehicles, green cars, clean cars, new energy vehicles, urban freight, urban goods, city logistics and urban logistics. There are total 28 combinations searched in these four database and finally 25 papers studied electric vehicles in the urban freight transport.

According to the contents, the selected papers are able to be classified in six categories. First of all are review papers. They summarized the rapid changes and developments during the past few years and analyzed the current issues and opportunities for employing ECVs in city areas (Taefi et al. 2014; Iwan et al. 2014; Foltyński 2014; Visser et al. 2014). The second category is survey papers. They discussed the needs, requirements and acceptance with questionnaires or interviews and illustrated drivers and barriers when adopt ECVs (Klauenberg et al. 2014; Ehrler and Hebes 2012; Sierzchula 2014; Klumpp et al. 2014; Ablola et al. 2014). The third category is about electric vehicle routing problem. They proposed different algorithms to minimize travel distance, service time and total costs with considering charging stations (Schneider et al. 2012; Yang et al. 2015; Conrad and Figliozzi 2011; Afroditi et al. 2014; Van Duin et al. 2013; Schau et al. 2015). Replaced ICE commercial vehicles with ECVs are the fourth category. The researchers tested feasibility of battery capacity, showed changes of emissions and analyzed potential problems in real life (Feng and Figliozzi 2013; Browne et al.

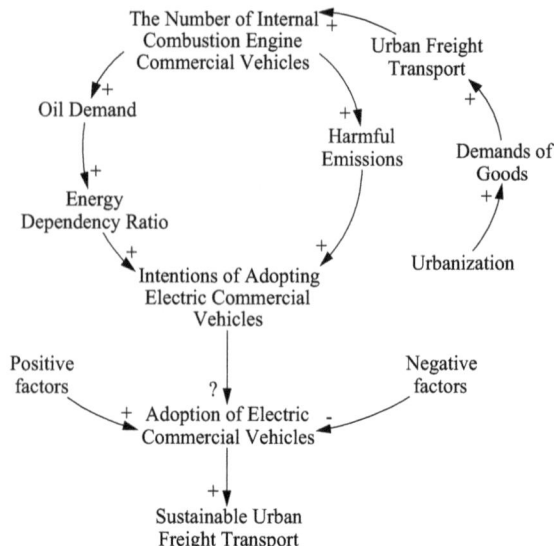

Fig. 23.1 The research framework

2011; Leonardi et al. 2012; Vonolfen et al. 2011; Lebeau et al. 2013; Melo et al. 2014; Tipagornwong and Figliozzi 2014). The fifth category is about new methodology. The researchers from different perspective examined and quantified the adoption of ECVs (Roumboutsos et al. 2014: Davis and Figliozzi 2013). Finally, the sixth category discussed the relationship between total costs of ownership and payload with a case study (Macharis et al. 2013).

In order to illustrate the motivation and the research question clearly, we developed a research framework (Fig. 23.1) with system dynamics. The model describes why the urban freight transport intends to employ ECVs, how the factors influence on the adoption, what is the objective of the research. Then, we proposed the research question: why the intention of the adoption is high, while the amount of employment is low.

23.4 Analysis of Factors

To answer the research question, the paper collected factors from above-selected literatures and classified them into the three pillars of sustainability (economic, social, and environmental) with integrating technological dimension. According to the definition of urban freight transport, the UFT is the commercial entities existing in the physical reality called environment. It is organized and operated in the society and supports social development. Furthermore, the UFT is a fundamental element for economic vitality and impacts on the efficiency of the economy. Therefore, when we discuss the factors in the UFT, the economic, social and environmental dimensions are essential roles. On the other hand, the electric commercial vehicle is

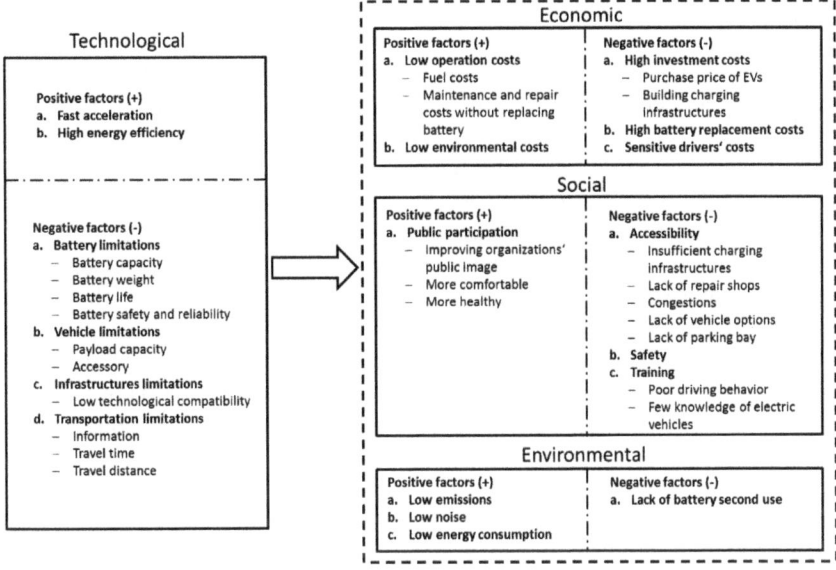

Fig. 23.2 Factors affecting employment of ECVs in the UFT

a technology, which is one of pathways providing innovations to overcome obstacles and to pursue the goal of sustainable development. In this research, when we study on the employment problem of the ECVs in the UFT, we have regarded them as an entirety. Thus, the research applied the three pillars as the basic dimensions and the technology as an external dimension to classify and analyze the factors.

Figure 23.2 shows the result of classification. The positive and the negative factors explain what are promoting ECVs in real life and what are barriers preventing ECVs from being adopted in the urban freight transport. The arrow in the diagram illustrates the technological dimension is an external element of the UFT and the dimension will be considered when the ECVs is employed into the UFT. In addition, the research proposed an ECV-UFT system. It is an urban freight transport system, which employs electric commercial vehicles to deliver goods.

23.4.1 Economic Dimension

The economic dimension in this paper describes a group of factors, which concerns total costs of ownership (TCO) in the ECV-UFT system. The TCO is defined as a purchasing tool and philosophy which is aimed at understanding the true cost of buying a particular good or service from a particular supplier (Ellram 1995). From the perspective of electric vehicles, TCO includes investment costs (purchase price and infrastructures), total operating costs (fuel costs, electricity costs, maintenance, and repair costs), and ownership costs (insurance costs, license and registration costs, depreciation costs, road and vehicle tax) (Taefi et al. 2014; American Automobile Association 2015). From the perspective of transportation, TCO contains vehicles costs, travel time costs, road and parking facility costs, congestion costs, traffic cashes costs, environmental costs and so on (Victoria Transport Policy Institute 2014). This paper illustrates the opportunities and the barriers of the employment in terms of both perspectives.

There are two categories of factors. The positive factors show that the operating costs and environmental costs are low, since electric commercial vehicles are powered mainly by electricity. It reduces the consumption of fossil fuels and harmful emissions. Moreover, due to the simplified structure of ECVs, the maintenance and repair costs are low as well when it doesn't include costs of replacing battery. However, because of high battery price and demands of recharging, when a logistics company replaces ICE vehicles with electric vehicles in their fleets, the high investment costs will be a barrier to prevent them from purchasing EVs and building charging infrastructures. Likewise, high price of battery leads to high battery replacement costs. Furthermore, because EVs are a new technology for drivers, the logistics companies need to pay the training fee for drivers to adapt EVs and drive efficiently. In addition, drivers' salary is related to working time. When the efficiency is improved by employing ECVs, the working time will be decreased so that the drivers' salary will be reduced too. Therefore, the drivers' cost is a sensitive factor in the economic dimension.

23.4.2 Social Dimension

The social dimension in this paper describes a group of factors, which concerns a livable community in the ECV-UFT system. Normally, equity, human health, education, community, accessibility, quality of life, and public participation are sustainability issues in the social dimension (Victoria Transport Policy Institute 2015; McKenzie 2004; Basiago 1999). Therefore, the paper collected and analyzed the factors according to these categories of the issues. First of all, the positive factor is public participation. When an organization employs electric vehicles in their fleet, it will improve their public image, since the environmental friendly features of EVs. Moreover, because of more comfortable and healthy working environment, it leads to a positive attitude from drivers when they have the experience of driving ECVs. Thus, they encourage the public participation.

In contrast, when ECVs are adopted in real life, there are three aspects negative factors. First, accessibility is a problem to support the employment, since congestions, insufficient charging infrastructures, short of repair shops, few vehicle options and limited parking bays. Congestions and lack of parking bays affect the distribution efficiency, because they consume time and battery energy. Insufficient charging infrastructures and short of repair shops are worries from customers, because they are necessaries when the customers consider the purchase of ECVs. More vehicle options are able to satisfy different requirements from different companies and driving conditions. Second, safety is a problem for pedestrians, because electric commercial vehicles are so quiet that the pedestrians are hard to hear them. Finally, poor driving behavior and little knowledge of electric vehicles decrease efficiency and acceptance. Therefore, training drivers is a way to improve current status.

23.4.3 Environmental Dimension

The environmental dimension in this paper describes a group of factors, which concerns environmental quality and natural resources in the ECV-UFT system. The environmental quality is related to the emissions and noise. The natural resources are referred to the energy consumption and the reuse. According to the above studies, most of papers focused on calculating CO_2 emissions and energy consumption from well to wheel when ICE vehicles are replaced with electric vehicles. The objective is to know how many emissions and how much energy are saved and evaluate the advantages of the replacement. However, noise is a neglected part in above search results. Most of the articles mentioned low noise as an advantage of employing ECVs, but apparently, noise has not priority over emissions and energy consumption. Even so, the low emissions, noise and energy consumption are triggers to research and develop electric vehicles. On the other hand, lack of battery second use is a negative factor in the environmental dimension. It means that there are problems in recycling and reusing of electric vehicles and their batteries. Nevertheless, same as noise, recycling, and reusing are still not the first place to be considered when ECVs are employed in real life.

23.4.4 Technological Dimension

The technological dimension in this paper describes a group of factors, which concerns current technological drivers and barriers in the ECV-UFT system. Apparently, when ECVs are employed in the urban freight transport, the energy efficiency is improved, because the electric vehicle operates without internal combustion engines or with them partly. When the engines converting chemical energy to mechanical energy, they lose more than 50 % energy to friction, pumping air and wasted heat. Likewise, because of high energy efficiency and smooth shifting of electric motors, EVs accelerate faster than ICE vehicles. Therefore, they are advantages attracting people to consider using ECVs. However, there are four limitations preventing the employment.

First of all are battery limitations. Customers expect long driving range, which require high battery capacity. Actually, it is possible to obtain a large battery capacity, but it results in heavier battery weight thereby increasing the gross vehicles weight. Therefore, a light, small size and high capacity battery is the goal of researchers, but the total costs limits the development and employment. Moreover, the battery life, battery safety and reliability are parameters to evaluate the value of replacing ICE vehicles and the timing to reuse and recycle. Secondly, the payload capacity and accessories are vehicle limitations. The battery size has an influence on volume of payload thereby affecting the efficiency of distribution. Furthermore, air conditions as one component of accessories consume a deal of electricity in summer or winter and reduce the driving range. Thirdly, low technological compatibility is a charging infrastructure limitation. Different connectors, sockets and payment systems complicate the process of recharging thereby reducing the desire of organizations to purchase ECVs in their fleets. Finally, transportation limitations include information, travel time and travel distance. With a suitable information system is able to minimize the travel time and travel distance in one planning route and increase frequencies of delivering goods so that improve the utilization of ECVs.

23.5 Conclusion

The research reviewed and described the state of the art in the research field of electric commercial vehicles, urban freight transport and the combination of these two fields. The papers of the combination are classified in six categories. The research collected and analyzed factors to answer the research question: why the intention of the adoption is high, while the amount of employment is low. There are three basic dimensions from the sustainability (economic, social and environmental) and an external dimension (technological) to categorize and record the collected factors. The results show that low operation and environmental costs, public participation, low emissions, noise and energy consumption, fast acceleration and high energy efficiency are positive factors promoting the employment of ECVs. On the other hand, investment costs, battery replacement costs, drivers' costs, accessibility, safety, training, battery second use, technological limitations of batteries, vehicles, recharging infrastructures and transportation are problems preventing the employment. The future work will focus on ranking the selected factors and finding the priorities to alleviate the impacts of the negative factors.

References

Ablola M, Plant E, Lee C (2014) The future of sustainable urban freight distribution – a Delphi study of the drivers and barriers of electric vehicles in London. Hybrid and electric vehicles conference; 5th IET, 1–7

Afroditi A, Boile M, Theofanis S, Sdoukopoulos E, Margaritis D (2014) Electric vehicle routing problem with industry constraints: trends and insights for future research. Transp Res Procedia 3:452–459

Allen J, Browne M, Woodburn A, Leonardi J (2012) The role of urban consolidation centres in sustainable freight transport. Transp Rev 32(4):473–490

American Automobile Association (2015) Your driving costs how much are you really paying to drive. http://exchange.aaa.com/wp-content/uploads/2015/04/Your-Driving-Costs-2015.pdf

Basiago AD (1999) Economic, social, and environmental sustainability in development theory and urban planning practice. Environmentalist 19:145–161

Browne M, Allen J, Leonardi J (2011) Evaluating the use of an urban consolidation centre and electric vehicles in central London. Int Assoc Traffic Saf Sci Res 35:1–6

Brundtland GH (1987) Report of the world commission on environment and development: our common future. Oslo

Chau KT, Chan CC, Liu CH (2008) Overview of permanent-magnet brushless drives for electric and hybrid electric vehicles. IEEE Trans Ind Electron 55(6):2246–2257

Comi A, Donnelly R, Russo F (2013) Modelling freight transport: chapter 8 urban freight models. London

Conrad RG, Filiozzi MA (2011) The recharging vehicle routing problem. In: Proceedings of the 2011 industrial engineering research conference

Dablanc L (2009) Freight transport for development toolkit: urban freight. The International Bank for Reconstruction and Development, Washington DC

Davis BA, Figliozzi MA (2013) A methodology to evaluate the competitiveness of electric delivery trucks. Transp Res Part E 49:8–23

Ehrler V, Hebes P (2012) Electromobility for city logistics- the solution to urban transport collapse? An analysis beyond theory. Procedia—Soc Behav Sci 48:786–795

Ellram LM (1995) Total cost of ownership an analysis approach for purchasing. Int J Phys Distrib Logist Manage 25(8):4–23

European Commission (2011) White paper: roadmap to a single European transport area—towards a competitive and resource efficient transport system. Brussels

Feng W, Figliozzi MA (2013) An economic and technological analysis of the key factors affecting the competitiveness of electric commercial vehicles: A case study from the USA market. Transp Res Part C 26:135–145

Foltyński M (2014) Electric fleets in urban logistics. Procedia—Soc Behav Sci 151:48–59

Granovskii M, Dincer I, Rosen MA (2006) Economic and environmental comparison of conventional, hybrid, electric and hydrogen fuel cell vehicles. J Power Sources 159:1186–1193

Hackbarth A, Madlener R (2013) Consumer preferences for alternative fuel vehicles: a discrete choice analysis. Transp Res Part D 25:5–17

Hao H, Wang HW, Ouyang MG (2012) Fuel consumption and life cycle GHG emissions by China's on-road trucks: future trends through 2050 and evaluation of mitigation measures. Energy Policy 43:244–251

Iwan S, Kijewska K, Kijewski D (2014) Possibilities of applying electrically powered vehicles in urban freight transport. Procedia—Soc Behav Sci 151:87–101

Klauenberg J, Gruber J, Frenzel I, Zajicek J, Kaplan S (2014) Needs, requirements and attitudes of specific commercial sectors in Denmark, Austria and Germany with respect to the use of electric vehicles in commercial transport. In: European electric vehicle congress, Brussels, Belgium

Klumpp M, Witte C, Zelewski S (2014) Information and process requirements for electric mobility in last mile logistics. Information technology in environmental engineering, selected contributions to the sixth international conference on information technologies in environmental engineering (ITEE2013). Springer, Berlin, Heidelberg

Lebeau P, Macharis C, Van Mierlo J, Maes G (2013) Implementing electric vehicles in urban distribution: a discrete event simulation. In: EVS27 international battery, hybrid and fuel cell electric vehicle symposium, Barcelona, Spain

Leonardi J, Browne M, Allen J (2012) Before-after assessment of a logistics trial with clean urban freight vehicles: A case study in London. Procedia—Soc Behav Sci 39:146–157

Lu LG, Han XB, Li JQ, Hua JF, Ouyang MG (2013) A review on the key issues for lithium-ion battery management in electric vehicles. J Power Sources 226:272–288

Macharis C, Lebeau P, Van Mierlo J, Lebeau K (2013) Electric versus conventional vehicles for logistics: a total cost of ownership. In: EVS27 international battery, hybrid and fuel cell electric vehicle symposium, Barcelona, Spain

Maden W, Eglese R, Black D (2010) Vehicle routing and scheduling with time-varying data: a case study. J Oper Res Soc 61:515–522

McKenzie S (2004) Social sustainability: towards some definitions. Hawke Research Institute Working Paper Series No 27

MDS Transmodal Limited (2012) DG MOVE European commission: study on urban freight transport final report

Melo S, Baptista P, Costa A (2014) Comparing the use of small sized electric vehicles with diesel vans on city logistics. Procedia—Soc Behav Sci 111:1265–1274

Moreno J, Ortúzar ME, Dixon JW (2006) Energy-management system for a hybrid electric vehicle, using ultracapacitors and neural networks. IEEE Trans Ind Electron 53(2):614–623

Roumboutsos A, Kapros S, Vanelslander T (2014) Green city logistics: systems of Innovation to assess the potential of E-vehicles. Res Transp Bus Manage 11:43–52

Schau V, Rossak W, Hempel H, Späthe S (2015) Smart City Logistik Erfurt (SCL): ICT-support for managing fully electric vehicles in the domain of inner city freight traffic: a look at an ongoing federal project in the city of Erfurt, Germany. In: Proceedings of the 2015 international conference on industrial engineering and operations management, Dubai, United Arab Emirates (UAE)

Schneider M, Stenger A, Goeke D (2012) The electric vehicle routing problem with time windows and recharging stations. Technical report

Sierzchula W (2014) Factors influencing fleet manager adoption of electric vehicles. Transp Res Part D 31:126–134

Statista (2014) Anzahl der Lastkraftwagen (Lkw) mit alternativen Antrieben in Deutschland (Strand: 1.Januar 2014) KBA. http://de.statista.com/statistik/daten/studie/259803/umfrage/lkw-bestand-mit-alternativen-antrieben-in-deutschland/

Taefi T, Kreutzfeldt J, Held T, Konings R, Kotter R, Lilley S, Baster H, Green N, Laugesen MS, Jacobsson S, Borgqvist M, Nyquist C (2014) Comparative analysis of European examples of freight electric vehicles schemes. In: 4th international conference on dynamics in logistics, Bremen

Taniguchi E (2013) Urban freight transport management for sustainable and livable cities. Global challenges in smart logistics—innovation driving supply chain control, Utrecht, Netherland. http://www.rvo.nl/sites/default/files/2014/09/Urban%20Freight%20Transport%20Management-Taniguchi.pdf

Taniguchi E, Thompson RG, Yamada T, Van Duin R (2001) City Logistics: Network modelling and intelligent transport systems. Pergamon, Amsterdam

The international council on clean transportation (2014) European vehicle market statistics pocketbook. Berlin

Tipagornwong C, Figliozzi MA (2014) An analysis of the competitiveness of freight tricycle delivery services in urban areas. In: The 93rd annual meeting of the transportation research board, Washington, D.C

Van Duin JHR, Tavasszy LA, Quak HJ (2013) Towards E(lectric)-urban freight: first promising steps in the electric vehicle revolution. European Transport\Trasporti Europei Issue 54, Paper n ° 9. ISSN 1825–3997

Van Vliet O, Brouwer AS, Kuramochi T, Van den Broek M, Faaij A (2011) Energy use, cost and CO_2 emissions of electric cars. J Power Sources 196:2298–2310

Victoria Transport Policy Institute (2014) Transportation costs and benefits. http://www.vtpi.org/tdm/tdm66.htm

Victoria Transport Policy Institute (2015) Sustainable transportation and TDM. http://www.vtpi.org/tdm/tdm67.htm

Visser J, Nemoto T, Browne M (2014) Home delivery and the impacts on urban freight transport: a review. Procedia—Soc Behav Sci 125:15–27

Vonolfen S, Affenzeller M, Beham A, Wagner S (2011) Simulation-based evolution of municipal glass-waste collection strategies utilizing electric trucks. In: LINDI 2011 3rd IEEE international symposium on logistics and industrial informatics, Budapest, Hungary

Yang HM, Yang SP, Xu Y, Cao EB, Lai MY, Dong ZY (2015) Electric vehicle route optimization considering time-of-use electricity price by Learnable Partheno-Genetic algorithm. IEEE Trans Smart Grid 6(2):657–666

Zhang Y, Yu YF, Zou B (2011) Analyzing public awareness and acceptance of alternative fuel vehicles in China: the case of EV. Energy Policy 39:7015–7024

Chapter 24
Evaluation of Practically Oriented Approaches for Operational Transportation Planning

Heiko Kopfer, Herbert Kopfer, Benedikt Vornhusen and Dong-Won Jang

Abstract Integrated operational transportation planning (IOTP) refers to the extension of vehicle routing by the option of subcontraction. In practice, planners often use semi-manual strategies for IOTP. Two wide spread semi-manual strategies for IOTP are presented. The strategies are evaluated by computational experiments comparing the results achieved by these strategies with the exact solutions which have been provided by MIP approach for IOTP.

Keywords Integrated operational transportation planning · Cherry-Picking · Bundle-Oriented strategy

24.1 Introduction

Due to fierce competition, freight forwarders are forced to increase the efficiency of their transportation processes. One important remedy for efficiency increase is realized by subcontraction; i.e., forwarders enlarge their transportation capacity and decrease costs by combining self-fulfillment (using own vehicles) and subcontraction (charging external carriers). Extending the traditional vehicle routing problem by adding the option of subcontraction is called integrated operational transportation planning (IOTP). The IOTP, including different tariffs for carrier payments, has been introduced in (Krajewska and Kopfer 2009). Several papers in the literature have

---

H. Kopfer (✉)
BMW Group, Entwicklung Antrieb, EA-202, Munich, Germany
e-mail: heiko.kopfer@bmw.de

H. Kopfer · B. Vornhusen · D.-W. Jang
Chair of Logistics, University of Bremen, Bremen, Germany
e-mail: kopfer@uni-bremen.de

B. Vornhusen
e-mail: bvornhusen@uni-bremen.de

D.-W. Jang
e-mail: dwjang@uni-bremen.de

shown that IOTP has tremendous advantages in comparison to traditional vehicle routing (e.g. Kopfer and Schönberger 2009; Kopfer and Wang 2009). First, IOTP can reduce fulfillment costs since the vehicles of the own fleet are deployed for efficient routes only, while the more or less incompatible requests, which cannot be assigned to such efficient routes are subcontracted to carriers. Second, IOTP enables forwarders to operate a fleet of reduced size. Operating a relatively small fleet, which is only able to carry out a part of the usually incoming amount of transportation requests, is valuable for guaranteeing a high degree of capacity utilization and efficient operations even in case of fluctuating order situations for transportation requests. On the one hand, IOTP is useful for increasing the transport efficiency of forwarders. On the other hand, the complexity of planning is essentially increased in comparison to the problem of traditional vehicle routing.

In literature, integrating subcontraction into vehicle routing has been solved by using different optimization methods (e.g. Pankratz 2002; Savelsbergh and Sol 1998; Bolduc et al. 2007; Chu 2005). Since the resulting planning problems are highly complex, almost all known optimization methods proposed for IOTP are heuristic approaches, which do not claim to guarantee the optimality of the generated solutions. An exception which is not based on heuristics is given by (Kopfer and Wang 2009), who analyze the benefits of subcontraction for small instances of the Vehicle Routing Problem (VRP).

In contrast to most approaches in the literature, this paper strives for exact IOTP. This is realized by applying a commercial solver to handle a mathematical model for IOTP. Of course, exact IOTP can only be reached for relatively small problem instances. A unique feature of this paper is that strategies for manual IOTP, which are actually used in practice, are considered and evaluated. Two manual strategies are emulated by using mathematical models for IOTP. The results achieved by the emulation of the two manual IOTP strategies are compared with the results of exact IOTP. In the following, IOTP is considered for the planning situation that all transportation requests are of the type "Pick-up and Delivery" and "Full-Truckload". Consequently, the basic planning problem of the considered scenario of IOTP is a Pick-up and Delivery Problem (PDP) with Full-Truckload (FTL), i.e., we consider FTL-PDP-based IOTP. Further on, it is assumed that the maximum length of tours generated by IOTP is restricted to an upper limit so that all tours can be performed by one driver within a single day.

24.2 Fulfillment Modes of IOTP

There are three fulfillment modes for IOTP: self-fulfillment, tour-based subcontraction, and daily-based subcontraction (Krajewska and Kopfer 2009; Kopfer and Wang 2009). Self-fulfillment means that vehicles of the own fleet are used for request fulfillment. The costs of using the own fleet are composed of fixed costs and variable cost. The fixed costs of the own fleet are not decision relevant for short-term planning on a daily planning interval. The variable costs for employing an own vehicle depends on the distance traveled by that vehicle and are assumed to be linear dependent on the travel distance. Subcontraction means that transportation requests are combined to complete tours which are forwarded to external carriers. The freight costs of subcontraction depend on the type of subcontraction. Applying the tour-based type of subcontraction, the payment for the execution of the tour depends on the length of the route to be performed. The calculation of the transportation fee is based on a fixed tariff rate per distance unit, i.e., the amount of payment is calculated by multiplying the length of the entrusted route with the agreed tariff rate. For the forwarding company there are no fixed costs incurred by the usage of external vehicles if subcontraction on tour basis is applied. But compared to the usage of own vehicles the variable costs for tour-based subcontraction are higher than those for self-fulfillment because the payment of the forwarder has to cover a part of the fixed costs of the carrier.

Daily-based subcontraction refers to paying the subcontractors on a daily basis independent of the size or length of the forwarded tours. In this case an external carrier gets a daily flat-rate and has to fulfill all the received requests of a single day up to an agreed distance and time limit. Costs for all three types of subcontraction are schematically shown in Fig. 24.1.

If only one single fulfillment mode is considered, IOTP reduces to a vehicle routing problem with homogeneous fleets, while the IOTP itself refers to a planning situation with heterogeneous fleets whose vehicles differ with respect to cost structures.

24.3 Commonly Used Planning Strategies for IOTP

Two planning strategies which are widely spread and commonly used in practice are presented in this section. The presented strategies are suitable for semi-manual planning performed by dispatchers who use a customary planning tool for traditional vehicle routing problems. The first strategy is called Cherry-Picking and the second one is called Bundle-Oriented strategy. These planning strategies pursue a sequential approach for IOTP. They reduce the heterogeneous vehicle routing

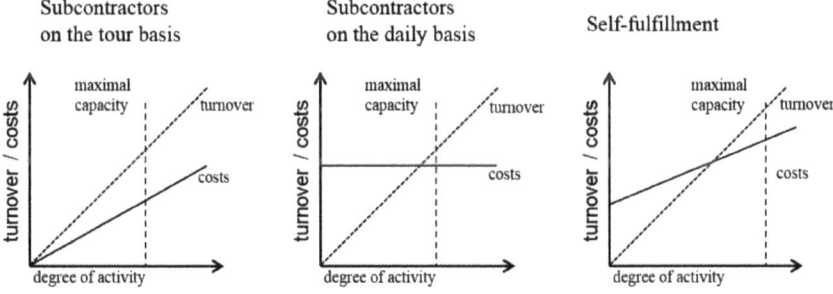

Fig. 24.1 Comparison of costs for different types of subcontraction and self-fulfillment (cf. Kopfer and Krajewska 2007)

problem associated with IOTP to related homogeneous vehicle routing problems, and then they build feasible solutions for IOTP based on the optimal solutions of the reduced homogeneous problems. Of course, solving the original IOTP in such a sequential way is a method for reducing the complexity of the problem but it does not guarantee the optimality of the generated solutions.

In principle, Cherry-Picking means that the best (i.e. most attractive) requests should be carried out by vehicles of the own fleet. Then, from the set of all requests which are not earmarked for self-fulfillment, the most attractive ones should be selected for tour-based subcontraction. Finally, all remaining requests are left to be executed by daily-based subcontraction. The aim of Cherry-Picking is composed of the following targets:

- First, for the own fleet a high degree of utilization as well as high profit contributions are aspired. The aim is to cover the high fixed costs of own vehicles and to achieve revenues which exceed the fixed cost pool of the own fleet as far as possible.
- Second, it is aspired that the requests which are subcontracted on tour basis are generating as much profit as possible. Tour-based subcontraction is planned with a lower priority than self-fulfillment since, in case of short-term planning, self-fulfillment is cheaper and more effective than tour-based subcontraction. The freight costs caused by requests of a tour-based subcontracted tour depend on the length of that tour, while the revenues for requests are fixed by the payment of the forwarder's shippers and are independent of the characteristics of the tours they are assigned to. Similar to planning for self-fulfillment, tour-based subcontraction also seeks a high degree of transportation service (respectively, revenue) opposed to relative low (freight) costs. This target is reached by constructing well-planned tours for attractive transportation requests with high profit contributions.
- Third, it is tried to keep the costs for daily-based subcontraction low. This transportation mode is relevant for all remaining requests (i.e., those request which have not been assigned to self-fulfillment or to tour-based subcontraction). Keeping the cost low is equivalent to minimizing the number of vehicles which are entrusted with requests. Since the remaining requests which are left over for daily-based subcontraction tend to have low profit margins and to be incompatible, it has to be accepted that no good bundling with efficient routes will be possible for daily-based subcontraction. The carriers paid on daily-based tariffs are only occasionally hired in case that it is not possible to fulfill all incoming requests by the modes of self-fulfillment and tour-based subcontraction. That is why daily-based subcontraction is planned with a lower priority than tour-based subcontraction.

According to the above targets for Cherry-Picking, a dispatcher will first try to build an efficient transportation plan for the own fleet. Then, he will try to build optimal routes for entrusting carriers on tour basis. Finally he will try to subcontract the remaining requests on a daily basis using a minimal number of vehicles.

The Bundle-Oriented strategy consists of two steps. The first step aims to build efficient routes by minimizing the travel distances and the second step tries to find the best way for assigning these routes to transportation modes.

- The first step (building efficient routes) is realized by solving the vehicle routing problem for the entire set of available requests and all available vehicles of any fulfillment mode. The objective criterion is the minimization of the total travel distance. Of course, there are some restrictions for the length of single tours, since no tour is allowed to exceed the agreed travel distance which can be reached by a single driver during one day. In our case this is assumed to amount to 630 km.
- The second step (assigning tours optimally to transportation modes) can easily be done. The longest tours are assigned to vehicles out of the own fleet so that all own vehicles are occupied by tours assigned to them. All remaining tours whose travel distance exceeds the upper limit for daily-based subcontraction are assigned to tour-based subcontraction. The residual tours are also assigned to tour-based subcontraction if the following two conditions are fulfilled. First, there are still unoccupied vehicles available for tour-based subcontraction and, second, tour-based subcontraction is cheaper than the flat-rate of daily-based subcontraction. Otherwise (i.e. one of the above two conditions is not fulfilled) the relevant tour is assigned to daily-based subcontraction.

The above two strategies differ with respect to the sequence of decision making. In case of the Bundle-Oriented strategy, a dispatcher first solves the vehicle routing problem in total without regarding fulfillment modes. Then he makes a decision on assigning tours to modes for the generated tours. On the opposite, a dispatcher following the Cherry-Picking strategy involves mode decisions already at the beginning of his planning process, while he does vehicle routing for all modes separately and in a sequential way. The strategies additionally differ with respect to the level of making decisions on fulfillment modes. Cherry-Picking makes the mode decision on the level of requests which are well fitting into routes. Contrary, the Bundle-Oriented strategy makes this decision on the level of previously generated bundles.

For both strategies, the dispatcher is assumed to play out his strategy semi-manually by applying planning software, which is able to handle the homogeneous vehicle routing problems within the different steps of the above strategies.

24.4 Mathematical Models for Common Planning Strategies and for Simultaneous Planning

In this section MIP formulation for FTL-PDP-based IOTP is presented. This model formulation can be used for generating exact solutions for small problem instances. Additionally, an optimization model useful for Cherry-Picking and a model for performing the Bundle-Oriented strategy are proposed.

Let $LOC = \{0, 1, 2n\}$ denote a set of locations. The identifier $loc = 1, \ldots, 2n$ is used for denoting customer locations and $loc = 0$ denotes the depot of the FTL-PDP-based IOTP. The distance matrix for locations from LOC is denoted by $DIST_{loc1,loc2}$ with $loc1, loc2 \in LOC$. Consider J to be a set which is composed of the depot $j = 0$ and of FTL-PDP requests $(j, j + n)$ with $1 \leq j \leq n$ and $j \in LOC$. Each FTL-PDP has assigned a freight rate Fr_j $(1 \leq j \leq n)$. The freight rate Fr_j specifies the amount of freight per transportation distance, which is obtained by the forwarder from its shipper for fulfilling request j. The transportation distance d_j^T of request j is equal to $d_j^T = DIST_{j,j+n}$. The FTL-PDP can be considered as an mTSP if the set $J = \{0, 1, \ldots, n\}$ consisting of the depot and the requests of the FTL-PDP are taken as the customer locations of the mTSP. The distance from the depot to a request j can be calculated as $d_{0j}^R = DIST_{0,j}$, and the distance from j to the depot is $d_{j0}^R = DIST_{j+n,0}$, respectively. The distance between two requests i and j is given by $d_{ij}^R = DIST_{j,j+n}$ for all $i, j \in J$. Assume that there is a set K of vehicles $k \in K = \{1, \ldots, m\}$. The set $K = K1 \cup K2 \cup K3$ is composed of $m = m_1 + m_2 + m_3$ vehicles with m_1 vehicles from $K_1 = \{1, \ldots, m_1\}$ for self-fulfillment; m_2 vehicles from $K_2 = \{m_1 + 1, \ldots, m_1 + m_2\}$ for subcontraction on tour basis; and m_3 vehicles from $K_3 = \{m_1 + m_2 + 1, \ldots, m_1 + m_2 + m_3\}$ for daily-based subcontraction. The constant t_k is used for specifying the maximal allowed travel distance of vehicle k. Let a_k denote for vehicle k the variable costs per travel distance and let b_k denote the fixed costs for employing vehicle k. The binary variable x_{ijk} is one if vehicle k is fulfilling request j straight after request i, and otherwise x_{ijk} is zero. The binary variable $y_{jk} = 1$ if request j is fulfilled by vehicle k, otherwise $y_{jk} = 0$.

The FTL-PDP based IOTP can be modeled as follows:

$$\max \sum_{i=1}^{n} d_i^T \cdot Fr_i - \left[\sum_{i=0}^{n} \sum_{j=0}^{n} \sum_{k=1}^{m} a_k \cdot \left(d_i^T + d_{ij}^R \right) \cdot x_{ijk} + \sum_{i=1}^{n} \sum_{k=1}^{m} b_k \cdot x_{0ik} \right] \quad (24.1)$$

Subject to

$$\sum_{i=0}^{n} x_{ijk} = \sum_{i=0}^{n} x_{jik} \quad \forall j \in J, \forall k \in K \quad (24.2)$$

$$\sum_{i=0}^{n} x_{ijk} = y_{jk} \quad \forall j \in J, \forall k \in K \quad (24.3)$$

$$\sum_{k=1}^{m} y_{jk} = 1 \quad \forall j \in J\setminus\{0\} \quad (24.4)$$

$$\sum_{j=1}^{n} x_{0jk} \leq 1 \quad \forall k \in K \quad (24.5)$$

$$\sum_{i=0}^{n} \sum_{j=0}^{n} (d_i^T + d_{ij}^R) \cdot x_{ijk} \leq t_k \quad \forall k \in K \quad (24.6)$$

$$u_i - u_j + n \cdot \sum_{k=1}^{m} x_{ijk} \leq n - 1 \quad \forall i \in J, \forall j \in J \setminus \{0\} \tag{24.7}$$

$$u_i \geq 0 \quad \forall i \in J \tag{24.8}$$

$$x_{ijk} \in \{0, 1\} \quad \forall i \in J, \forall j \in J, \forall k \in K \tag{24.9}$$

$$y_{jk} \in \{0, 1\} \quad \forall j \in J, \forall k \in K \tag{24.10}$$

The objective function (24.1) maximizes the total profit contribution which is achieved together by all three fulfillment modes. Equation (24.2) guarantees the balance of flow for the vehicles. Equation (24.3) establishes the connections between vehicles and requests served by them. Equation (24.4) ensures that each request is served exactly once. Constraints (24.5) require that each vehicle is used at most once. The maximum tour length is limited by constraints (24.6). Constraints (24.7) and (24.8) are needed for the sub-tour elimination. Finally, constraints (24.9) and (24.10) specify the integral conditions for the variables x_{ijk} and y_{jk}. The model (24.1)–(24.10) simultaneously optimizes for all three fulfillment modes the corresponding transportation plans. Thus, it is able to generate exact optimal solutions for IOTP, provided that the model can be solved to optimality within reasonable time by the used solver.

The strategy of Cherry-Picking is conducted by sequentially generating transportation plans for three different fulfillment modes: first for self-fulfillment, then for tour-based subcontraction, and finally for daily-based subcontraction. For self-fulfillment and tour-based subcontracting, the task of generating optimal plans yields a selection vehicle routing problem since there are not enough vehicles for serving all requests. In this case, a subset of requests and a transportation plan for serving these requests have to be determined such that the achieved contribution margin is maximized. Equation (24.11) shows the objective function of a model for the selection FTL-PDP which has to be solved in case of self-fulfillment.

$$\max \sum_{i=1}^{n} \sum_{k \in K_1} d_i^T \cdot Fr_i \cdot y_{ik} - \left[\sum_{i=0}^{n} \sum_{j=0}^{n} \sum_{k \in K_1} a_k \cdot \left(d_i^T + d_{ij}^R \right) \cdot x_{ijk} + \sum_{i=1}^{n} \sum_{k \in K_1} b_k \cdot x_{0ik} \right] \tag{24.11}$$

For tour-based and daily-based subcontraction, the objective function is identical to (24.11) except that K_1 has to be substituted by K_2 and K_3, respectively. Apart from the following minor modifications, the constraints of a model for planning self-fulfillment, tour-based subcontraction, and daily-based subcontraction within the Cherry-Picking strategy are identical to the Eqs. (24.2) to (24.10). For self-fulfillment and tour-based subcontraction Eq. (24.4) have to be modified to Eq. (24.4′) since selection vehicle routing problems have to be solved for these fulfillment modes.

$$\sum_{k \in K'}^{m} y_{jk} \leq 1 \quad \forall j \in J_1 \backslash \{0\} \wedge J_2 \backslash \{0\} \tag{24.4'}$$

K' in Eq. (24.4') is equal to K_1, respectively, K_2 for self-fulfillment, respectively, tour-based subcontraction. The set K has to be replaced by K_1, K_2 or K_3 for self-fulfillment, tour-based subcontraction or daily-based subcontraction, respectively. The set J of available requests also has to be adjusted to J_1, J_2 or J_3 for the corresponding fulfillment mode. In case of self-fulfillment, J_1 is equal to J. For tour-based subcontraction J_2 consists of the subset of all originally available requests which have not been selected for self-fulfillment. In the model for daily-based subcontraction the set J_3 has to reduce to the subset of requests which have been left over by the other two fulfillment modes. In the transportation plan for subcontraction on daily basis, all of the remaining requests have to be served. That is why the optimization problem related to daily-based subcontracting is not a selection vehicle routing problem but a regular vehicle routing problem.

After having solved the optimization models for all three fulfillment modes in the prescribed sequence, the value of the objective function of the IOTP-solution derived by Cherry-Picking is equal to the sum of the three objective values obtained by the three models.

The tours for the Bundle-Oriented strategy are constructed according to the same rules which are applied for self-fulfillment, only the objective function is different. The Bundle-Oriented strategy strives for short travel distances. That is why the objective function is represented by

$$\min \sum_{i=0}^{n} \sum_{j=0}^{n} \sum_{k=1}^{m} d_{ij}^{R} \cdot x_{ijk} \tag{24.12}$$

The constraints of the Bundle-Oriented model consist of the Eqs. (24.2) to (24.10).

After having solved the Bundle-Oriented model, the optimal solution $\bar{x}_{ijk}$ obtained for that model is taken as value for the variable x_{ijk} and for calculating the value of the objective function in Eq. (24.1).

24.5 Computational Experiments

Due to the lack of powerful optimization software for IOTP, Cherry-Picking and the Bundle-Oriented strategy are widely used in practice. In this section both strategies are evaluated by computational experiments on problems of a size that can be handled by MIP-solvers.

For evaluating the two strategies, three test sets with differently sized problems are generated: a test set T_1 containing small problem instances, another one T_2 containing medium sized problem instances, and the last one T_3 containing relatively large instances with 8, 10, and 11 transportation requests, respectively. For each test set, 10 problem instances are randomly generated. Each instance is solved by any of the three proposed approaches for IOTP (simultaneous IOTP by model (24.1) to (24.10), model-based Cherry-Picking, model-based Bundle-Oriented strategy). Altogether, $3 \times 3 \times 10 = 90$ test cases have been solved for strategy evaluation.

The randomly generated problem instances, which are used for constructing test cases, are FTL-PDP requests within a grid square of 200×200 km. The depot is located in the center of the grid square. For generating a single instance, $2n$ customer locations loc_j ($1 \leq j \leq 2n$) are generated by distributing them equally over the grid square. The matrix $DIST$ for the distances between the locations is calculated based on the coordinates of the locations and the Euclidean distance metric. Then, these locations are used for building n FTL-PDP requests $j = (loc_j, loc_{j+n})$ with loc_j as pick-up location and loc_{j+n} as delivery location of request j. The freight rates for requests randomly vary between 2.00 and 2.50 €/km.

The number of vehicles available for request fulfillment is $m = 8$, with $m_1 = 2$ vehicles for self-fulfillment, $m_2 = 2$ vehicles for tour-based subcontraction, and $m_3 = 4$ vehicles for daily-based subcontraction. The maximum tour length t_k is set to $t_k = 630$ for $k \in K_1 \cup K_2$ and to $t_k = 450$ for $k \in K_3$. The variable costs per travel distance a_k are set to 0.41 €/km, 1.04 €/km, and 0 €/km for K_1, K_2, K_3, respectively, and the fixed costs b_k assumed for the computational experiments are 0 €/day, 0 €/day, 460 €/day, respectively.

All 90 test cases could be solved by CPLEX to optimality (i.e. without any optimality gap). For a problem size of $n = 8$, simultaneous IOTP needs 4.7 s cpu-time on average; 205 s for $n = 10$; and 971 s for $n = 11$. In case of the Bundle-Oriented strategy the averaged cpu-time amounts to 1.26 s (for $n = 8$), 16 s (for $n = 10$), and 130 s (for $n = 11$). In case of Cherry-Picking, the averaged cpu-time is 0.56 s, 0.83 s, and 0.98 s for problems of the size $n = 8$, $n = 10$, $n = 11$, respectively. On average over all 90 test cases, simultaneous IOTP needs 6.5 min cpu-time per instance, the Bundle-Oriented strategy needs about 50 s per instance, and Cherry-Picking averagely needs only 0.8 s per instance.

Table 24.1 shows the total profit contribution (TB), which could averagely be achieved for test cases of T_1, T_2 and T_3 by Simultaneous Planning (SP), Bundle-Oriented strategy (BS), and Cherry-Picking (CP). Additionally, Table 24.1 shows the values for the total travel distance of the transportation plans generated by the three different proposed approaches for IOTP. The results of Table 24.1 demonstrate that BS is able to generate high quality solutions with very small deviations from the optimal solutions derived by SP (0.5 % for T_1; 1.6 % for T_2; 0.3 % for T_3) and with values for the travel distances which differ only slightly from those generated by SP. Contrarily, the values for the objective function and the travel distance generated by CP differ a lot from the values obtained by the other two approaches. Compared to SP, the objective value is worsened by 11 % (T_1), 25 % (T_2), and 19 % (T_3); the travel distances are increased by 3 %, 9 %, and 7 %, respectively. The experiments performed on T_1, T_2, and T_3 have shown that the solutions obtained by SP and BS are equal with respect to the number of used

Table 24.1 Total profit contribution (TB) and travel distances (D) for the test cases of T_1, T_2, and T_3

	$T_1(n=8)$		$T_2(n=10)$		$T_3(n=11)$	
	D (km)	TB (€)	D (km)	TB (€)	D (km)	TB (€)
SP	1488.89	1129.63	1850.97	1306.71	1871.85	1403.26
BS	1480.37	1123.72	1836.72	1287.54	1867.24	1399.46
CP	1537.08	997.83	2002.91	1008.36	1990.45	1137.88

vehicles, in all 90 test cases (with one single exception). CP mainly differs from that with respect to the number of vehicles charged for subcontraction. Compared to SP and BS, CP uses more vehicles on daily basis and fewer vehicles for tour-based subcontraction.

24.6 Future Research/Conclusions

The experiments have shown that BS outperforms CP by far. This is a surprising and important insight, since a survey in freight forwarding companies has shown that practitioners assume CP to be superior and consequently CP is used more commonly than BS in practice. Actually, BS has shown in our experiments that it could generate solutions which are very close to the optimum within a short time.

Acknowledgments This research was supported by Basic Science Research Program through the National Research Foundation of Korea (NRF) funded by the Ministry of Education, Science and Technology (2015R1A6A3A03019652).

References

Chu C (2005) A heuristic algorithm for the truckload and less than truckload problem. Eur J Oper Res 165:657–667

Bolduc M, Renaud J, Boctor F, Laporte G (2007) A perturbation metaheuristic for the vehicle routing problem with private fleet and common carriers. J Oper Res Soc (JORS) 59(6):776–787

Kopfer H, Krajewska M (2007) Approaches for modelling and solving the integrated transportation and forwarding problem. In: Corsten H, Missbauer H (Hrsg) Produktions- und Logistikmanagement. Verlag Franz Vahlen, S 439–458

Kopfer H, Schönberger J (2009) Logistics: the complexity of operational transport optimization. In: Lucas P, Roosen P (eds) Emerge, analysis and optimization of structures—concepts and strategies across disciplines. Springer, pp 175–187

Kopfer H, Wang X (2009) Combining vehicle routing with forwarding—extension of the vehicle routing problem by different types of sub-contraction*. J Korean Inst Ind Eng (JIKIIE) 35(1): 1–14

Krajewska M, Kopfer H (2009) Transportation planning in freight forwarding companies—Tabu Search algorithm for the integrated operational transportation planning problem. Eur J Oper Res (EJOR) 197(2):741–751

Pankratz G (2002) Speditionelle transportdisposition. Ph.D. thesis, University of Hagen, Wiesbaden

Savelsbergh M, Sol M (1998) DRIVE: dynamic routing of independent vehicles. Oper Res 46:474–490

Chapter 25
Intelligent Transport Systems for Road Freight Transport—An Overview

Ilja Bäumler and Herbert Kotzab

Abstract This paper presents a state-of-the-art analysis of Intelligent Transport Systems (ITS) for road freight transport including an overview of telematics applications for road freight transport. Furthermore, an analysis on how different actors of a transport chain perceive the developments of ITS is given. The paper also presents selected examples of practical ITS usage.

Keywords Intelligent Transport Systems (ITS) · Road freight transport · Telematics

25.1 Problem Background and Research Questions

The majority of freight transport in Germany is carried out by trucks using the road network (Mondragon et al. 2009) and it is expected that this dominance will continue in the future. However, investments into road infrastructure will not increase in the future thus there is a need for technological systems which control the traffic flows on roads, try to avoid traffic disturbances, as well as to ensure increased road safety. In addition, such systems could also improve the efficiency and productivity of public and private fleets. These systems are known as Intelligent Transport Systems (ITS) or telematics and are defined by Müller (2012) as the process of data collection, processing, and output which achieves goals by using sensors, information and communication technology, and mathematical models. As ITS also allows autonomous decision making on a vehicle by vehicle basis, it is important to know how different user groups of a transport chain accept ITS and its consequences.

In this paper, we want to present a state-of-the-art analysis of the current progress in the field of road freight related ITS as well as a critical reflection on these developments from the perspective of different user groups. This includes a

---

I. Bäumler (✉) · H. Kotzab
Department of Logistic Management, University Bremen, Bremen, Germany

presentation of the advantages and disadvantages of ITS and an elaboration into potential future developments. The paper is driven by the two following research questions, which are answered by using secondary data, as well as on the findings of three expert interviews representing the view of a transport provider, a professional truck driver, and a software programmer:

(a) What are the application areas for ITS in the field of freight road transport and which kind of systems are utilized?
(b) How do different user groups of a transport chain evaluate current ITS developments?

The remainder of the paper is as follows: After a short introduction, we present a short description of the basic set up of ITS based on the notions of Sussman (2005). Then, we show the main application areas of ITS for road freight transport including some examples of how ITS is currently used. Afterwards, we discuss these developments from the perspective of the different user groups. The paper closes with a critical reflection and an outlook for future research.

25.2 Fundamentals on ITS

Any ITS consists of several system components including a communication infrastructure, a service provider, positioning systems, end user devices, and systems for external information provision (Sussman 2005). Global navigation satellite systems (GNSS) are a fundamental part of ITS as they allow positioning while the vehicle moves. At least four satellites are required for exact positioning for route navigation (Evers 1998). The most prominent GNSS are the US-American GPS, the Russian GLONASS, and the European Galileo system. Communication systems refer to monodirectional and bidirectional communication networks based on radio, mobile, and satellite frequency (ESA 2011). Furthermore, telematics standards include Open Service Gateway initiative Alliance (OSGi Alliance), The Motor Industry Software Reliability Association (MISRA), open systems and their interfaces for electronics inside motor vehicles in cooperation with the Vehicle Distributed Executive (OSEK/VDX), and ERTICO-ITS Europe. Sussman (2005) divides ITS into the following six categories: (1) Advanced Traffic Management Systems (ATMS), (2) Advanced Traveller Information Systems (ATIS), (3) Advanced Vehicle Control System (AVCS), (4) Commercial Vehicle Operations (CVO), (5) Advanced Public Transportation Systems (APTS) and (6) Advanced Rural Transportation Systems (ARTS). For freight road transport, AVCS as well as CVO systems are of special importance. AVCS helps drivers to keep control over the vehicle while CVO systems ensure the improvement of efficiency, safety, and other controls of the transport.

25.3 ITS for Road Freight Transport

Access to current information on the traffic situation, weather, as well as cargo conditions is crucial for operating road freight transport in an efficient way. A continuous surveillance of the involved road freight transport processes further allow for a better foundation for long-term as well as short-term planning, thus leading to improved profitability as well as savings potentials due to reduced transport efforts. In the following section, we discuss eight ITS areas as suggested by Sussman (2005) and the German Research Information System (FIS) (2015). These ITS application areas for freight road transport refer to (1) fleet management, (2) toll collection and control, (3) tracking and tracing, (4) emergency and disturbance management, (5) control of public traffic systems, (6) provision of traffic information, (7) control of hazardous goods and heavy bulk transports, and (8) primary traffic control units (Baumann 2011). Most of these application areas belong to the field of Commercial Vehicle Operations (CVO) (see Sussman 2005). Fleet management can be considered to be the most prominent application area within road freight transport as it includes the processing of order and traffic information and the adequate provision of data for drivers. Permanent surveillance, securing and control of hazardous goods or heavy load transports is also an important ITS application area. The use of sensors may refer to Advanced Vehicle Control Systems (AVCS) (see Sussman 2005), which are seen as to be of major importance for road freight transport (Kortüm et al. 1998).

Maurer (2012) differentiates between conventional driver assistance systems and driver assistance systems with automatic detection. The first group includes systems which simply assist drivers in controlling their vehicles with easy to measure key performance indicators including Antilock Braking Systems (ABS) or Tire-Pressure Monitoring Systems (TPMS). The second group is also known as Advanced Driver Assistance Systems (ADAS) as they include Automated Data Processing Systems (ADAS). ADAS assists drivers in difficult traffic situations or in other unobservable situations with predetermined ranges of performance indicators. This refers to situations such as emergency brake systems, Lane Departure Warning systems (LDW), anti-drifting systems or systems which are able to recognize driver fatigue. Within the context of the European Union, CVO as well as AVCS are of increasing importance for road freight transport as many vehicles cross multiple country borders and are then confronted with different toll systems.

Given the increasing number of goods transports within urban areas, the control of such transports is becoming more important too. To date, traffic control systems have been used to stabilize, harmonize, and allocate traffic allowing for better traffic flow within an urban area. This is also relevant for road freight transport as, recently some ITS-related city logistics systems have been introduced (Anonymous 2015a; Dresdner Verkehrsbetriebe 2015).

The next section presents up-to-date examples for all ITS areas which should provide further insight into the topic matter.

25.4 Selected ITS Application Areas in Road Freight Transport

Table 25.1 presents selected examples of ITS where we identify the latest developments in fleet management, city logistics, Collaborative Adaptive Cruise Control (CACC), toll systems. The table also presents an example of an area-wide hazardous goods control.

The broad range of various examples shows that road freight transport provides for each of the identified ITS areas, adequate application possibilities. Applications for truck parking lot allocation, which is a sub area of fleet management, has progressed considerable in the past. Specific applications therefore refer to ITS-assisted compact parking (Kleine and Lehmann 2014) or the app-based parking lot search and reservation system with which the driver or his freight forwarder can book parking space for the desired truck service area. "Highway-Park" or "Systemparken" are such app-based systems. A successful utilization of ITS depends however on the acceptance of the various user groups in a transport chain. We are now going to show how different user groups assess the potentials of ITS for freight road transport. The results of this assessment are shown in the following section.

25.5 Critical Evaluation of ITS from the Perspective of Different Actors of a Transport Chain[1]

The main results of this assessment are summarized in Table 25.2. They refer to the main arguments of the various ITS user groups in regards to road freight transport.

25.5.1 A Transport Provider's View on Using ITS

So far, transport providers cannot recognize any economic advantages by using ITS in their fleet management. In accordance to Directive EC 561/2006 of the European Parliament, transport providers are allowed to control driving, working, standby, and idle time with electronic tachographs. Based on this, transport providers are already capable of identifying delays or failures.

ITS is capable of warning drivers in good time about traffic disturbances and suggest bypasses in order to improve adherence to delivery dates. This helps the transport provider avoid any surcharges due to delays. Consequently, the risk of losses is minimized or even avoided. However, these advantages are difficult to

---

[1]If not else indicated, this section refers to the expert interviews: Anonymous (2015b), Gieske (2015), Warkentin (2015).

Table 25.1 ITS categories with application and characteristics

ITS	Application and characteristics	
Fleet management	**"SKEYE Fleet" (Evers 1998)**	
	Drivers exchange with the control center in a wireless manner order cargo information as well as operating conditions	Operational conditions and other information are captured by additionally implemented sensors in an electronic logbook
	A geographic information system determines the position	Car theft protection by definition of a free travel zone
	SKEYE Fleet Software allows the implementation of an individual fleet management control center	Depending on the signal, the system differs between a simple logbook entry and an alert
	Rule-based real-time control (Kapsalis et al. 2010)	
	Hub-and-Spoke-system based network architecture	Remote configuration of the vehicle system with rule-based event activities possible
	Improved assistance of transport relevant to security by vehicle-installed sensors which are linked to the systems	Safety mechanisms with three levels for sound conditions, system independent firewall, and authenticity check
	Vehicle-installed gateway for connection with fleet control center	Little programming skills for rule development required
City logistics	**Computer assisted truck guidance system (Anonymous 2015a)**	
	Discharging of inner-city traffic network by redirecting heavy bulk load transports	Reduction of energy consuming deceleration and acceleration processes by cross road prioritization
	Faster routes due to priority traffic light circuits at cross roads	Reduction of pollution and noise emissions
	No need for further road network expansion	Reduction of inner-city rat run truck traffic
	"CarGo-Tram" (Dresdner Verkehrsbetriebe 2015)	
	Relief of inner-city traffic network by intermodal transport chains	Faster routes by priority traffic light circuits at cross roads
	Allows JIT-delivery within urban areas	Only efficient if track infrastructure exists
Advanced vehicle control systems	**Collaborative adaptive cruise control (Zambou et al. 2003)**	
	Shorter vehicle space due to automatic cruise control	Stable convoy of vehicles due to adjusting engine torque of all participants of the convoy
	Three possible communication structures: direct, bus, and direct with broadcasting	The smaller the vehicle space, the more complex the programming effort
	Efficient driving with fewer deceleration and acceleration processes	Uncertain legal security in case of accident due to software failure

(continued)

Table 25.1 (continued)

ITS	Application and characteristics	
Toll system	**"Toll-Collect" (D) (Toll-Collect** 2015)	
	Established since 2005	Current Limit weight of 12 t, by October 2015–7.5 t
	Nationwide (federal road and Autobahn)	Emission dependent environmental toll
	Generation of additional income sources	Easy expandability through reprogramming
	"GO Maut" (A) (Asfinag 2015)	
	Established since 2004	Limit weight of 3.5 t
	Nationwide (highways and express ways)	Emission dependent environmental toll
	Generation of additional income sources	Expansion only by further implementation of toll bridges
	"LSVA" (CH) (Eidgenössische Zollverwaltung 2013)	
	Established since 2001	Limit weight of 3.5 t
	Nationwide (all types of roads)	Emission dependent environmental toll
	Generation of additional income sources	No expansion required as all roads are tolled
Hazardous goods control	**"SHAFT" (Zajicek and Schechtner** 2005)	
	Area-wide control of hazardous goods and heavy bulk transports	Transmission of electronic shipping notification to On Board Unit (OBU) and database system
	System components: OBU, hazardous goods server, hazardous goods database	Automatic route control in emergency situations
	GPS and wireless transmission technology for permanent positioning	Simple cross-border expansion possibilities

validate as comparison values are lacking. Another potential advantage refers to an automated truck storage space allocation by feeding the system with order information. This allows the optimal usage of time and truck storage space in relation with the picking and receiving location. However, the replacement of shippers by automated systems could lead to system breakdowns as "computers" do not consider any human interaction.

25.5.2 A Professional Truck Driver's View on Using ITS

The job of a professional truck driver is not an easy task, and in today's world requires ever expanding skills including planning skills and technical know-how.

Table 25.2 Perspectives on ITS

Transport provider	• Links such systems with additional financial burden • Does not see any direct benefit • Requires comparable entities to analyze the cost/benefit ratio • Fears its replacement by intelligent systems
Professional truck driver	• Fears an increase of external control, thus loosing direct responsibility • Despite partly autonomous systems, failure is always connected with the driver • Needs help with route navigation and compliance with its driving and rest periods • Approves the importance of additional safety systems
Traffic control center	• Needs such systems for controlling and influencing traffic flow due to a steady increase in traffic • Sees high potential benefit • Has to be constantly adapted to the latest safety standards
Client and receiver	• Is mainly interested in a low-cost and short delivery process • Does not want any cost sharing for increasing transportation safety • Tracking and Tracing seems to be sufficient enough, does not demand further technologies
Public authorities	
Programmer	• Has to consider economy's and population's interest while applying ITS • Needs ITS for efficient control and for better law enforcing • Boosts widespread usage of ITS through bills and incentives • Has to worry about and ensure road safety
	• Extends every system with additional capabilities • Is responsible for a system's portability and expandability • Is responsible for better comprehensibility of the professional program code

Although the requirements for drivers are permanently increasing, drivers additionally face more and more surveillance.

The use of ITS is justified as drivers are subject to various bills and restrictions which need to be controlled. Furthermore, the position of the vehicle needs to be known in regards to cargo control. ADAS can assist drivers along the trip, which drivers do not always consider to be a positive issue as ITS may lead to diminishing self-dependent driving thus leading to decreased job satisfaction. This further leads to potential physical as well as mental stress. ITS is not able to interpret or to foresee these effects. Despite all technical possibilities, any failure will be attributed to the driver even though the driver may not be responsible. The realization of driving and idle times is, however, a positive application area for ITS. Even though drivers are aware of the legal situation, ITS can better assist drivers in their decision making, especially when to take a break. This will lead to increased traffic safety.

25.5.3 A Traffic Control Center's View on the Use of ITS

All traffic information from different systems are fed in the traffic control center. This includes real-time recordings of inner-city traffic surveillance cameras, weather and traffic sensors on motorways and highways which help to monitor the traffic situation, and state of the technical equipment in the traffic infrastructure. Based on the data analysis, traffic control centers can intervene in current systems by, e.g., changing light signals dependent on the traffic situation, which conveys information regarding changing conditions to road users who are then able to adapt to the situation. Any modern traffic control center has integrated user software allowing all users to access required and relevant data on individual work stations. Today's increased traffic density, increased road network, and other traffic related factors require systems which are able to cope with many interactions of these factors. ITS offer many positive advantages for traffic control centers regarding traffic control and planning. However, they require perpetual systems updates in order to include and to adapt to current safety standards (Landeshauptstadt München Baureferat 2012).

25.5.4 A Client's/Receiver's View on the Use of ITS

Any client wants products to be delivered to a receiver without damages. As long as no problems occur and as long as costs are low, the whole delivery process is of very low interest. A need for information is generated if products do not arrive as promised, either because of time delays or product characteristics. Tracking and Tracing helps to monitor the position of a delivery. However, delivery costs are required to be kept to a low level. Any ITS system which helps to keep delivery costs as low as possible and simultaneously keeps promised delivery times is of interest for clients/receivers.

25.5.5 A Public Authority's View on the Use of ITS

Any state represents the interest of the public but it also has to represent economic interests. This can lead to limitations in transport policy. Public interest includes urban areas free of emissions, reduction of noise pollution, reduction of CO_2 emissions, creation of high quality living conditions, and congestion-free roads.

City logistic concepts, such as the examples as shown in Table 25.1, consider all these issues in their concept.

Decreased acceleration and deceleration processes lead to positive gains as average speed is increased and emissions are reduced. Furthermore, goods freight transports are reduced in inner-city locations. However, public authorities are required to take over the installation of such concepts due to higher costs which private organizations are not willing to cover. Therefore, it is necessary to find a balance between legal obligation and positive incentives for companies to invest into better equipped and safer trucks.

25.5.6 A Programmer's View on ITS

A programmer views ITS as a closed system with partly blurred and partly clear determined boundaries. The project description by the customer is therefore crucial for the adaptability and expansion of the system. Thus, the requirement catalog includes the minimum requirements as well as 'nice-to-have' functionalities as well as the future vision of the project. The anticipatory thinking of a programmer plays a major role for the sustained and long-term advantage of ITS. It is the programmer's responsibility to develop the program code in such a manner so that later successors can adapt it or expand it easily. These findings show that ITS can only be used for guaranteeing safe and efficient freight road transports if the various interests of the transport providers, professional drivers, traffic control centers, customers and receivers, public authorities as well as programmers are equally considered. This requires the active involvement of these groups in defining the prerequisites for ITS.

25.6 Conclusion

The purpose of this paper was to present a state-of-the-art analysis of the current progress in the field of freight road related ITS as well as a critical reflection of these developments from the perspective of different user groups. Overall, we were able to see that within the analyzed literature there is a focus on the interplay between ITS, a vehicle and its environment (e.g. road infrastructure, road conditions, parking, etc.). For further research it would be interesting to differentiate further between vehicle and load units. First attempt of this can be found in the research of the SFB 637 with the intelligent container (e.g. Lang et al. 2011) or swap bodies (e.g. Podlich et al. 2009).

Based on the presented discussion in the previous sections, we are able to answer our two research questions as follows.

What are the application areas for ITS in the field of freight road transport and which kind of systems are utilized?

Based on the systematic ITS application areas in accordance to Sussman (2005), we were able to identify fleet management, city logistics, ADAS, toll systems, and hazardous goods and heavy bulk transport controls as the most important areas. For each of these areas, we have presented at least one application system (see Table 25.1).

How do different user groups of a transport chain evaluate current ITS developments?

We have provided in Table 25.2 a summary of the various views from different actors of a transport chain. We therefore recognized the various dilemmas between user groups depending on their individual goal settings.

As the field of ITS is significantly influenced by current technology development, it is necessary to see how these developments affect these systems.

Taking the ADAS perspective, we need to consider the future development of system architectures and sensor technology which allows for more precise measuring. Currently, ADAS is operated within rigid system architecture. In the future, system borders need to be opened allowing the interaction between different ADAS systems. This will lead to even more complex systems inside the vehicles, which will require different computer nodes compared to today (Reichart and Bielefeld 2012). Another wide area for future developments is urban logistics concepts, out of which city logistics has been positively developed in larger cities. Future intelligent systems need to be able to improve the efficient utilization of fleets as well as an efficient allocation of orders amongst various actors. One example is the Internet-based combinative auction trade of transport orders.

Considering the broad range of technologies which are able to realize ITS, we recognize a huge potential for future research. New ITS for road freight transport are seen to be integrated, intermodal, Internet-based and intelligent (Giannopoulos 2009) and will require more complex methods and processing algorithms in order to allow for an appropriate utilization of collected data (Crainic et al. 2009).

Finally, it is the will and motivation of all road users and their ability to adapt and change that will promote the use of any future ITS system.

References

Anonymous (2015a) Wirtschaftsverkehr in Ballungsräumen. Nienburg: Rechnergestütztes LKW-Leitsystem [Commercial trade in agglomeration areas. Nienburg: Computer assisted truck guidance system]. http://www.vsl.tu-harburg.de/gv/4/test?menu=4a&inhalt=4a2

ASFiNAG (2015) GO Maut. https://www.go-maut.at/portal/faces/pages/common/portal.xhtml

Baumann S (2011) Telematik für den Güterverkehr [Telematics for freight transport]. http://www.forschungsinformationssystem.de/servlet/is/339629/

Coronado Mondragon AE, Lalwani CS, Coronado Mondragon ES, Coronado Mondragon CE (2009) Facilitating multimodal logistics and enabling information systems connectivity through wireless vehicular networks. Eur Transp Res Rev 122:229–240

Crainic TG, Gendreau M, Potvin J-Y (2009) Intelligent freight-transportation systems: assessment and the contribution of operations research. Transp Res Part C: Emerg Technol 17:541–557

Dresdner Verkehrsbetriebe AG (2015): CarGoTram: Autoteile fahren Bahn [CarGoTram: automobile parts drive by tram]. https://www.dvb.de/de-de/die-dvb/technik/fahrzeuge/cargotram/

Eidgenössische Zollverwaltung (2013) LSVA-Übersicht. Leistungsabhängige Schwerver-kehrsabgabe [LSVA-overview. Performance-related heavy vehicle fee]. http://www.ezv.admin.ch/zollinfo_firmen/04020/04204/04208/04744/index.html?lang=de

Evers H (1998) Kompendium der Verkehrstelematik: Technologien, Applikationen, Per-spektiven [Compendium of transport telematics: technologies, applications, perspectives], eds. (Köln: Verl. TÜV Rheinland)

Giannopoulos GA (2009) Towards a European ITS for freight transport and logistics: results of current EU funded research and prospects for the future. Eur Transp Res Rev 1:147–161

Kapsalis V, Fidas C, Hadellis L, Karavasilis C, Galetakis M, Katsenos C (2010) A networking platform for real-time monitoring and rule-based control of transport fleets and transferred goods. In: 13th International IEEE conference on intelligent transportation systems—(ITSC 2010), pp 295–300

Kleine J, Lehmann R (2014) Intelligent controlled compact parking for modern parking management on German motorways. In: Transport Research Arena (TRA) 5th conference: transport solutions from research to deployment

Kortüm W, Goodall RM, Hedrick JK (1998) Mechatronics in ground transportation-current trends and future possibilities. Annu Rev Control 22:133–144

Landeshauptstadt München Baureferat (2012) Neubau Technisches Betriebszentrum mit Verkehrsleitzentrale [Reconstruction of a technical operations center with traffic control center]. http://www.muenchen.de/rathaus/Stadtverwaltung/baureferat/projekte/technisches-betriebszentrum.html

Lang W, Jedermann R, Mrugala D, Jabbari A, Krieg-Brückner B, Schill K (2011) The intelligent container—a cognitive sensor network for transport management. IEEE Sensors J Spec Issue Cogn Sensor Netw 11(3):688–698

Maurer M (2012) Entwurf und Test von Fahrerassistenzsystemen [Design and test of driver assistance systems], in Handbuch Fahrerassistenzsysteme. In: Winner H, Hakuli S, Wolf G (eds) (Vieweg + Teubner Verlag), pp 43–54

Müller S (2012) Makroskopische Verkehrsmodellierung mit der Einflussgröße Telematik [Macroscopic modeling of transport with the influence of telematics]. Dissertation, Technische Universität, Berlin

Podlich A, Weise T, Menze M, Gorldt C (2009) Intelligente Wechselbrückensteuerung für die Logistik von Morgen [Intelligent swap body control for tomorrow's logistics]. In: Electronic communications of the EASST, vol 17 (2009). http://journal.ub.tu-berlin.de/eceasst/article/download/205/207

Reichart, G. and Bielefeld, J. (2012); Einflüsse von Fahrerassistenzsystemen auf die Syste-marchitektur im Kraftfahrzeug [Influences of driver assistance systems on the system architecture in the automobile], in Handbuch Fahrerassistenzsysteme, H. Winner, S. Hakuli and G. Wolf, eds. (Vieweg + Teubner Verlag), 84–92

Sussman JS (2005) Perspectives on Intelligent Transportation Systems (ITS) (Boston. MA, Springer, US)

Toll-Collect (2015): Allgemeines [Miscellaneous], https://www.toll-collect.de/de/web/public/toll_collect/service/fragen___antworten/allgemeines_1/allgemeines.html

Zajicek J, Schechtner K (2005) Area wide hazardous goods monitoring on the TERN in Austria project SHAFT. In: 2005 IEEE intelligent transportation systems, pp 287–289

Zambou N, Enning M, Abel D (2003) Längsdynamikregelung eines Fahrzeugkonvois mit Hilfe der Modellgestützten Prädiktiven Regelung [longitudinal dynamic control of vehicle convoys using model-based predictive control]. In: Verein Deutscher Ingenieure eds.: Telematik 2003

Personal Interviews

Anonymous (2015b) Expert interview, 26.03.2015, 17:30–19:00, Bremen. The expert requested an anonymous treatment of his interview
Gieske J (2015) Expert interview, 03.04.2015, 12:30–13:30, Bremen
Warkentin A (2015) Expert interview, 10.04.2015, 19:30–20:30, Osnabrück

Chapter 26
Imposing Emission Trading Scheme on Supply Chain

Fang Li and Hans-Dietrich Haasis

Abstract Researchers intend to employ emission trading scheme (ETS) as one of the cost-effective policy instruments on the level of supply chains (SCs) to control SC emissions. This paper reviews literatures from recent 5 years in addressing ETS in the context of SCs. Through analyzing mathematical models adopted in available literatures to address the implementation, it identifies the intentions and approaches of imposing ETS in the context of SCs. This paper is believed to be one of the first literature reviews in addressing ETS in the context of SCs. It provides insights for supply chain managers and policy makers on whether including SCs into ETS.

Keywords Emission trading scheme · Supply chain

26.1 Introduction

In the background of global warming, firms are required or encouraged to be more environmental friendly by limiting their own emission and emission from its supply chains (SCs). And accurately SC emission account for around 75 % of the whole emission from an industry sector while companies' direct emission average only 14 % of their SC emission prior to use and disposal across all industries (Huang et al. 2009). Moreover, companies must take a global SC perspective in order to identify the most profitable means to reduce overall supply chain emission (Plambeck 2012). For example, non-European sources of emission remain linked to Europe via the international SCs of European companies (Skelton 2013). Legislation is identified as a key driver of proenvironmental behavior in organizations (Diabat and Govindan 2011). Among a range of policy responses, emission trading

F. Li (✉) · H.-D. Haasis
Chair of Maritime Business and Logistics, Faculty of Economics,
University of Bremen, Bremen, Germany
e-mail: lif@uni-bremen.de

H.-D. Haasis
e-mail: haasis@uni-bremen.de

scheme (ETS), also called cap and trade, is becoming one of the most cost-effective instruments and spreading around the world. The environmental impacts of life cycle products could be best managed through goal-oriented and market-based mechanisms like cap and trade that provide flexibility in choosing compliance levers to the targeted firms or industries (Gupta and Palsule-Desai 2011). With such intention in mind, researchers are interesting in employing ETS as one policy instrument on the level of SC in order to control SC emission (Chaabane et al. 2008, 2011, 2012).

In Sect. 26.2, this paper gives a graph introduction about the development of ETS. Section 26.3 analyzes the intentions and possible ways of imposing ETS in the context of SC, and it reviews mathematical models adopted in available literatures to address the implementation. Moreover, this paper identifies the drivers and barriers of imposing ETS on the level of SC in Sect. 26.4 and thereby provides insights for both policymaker and business organizations.

26.2 ETS

26.2.1 Working Principle

The ETS works on the 'cap and trade' principle. The 'cap', or limit, is set by selected officials on the total amount of certain greenhouse gases (GHGs) emission that can be emitted by the liable entities. The cap is reduced over time so that total emission fall. Entities are allocated with separate carbon permits summed up to the amount of cap at free (by free allocation) or a certain cost (by auction). One carbon permit gives the right to emit one ton of carbon dioxide (CO_2) or an equivalent amount of GHGs. The 'trade' creates a market for carbon permits, helping liable entities innovate in order to meet, or come in under their allocated limit. The price of permits is jointly decided by the demand and supply of them in the market. In a word, entities under ETS are allowed to exchange the carbon permits via carbon market with a certain carbon price as needed. At the end of certain period, entities that emit less than its allocated permits may bank the spare permits to cover its future needs or else sell them to other entities that are short of permits. Entities whose cumulative emission exceed allocated permits, after taking possible green potentials or not, may buy permits from carbon market in order to meet their excessive emission. Since the compliance cost usually is much higher than the carbon price, carbon market may offer one solution to meet cap before suffering heavy fines. On one side, by imposing a cap on firms' emission and thereby giving cost for each unit of emission, ETS puts environmental issues into companies' economic agenda in terms of internalizing the cost of external resources. On the other side, with the aim of trade, ETS helps companies reach their emission reduction goal by paying the least to do so.

Entities could be here in the name of country, firm, region, industry sector, or any business cluster. When entities are different regions or sectors within a country, it stands for national ETS. Sovereign governments control the types of emission units that are acceptable within the ETS that they oversee. For example, the EU ETS trades primarily in European Union Allowance (EUA), the Australia ETS in Australian Unit (AU), the California cap and trade scheme in California Carbon Allowance (CCA), and the New Zealand scheme in New Zealand Unit (NZU). When entities stand for countries, it corresponds to international ETS, such as the designed ETS under Kyoto Protocol in Assigned Amount Unit (AAU). Until now, some emission units between different schemes are actively linked and exchanges are allowed in certain proportion. For example, one EU Allowance (EUA) is equivalent to one Australian Unit (AU) by trading (Talberg and Swoboda 2013). Except the compulsive implementation of ETS on certain industry sectors, plants and installations, any individual institution is free willing to join ETS as well.

26.2.2 ETS in Practice

(1) EU ETS

ET is currently widely adopted by different countries as one of the most important instruments to realize their own emission reduction targets, for example, the European Union (EU), Australia, Canada, Norway, China, Japan, etc. Among them, the European Union—ETS (EU ETS), the first and still by far the largest multicountry, multisector system for trading GHG allowances, accounts for about 45 % of EU GHG emission, 40 % of EU CO_2, and covers all 27 EU member states, as well as other non-EU nations, Iceland, Liechtenstein, Norway, and Croatia. Up to now, more than 11,000 facilities covered by the EU ETS included (European Commission 2013).

The EU ETS runs in phases (seen Table 26.1). The first phase of EU ETS was intended as a stage of 'learning by doing,' and no strict cap was designed in the scheme except modest abatement aims. Too many allowances were allocated into the market and the allowance price was zero as a result. In the later phases emission reduction targets were designed strictly. However, due to the financial crisis in 2007, emission decreased as the production activities were decreased. Allowance price was still zero since there lacked demand for it.

(2) ETS in China

As one of contributions to the international efforts addressing climate change, China commits to lower CO_2 emission per unit of GDP by 40–45 % by 2020 compared to 2005 (Zhang et al. 2014). To approach the emission reduction target, cap and trade scheme was approved first by the National Development and Reform Commission (NDRC) of China in 2011. Supported by the 12th Five Year Plan (2011–2015) endorsed by the National People's Congress, a mandatory 'cap and

Table 26.1 EU ETS phases

	Phase I	Phase II	Phase III
Period	2005–2007	2008–2012	2013–2020
Target	Modest abatement aims	8 % below 1990 levels	21 % on 2005 levels
Cap setting	Bottom up by National allocation plan (NAP)	NAP	EU-wide cap
Participant country	25 EU States	27 EU States	27 EU States + Norway, Iceland, Liechtenstein
Participant industry	Power generator; Energy-intensive industry	Power generator; Combustion plants; Oil refineries; Coke evens; Iron and steel plants; Factories making cement, glass, lime, bricks, ceramics, pulp, paper, and board	Aviation; Petrochemicals, ammonia, aluminum; Nitric, adipic, and glycolic acid production
Emission type	CO_2	$CO_2 + N_2O$	$CO_2 + N_2O + PFC$
Covered emission in per cent (%)	40	40	43
Permits allocation	95 % free	90 % free	Less than 50 % free
			None for power plants
			Up to 80 % of benchmark
			Free for heavy emitters
			Max. 10 % auctioning
Compliance cost (€/ton)	40	100	100
Permits price (€/ton)	0	0	3.9

Source Talberg and Swoboda 2013

trade' emission trading pilot scheme is adopted in seven provincial regions by 2013, including Beijing, Shanghai, Tianjin, Hubei, Chongqing, Guangdong, and Shenzhen. And, Chinese government aims to establish a national ETS during the 13th Five Year Plan (2016–2020).

Focusing only on carbon dioxide, the seven pilots cover roughly 40–60 % of a city or province's total emission, and apply to power and other heavy manufacturing sectors such as steel, cement, and petrochemicals. These seven pilot schemes are collectively expected to cover 700 MtCO2e, a quantity only behind the 2.1 GtCO2e that the EU ETS covers by 2014 (Zhang et al. 2014). It is reported that China national ETS would be then the largest one around the world.

The pilot program across China serves as basic foundation for a national-level and unified emission trading market. However, by putting it into reality, China is experiencing challenges to establish the national ETS.

- Experiences from existed ETS show no implication for China at provincial- and city scale pilot level. Other ETSs are directly carried out at national level and do not cover so large spectrum of economy development. The economic gap among China provinces or cities is larger than that of EU ETS participant countries.
- No experiences can be derived from developing countries. China is the largest developing country, and among developing countries China is the first trial to build ETS.
- Regional pilots experiences cannot be easily transferred into useful knowledge to benefit establishing the nationwide ETS. Permits price in the initial stage is of big difference in different pilots region where Shenzhen has the highest price. The different price levels show it is still early days for market participants and that there is great deal of uncertainty about whether permits are scare.
- The complicated administrative structure in China results in the particularity to tackle with planning and operation processes of establishing China ETS. For example, central governance is efficient in decision-making on one side; however, it might bring conflicts between central control and local flexibility on the other side.

26.3 ETS in Line with SC

26.3.1 Imposing ETS on SC—Intentions

So far ETS is only compulsive on individual firms in energy-intensive industries and other firms in other sectors are free to join. Although SC is not subject to ETS directly so far, pre-research in addressing ETS in-line with SC bases on two scenarios. First, firms subject to ETS could reach their emission reduction targets by reducing emission from both within and beyond their own operations. Second, SC is devoted to control the whole emission under a compulsive or voluntary target in order to reduce the emission per product. Four cases are together illustrated from both scenarios in below. In all cases, ETS implementation affects the management of SC by inducing an emission cost and by leveraging the opportunities provided by

ETS as well as other operational adjustments; SC could be optimized for a minimum cost.

Scenario 1: ETS targeted on a single node in SC.

Case 1 (internal emission abatement): The leading firm (i.e., manufacturer) in a SC is directly subject to ETS. There is a limit on its own emission (scope 1 and 2 emission), and it can trade permits in market according to its need. SC partners agree to reduce emission cooperatively along the SC in order to satisfy customers' requirements for low emission products. This case could be found by Jaber et al. (2013). Emission is assumed to come from only production activities. The total SC cost induced by production–inventory cost as well as emission cost is minimized by deciding the production rate and coordination multiplier between SC partners.

Case 2 (external emission abatement): The leading firm (i.e., manufacturer) in a SC is directly subject to ETS. There is a limit on its own emission (scope 1 and 2 emission) and it can permit trade in market according to its need. To reach the goal, the firm can take both internal (i.e., product output reduction, green technologies, production processes improvement) and external measures (i.e., green supplier, green customer, green logistics). The leading firm can earn credits by adopting external measures and uses them to offset its emission reduction target in return. When the leading firm lies in a global SC, such measures earn CDM^1 and JI^2 credits. In this way, SC could be emission credit seller and the leading firm is the credit buyer. They trade credits within the scope of SC. Researchers do not yet study this case, but Long and Yong (2015) mention it as one of potential intervention options, so-called SC credit scheme. SC leading organizations who conduct work within a supplier to reduce GHGs would gain credits, which could be used to offset linked tax liabilities or as additional credits within existing cap and trade schemes.

Scenario 2: ETS targeted on the whole SC nodes.

Case 3 (Single stakeholder): It is assumed in this case that one big single firm owns the whole SC. The leading firm in a SC is motivated to estimate and control emission from its whole SC. It sets up a limit for the whole SC emission either by customers' requirement or government regulations. ETS offers one option for the SC to reach its goal when it is free willing to join ETS. This case is analyzed by Ramudhin et al. (2008), Chaabane et al. (2011), Fahimnia et al. (2013), Jin et al. (2014), etc. ETS could be leveraged in this case by setting one common cap on the whole SC. It means, emission from raw materials production, products manufacturing, transportation, warehousing, and even reverse logistics are subjective to a common cap. The firm could minimize its total logistics cost including emission cost by operational adjustment (i.e., facility allocation, transportation configuration,

---

[1]CDM: Clean Development Mechanism, one of market-based instruments proposed along with ETS in Kyoto Protocol, allows Annex I countries to invest into emissions reduction projects in developing countries.

[2]JI: Jointly Investment, another market-based instrument proposed along with ETS in Kyoto Protocol, allows Annex I countries to invest into emissions reduction projects in Annex I countries.

supplier selection) and green investment (i.e., green technologies adoption) and permit purchase.

Case 4 (Multistakeholder): It is assumed in this case that SC is composed of multiplayers, including raw material supplier, manufacturer, wholesaler, broker, and retailor, etc. All SC partners are separately or jointly subjective to ETS. This case is illustrated by Benjaafar et al. (2013). SC emission is considered with regard to procurement, production, and inventory. SC cost including emission cost could be minimized by deciding variables such as order and inventory quantity with or without collaboration among SC partners. Imposing SC-wide emission caps is more cost-effective than individual cap installation on each firm, and it also increases the value of collaboration (Benjaafar et al. 2013).

Green SC design will gain in richness and mind share if they leverage the opportunities offered by carbon trading markets for those companies pursuing a green strategy or having to regulate their GHGs emission (Ramudhin et al. 2008). While different firms have different emission marginal abatement cost, a broader ETS allows emission is reduced in a more cost-effective way due to the participation of agents with low marginal cost in emission reduction. Besides, in order to maintain competitive and flexible under a wide range of future scenarios, SC managers should consider possible decreased permits allocation in the future when they set emission reduction targets (Diabat and Simchi-Levi 2009).

26.3.2 SC Design Under ETS—Mathematical Models

It is not surprising to see integrated logistics mathematical modeling based methodologies are the most common approaches used to tackle green SC design problems with environmental considerations, as seen in Table 26.2.

26.4 Discussion

Previous research focuses mostly on the mathematical application in imposing ETS on the level of SCs, and there lack qualitative analysis in this area to draw drivers and barriers behind such implementation. Long and Young (2015) study the intervention options to enhance the management of SC GHGs in UK. They mention SC tax, emission trading, and credit schemes as economic instruments among others to control SC emission. However, they do not analyze the drivers and barriers in detail in imposing SC ETS. In this section, this paper addresses the gap from both administrative and corporate perspectives, as seen in Table 26.3.

Existing ETSs could be feasibly expanded to include SCs by including the SCs of targeted companies as illustrated in Sect. 26.3.1 Case 2. However, policymakers may face problems in setting the scope and scale of SC ETS, especially in a global case. Firms may allocate their plants to other regions which are absolvent from

Table 26.2 Literatures summary—GSCM under emission trading

	Authors	Field	Methods	Contributions
1	Jaber et al. (2013)	Production–inventory management	Mathematical programming	The vendor is subjective to ETS and considers only production emission
2	Ramudhin et al. (2008)	SC network design (SCND)	Mixed integer programming (MIP) and goal programming (GI)	It provides decision-makers with the ability to evaluate different strategic decision alternatives, such as supplier and subcontractor selection, product allocation, capacity utilization, and transportation configuration. Emission is limited to production and transportation activities.
3	Chaabane et al. (2011)	SCND	Multiobjective mixed integer linear programming (MILP)	The proposed approach helps SC managers to evaluate the average abatement cost (technology acquisition) as a function of carbon emission reduction targets. Emission is limited to production and transportation activities
4	Bing et al. (2015)	Reverse global SC design	Integer programming	It considers reprocessors' allocation under two different ETSs in two countries in a reverse SC. Emission is considered from only reprocessing
5	Fahimnia et al. (2013)	Close loop SC (CLSC)	Mathematical programming	It takes reverse logistics into account and indicates that reverse logistics should be subsidized in order to reduce the whole SC carbon footprint
6	Chaabane et al. (2012)	CLSC	MILP-LCA	It takes reverse logistics into account and distinguishes between solid, liquid wastes, and gaseous emission due to various production processes and transportation systems
7	Abdallah et al. (2012)	Green procurement	MIP-LCA	Emission is considered from raw materials, shipping, assembly, and warehousing
8	Zakeri et al. (2014)	SCND	MILP	It conducts a comparative analysis between impacts of carbon tax and emission trading on the SC performance

(continued)

Table 26.2 (continued)

	Authors	Field	Methods	Contributions
9	Jin et al. (2014)	SCND	MILP	It performs impacts analysis of carbon pricing policies on a major retailer (i.e., Walmart) in terms of SC design and transportation modes choice
10	Drake et al. (2010)	SCND	A 2-stage, stochastic model	It takes two alternative technologies as compliance strategies into consideration
11	Benjaafar et al. (2013)	Procurement, production, and inventory management	Lot sizing model	It employs ETS on all of SC nodes with a single cap or separate caps

Table 26.3 Drivers and barriers of imposing ETS on SC

	Policymaker	Business organization
Drivers	• Existed ETS • Existed parties for emission verification, measure, and report	• Increasing direct benefit from SC emission reduction • Decreasing relative cost of compliance measures • Increasing competitiveness
Barriers	• Verification, measure, and report of emission • Setting scale and scope • Lack of understanding • Increasing administrative cost	• Conflicting objective among SC partners • Lack of SC environmental management system • Lack of management standards • Lack of consumer demand • Lack of understanding • Increasing management cost

ETSs. It is also possible that firms in a global SC may be subjective to different ETSs, and the exchange of these ETSs is necessary for implementing a whole SC ETS. Moreover, government has to provide or motivate to build an accurate system for emission verification, measure, and report.

As the ETS working principle explains, firms joining ETS could benefit from reducing emission in a cost-effective manner, and this principle applies to SCs as well. Besides, by setting an emission target on the SC therefore on SC firms, they could earn market competitiveness by providing emission information on products (carbon labels). However, supposing SC is composed of different stakeholders, for example, medium-sized enterprises and each of them has different objectives. This would be a barrier for SC optimization. In addition, although there exist extra saved emission or profits for SC as a whole subjective to ETS, it is still difficult to allocate such benefits among different SC stakeholders considering such implementation would decrease cost or emission for some and increase for others. What's more, a management standard for SC emission control under ETS is needed to provide practitioners with instructions. And, environmental management system must be installed in SC firms in order to share update information of the SC emission reduction. Last but not least, customers' awareness which affect customers' behavior in the end market is still not high enough to motivate SC emission control.

26.5 Conclusion

This paper stresses the importance of reducing GHG emissions from the SC perspective, and it investigates how ETS could be employed as one of policy instruments to achieve emission reduction. It is derived that ETS could be applied to single node of SCs as well as whole SCs in different scenarios. This paper also reviews literatures addressing ETS in the context of SCs and summarizes most commonly used mathematical methods in this field. It is believed to be one of the first in conceptualizing the application of ETS in SCs. However, it is necessary to apply ETS to different structures of SCs and analyze the effects from the quantitative perspective.

References

Journal Article

Benjaafar S, Li Y, Daskin M (2013) Carbon footprint and the management of SCs: insights from simple models. IEEE Trans Autom Sci Eng 10(1):99–116

Chaabane A, Ramudhin A, Paquet M (2011) Designing SCs with sustainability considerations. Prod Plann Control 22(8):727–741

Chaabane A, Ramudhin A, Paquet M (2012) Design of sustainable SCs under the ETS. Int J Prod Econ 135(1):37–49

Diabat A, Govindan K (2011) An analysis of the drivers affecting the implementation of green SC management. Resour Conserv Recycl 55(6):659–667

Diabat A, Simchi-Levi D (2009) A carbon-capped SC network problem. In: IEEE International conference on industrial engineering and engineering management, 2009. IEEM 2009. IEEE, pp 523–527

Drake D, Kleindorfer PR, Van Wassenhove LN (2010) Technology choice and capacity investment under emission regulation. Fac Res 93(10)

European Commission (2013)

Fahimnia B, Sarkis J, Dehghanian F, Banihashemi N, Rahman S (2013) The impact of carbon pricing on a closed-loop SC: an Australian case study. J Clean Prod 59:210–225

Gupta S, Palsule-Desai OD (2011) Sustainable SC management: review and research opportunities. IIMB Manage Rev 23(4):234–245

Huang YA, Weber CL, Matthews HS (2009) Categorization of scope 3 emission for streamlined enterprise carbon footprinting. Environ Sci Technol 43(22):8509–8515

Jaber MY, Glock CH, El Saadany AM (2013) SC coordination with emission reduction incentives. Int J Prod Res 51(1):69–82

Jin M, Granda-Marulanda NA, Down I (2014) The impact of carbon policies on SC design and logistics of a major retailer. J Clean Prod 85:453–461

Long TB, Young W (2015) An exploration of intervention options to enhance the management of SC greenhouse gas emission in the UK. J Clean Prod

Plambeck EL (2012) Reducing greenhouse gas emission through operations and SC management. Energy Economics 34:S64–S74

Ramudhin A, Chaabane A, Kharoune M, Paquet M (2008) Carbon market sensitive green SC network design. In: IEEE International conference on Industrial engineering and engineering management, 2008. IEEM 2008. IEEE, pp 1093–1097

Skelton A (2013) EU corporate action as a driver for global emission abatement: a structural analysis of EU international SC carbon dioxide emission. Glob Environ Change 23(6):1795–1806

Talberg A, Swoboda K (2013) Emission trading schemes around the world. Background Note. Parliamentary Library. Department of Parliamentary Services, Parliament of Australia, Canberra, ACT. http://www.aph.gov.au/About_Parliament/Parliamentary_Departments/Parliamentary_Library/pubs/BN/2012-2013/EmissionTradingSchemes#_ftn11. Accessed 29 Nov 2013

Zakeri A, Dehghanian F, Fahimnia B, Sarkis J (2014) Carbon pricing versus emission trading: a SC planning perspective. Int J Prod Econ

Zhang D, Karplus VJ, Cassisa C, Zhang X (2014) Emission trading in China: progress and prospects. Energy policy 75:9–16

Chapter 27
Planning of Maintenance Resources for the Service of Offshore Wind Turbines by Means of Simulation

Stephan Oelker, Abderrahim Ait Alla, Marco Lewandowski and Michael Freitag

Abstract In the last decade, the erection of offshore wind turbines, especially in the northern sea, has been showing a significant growth and this trend will continue further in the next years. In order to make the offshore wind energy competitive and attractive, the different processes related to the overall life cycle cost of an offshore wind farm have to be optimized. In this context, the cost for operation and maintenance (O&M) is estimated between 15 and 30 % of the total costs generated by offshore wind farms. Thereby, the efficiency of the maintenance processes is a crucial factor to guarantee sustainable energy and improve the reliability and availability of an offshore wind turbine. Indeed, the maintenance activities in the offshore field are a challenging task especially due to the harsh maritime environment which leads to high material stress and low resource utilization. In this paper, we model the maintenance processes of an offshore wind farm by means of a discrete event and agent - based simulation model. The objective is to schedule the maintenance tasks taking into account all real restrictions based on historical data in order to determine important factors and potential operational improvement. As an example, the simulation will be used to determine the optimal number of resources needed to perform maintenance activities by keeping the resource utilization in an acceptable level.

Keywords Operation and maintenance · Offshore wind park · Simulation

---

S. Oelker (✉) · A.A. Alla · M. Lewandowski · M. Freitag
BIBA – Bremer Institut für Produktion und Logistik GmbH at the
University of Bremen, Bremen, Germany
e-mail: oel@biba.uni-bremen.de

A.A. Alla
e-mail: ait@biba.uni-bremen.de

M. Lewandowski
e-mail: lew@biba.uni-bremen.de

M. Freitag
e-mail: fre@biba.uni-bremen.de

M. Freitag
Faculty of Production Engineering, University of Bremen, Bremen, Germany

© Springer International Publishing Switzerland 2017
M. Freitag et al. (eds.), *Dynamics in Logistics*, Lecture Notes in Logistics,
DOI 10.1007/978-3-319-45117-6_27

27.1 Introduction and Problem Description

The offshore wind energy (OWE) is considered as an important factor of renewable energy, which can provide a major contribution to the objectives of the green energy production by considerably reducing the carbon dioxide emissions. In this regard, there has been a significant growth in number of built offshore wind farms in the last years. However, this growth implies different challenges that lead to debates on the profitability of OWE. Indeed, in comparison to onshore wind energy, the OWE faces more challenges and restrictions. Maintenance activities for the OWE have to cope with the harsh maritime conditions. Due to that, in the case of failures, accessibility of the offshore wind turbines (OWT) is very difficult and has an effect of the availability of offshore wind turbines. In addition, the availability of an offshore wind farm (OWF) depends on the distance from the coast to the OWF. In this context, the availability of an offshore wind turbine in some literatures is estimated to be between 60 and 90 % (Scheu et al. 2012; Netland et al. 2014; Rohing 2014), which is very low in comparison to onshore wind turbines.

Another problem is related to the size of offshore wind turbines. These are taking on ever-increasing dimensions, and necessitate special vessels to carry out spare part needed for maintenance activities. In order to overcome these challenges and to make the OWE attractive and competitive, different improvements have to be made during the offshore wind turbine lifecycle. Different works have dealt with the investigation and identification of the improvement factors in the OWE (cp. Martin et al. 2016). In this context, several published research articles have paid much attention to the operation and maintenance (O&M) of offshore wind turbines. Indeed, the costs for O&M during the overall life cycle of the OWF is estimated to be between 15 and 30 % of the total costs resulting from offshore wind farms (Besnard et al. 2013). Therefore, the reliability of the offshore wind turbines has become an important factor. For instance, there are efforts to implement condition monitoring systems as an extension to the established SCADA systems, in order to permanently monitor the different components of a wind turbine, and detect possible degradation in a predictive way. In this context, the planning and scheduling of maintenance activities on monitored parts is proactively performed, which can avoid unexpected breakdowns, and maximize power generation.

Therefore, nowadays wind turbines are equipped with various sensors, which monitor the behavior of the wind turbine and detect incipient failures. In this regard, the shift to the condition-based monitoring (CBM) is possible, which can avoid unnecessary visits triggered from the time-based maintenance strategy and avoid unanticipated breakdowns from failure-based maintenance strategy. Nevertheless, maintenance practices show that the time-based and failure-based maintenance strategy are still the popular maintenance strategies adopted widely by the OWE (Andrawus et al. 2007). This is on the one hand due to the limited knowledge concerning failure modes and their relationship with observed data, and on the other hand due to the difficulty to guarantee reliable sensor data of all wind turbine components. To address this issue, there have been several works regarding the benefits of

CBM in the OWE (Byon et al. 2011). For these reasons, it is challenging to manage the maintenance logistics for OWE efficiently, so that simulation modeling seems to be a powerful approach to cope with the maintenance logistic complexity for OWE.

In this regard, various simulation models have been developed to model the O&M of OWF (cp. Martin et al. 2016). The simulation models are used to evaluate the suggested O&M, and through assessment of predefined key performance indicators (KPI) they can help in the decision-making process. The most proposed KPIs are the O&M costs and the wind farm availability. In this paper we propose a simulation model which considers critical stochastic aspects of OWE, like weather restrictions, resource restrictions, resource availability, and wind turbine reliability data. To this end, we use the agent-based and discrete event simulation software, AnyLogic, to build a model simulating the O&M of an OWF (AnyLogic).

27.2 State of the Art

Generally accepted, maintenance activities can be assigned to two major maintenance strategies: corrective maintenance and preventive maintenance (Fig. 27.1).

Corrective maintenance strategies renounce of inspections and a maintenance task is only carried out after a failure or breakdown of a component. This requires spontaneous and fast working, which is often not possible at sea. Preventive maintenance strategies can be subdivided further into time based, condition based, and predictive maintenance. This subdivision is done with respect to the different triggers for inspection measures. Within time-based maintenance the service and inspection will be carried out according to defined intervals, while condition-based maintenance strategies use the condition of the system as trigger, e.g., by using data collected by a condition monitoring systems. Predictive maintenance represents a further development of condition-based maintenance. In this case, the aim is the early detection of potential faults and the prevention of the occurrence of a fault (Schenk 2010).

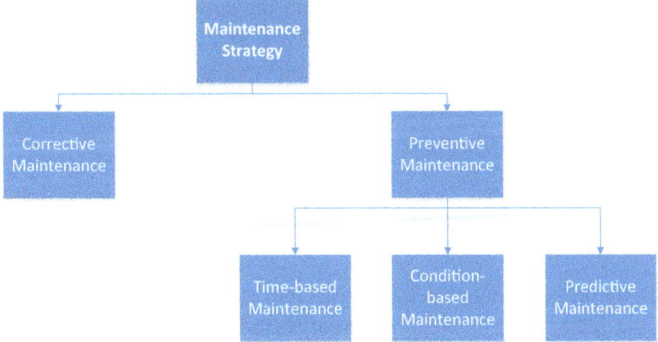

Fig. 27.1 Systematics of maintenance strategies (*source* Matyas 2002)

Simulation approaches are widely employed in order to analyze different complex systems. There are a number of works in which the simulation models are used in order to identify important factors in the O&M and to investigate the benefits of CBM strategies.

Regarding the maintenance logistics in OWE a state of the art is reviewed in (Shafiee 2015). The authors proposed a classification scheme involving three echelons of strategic, tactical and operational decision-making. Despite the number of published studies, they found out that there is still a huge potential for future developments in the O&M logistics. Martin et al. (2016) investigated the important factors affecting operational costs and availability for wind farm operators. They applied a sensitivity analysis approach to identify these factors. To this end a simulation model was developed which is capable of simulating corrective and time-based maintenance strategy, but not condition based. They pointed out that the important factors for total O&M cost were access and repair costs along with failure rates for both minor and major repairs. For time-based strategy, the important factors identified were those related to the length of time conducting the maintenance tasks, i.e., the operation duration and the working day length. Byon et al. (2011) developed a discrete event simulation model for wind farm operations and maintenance. The simulation platform represents actual wind farm operations with sufficient details, such as weather disruptions, turbine failure and spare parts management. It provides a tool for wind farm operators to select the most cost-effective O&M strategy between time-based maintenance and condition-based maintenance. Their results show that condition-based maintenance enables more wind power generation by reducing wind turbine failure rates and thus increasing wind turbine available. Scheu et al. (2012) developed a simulation model in order to simulate the operating phase of wind farm. The objective was to investigate the influence of certain parameters like the change in wave height limit for the utilized vessels, and changing the accuracy of weather forecasts. The results show that influencing these parameters enable to increase availability and the economic efficiency of an OWF.

Hofmann and Sperstad (2013) developed a time-sequential event-based Monte Carlo technique, which serves as a decision support tool "NOWIcob" that simulates the operational phase of an OWF. In addition, the simulation model takes weather uncertainty, vessel concepts for accessing OWT and different maintenance strategies into consideration. The simulation results are availability, life cycle profit, operation and maintenance cost, as well as, produced electricity.

27.3 Simulation Model and Performance Measures

In this paper we propose an agent-based, discrete event simulation model in order to understand the interaction between different elements of the O&M of an OWF. In this context, different actors involved in the O&M of the OWF are represented as agents who communicate with each other. The simulation approach is capable to integrate different planning level parameters from strategic, tactic, to the operative nature into the model, and to evaluate changes of these parameters in different planning levels.

From the strategic point of view, the model enables the change of the OWF structure and location, change of the port location, the number of the OWT and the choice between different maintenance strategies.

Regarding tactical issues, the model supports the change of resource numbers, resource types, the assignment of OWF to a specific maintenance service, and adaptation of different failure modes. The operational characteristics of OWF are taking into account in the simulation model, for example it is possible to change the lead time of different tasks as well as the resource and operation restriction like wind and wave limits, etc.

The process flow adopted in the simulation model is depicted in Fig. 27.2. Based on the failure analysis, the wind turbine releases a failure notification. The control center receives this notification and evaluates the failure. If it is a wrong failure a remote reset is performed in the wind turbine. Otherwise, based on the failure evaluation, a repair order is created and sent to the maintenance service. The maintenance service defines the maintenance requirement such as equipment and personnel needed for the particular O&M tasks. Since the weather forecast is considered as the main source of uncertainty, meteorological data is retrieved from a weather operator. If the weather condition is good and the needed resources are available for the repair order, the service crew leaves the harbor to carry out the maintenance activities. Since these decisions are based on forecasted weather data, the possibility that the real data do not match the forecasted data leads to the fact that the maintenance crew operating in the wind park has to interrupt repair activities and has to return to the harbor. In this case the repair order has to be rescheduled. This situation leads to undesirable efforts with more costs. After the maintenance crew carried out the maintenance activities the technicians and the deployment resources are released for next activities. It will be appreciated that for financial reasons, the deployment of crew transfer vessels is preferred compared to the use of helicopters.

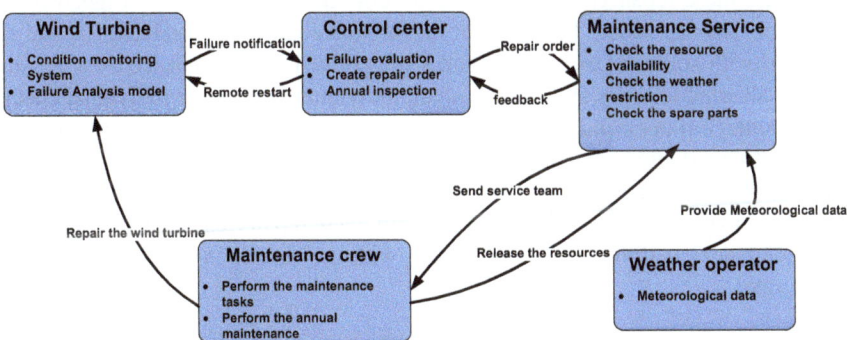

Fig. 27.2 Process flow described in the developed simulation model

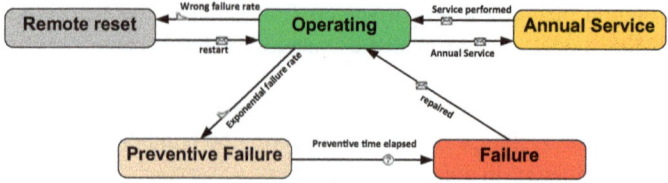

Fig. 27.3 State transition diagram of offshore wind turbine

Regarding the wind turbine behavior, the failures in the wind turbine are modeled based on exponential distributed failure rates assumed according to literature. The wind turbine is represented with a state chart, which describes the event- and time-driven behavior of the wind turbine according to Fig. 27.3. The shift from an operating state to a failure state is triggered by failure events that are determined by corresponding failure rate. The transition from failure modes to an operating state is triggered by a message event from the maintenance crew who performed the maintenance activities. Furthermore, the wind turbine has a preventive failure state which models the CBM strategy. In this case the wind turbine changes its state to the preventive failure mode when the condition monitoring system detects possible malfunctions a priori. Consequently, if the wind turbine has not been repaired for a certain time, its state will be changed to a failure state to trigger inspection as a precaution.

In principle, there are four types of repairs depending on the failure severities. First, marginal repair which need a remote manual reset to restart the wind turbine. In this case, the remote restart is executed on average once each 2 weeks and the wind turbine is off for about 2 h. The second type of repair is small repair which requires the use of helicopter or crew transfer vessel (CTV) and the maintenance activities takes about 6 h. The third type of repair is major repair which can be conducted by means of Jack-up vessels and consists of changing a big component in the OWT. This is executed exponentially once each 6 months and takes about 12 h. Finally, fatal repair usually needs the deployment of installation vessels to carry out maintenance activities and takes about 24 h.

Besides these repair types, an annual control takes place which is trigged from the control service room. To carry out this annual control an installation vessel is deployed with sufficient technicians and spare parts. The maintenance crew passes through all wind turbines and carries out an individual check. The individual check takes about 72 h per wind turbine. Failures are assumed to trigger with exponentially distributed time intervals with the parameter mean time between failures (MTBF). In this context it will be assumed that minor repairs take place each 3 months, major repairs each 6 months, and fatal repairs each 18 months. The Table 27.1 depicts different type of repairs with the correspondent restriction and failure rate.

Table 27.1 List of repair types with the corresponding repair restriction and failure rate considered in this paper (adopted from Krokoszinski 2003 and Rademakers et al. 2008)

Type of repairs	Repair restriction	Corresponding failure rate
Remote restart	Repair time = 2 h	Each 2 weeks
Small repairs	Repair time = 6 h – helicopter or crew transfer vessel and 3 technicians required	Each 3 months
Major repairs	Repair time = 12 h – crew transfer vessel with crane and 3 technicians required	Each 6 months
Fatal repairs	Repair time = 24 h – installation vessel with huge crane and 6 technicians required	Each 18 months
Annual maintenance	Control time = 72 h per OWT – installation vessel and 6 technicians required	Once a year

27.4 Simulation Results and Discussion

We conducted different simulation scenarios for the scheduling time horizon of 4 years in an OWF consisting of 60 wind turbines. We changed parameters like the number of different resources, and the selected maintenance strategy in order to analyze the importance of these parameters in O&M of OWE. Regarding the meteorological data, we consider real weather data from the year 2000 till the end of 2003 from a wind park located in the north of Germany (compare Fig. 27.4).

The performance measures used to assess the results are the wind park's availability, the generated power in the observed simulated time horizon and the

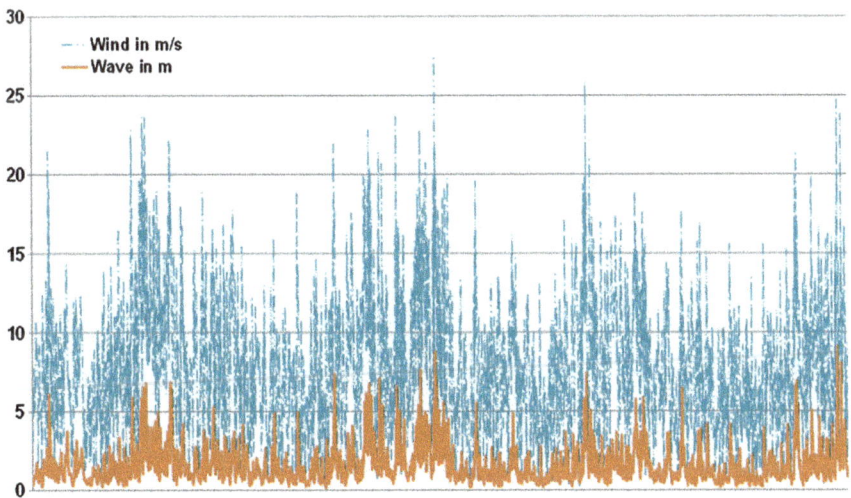

Fig. 27.4 Wave and wind speed for the scheduling time horizon between 2000 and 2003

utilization of different resources. Besides these key parameters, the mean time to repair (MTTR) of different repair type is collected. For the calculation of the electrical power of a wind turbine P_w, the formula of (Barth et al. 2012) is used:

$$P_w = c_p \frac{1}{2} \rho A v^3 \qquad (1)$$

where C_p is a coefficient of performance related to the OWE (0.79), ρ is air density, A is the rotor surface, and v is the wind speed.

The proposed simulation model has been validated by comparing the OWF availability obtained in the simulation model with the availability achieved in a OWF in the north of Germany and the availability cited in the literature (Scheu et al. 2012; Netland et al. 2014; Rohing 2014).

The results of the simulation are reported in the Tables 27.2 and 27.3. In Table 27.2, the results show that an increased use of resources leads to high OWF availability and more generated energy, which is like expected. However, this increase leads to a low utilization of the resources. Moreover, the results demonstrate that in our case the MTTR of fatal repairs is still high despite the use of two installation vessels. This is essentially due to weather restrictions of the deployment of installation vessels. In this context, the deployment of vessels with more resistance to wave height can decrease the MTTR of fatal repair and increase the OWF availability.

The results in Table 27.3 show that the CBM for the case of the deployment of one helicopter, one CTV, and one installation vessel outperforms the case without CBM. In addition, the preventive time referred to the time between the detection of failures and its occurrence, plays an important role in the increase of the availability and the energy generated.

Moreover, the results show that, even with the use of a few resources in the CBM scenario, the result is better than the scenarios without CBM strategy but with more resources.

Table 27.2 Simulation results of different number of resources without CBM strategy

Number of resources			Availability	Energy	MTTR (repair types) in hours			Workload (%)		
Heli	CTV	IV	%	GWH	Small	Major	Fatal	Heli	CTV	IV
1	1	1	84.3	9214.94	26	122	290	22	31	50
1	1	2	84.7	9197.09	25	140	280	21	32	24
1	2	1	85.5	9323.87	27	88	327	15	19	51
1	2	2	85.6	9339.12	23	82	239	15	20	25
2	1	1	84.2	9134.34	22	152	359	11	31	49
2	1	2	85.9	9327.00	21	123	323	12	31	25
2	2	1	86.1	9350.14	21	88	404	8	18	48
2	2	2	86.1	9446.09	19	74	283	7	18	25

Table 27.3 System performances comparison between CBM strategies with different preventive times

Number of resources			CBM preventive time	Availability	Energy	MTTR (repair types) in hours			Workload (%)		
Heli	CTV	IV	Days	%	GWH	Small	Major	Fatal	Heli	CTV	IV
1	1	1	0	84.3	9214.94	26	122	290	22	31	50
1	1	1	3	90.1	9600.08	18	111	367	6	17	49
1	1	1	7	91.1	9592.82	18	109	290	6	18	51
1	1	1	14	91.6	9711.77	17	62	312	6	17	49

27.5 Conclusion

In this paper, an agent-based discrete simulation model is proposed that simulates the O&M processes. The simulation model can help to understand which factors are important in the field of OWE. The results of the simulation show that the influence of increasing the number of utilized resources on OWF availability is limited due to weather limitations of the deployment of resources. In addition, the implementation results show the benefit of CBM. The time between the detection of the failures and its occurrence plays also an important role in enhancing the availability of the OWF. Further works therefore will focus on the issue, determining the optimal preventive time, leading to better OWF availability. Furthermore, other operative issues will be integrated as well in the simulation model in order to analyze different operative maintenance strategies like opportunistic maintenance. Finally, the proposed simulation model can be extended to incorporate more tactical issues to improve the availability and the generated power of the OWF. In this context it is interesting to investigate the increase of the number of resources in predefined periods, e.g., annual or seasonal maintenance.

Acknowledgments This work is part of the preInO project, funded by the Federal Ministry of Economic Affairs and Energy (BMWi; funding code 0325587A).

References

Andrawus JA, Watson J, Kishk M (2007) Wind turbine maintenance optimisation: principles of quantitative maintenance optimisation. Wind Eng 31(2):101–110
AnyLogic (2015) http://www.anylogic.de/. Accessed 24 Oct 2015
Barth V, Canadillas B, Neumann T, Westerhellweg A, Neddermann B (2012) Abschätzung des Energieangebotes. In: Handbuch Offshore-Windenergie. Rechtliche, technische und wirtschaftliche Aspekte. München: Oldenbourg, R, pp 396–422
Besnard F, Fischer K, Tjernberg LB (2013) A model for the optimization of the maintenance support organization for offshore wind farms. IEEE Trans Sustain Energy 4(2):443–450
Byon E, Pérez E, Ding Y, Ntaimo L (2011) Simulation of wind farm operations and maintenance using discrete event system specification. Simulation 87(12):1093–1117

Hofmann M, Sperstad IB (2013) NOWIcob—a tool for reducing the maintenance costs of offshore wind farms. Energy Procedia 35:177–186

Krokoszinski HJ (2003) Efficiency and effectiveness of wind farms—keys to cost optimized operation and maintenance. Renew Energy 28(14):2165–2178

Martin R, Lazakis I, Barbouchi S, Johanning L (2016) Sensitivity analysis of offshore wind farm operation and maintenance cost and availability. Renew Energy 85:1226–1236

Matyas K (2002) Ganzheitliche Optimierung durch individuelle Instandhaltungsstrategien. Industrie Management 18(2):13–16

Netland Ø, Sperstad IB, Hofmann M, Skavhaug A (2014) Cost-benefit evaluation of remote inspection of offshore wind farms by simulating the operation and maintenance phase. Energy Procedia 53:239–247

Rademakers L, Braam H, Obdam T, Frohböse P, Kruse N (2008) Tools for estimating operation and maintenance costs of offshore wind farms: state of the art. In: Proceedings of EWEC 2008

Rohing K (2014) Windenergie Report Deutschland 2013

Schenk M (2010) Instandhaltung technischer Systeme. Methoden und Werkzeuge zur Gewährleistung eines sicheren und wirtschaftlichen Anlagenbetriebs. Springer, Berlin, Heidelberg

Scheu M, Matha D, Hofmann M, Muskulus M (2012) Maintenance strategies for large offshore wind farms. Energy Procedia 24:281–288

Shafiee M (2015) Maintenance logistics organization for offshore wind energy: current progress and future perspectives. Renew Energy 77:182–193

Chapter 28
Shopper Logistics Processes in a Store-Based Grocery-Shopping Environment

Jon Meyer, Herbert Kotzab and Christoph Teller

Abstract This paper discusses the process model of consumer logistics by Granzin and Bahn (1989) from the perspective of the shopper. Thereby, we propose a model of shopper logistics and provide further insight into the planning and execution of shopper logistics based on an empirical study amongst private households. Our findings identify a distinction between planning and executing logistics activities which are not performed in a sequential but in a simultaneous manner.

Keywords Consumer logistics · Shopper logistics · Physical distribution · Household management

28.1 Introduction

When it comes to the distribution of fast moving consumer goods (FMCG), we observe a low level of involvement consumers in logistics as these goods influence daily routines the most as they represent the majority of consumers' basic needs (Baxter and Moosa 1996). However, according to Nielsen (2013) more than half of all consumers thoroughly plan their shopping trips to grocery stores by using a shopping list. 60 % of consumers like to shop at physical stores since shopping can be done quickly and conveniently and 69 % of consumers pay attention to price

---

J. Meyer (✉) · H. Kotzab
Lehrstuhl für ABWL und Logistikmanagement, Wilhelm-Herbst-Str. 5,
28359 Bremen, Germany
e-mail: jon.meyer@uni-bremen.de

H. Kotzab
e-mail: kotzab@uni-bremen.de

C. Teller
Department of Marketing and Retail Management (The Surrey Business School),
University of Surrey, Surrey, Guildford GU2 7XH, UK
e-mail: c.teller@surrey.ac.uk

promoted products. These figures indicate that consumers aim to perform shopping somehow efficiently, effectively, and rationally (Nielsen 2013).

While the marketing channel literature (e.g., Coughlan et al. 2006) has recognized consumers as active members of distribution channels, logistics literature has so far not focused on these members of a logistics chain. Worthwhile exceptions are the papers by of Granzin and colleagues where Granzin and Bahn (1989) were one of the first who recognized these issues and introduced the notions of consumer logistics (CL). Granzin (1990) defined CL as the efficient management and procurement of final products, carried out by the consumer in order to fulfill the household's consumption needs. Overall research on CL features conceptual and deductive approaches and is mainly descriptive in its nature (see, e.g., Granzin and Bahn 1989; Granzin 1990; Teller and Kotzab 2004; Teller et al. 2012). Prescriptive research refers only to a few contributions, which mainly deal with the optimization of CL in terms of efficiency and effectiveness (see, e.g., Granzin et al. 1997).

The aim of this research is to critically assess the CL-model by Granzin and Bahn (1989) by proposing the following research question: How does the CL-model by Granzin and Bahn (1989) differ if focusing on the logistics activities from the perspective of the actor who executes this task? By introducing the notion of shopper logistics (SL), we expect differences to the more general CL-model of Granzin and Bahn (1989) in all phases of CL.

The paper is structured as follows: After a brief description of the state of the art of CL research we present the basic conceptualization standing behind the notions of our understanding of SL. Thereafter we present the methodology of our research and provide an overview on the results. A discussion and an outlook for further research conclude this paper.

28.2 Conceptualizing Shopper Logistics

Overall the aim of logistics is to "provide [...] the right quantities of goods most efficiently at the right place in the right order within the right time" (Gudehus and Kotzab 2012, p. 3; see also Mangan et al. 2008, p. 9). This also holds for consumers and/or private households. Moreover, by providing space and time, households cover their needs of consumption. From this perspective, Granzin (1990, p. 239) defines CL as a complement to business logistics, which includes all activities of the consumers that "[...] support the handling of goods from the point of acquisition [...] to the point of consumption, or other disposition". Although logistics comprises the accomplishment of objectives in regards to effectiveness and efficiency, household organizations do the same with reduced scope (Granzin 1990). The consideration of a household's decisions and activities due to the movement and storage of goods, led Granzin and Bahn (1989) to build a descriptive model of CL based on the systems approach of business logistics (see Bowersox 1978). Their CL-model included the following sequential decision areas (see also Fig. 28.1): Setting, pre-trip stock assessment, trip planning, outbound travel, in-store selection, inbound travel, post-trip stock management, disposal, and post-trip communication. Based on this model, Granzin (1990) identified specific functions of CL which are presented in following Table 28.1.

In addition to these activities, Barth et al. (2007) found out that consumers would consider it necessary to calculate and plan their CL activities, just as a professional organization would do. When focusing on CL for grocery items, Teller et al. (2012) found out that this task is as compared to other product categories frequently executed in a habitual manner. Especially for those consumers who actually execute the shopping activity Teller and Kotzab (2004) revealed that consumers perceive CL as a challenging task but as an inevitable evil.

When examining a household's participation in CL, demographics and shopping behavior, Granzin et al. (1997) identified six consumer segments that represent CL patterns of shoppers, which were labeled, "Household Captain, Minimizers, Extended Shoppers, Finally Supporters, Flexible Shoppers, and Helpers". For example, the segment "Household Captain" represents multi-member households, who are living in a house, and primarily engaged in storage management, inventory acquisition, and logistics coordination. Overall we can see a difference between those household members who are actively engaged in the shopping activities (= shoppers) and those who are supporting consumer logistics activities by other means than active involvement in the shopping process. This aspect supports the

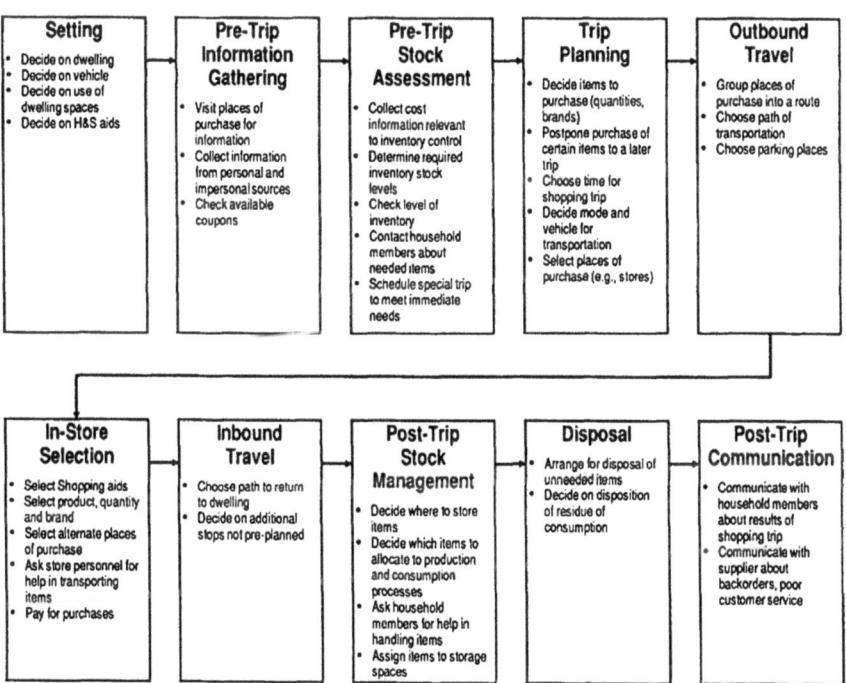

Fig. 28.1 A process model of CL decisions (Granzin and Bahn 1989, p. 93)

Table 28.1 Identified consumer logistic functions (adapted from Granzin 1990, pp. 248–251)

In-home stock management (inventory): *Includes activities/decisions on: where items to store at home, when to buy a product, the discard of items, the removal of items from stock, the management of the household's supply, determination of the needs, control of available items at home, and the formalization of a shopping list*
Selection of trip origin (location): *Includes activities/decision on: from where to start the shopping trip*
Trip management (transportation): *Includes activities/decisions on: which vehicle, means of transport to use, which route to follow, when to shop (time), persons who has to shop with whom*
Trip rescheduling (transportation): *Includes activities/decisions on: how to substitute items if the shopping trip is not possible, postponing, or canceling the shopping trip*
Nature of travel (transportation): *Includes activities/decisions on: taking a long trip or a short trip, how much time should be spent, how many and which stores are visited, involvement of carrying purchased items*
Store selection and usage (location): *Includes activities/decisions on: the type of store, how many stores to visit in regard to time limit, comparison between stores in regard to prices and products*
In-store information gathering (communication): *Includes activities/decision on: what products to buy due to information from store personnel*
In-store substitution (inventory): *Includes activities/decision on: substitution of the item by another article, product, or brand, in regard to out-of-stock (OOS), or by switching the store*
Transport-related materials handling (handling and storage): *Includes activities/decisions on: how to manage conveyance of items within the store, home, and from the mode of transportation*
In-home customer service communication (communication): *Includes activities/decisions on: arrangement of joint needs, discussions about feedback and satisfaction of the shopping trip*
External food supply (location): *Includes activities/decisions on: the supply of food, eating in a restaurant, or food delivery*
Household supporting operation (handling and storage): *Includes activities/decisions on: Maintenance of equipment (vehicle), disposing of garbage, supporting the household*

assumption of a shopper related aspect of CL, which might correlate to categories of household demographics and shopping habits, thus we label this aspect of CL as SL.

Also Teller et al. (2006) explored consumers' CL participation related to different store-based retail formats. They assumed that consumers engage in CL with different intensity, as store formats cater to for different clienteles. The consumer' CL participation is differentiated between management and operation. The results show that planning intensity seem to be quite homogenous, while the dimensions of operation (number of procurement actions, transport effort, and commissioning effort) significantly differed between store types. Kotzab et al. (2006) could further show that consumers evaluate their undertaken CL performance as being efficient.

In conclusion we distinguish between SL and CL depending on who is involved in the logistics activities as households represent—such as industrial organizations—buying centers consisting of members with different roles (see Webster and Wind 1969). SL patterns may be categorized by the demographics of a household and its shopping behavior (Teller and Kotzab 2004; Granzin et al. 1997). In addition, store formats may also affect the intensity of SL (Teller et al. 2006). Overall, existing knowledge on SL is limited thus we want to gain more insight into this subject matter.

28.3 Methodology

The aim of this study is to gain further insight into the household logistics activities. Especially we were interested to get empirical support on our notions on SL. The unit of analysis in our empirical study is the shopper's engagement in planning, processing, and management of logistics activities and decisions within the consumption process of groceries. Based on the notions of Granzin et al. (1997) we propose that grocery shopping is conducted in a habitual manner. Based on Teller and Kotzab (2004) we further propose that shoppers perceive logistics activities and decisions only to a certain degree. Based on these propositions/assumptions and due to the lack of existing theories, a "grounded approach" was chosen to explore this phenomenon. While retaining the principles of the Grounded Theory approach (see, e.g., Glaser and Strauss 1967 or Charmaz 2013), we applied the ethnographic method "Participant Observation" (see, e.g., Mathews and Kaltenbach 2007 or Bachmann 2009) in four different households. There we specifically focused on the processes related to grocery shopping (see Table 28.2).

Table 28.2 Characterization of the sample

	Household I	Household II	Household III	Household IV
Description	Young couple, male at the age of 26 years, female at the age of 28 years, not married, without children	4 household members, parents (50–55 years) and grown up daughters at the age of 21 and 24 years	Single household, male at the age of 70 years	A family of three members, male at the age of 41 years, female at the age of 40 years and female child at the age 5 years, one dog, two available cars
Properties of dwelling	50 m² living space, two-room department on third floor, integrated kitchen, no lift inside the building	Ca. 200 m² living space, house on a farm, kitchen in the first upper floor	Ca. 60 m² living space, apartment on the fourth floor, kitchen in a separate room, no lift inside the building	Detached house, ca. 400 m² living space, kitchen in a separate room
Monthly expenditures on groceries	220 EUR	550 EUR	No comment	800–1,000 EUR

The selected households were little known by the researcher, as they were distant acquaintances or interceded persons. This secured a comfortable interaction between researcher and participant. Each household's representative was informed about the study's objective and data privacy before the investigation started. The researcher would accompany the shopping household member on their shopping trips. The observation focused on three shopping process sequences, i.e., pre-trip section, operation section, and post-trip section. Alongside the participant observation qualitative interviews were conducted before and after the shopping trips. The participants are visited in their dwellings an hour before the shopping trip begins. The interviewer, carries on a conversation, and observes the situation in the pre-trip section. Subsequently the researcher accompanies the participant on her. Simultaneously, the researcher observed the situation, took notes, and pictures. When the shopping trip had ended, the researcher accompanied the participant back to their dwellings. Finally, the participant was interviewed and an interview was conducted, reflecting on the shopping trip. After gathering the data, the sound files and notes were transcribed according to the rules of easy transcription by using the F4 software (Dresing and Pehl 2011). The text data included three types of documents: Pre-trip interview transcripts, post-trip interview transcripts and participant observation protocol. Content analysis including an individual analysis as well as cross-analysis was conducted in the second stage (e.g., Hsieh et al. 2005).

28.4 Results

28.4.1 Four Areas Determine Shopper Logistics

One major finding of our research was the identification of four general areas, which determine the way how SL is performed. These areas refer to household, event, supply, and shopping management follows:

- **Household management** includes all taken decisions and performed activities by household members. The overall goal of household management is to ensure the flows of daily routines without any disturbances;
- **Event management** refers to all decisions and activities to be performed in regards to specific and/or unexpected incidents influencing the daily routines of a household;
- **Supply management** includes the identification of offerings, products and services which fit to the needs and wants of a household
- **Shopping management** includes all taken decisions and performed activities by household members regarding the procurement of required goods and services.

Household and event management determine the demand of a household. Supply management selects the adequate supply while shopping management determines the procurement mode. Any SL decision is based on the interaction of these four areas. All decisions that are taken within these areas result in activities that determine the individual grocery-shopping process Household management takes purchasing decisions, which are either general or specific depending on external developments which event management is aware of. Typical example for this is the 'weekly purchase' against the 'birthday party'. These two areas determine the basic needs and wants and translate them into shopping requirements. Supply management is concerned with the comparison of shopping requirements with existing stock levels of goods in the household. After this comparison, the specification of the shopping list determines the requirements for shopping management including the choice of products, the choice for the store(s) as well as potential time constraints, which need to be considered.

28.4.2 Planning and Execution of Shopper Logistics

Another finding of our research refers to the explicit identification of a separation between 'strategic' planning activities for a shopping trip and the subsequent 'operational' execution of the shopping trip. While planning of SL usually takes place in a nonstore environment (e.g., at home), the execution of SL is a pure point-of-sale activity. The planning of SL determines its execution, meaning the better the planning, the easier the shopping trip. Contrary to Granzin and Bahn (1989) we were not able to identify a clear sequence of decision and activities. However, we observed decisions and activities that can be associated with two succeeding levels referring to a strategic and an operative level. The strategic level includes the planning of the shopping trip which is done outside the shopping location. The operative level refers to the execution of the shopping process and occurs at the point-of-sale.

28.4.3 Simultaneous Planning and Execution of Shopper Logistics

A further finding of our research was the identification of simultaneous planning and execution activities, which is in contrast to the sequential mode of activities as suggested by Granzin and Bahn (1989). In the planning phase, the shopper and the household members recognize their needs and translate them into a more or less specific shopping list. In the execution phase, the plan is translated into 'shopping action' where depending on the specificity of the planning; further planning activities need to take place. Furthermore, the shopper is confronted with a 'store reality', which also forces the shopper to adapt the plans to the given environment. This means that there is a sequence in planning and execution, however, certain internal and external disturbances affect execution and forces shoppers to plan in the

Table 28.3 Simultaneous planning and execution

In-store selection			
Household I	Household II	Household III	Household IV
• Plans further items to be purchased • Substituting items due to out-of-stock	• Get influenced by promotion • Substituting items due to better alternatives • Recalculating the needs by thinking of home's inventory levels • Reacting on OOS	• Searching for price promotion • Rescheduling meals due to supply • Consideration of forthcoming events to decide on better alternatives	• Shelves are passed by checking if there are items needed • Thinking of household member's needs

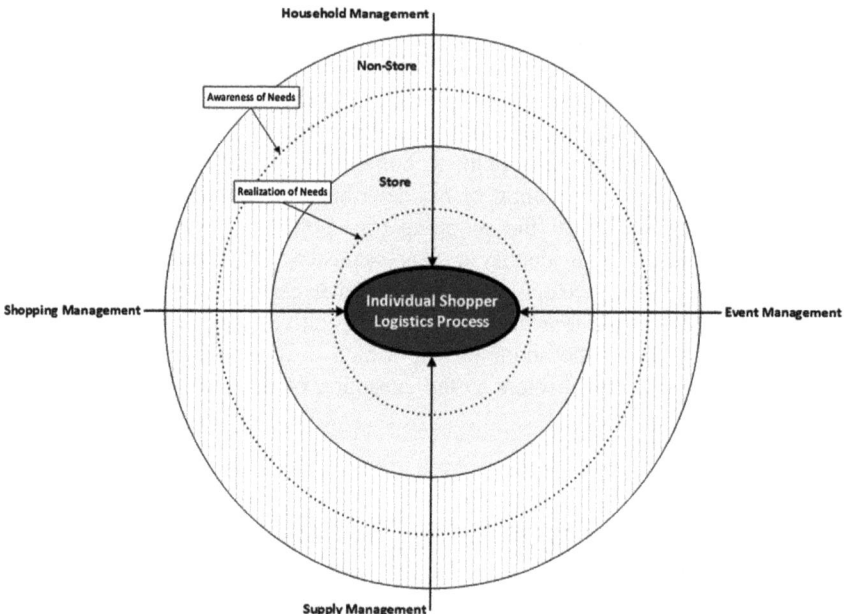

Fig. 28.2 A different approach to consumer logistics

execution phase. Table 28.3 views our identified categories, which include the respondent's planning activity due to disturbances, while shopping in store. The typical case for this refers to the out-of-stock-situation that cannot be considered at a planning stage at home and thus makes a shopper switch the shopping format.

28.4.4 Deriving a Different Model of Consumer Logistics

Based on our findings, we propose a different model of CL as opposed to the notions of Granzin and Bahn (1989) that is shown in Fig. 28.2. Our model consists of four determination areas (household, event, supply and shopping management) which influence the planning and execution of CL and consequently SL.

We distinguish furthermore between planning and execution of SL that takes typically place at different locations. However, depending on the quality of planning (= internal disturbance) and the store offering (= external disturbance), planning and execution is not performed sequentially but simultaneously. Thereby communication processes between household members and shoppers happen.

28.5 Conclusions and Outlook for Further Research

The goal of our empirical study was to gain insights into the SL processes in four different demographic household settings. There we empirically investigated different types of households and their planning and operation of shopping for groceries. Each case offered insights in regards to their individually conducted shopping trips. The cross-case analysis thus led to new ways to look at this phenomenon. Overall we were able to identify major differences to the existing knowledge on CL. We found several SL activities and decisions the households conducted in order to satisfy their needs, which were determined by overarching decision areas. We also found SL tasks to be distinguished between planning and operation, which were not performed solely in a sequence.

We see the main reason for that in the fact that due to the individual needs, individual structures, and individual daily routines, shopping and related SL tasks were conducted habitually and individually. This result also holds for the household's consumption goals, which result in a shopping list representing the potential customer order. The way and how this order is realized depends on the shopper's SL planning and its execution as well as on external factors such as store offerings. Thereby we were able to observe that consumers obviously conduct logistical planning activities during shopping. This insight let to the conclusion that SL planning overlaps with SL operation. Whenever relevant needs are not considered in the planning phase, the order becomes incomplete, and the consumer is engaged in further planning activities during operation. This phenomenon was labeled "overlap". The consideration of variables that influence the SL process gained further insights into how activities and decisions interrelate.

Our analysis also showed a limitation of the Granzin-model itself, as this approach does not include any reverse logistics activities such as disposal, returning of empties or dealing with packaging material For further studies, the inclusion of disposal logistics needs to be considered.

Further limitations of our study refer mainly to the sample that offered quite a wide range of grocery shopping and its related activities and decisions. Despite these limitations, several aspects of' SL activities and decisions were discovered and examined. Second, the individual shopping trips were captured in categories. Initial coding captured content due to the household's engagement in CL in terms of conspicuous activities, decisions, statements, etc., the participated consumer made. Codes were then classified to categories. During further procedures, these categories were aggregated into core categories, whenever saturation had been reached. Because, saturation of categories was evaluated subjectively, and intercoder-consistency matrixes were not considered due to the scale and scope of this study, it has to be acknowledged that the classified categories might lack saturation. The proposed CL-model is strongly dependent on subjective interpretations. Further research on CL should repeat the applied research procedure to verify and identify further categories and variables. Intercoder-consistency matrixes would reduce the problem of subjectivity. Bigger sample sizes would present a more comprehensive understanding of the household's engagement in CL.

References

Bachmann (2009): Teilnehmende Beobachtung. In: Handbuch Methoden der Organisationsforschung: quantitative und qualitative Methoden. VS, Verlag für Sozialwissenschaften, Wiesbaden
Barth et al (2007) Betriebswirtschaftlehre des Handels. Gabler, Wiesbaden
Baxter, Moosa (1996) The consumption function: a basic needs hypothesis*. J Econ Behav Organ 1(31):85–100
Bowersox (1978) Logistical management, 2nd edn. MacMillan, New York
Charmaz (2013) Constructing grounded theory, 2nd edn. Sage Publications, Thousand Oaks, CA
Coughlan et al (2006) Marketing Channels. Prentice-Hall, Upper Saddle River, NJ
Dresing, Pehl (2011) Praxisbuch Transkription: Regelsysteme, Software und praktische Anleitungen for qualitative ForscherInnen. Eigenverlag, Marburg
Glaser, Straus (1967) The discovery of grounded theory: strategies for qualitative research. Aldine Pub. Co, Chicago
Granzin (1990) The consumer logistics system: a focal point for study of household-consumption process. J Consum Stud Home Econ 14(3):239–256
Granzin, Bahn (1989) Consumer logistics: conceptualization, pertinent issues and a proposed program for research. J Acad Mark Sci 17(1):91–101
Granzin et al (1997) Consumer logistics as a basis for segmenting retail markets. J Retail Consum Serv 4(2):99–107
Gudehus and Kotzab (2012) Comprehensive logistics. Springer, Berlin, Heidelberg and New York
Hsieh et al (2005) Three approaches to qualitative content analysis. Qual Health Res 15(9):1277–1288
Mangan, John, Chandra Lalwani, and Tim Butcher (2008). Global logistics and supply chain management. John Wiley & Sons
Mathews, Kaltenbach (2007) Ethnographie: Auf den Spuren des täglichen Verhaltens. In: Qualitative Marktforschung in Theorie und Praxis: Grundlagen, Methoden und Anwendungen. Gabler, Wiesbaden, pp 139–149

Nielsen (2013) Deutschland 2013: Handel, Verbraucher, Werbung. http://www.nielsen.com/content/corporate/de/de/insights/reports-downloads/_jcr_content/par/download_0/file.res/Nielsen_Universen_D_Internet.pdf Accessed 24 June 2014; [MEZ] 16:22

Teller, Kotzab (2004) Proposing a model of consumer logistics. In: Enhancing competitive advantages through supply chain innovation. Presented at: logistic research network 2004 conference proceedings, Dublin, pp 1–7

Teller et al (2006) Betriebstypen und Konsumentenlogistik. In: Schnedlitz, Peter et al (eds) Innovationen in Marketing und Handel. Linde, Wien, pp 214–232

Teller et al (2012) The relevance of shopper logistics for consumers of store-based retail formats. J Retail Consum Serv 19(1):59–66

Webster and Wind (1969) A general model for understanding organizational buying behavior. J Mark 36:12–19

Chapter 29
Empty Container Repositioning from a Theoretical Point of View

Stephanie Finke

Abstract The repositioning of empty containers is one of the most important tasks in container shipping. The paper discusses the repositioning with non-logistic theories to explain problems between different actors in the supply chain of empty container repositioning. The transaction cost theory explains the conditions under which ocean carriers or exporters have to find a transaction partner or the choice between hierarchy and market. The principal agent theory explains how different actors react in a supply chain and gives advices for contract design. By means of these theories one specific relationship in empty container management is analyzed. With the help of this analysis, advices for later research could be given and a better understanding for problems between the actors could be shown up. The fact that only the relationship between exporter-ocean carrier-railway company is analyzed and not the whole supply chain limit the research. The main contribution is that a research gap in empty container repositioning could be closed, because there is no paper with a theoretical background of new institutional economics.

Keywords Empty container management · Intermodal · Transaction cost theory · Principal agent theory

29.1 Introduction

Although the repositioning of empty containers is very important and often discussed in the literature, there are only a few papers that analyze the act of repositioning from a theoretical point of view (Halldorsson and Kotzab 2007). Mostly the problem is analyzed and solved with operations research (OR), but in empty container repositioning, some problems occur, which cannot be solved with the help of OR, but from a socioeconomic perspective. The character of this paper is of

S. Finke (✉)
Lehrstuhl für ABWL und Logistikmanagement, Wilhelm-Herbst-Str. 12,
28359 Bremen, Germany
e-mail: stephanie.finke1@ewetel.net

conceptual nature, so the main purpose of this paper is to look at one particular problem of empty container management, which is the repositioning through the lenses of principal agent and transaction cost theory. This should lead to a better understanding of the relationships between the different actors within the supply chain. Therefore, one typical problem in empty container management, the need of repositioning containers, is shown. In addition this scenario is common in container management. In addition the main actors in empty container repositioning are described explained in Chap. 2. In a second step, the theory of principal agent and transaction costs will be presented in Chap. 3. In Chap. 4, the described situation and the relationships between the actors within the supply chain are analyzed under the determinants of the theories. The paper closes with a conclusion and implications for future research.

29.2 Typical Example in Empty Container Repositioning and Main Actors in the Supply Chain

First of all, in this example, the relationship between an ocean carrier and a forwarder and an exporter is regarded. This scenario is addressed to the problem of trade imbalances and its problem of less empty containers at the right place.

In this example an ocean carrier gets the assignment to deliver some empty containers to the exporters. The exporting company operates in the car producing sector and often needs empty containers. Due to this fact, the exporter and the ocean carrier know each other very well, because they often cooperated in the past. The ocean carrier has no own depots in the hinterland and its container stock consists of leasing and own containers. Additionally, the exporter has access to rail connection, so trains always transport the containers. Based on the frequency of empty container deliveries, the ocean carrier deliberates whether to found its own railway shuttle company, or not.

In this case we have carriers haulage, so the carrier has to organize the pre-run and post-run of the empty containers. It is assumed that the ocean carrier has not enough empty containers next to exporters place, due to the trade imbalances. So it has to reposition empty containers. A railway company should transport these containers over landside in order to meet the demand of the exporter and to work against the mismatch of empty containers. So, the ocean carrier instructs a railway company to transport a given number of containers to exporters.

For a better understanding of the interaction of different actors in empty container repositioning, the main actors in the supply chain are: container owners, ocean carrier, empty container depot operators and service companies, terminal operators, port operators, forwarders, carriers, container leasing companies, operators and shippers and consignees. Only the actors, which are need in the scenario will be characterized in detail:

- Ocean carrier—They control a large part of the logistic chain. Ocean carriers perform the container transport by sea, own or lease containers, and operate ships for the container transport (The Tioga Group 2002; Hildebrand 2008; Schieck 2008). In addition in some cases they conduct the hinterland transport (Pawlik 1999).
- Forwarders—Forwarders have contracts with shippers, consignees and sometimes other forwarders. In case of a contract with another shipper, the second shipper organizes the transport chain by order of the forwarder (Schwarz 2006; Hildebrand 2008). Forwarders usually do the calculation of the costs for transport and handling, preparation of all needed documents like transportation and shipping documents. In addition they do all duty transactions for import and export goods (Bischof 1995). The accomplishment of the container transport can be done by the shippers self-contained or outsourced. Also other transport operators or stevedore companies can act as shippers. In reality, this fact often resulted in a fusion of shippers and forwarders, so a clear separation between both groups is difficult. Additionally, one can distinguish between overseas shippers and hinterland shippers (Hildebrand 2008).
- Carriers—This group consists of barge operators, feeder shippers, railway companies, trucking companies and independent ship owners. Independent ship owners, so called Partikuliere, are independent, work with a small barge fleet and have contracts to other barge operators (Hildebrand 2008).
- Shippers and consignees—This group represents the primary purchaser. By demanding transport activities and defining the sort of goods to be transported, they set the key requirements for the design of the transport chain (Hildebrand 2008).

The design of the hinterland chain can be selected by the actors. In general, it can be distinguished between carriers haulage, merchants haulage and mixed arrangements. In the first case, the ocean carrier assumes the arising expenses and the complete organization of pre-run and post-run of full and empty containers. If forwarders or shippers carry out these tasks, we have the case of merchants haulage. The concept of mixed arrangements applies when the pre-run and post-run of the empty container is organized by the ocean carrier and therefor the carrier takes a fee. The forwarder bears the costs for the transportation of the full container and he organized it (Schwarz 2006).

In this chapter a supply chain in empty container management in case of an import is described in the next sentences. First of all, a container is owned by a leasing company, a shipping company, or third parties (Theofanis and Boile 2009; The Tioga Group 2002). The box should be shipped from point A, a foreign port, to point B, a consignee in hinterland. In the example, the ocean carrier transports the container to the port from sea side. When the container arrives at the port there are three actors in the SCM who handle the container. First of all there are the port operator and the terminal operator. In addition we have container service companies in the port. The hinterland transport of the container can be done in three different ways. First, the container is transported to the consignee by truck. Second the box is transported to an inland terminal or so called GVZ by train and later to the consignee by truck. In the third scenario the container is transported to an inland port by barge first and second directly to the consignee by truck. Or the container is transported afterwards to an inland terminal by train and to the final destination by truck.

29.3 The Theory Applied to the Repositioning Problem

29.3.1 The Principal Agent Theory

In principal agent theory a principal (purchaser) and an agent (contractor) react in an environment with uncertainty and asymmetric information (Mensch 1999). In general, "whenever one individual depends on an action of another, an agency relationship arises" (Pratt and Zeckenhauser 1985, 2). This definition is very common, so all negotiations between two parties that interact with each other can be referred to a principal agent problem. The definition of Jensen and Meckling (1976) is more precise. They point out that a principal agent relationship is "a contract under which one or more persons (the principal(s)) engage another person (the agent) to perform some service on their behalf which involves delegating some decision-making authority to the agent" (Jensen and Meckling 1976, 308). Typical examples for this are the relationship between a buyer and a seller of goods or between an employer and an employee (Göbel 2002).

The principal agency relationship is characterized by the following assumptions:

- the behavior of the agents can have a positive or negative impact on the well-being of the principal
- the principal and the agent are rational and want to maximize their own profit
- principal and agent have different expectations of the use, because of their different rights at the goods
- between principal and agent exist asymmetric information. The agent has an information related advance against the principal.

Due to these assumptions, the following problems can occur: hidden characteristics, hidden action, hidden information, and hidden intention. Hidden characteristics occur ex ante the contract is signed, as the principal cannot get the hidden properties of the agent, so adverse selection (selection of a poor contractor) is possible (Milgrom and Roberts 1992). In case of hidden action, the principal cannot observe all actions of the agent and due to asymmetric information, the principal can exploit this situation. This and the following problem arise ex post. The problem of hidden information appears, when the principal can observe the actions of the agent, but he cannot judge the quality of the action due to lack of expertise (Göbel 2002). The last problem, hidden intention, deals with the situation that the principal does not know everything about the accommodating of the agent in case of a conflict, before signing the contract. Another problem, which also belongs to hidden intention, is the risk of hold up. In case of hold up, the agent exploit the dependence of the principal for his own use (Pratt and Zeckenhauser 1985).

This theory fits perfectly in case of empty container repositioning, because it offers a model to explain the actions of people and institutions in hierarchy. In addition it gives advices for the design of contracts.

29.3.2 The Transaction Cost Theory

Firstly the term of transaction costs was established by Coase 1937. He was the first scientist, who argued that a transaction can be more efficient, if coordination forms like corporate hierarchy are used, instead of coordination through markets (Coase 1937, 1960). Later, this theory has been mainly expanded by Williamson (1975) and others (e.g., Teece 1986). The analysis unit of the theory is the transaction, but in literature, there is no integrative definition of a transaction (Göbel 2002). The most commonly used definition derives from Williamson. He says that a transaction occurs, when a good or a performance is transmitted across a technological interface (Williamson 1975, 1985). So transaction costs encompass costs of market-use (Coase 1960) and costs which arise in organizations. Based on the pioneer work by Coase, Williamson introduced two behavioral assumptions regarding the actors: bounded rationality and opportunism (Williamson 1975). "Bounded rationality may result from insufficient information, limits in management perception or limited capacity for information processing" (Skjoett-Larsen 2000). Opportunism is characterized by the fact that actors operate in their own interests and they are trying to maximize their own profit (Williamson 1996).

Additionally, Williamson classified the transaction costs into three main categories: information costs, bargaining costs and enforcement costs. An example for information costs is costs which result from finding a transaction partner. Costs of preparing and drawing a contract belong to the group of bargaining costs. To the enforcement costs amongst costs, which origin is in monitoring the contract execution (Williamson 1985).

According to Williamson (1985), there are three main dimensions, which characterize a transaction: frequency, uncertainty and specificity. Frequency is a very simple attribute. It states, if a transaction is often realized or not. The term of uncertainty can be divided into environmental uncertainty and behavioral uncertainty. The first type of uncertainty can lead to external errors regarding the transaction, so that the agreement between the two actors has to be adapted to the new situation. Behavioral uncertainty means that the contractual partners do not know the behavior of the other (Williamson 1985). Williamson divided the last dimension of transaction costs into physical asset specificity, site specificity, human asset specificity and dedicated asset specificity. Generally spoken, specificity means uniqueness and non-exchangeability of a good or performance. Asset specific means that one transaction partner does investments for a specific transaction with a low value when used in other transactions. In case of site specificity, a factor differs from others by its specific location. An example for human asset specificity is an employee, who has gained certain skills and knowledge for his work through years. For instance an investment into a special machine that can only produce components for a specific customer belongs to dedicated asset specificity.

"Transaction cost economics has been widely used for assessing strategic changes in organizational forms and organizational innovations and in particular for determining optimal structures of governance" (Panayides 2002; Williamson 1996). These insights are relevant for the analysis of the supply chain in empty container repositioning and this theory could be a useful framework for changes in container repositioning.

29.3.3 Analysis of the Repositioning Problem

In the presented scenario, a trade relation has been introduced. At first, the example will be analyzed by disposing the principal agent theory and second by transaction cost theory.

29.3.3.1 Scenario Applied to the Principal Agent Theory

In the sample, we have two principals and multiple agents. For example, in the relation between the exporter and the ocean carrier, the principal, is presented by the exporting company and the agent by the ocean carrier. Due to this, the exporter is the principal, because he needs an empty container and he instructed the ocean carrier to get one. Additionally, the ocean carrier is a principal as well it is principal compared to the container leasing company in case of a leased container and also to depot operators and to the railway company.

To have a clear overview, the paper only focuses on the principal agent relation between the exporter and the ocean carrier and additionally between the ocean carrier and forwarder, represented by a railway company. As mentioned in the third chapter, there are four different problems in principal agent theory: hidden characteristics, hidden action, hidden information and hidden intention. Not every relationship conducts all four theory related problems. Due to this, the relationships between the two different trade relations, on the one hand exporter-ocean carrier and on the other hand ocean carrier-railway company, are simultaneously analyzed.

The example has two principals and multi agents. There are two different relationships: on the one hand there is a relationship between the exporter and the ocean carrier, who is tasked to transport an empty container to exporters place. On the other hand the relationship between the ocean carrier and the railway company is mentioned. The relations between ocean carrier and depot operator or leasing companies are disregarded.

For instance, the ocean carrier is principal and agent at the same time. It is agent in the relationship with the exporter, while the exporting company is the agent. The ocean carrier takes the role of principal in the relation between itself and the railway company.

The case of hidden action is given by the fact that the exporter does not know, if the ocean carrier lies or not, when they transport an empty container over a far distance, because there was no container close by. Furthermore, shirking is possible between the ocean carrier and the forwarder of the railway company. It is not obvious, if the forwarder loads the containers as fast as he can, or if he is working slowly.

The problem of hidden information exists between ocean carrier and railway company. For instance, the railway company only knows, if they can use all tracks in the railway network, or if there are some tracks which are too dangerous for the containers. Another example for hidden information is that only the railway company knows, whether all necessary protections are done for the container transport or not.

The risk of hold up exists between the exporter and the ocean carrier, because, there are not as many shipping companies in Germany. Additionally hidden intention and hold up are also given between the ocean carrier and the railway company, since only a few railway companies transport containers in Germany.

29.3.3.2 Scenario Applied to the Transaction Cost Theory

As shown in the theory chapter, transaction costs can be divided into information costs, bargaining costs, and enforcement costs. Furthermore, transactions are characterized by frequency, uncertainty, and specificity.

The object of the given scenario is the relationship between the ocean carrier and the railway company and additionally the relationship between the ocean carrier and the exporter. We will start with the analysis of the relationship between the ocean carrier and the railway company.

First, we have a transaction according to the transaction cost theory.

The transfer of property rights of empty containers has been done by transferring the rights from the ocean carrier to the carrier (railway company) in order to reposition the containers. Thereby we have a transfer of a material resource, the empty container, and a transfer of service in case of the transport.

In this scenario the specificity according to the transaction cost theory is defined as the repositioning of empty containers. There are many different types of containers, which can be repositioned and some types need special transport nodes, because they cannot be transported with a standard truck for example. As shown in the theory chapter the main task of a forwarder is the transportation of containers. In this scenario the empty containers and the transportation by a forwarder characterize the specificity.

Uncertainty refers to the empty container reposition by the railway company and its timeframe. If there is a delay, it could have negative aspects for the ocean carrier and its plans. For example, additional services of the empty container management will be too late or cannot be done. The result is that additional costs for the ocean carrier arise. In this case, the transport of empty containers has a high priority for the ocean carrier, while the priority is lower for the railway company, because the profit of the transportation of an empty container is lower than a full container. So the forwarder prefers to transport full containers. The named for aspects of uncertainty refer to the bounded rationality of the ocean carrier to the railway company. On the basis of bounded rationality, the ocean carrier has not all the information of the container transport, which is done by the forwarder and the forwarder as an actor. Due to restricted information, it is impossible for the ocean carrier to remove all uncertainties, out of which coordination problem arise. Because of these problems, the ocean carrier wants to control the performance and with this the railway company. In case of nonconformity, the ocean carrier will impose sanctions. These measures have an effect in the resulting transaction costs. In addition, due to bounded rationality, it is assumed that the railway company has an opportunistic behavior.

Opportunism, bounded rationality, and uncertainty will decline when the transaction will be done more often and cooperation arises. Finally, the transaction costs will also go down.

Now the relationship between the ocean carrier and the exporter will be analyzed shortly. The information costs are low for the exporter when he hires an ocean carrier for its assignments, because in Germany there are not many carriers which handle containers. From the perspective of the ocean carrier, the information costs are low as well, since only a few railway companies may transport containers.

In this scenario, the transaction frequency is very high, because the exporter often needs empty containers and the transaction partners have a common history. Therefore, the ocean carrier has long-term contracts with the exporter and the railway company. In both cases, bargaining and enforcement costs are low, because of the long-term agreements.

In general, the specificity of this transaction is low. As in scenario A, the railway company can be replaced by another company or the container can be transported by another transportation mode like truck or barge.

Due to the fact that the railway company also can transport full and empty containers from other customers, the asset specificity is low. In contrast, the specificity is high, if the ocean carrier founded its own railway company.

The human asset specificity is quite high, since the individual members of the railway company got to know the exact process with the exporter and ocean carrier very well in the course of all the transactions.

29.4 Conclusion and Managerial Implications

The paper contributed the research in empty container repositioning as it analyzed the relationships of the various actors who are involved in more detail and under a theoretical point of view. The analysis has shown that different principal agent problems like the risk of hold up or hidden action exist in empty container repositioning. Furthermore, transaction costs also apply in repositioning. This is important because in shipping the ocean carries and other actors need to keep the costs low, including transaction costs. Analyzing different relationships between actors of a supply chain has showed the utility of the theories for understanding problems. In addition, one example is given and analyzed under theoretical determinants. The research is limited by missing links in the whole supply chain applied to the theory. Only two relationships were analyzed in detail.

The analysis has shown that the two presented theories fits perfectly to the empty container management and gives some advices for solving typical problems. First of all, the principal agent theory has shown that a contract between the acting companies will help to reduce typical principal and agent problems. A contract is also a solution in case of transaction costs. All in all, certain conditions related to the specific services shall be recorded in the contract. Due to this, the actors would have a better planning basis und uncertainty will be reduced. According to the transaction cost theory opportunistic behavior, bounded rationality and uncertainty can be reduced by frequent transactions in the form of a cooperative collaboration. Through a long-term cooperation, cooperation problems and transaction costs can be reduced. In addition the actors have the option to note certain conditions related to the specific service in the contract. This allows a better planning basis for the actors and also reduces uncertainty.

The next step in future research could be to use these findings in order to optimize empty container repositioning and to use the insights by better design of contracts. An example for this is that the contracts should be more long term in order to reduce transaction costs like information costs. Additionally the contracts could be clearer regarding to contractual penalties or time of delivery date.

References

Bischof KD (1995) Speditionsbetriebslehre, 4th edn. Stam Verlag, Köln
Coase RH (1937) The nature of the firm. Economica 4(16):386–405
Coase RH (1960) The problem of social cost. J Law Econ 3:1–44
Göbel E (2002) Neue Institutionenökonomik. Lucius und Lucius Verlagsgesellschaft mbH, Stuttgart
Halldorsson Arni, Kotzab Herbert (2007) Complementary theories to supply chain management. Supply Chain Manage: An Int J 12(4):284–296
Hildebrand W-C (2008) Management von Transportnetzwerken im containerisierten Seehafen-hinterlandverkehr: Ein Gestaltungsmodell zur Effizienzsteigerung von Transportprozessen in der Verkehrslogistik. (Schriftenreihe Logistik der Technischen Universität Berlin, 6, Baumgarten, H., Straube, F., Klinkner, R. (eds)) Universitätsverlag der TU Berlin, Berlin
Jensen MC, Meckling WH (1976) Theory of the firm: managerial behaviour, agency costs and ownership structure. J Finance Econ 305–360
Mensch G (1999) Grundlagen der Agency-Theorie, in: WISU – das Wirtschaftsstudium 28 (5):686–688
Milgrom P, Roberts J (1992) Economics, organization and management. Prentice Hall, Englewood Cliffs

Panayides PM (2002) Economic organization of intermodal transport. Transp Rev: Transnational Transdisciplinary J 22(4):401–414

Pawlik T (1999) Seeverkehrswirtschaft: Internationale Containerlinienschifffahrt; eine betriebswirtschaftliche Einführung. Betriebswirtschaftlicher Verlag Dr. Th. Gabler GmbH, Wiesbaden

Pratt JW, Zeckenhauser RJ (1985) Principals and agents: an overview, In: Pratt JW, Zeckenhauser RJ (eds) Principals and agents: the structure of business. Boston 1–35

Schieck A (2008) Internationale Logistik: Objekte, Prozesse und Infrastrukturen grenzüberschreitender Güterströme. Oldenbourg Wissenschaftsverlag GmbH, München

Schwarz F (2006) Modellierung und Analyse trimodaler Transportketten für Seehafenhinterlandverkehre. (Logistik, Verkehr und Umwelt, Hrsg.: Clausen, U.) 1. Aufl. Dortmund: Verlag Praxiswissen

Skjoett-Larsen T (2000) Third party logistics—from an interorganizational point of view. Int J Phys Distrib Logistics Manage 30(2):112–127

Teece DJ (1986) Transaction cost economics and the multinational enterprise: an assessment. J Econ Behav Organ 7:21–45

The Tioga Group (2002) Empty ocean container logistics study. Technical Report, Submitted to the Gateway Cities Council of Governments, CA

Theofanis S, Boile M (2009) Empty marine container logistics: facs, issues and management strategie. GeoJournal 74(1):51–65

Williamson OE (1975) Markets and hierarchies: analysis and anti-trust implications. a study in the economics of organization. Free Press, New York

Williamson OE (1985) The economic institutes of capitalism. Free Press, New York

Williamson OE (1996) The mechanisms of governance. Oxford University Press, New York

Chapter 30
Frugal and Lean Engineering: A Critical Comparison and Implications for Logistics Processes

Eugenia Rosca and Julia Bendul

Abstract Frugal innovation gains momentum in literature and practice as the next important management approach transferred from East to West after lean. The two paradigms have launched powerful ideas challenging the traditional Western approaches, with frugal innovation being the most recent. This paper focuses on the application of lean management and frugal innovation in the field of engineering. The principles employed during the design stages are responsible for a high percentage of logistics costs and consequently influence significantly logistics processes. Therefore, we highlight similarities and differences of the two approaches, and show how to combine the underlying principles in order to develop more efficient products with efficient production processes especially for cost-sensitive consumers in both developing and developed countries. The implications of lean and frugal engineering for procurement, production, distribution and disposal logistics are relevant not only for sustainability managers, product or logistics managers concerned with sustainability aspects but also for executives aiming to expand their operations in developing countries.

Keywords Frugal innovation · Lean management · Logistics processes

30.1 Introduction

Many scholars regard frugal philosophy as the next important philosophy after lean management to migrate from East to West (Bhatti et al. 2013). Lean thinking migrated as a management philosophy in the late 1980s and is regarded by Japanese companies as a way of working, rather than specific tools and techniques (McManus et al. 2007). In the West, lean ideas translated into principles and tools

E. Rosca (✉) · J. Bendul
Jacobs University, Bremen, Germany
e-mail: e.rosca@jacobs-university.de

J. Bendul
e-mail: j.bendul@jacobs-university.de

for manufacturing processes and also gave birth to the concept of lean engineering for product development activities. Similarly, the concept of frugal innovation diffuses as a management philosophy and induces frugal engineering principles, however, it still lacks solid operational tools and methods (Sehgal et al. 2010).

Using lean or frugal management principles during early product and process development activities facilitates the diffusion of lean and frugal philosophy across the organization. Therefore, understanding and applying lean and frugal engineering enables the dissemination of both approaches across the value chain activities as originally intended. Moreover, since product design principles influence procurement, production, distribution, and disposal costs, understanding both lean and frugal design principles and their impact on logistics processes is crucial. Several authors briefly associate the two constructs and highlight the similarities between lean and frugal (Tiwari and Herstatt 2014, Soni and Krishnan 2014). Lean and frugal engineering are rather complementary approaches and therefore synergies can be formed through the integration of both concepts. In order to evaluate synergies and their implications for logistics processes, similarities, and differences between the two concepts should be evaluated. Therefore, this work extends and complements existing research aiming to critically compare lean and frugal approaches by highlighting similarities and differences. This work's novel contribution is the discussion of implications that the emerging paradigm of frugal innovation bares upon traditional logistics processes in order to determine necessary changes. The understanding of the two design principles and implications for logistics processes is crucial for engineers and product managers.

This article is structured in six sections. After introduction, a comprehensive review of the frugal engineering concept follows in Sect. 30.2. Section 30.3 describes lean engineering and Sect. 30.4 compares both concepts in terms of similarities and differences. Section 30.5 discusses the implications of frugal and lean engineering for logistics processes, and Sect. 30.6 concludes with future implications for researchers and managers.

30.2 Frugal Engineering: Meaning and Principles

The term *'frugal engineering'* was first coined in 2006 by Carlos Ghosn, the CEO of Renault, who referred to it as the process of designing and developing the world's cheapest car that targets the base of the economic pyramid (BOP) customers in India— TATA Nano (Sehgal et al. 2010). Frugal engineering concept is rooted in the frugal innovation philosophy and denotes the actual process of developing frugal products. Therefore, frugal innovation is a philosophy, while frugal engineering refers to the actual product development practices (E Cunha et al. 2014) and the results of frugal engineering are frugal products (Soni and Krishnan 2014). Frugal engineering comes as an alternative approach to product development which highlights the wastefulness of over-engineered products, and challenges the traditional business model for R&D of Western companies (Radjou and Prabhu 2015).

Frugal engineering refers to a set of principles and methods used to design and develop low-cost, high-quality products in order to satisfy the needs of poor customers in developing markets (Kumar and Puranam 2012). Therefore, the inherent focus of frugal engineering are BOP markets which are defined as the largest and poorest socioeconomic group in terms of population. Frugal engineering is a 'clean sheet,' bottom-up approach to product development (Sehgal et al. 2010). While cost cutting is an important part of frugal engineering, the focus is to avoid unnecessary costs in order to ensure that the product is functional in resource constrained environments. In order to achieve this, new ways of thinking about customers, innovation, and organization are needed. Furthermore, organizational agility as a whole is needed for frugal engineering, in terms of cross-functional teams, non-traditional supply chains and top-down support for product development (Sehgal et al. 2010).

The value proposition of frugal products and services is defined by Tiwari and Herstatt (2014) as decreased total ownership cost, robustness, user-friendliness, and economies of scale. One critical aspect of this value proposition is the ability to cope with infrastructural barriers and challenges, and severe affordability constraints (Tiwari and Herstatt 2014). Kumar and Puranam (2012) studies a large set of frugal products and identifies several principles of frugal engineering. They include product features such as robustness and portability, process features such as de-featuring, leapfrog technology, mega-scale production and service ecosystems. Reducing costs significantly in order to account for affordability constraints while offering attractive value propositions is a central aspect of frugal engineering. Rao (2013) provides an overview of existing products and services which cost between 50 and 97 % less than corresponding non-frugal outcomes. For instance, one of the most popular frugal engineering outcomes is the world's cheapest car—TATA Nano. Table 30.1 includes several important design features of the car and highlights rationale for the decisions taken according to a study performed by Ray and Ray (2011). One can easily observe that the overarching aim of the design features presented is to reduce complexity of procurement and production processes, enable distribution to remote locations and considerably reduce costs.

Frugal engineering proposes a new way of approaching product development processes. Since frugal products allow for simplification of logistics processes, it is a part of frugal engineering to develop new ways of organizing production and logistics processes (Sharma and Iyer 2012). Yet, the literature has not emphasized so far the implications of logistics processes of frugal engineering.

30.3 Lean Engineering

Lean philosophy originated from Japan as a management philosophy, and it is an approach that emphasizes the identification and removal of waste, and the focus on value adding activities in order to satisfy customer needs (Womack and Daniel 2003). Originally, waste was defined for manufacturing environments, but recently

Table 30.1 Frugal engineering design features and rationale—The case of TATA Nano based on Ray and Ray (2011)

Frugal design feature	Rationale
One windshield wiper instead of two	Lower cost per unit due to fewer resources
Small 65R12 tires and 12 inch wheels	Lower cost per unit due to fewer material used
Three lug nuts instead of four in the twelve inch wheel mounting	Lower cost per unit due to fewer components used
Standard-life bulb in car lights instead of a long-life bulb	Lower costs per unit
Reduced thickness of bumper	More compact Lower costs per unit due to fewer materials used
Small (624 cc) rear-mounted engine with two cylinders and a single balance shaft instead of one for each cylinder	Small engine lowers fuel consumption and lowers car weight requires so that less equipment to operate Lowers cost per unit and reduces maintenance required Rear engine allows optimal layout for a compact car Lower emissions
Hollow steering column	Lower cost per unit due to fewer material required
Additional opening on the rear floor of car	Provide easier access to the intake manifold and starter of car

the concept of waste is translated in many other sectors. Lean evolved as a multidimensional, integrated system approach incorporating practices such as Just-in-Time (JIT), quality systems, cellular manufacturing and Kanban (Baines et al. 2006). The main pillars of lean management include waste elimination, identification of value streams, achievement of flows through the processes, adoption of a pull system, and most importantly the creation of a culture of continuous improvement (Bendell 2006). Because of the large influence lean principles has in the manufacturing sector, many other sectors such as product development, marketing, and sales started to implement lean practices (Stone 2012).

The lean philosophy has revolutionized the manufacturing processes in the West, however, reaching transformational change involves following the original Japanese model of incorporating lean thinking across the entire organizational processes. In this sense, lean engineering developed as a powerful concept in order to ensure that lean principles span the organization as early as possible during the product and process design (McManus et al. 2007). The principles of Lean Engineering or Lean Product Development Model (see Fig. 30.1) include the concurrent engineering process, a value-focused planning and development, knowledge-based environment, continuous improvement culture (kaizen) and the chief engineer technical and entrepreneurial leadership (Khan et al. 2013). In this model the focus on value, the kaizen culture and the knowledge environment are embedded in the

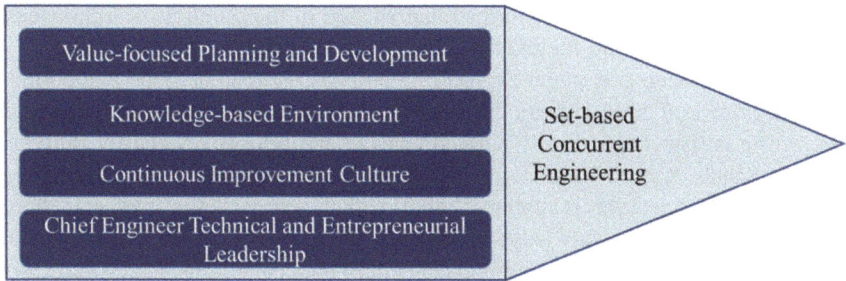

Fig. 30.1 Lean engineering model (Khan et al. 2013)

process, while the technical and entrepreneurial skills of the chief engineer guide it. The main enabler of the lean engineering processes is the set-based concurrent engineering, which ensures that several solutions are developed and analyzed in parallel (Ward 2007).

The proposed model incorporates some aspects of lean philosophy, namely focus on value, knowledge, continuous improvement and human competence. However, incorporating main lean principles from the manufacturing environment in engineering proves to be more challenging. First, the distinction between value adding services, evident and concealed waste are easy to depict in manufacturing settings. However, in engineering, especially early stages, the value is more difficult to define. Also, the large degree of uncertainty makes it difficult to distinguish between waste and value adding activities. Second, the value stream mapping is mainly composed of knowledge and information flows in product development stages, and therefore, more difficult to track compared to manufacturing flows (McManus et al. 2007). Third, implementing the 'pull' system which enables the customer demand to be the main determinant of takt time is not feasible since engineering processes are rather small steps of collective enterprise efforts to create more value (McManus et al. 2007).

30.4 Lean and Frugal Engineering: A Critical Comparison

Lean and frugal engineering both follow a reverse path from East to West and the similarities between them enable a better understanding of how synergistic effects can be developed through the combination of both principles.

First, both lean and frugal are management philosophies which span entire organizational processes, at least according to the original understanding. Even now, lean is perceived by Japanese companies as a culture and philosophy, rather than specific tools and methods which is the approach preferred by Western companies (Tiyjabi 2015). This implies that successful implementation of lean or frugal philosophy requires transformational change in the mindsets of employees, organizational culture and the overall management approach (Agnihotri 2015).

Second, both lean and frugal engineering emphasize creating value while minimizing waste (Soni and Krishnan 2014). One of the most important aspects of lean engineering is that it specifically aims to analyze and prioritize value for different projects (Schuh et al. 2011). The frugal principle of identifying and prioritizing customer essential features is similar. Therefore, both approaches provide guidelines on how to use existing knowledge in order to convert it into value (Sehested and Sonnenberg 2010). Waste is a common element in both paradigms: lean specifically emphasizes identifying and eliminating waste in terms of unused value, while frugal focuses on identifying waste in the form of features customers do not want.

Third, both philosophies have an indirect ecological focus. In both cases, the aim is not to develop green products, but rather green aspects come as spill-over effects. Lean product development has been found to have synergistic relationship with green product development due to its continuous improvement focus (Johansson and Sundin 2014). Similarly, some authors investigate the ecological promise of frugal engineering due to its focus on simple solutions, affordability and lower resource usage during product development, production, distribution, operation and disposal (Brem and Ivens 2013).

Fourth, both lean and frugal engineering approaches emphasize the importance of supplier relationship and partners. Lean engineering principles entail that suppliers are treated as partners since they have valuable knowledge which helps to design products which meet customer expectations (Womack et al. 1991). Similarly, Ray and Ray (2011) emphasize the role of suppliers for frugal engineering processes.

There are also several important differences between lean and frugal which need to be understood in order to better highlight irreconcilable aspects.

First, while frugal and lean engineering both attempt to improve and optimize value chain activities, their goals might be different because frugal engineering aims specifically to develop cheap and high value products and services which requires a much more creative and innovative process (Tiwari and Herstatt 2014).

Second, frugal approach entails a BOP focus and therefore severe resource constraints orientation. The engineering process is guided by the customer needs, but frugal philosophy regards customer needs in a broader sense, including also institutional and local constraints. Frugal engineering is mostly applied for products and services of utilitarian nature rather than hedonistic (Agnihotri 2015). Indeed, the lean philosophy also developed as driven by the lack of natural resources in Japan as well as limited space and international access (Womack et al. 1991) which means that it originally intended to also create more value with fewer resources. The severe resource constraint focus during the engineering process is marginally mentioned in lean engineering, however, it is a hallmark for the frugal engineering approaches. Moreover, resource constraints are perceived as innovation triggers for frugal engineering, while for lean engineering the triggers are related to strategies to avoid competition and capture new market share (Agnihotri 2015).

Third, while lean thinking focuses largely on process efficiency, frugal approach adopts an outcome efficiency orientation. While lean also considers customer value during engineering processes, frugal products need to be designed, produced, and delivered in order to fit specific constraints related to customers, such as affordability, acceptability, availability, and awareness, but also the general environment in terms of socioeconomic, institutional, and environmental requirements of developing countries (Anderson and Markides 2007).

This process efficiency and outcome efficiency orientation suggest that both approaches are complementary, and through combination those significant benefits can be achieved not only for developing countries but also for Western cost-sensitive consumers (Bhatti et al. 2013). Lean techniques could and should be used to achieve lower costs so that the frugal engineered products can be more affordable. For example, the value system framework (Schuh et al. 2011) helps map customer value during the development process in order to guide the engineering activities. Such tools could greatly advance the frugal engineering approach and enhance its theoretical and practical relevance.

30.5 Synthesis: Implications for Logistics Processes

It is important to understand the implications of both the lean and frugal approaches for logistics processes because up to 80 % of product lifecycle costs are pre-determined during the engineering phase (McManus et al. 2007). Thus, the applied engineering principles influence significantly the production, distribution, operation and maintenance costs associated with a product. Additionally, employing lean or frugal engineering facilitates the diffusion of lean or frugal activities across the complete chain of logistics activities. As a result of the differences and similarities as well as the orientation of each of the paradigms, namely lean as process orientation and frugal as outcome orientation, implications for logistics processes arise (Table 30.2). In this sense, implications are discussed according to the four main logistics processes, namely procurement, production, distribution, and disposal.

First, lean entails a large spectrum of methods and tools used to optimize the different stages of the value chain, while the frugal approach emphasizes the design and development of simple and basic products and services for the BOP markets, and therefore the logistics processes involved to produce and deliver them also focus on simplified solutions. Lean techniques in developed countries usually attempt to solve problems such as high inventory levels, high delivery times, and increased complexity due to changing customer preferences. However, the aim of frugal logistics goes back to more basic problems such as lack of appropriate infrastructure, weak regulatory and legal systems, and low education levels. This difference suggest that while lean engineered products involve complex logistics processes, frugal solutions bring in simplified logistics processes. The simplification of logistics processes supporting frugal products is based on the localization of procurement, production and distribution activities in order to manage existing gaps in developing countries (Gold et al. 2013).

Table 30.2 Lean and frugal engineering—Implications for logistics processes

Logistics Process	Lean engineering	Frugal engineering
Procurement	Early involvement of suppliers Standardization and reduction of complex parts which enables economies of scale Cost reduction, differentiation and the security of supply	Use of locally available materials Local suppliers Use of waste/recycled products as raw materials Reduction of complex parts Early involvement of suppliers Use of off-the-shelf components
Production	Kanban, Just-in-Time/Just-in-Sequence techniques Automatization Modularization Low inventory levels	Higher resource productivity Use of local available tools and skilled labor Modularization Labor-intensive manufacturing
Distribution	Milk-run concepts in order to achieve shorter lead times and lower costs Customer relationship management Multichannel management Integrated services and products	Locally available, existing distribution channels Robust and resistant packaging to withstand infrastructural problems during transport in remote areas
Disposal	Easy to disassembly modular design	Easy recycling and disposal of products Easy to disassemble designs

From a procurement perspective, lean processes focus mostly on cost reduction, differentiation and the security of supply (Harrison and van Hoek 2008), while sourcing for frugal solutions is driven mainly by limited financial resources. Fluctuating customer demands require standardized tools such as the vendor managed inventory (VMI) or collaborative product forecasting and replenishment (CPFR) concepts (Harrison and van Hoek 2008). On the other side, frugal procurement focuses on acquiring the best quality at lowest possible cost available and produced locally and also the procurement of reusable materials. Similar to frugal, lean procurement suggests the reduction of supplier base to several key partnerships and collaborators. Use of off-the-shelf components as frequently done in frugal engineering simplifies procurement repair and maintenance for product lifecycle.

From a production perspective, lean engineering results in production processes guided by the following key performance indicators: short lead times, low inventory level, high utilization of production resources and delivery reliability (Nyhuis and Wiendahl 2008). Under these performance indicators, lean techniques such as Kanban and JIT are used. However, frugal manufacturing focuses on reduced complexity of production activities and local solutions. As such, the focus is on labor intensive manufacturing, the use of locally available tools and techniques.

From a distribution perspective, lean processes focus on reduced costs, short distribution times, and customer satisfaction. Frugal distribution processes need to integrate various channels in order to cope with severe infrastructural deficits. They need locally embedded distribution channels to ensure the reach of geographically dispersed individuals in rural remote areas or condensed urban slums.

Overall, frugal engineering enables the design of logistics processes aligned with a sustainability agenda both from ecological and social perspective. Due to affordability constraints, frugal engineering implies lower resource usage during logistics processes as well as product lifecycle. Therefore, the combination of lean engineering in order to optimize and continuously improve processes and frugal engineering in order to reduce resource usage and support ecological process and product orientation is recommended.

30.6 Conclusion

Frugal innovation has been a highly influential management approach in the past decade, but similar to the early beginnings of lean it provides little operational methods to guide the application of the strategic goals it sets forth. The lack of clear guidelines may result similarly to the confusion created by lean, pull and Kanban terms (Hopp and Spearman 2004). Thus, this study suggests the adoption of lean engineering techniques which are more mature in order to guide the development of operational tools for frugal engineering. Since lean and frugal are complementary approaches, the adoption of lean methods may serve as a good starting point. The combination of the two approaches promises optimized processes as well as ecological processes and products. Therefore, since frugal engineering appears to have significant ecological effects on logistics processes, companies could potentially improve the sustainability of their operations by employing it. However, this research only opens up the topic by suggesting a conceptual symbiosis between the two concepts. This means that further research should investigate specific case studies of frugal and lean engineering in order to develop propositions for the contributions and limitations of the two approaches.

From a managerial perspective, frugal principles and implications are highly relevant for sustainability managers or logistics managers concerned with sustainability aspects and also for executives aiming to expand their operations in developing countries targeting the BOP customers.

It will be interesting to observe how this new approach will diffuse into the operations of Western companies and the impact it will have on various management disciplines and industry sectors.

References

Agnihotri A (2015) Low-cost innovation in emerging markets. J Strateg Mark 23(5):399–411
Anderson J, Markides C (2007) Strategic innovation at the base of the pyramid. MIT Sloan Manag Rev 49(1):83–88
Baines T, Lightfoot H, Williams GM, Greenough R (2006) State-of-the-art in lean design engineering: a literature review on white collar lean. Proc Inst Mech Eng Part B J Eng Manuf 220:1539–1547
Bendell T (2006) A review and comparison of six sigma and the lean organizations. TQM Mag 18:255–262
Bhatti YA, Khilji SE, Basu R (2013) Frugal innovation. Globalization, Change and Learning in South Asia. Oxford, UK: Chandos Publishing
Brem A, Ivens BS (2013) Do frugal and reverse innovation foster sustainability? Introduction of a conceptual framework. J Technol Manag Grow Econ 4(2):31–50
E Cunha MP, Rego A, Oliveira P, Rosado P, Habib N (2014) Product innovation in resource-poor environments: three research streams. J Prod Innov Manag 31(2):202–210
Gold S, Hahn R, Seuring S (2013) Sustainable supply chain management in "Base of the Pyramid" foodprojects—A path to triple bottom line approaches for multinationals? Int Bus Rev 22 (5):784–799
Harrison A, Van Hoek RI (2008) Logistics management and strategy: competing through the supply chain. Pearson Education
Hopp WJ, Spearman ML (2004) To pull or not to pull: what is the question? Manuf Serv Oper Manag 6:133–148
Johansson G, Sundin E (2014) Lean and green product development: two sides of the same coin? J Clean Prod 85:104–121
Khan MS, Al-Ashaab A, Shehab E, Haque B, Ewers P, Sorli M, Sopelana A (2013) Towards lean product and process development. Int J Comput Integr Manuf 26:1105–1116
Kumar N, Puranam P (2012) Frugal engineering: an emerging innovation paradigm. Ivey Bus J 76 (2):14–16
McManus HL, Haggerty A, Murman E (2007) Lean engineering: a framework for doing the right thing right. Aeronaut J 111:105–114
Nyhuis P, Wiendahl HP (2008) Fundamentals of production logistics: theory, tools and applications. Springer Science & Business Media
Radjou N, Prabhu J (2015) Frugal innovation: how to do more with less. Profile Books
Rao BC (2013) How disruptive is frugal? Technol Soc 35:65–73
Ray S, Ray PK (2011) Product innovation for the people's car in an emerging economy. Technovation 31:216–227
Schuh G, Lenders M, Hieber S (2011) Lean Innovation—introducing value systems to product development. Int J Innov Technol Manag 8:41–54
Sehested C, Sonnenberg H (2010) Lean innovation: a fast path from knowledge to value. Springer Science & Business Media
Sehgal V, Dehoff K, Panneer G (2010) The importance of frugal engineering. Strategy Bus 59:1–5
Sharma A, Iyer GR (2012) Resource-constrained product development: implications for green marketing and green supply chains. Ind Mark Manag 41:599–608
Soni P, Krishnan TR (2014) Frugal innovation: aligning theory, practice, and public policy. J Indian Bus Res 6:29–47
Stone KB (2012) Four decades of lean: a systematic literature review. Int J Lean Six Sigma 3:112–132
Tiwari R, Herstatt C (2014) Emergence of India as a lead market for frugal innovation. Opportunities for Participation and Avenues for Collaboration. Working Paper. Consulate General of India, Hamburg, Germany

Ward AC (2007) Lean product and process development. Cambridge, USA: Lean Enterprise Institute

Womack J, Jones D, Roos D (1991) The machine that changed the world. Harper-Collins, New York

Womack JP, Daniel T (2003) Jones. lean thinking: banish waste and create wealth in your corporation, 2nd edn. Free Press, New York

Chapter 31
Logistics Dynamics and Demographic Change

Matthias Klumpp, Hella Abidi, Sascha Bioly, Rüdiger Buchkremer, Stefan Ebener and Gregor Sandhaus

Abstract Change and dynamics in logistics are interestingly driven at the same time by external as well as internal forces. This contribution outlines a big data literature review methodology to overview recognizable external changes and analyzes the interaction of one major trend—demographic change—further in order to allow for change management and adaption concepts for successful logistics. Therefore, this may be a first blueprint of how to analyze and react to specific trends in a holistic manner embedded into a context and environment of trends and changes. This may allow logistics dynamics concepts also to be possibly more sustainable in terms of applicable for a longer period of time—and not to be overcome by other trends.

Keywords Demographic change · Logistic trends · Global trends · Interaction

31.1 Introduction

Trends and dynamic settings in logistics have a long tradition as well as importance still today (Craighead et al. 2007). In recent years, the notion of an internet of things —among other important ones like sustainability, agility as well as resilience and risk management—was one of the key trends implying change and dynamics for logistics concepts (Zijm and Klumpp 2016). As trend analysis is nothing new, the

M. Klumpp (✉)
University of Twente, Enschede, Netherlands
e-mail: matthias.klumpp@fom.de

M. Klumpp · H. Abidi · S. Bioly · R. Buchkremer · G. Sandhaus
FOM University of Applied Sciences, Leimkugelstr. 6, 45141 Essen, Germany

H. Abidi
Nottingham Trent University, NG1 Nottingham, UK

S. Ebener
NetApp Deutschland GmbH, Gladbecker Strasse 3, 40472 Düsseldorf, Germany

authors here propose a concept of trend *interaction* analysis with the example of demographic change as one trend which is tested and connected toward a range of other trends relevant for logistics. Therefore, Sect. 31.2 paves the ground by identifying relevant trends in the literature. Section 31.3 is elaborating on these and Sect. 31.4 is outlining the interaction analysis with the help of an expert analysis.

31.2 Trend Analysis: Research Approach

Importance as well as maturation of supply chain management and logistics as an academic discipline is confirmed by the high frequency of articles devoted to theory building, hypothesis testing, and exploratory research. In order to accommodate the evolution of supply chain management and logistics we aim to identify development trends—to do so we apply a literature review as this provides evidence for informing policy and practice on research topics in any discipline. As supply chain management and logistics topics are published in a broad array of journals (Grimm et al. 2014), we conducted an in-depth analysis using academic journals. Academic journals play a strategic role in developing and communicating disciplinary knowledge and are an important educational resource for knowledge dissemination (Fawcett et al. 1995). For our analysis we used the data platform Web of Science in order to retrieve supply chain management and logistics articles published from 2005 to 2015. In order to prioritize, we focused on the list of journals provided by McKinnon (2013). McKinnon (2013) highlighted the most recent ratings of supply chain management and logistics journals in the main listings used for business-related publications in Germany, UK, and Australia. To proceed, we developed key terms to pinpoint and analyze the literature and to avoid unbiased research (Buchkremer 2015). We collected articles focusing on keywords in titles and abstracts: "*supply chain*" and "*logistics*"; both terms were searched in following journals that are regarded as the core peer-reviewed literature in supply chain management and logistics field: *International Journal of Logistics Management, International Journal of Logistics: Research and Applications, International Journal of Physical Distribution & Logistics Management, International Journal of Shipping and Transport Logistics, Journal of Business Logistics, Journal of Supply Chain Management, Logistics Research, Maritime Economics and Logistics, Naval Research Logistics, Supply Chain Management: An International Journal, Transportation Research E: Logistic and Transportation Review*. We finally identified 3,469 articles—213 articles were deleted that were announcing for example a special issue or editorial or call for papers. Using the citation report provided by Web of Science we analyzed the high cited articles and read the abstracts. We identified a list of most cited focal themes as indicated in Table 31.1.

In order to identify the trends we conducted a further analysis as the second step using **SAS Enterprise Miner**. We created a list of further keywords using the focal themes from Table 31.1 and investigated the frequency of those themes in the abstract and title of 3,469 articles. In the third step we read the articles and identified four main trend areas from the selected articles (see Fig. 31.1).

Table 31.1 Focal themes in logistics and supply chain management

Risk management	Sustainability	Coordination
Operations management	Green SC/LOG	Collaboration
Return logistics	Information technology/industry 4.0	Cooperation
Vehicle routing	Demographic change	Supply chain integration
Distribution	Performance management and measurement	Relief/humanitarian operations
Purchasing/supply management	Closed loop	Packaging
Skills/competences	Human resources	Tracking and tracing
Organizational learning (education and training)	Inventory/warehouse	Agility/lean/flexibility

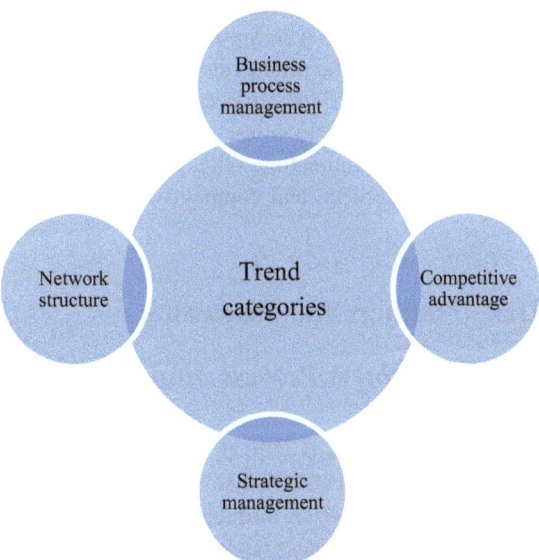

Fig. 31.1 Main areas on logistics and supply chain management trends

The area *strategic management* describes the identification and implementation of objectives based on assessment of internal and external environment factors in the organization considering efficient resource allocation. The second area *competitive advantage* is a major topic in academia and business practice. This includes innovative concepts and tools that support organizations to outperform its competitor. Furthermore, it is an element of strategic management. The area *business process management* deals with management of activities that produce a certain output based on the demand and request of customer (Cooper et al. 1997a, b; Davenport 2003). Moreover, it provides methods, techniques, and tools to support the design, management, and analysis of operational business processes (Van der Aalst et al. 2003). The last area deals with *network structure*: this gives insight on

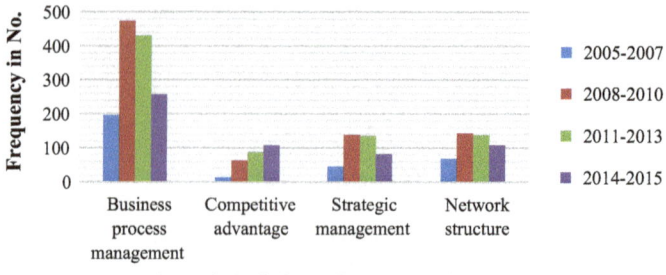

Fig. 31.2 Development of four areas in logistics and supply chain

the flow of materials and information from start to end. Furthermore, it shows value that has been created due to partner cooperation.

Finally, in order to gain an overview we summarized the results. Consequently, we have classified the articles in four *time phases* 2005–2007, 2008–2010, 2011–2013, and 2014–2015 to gain deeper insights into the development of the four areas and their related subcategories. As Fig. 31.2 indicates research on business process management is dominating—though declining—followed by network structure, strategic management, and competitive management.

31.3 Findings and Discussion

31.3.1 Business Process Management

The supply chain starts with the source of supply and ends at the point of consumption (Stevens 1989); it links each element of the production and supply processes from raw materials through to the end customer (Scott and Westbrook 1991). Supply chain management targets to synchronize the requirement of customers with the flow of material from suppliers even the goals of high customer service level, low inventory investment, and low unit cost (Stevens 1989; Scott and Westbrook 1991; Christopher and Towil 2001; Cooper and Ellram 1993; Christopher 1999). Over the last years different branches, in particular automotive branch, were confronted to increase their competition from their counterpart in the U.S. and Asian countries. This pressure leads the companies to improve quality, to reduce product development and manufacturing time as well as development and manufacturing costs. Therefore, the motivation on optimizing processes and increasing performance within and across actors of the supply chain was both to improve inventory and warehouse management and to promote distribution management while reducing costs. For example minimizing relevant **inventory and transportation** costs can be supported by using a network of inventory queue that incorporates an inventory replenishment policy for a store (Dong and Chen 2005). However, in the supply chain, it is an important issue for logistic managers to offset the replenishment cycles of multiple products sharing a warehouse so as to minimize the maximum warehouse space requirement (Yao and Chu 2008). A further area where more attention has been paid recently is the area of **e-business**. For example research has been executed on supporting retailers to increase customer satisfaction in fulfilling orders while minimizing logistics costs. Hereby each customer demand is fulfilled from the closest fulfillment center if there are enough inventories. Otherwise, the retailer would transship stock from a nearby facility or transfer the customer order so it is fulfilled from another facility, depending on the economics of transportation (Torabi et al. 2015). To address this gap Torabi et al. (2015) developed a mixed-integer programming model to efficiently find optimal solutions. **Last-mile distribution** is an essential logistics component for practice and fossil fuels vehicles "are polluting the world's cities, dumping increasing amounts of carbon dioxide and other greenhouse gases into the atmosphere, and consuming vast quantities of petroleum" (Sperling 1995, p. 1). Nevertheless, oil demand is increasing and will exceed production by 2035. This will have an impact on fuel availability as well as on cost of fuel (Charles et al. 2014). However, accomplishment of travel reduction or transportation of goods will be difficult therefore sustainable energy and transportation systems are needed to be established. According to this, the use of electric vehicles in the delivery process (last-mile distribution) would be an opportunity to decrease the economic and ecological key driver fuel. Another advantage would be decreasing noises and reducing carbon emission. In the logistics industry, for example Deutsche Post DHL has integrated electric vehicles in their fleet pool for urban delivery (Deutsche Post DHL 2014). To enhance the use of electric vehicles different research efforts have been driven by different disciplines such as engineering, economic, or business science research fields. Research on supply chain vulnerabilities due to macro and micro risks have been increased recently. For example since 1970 the total number of natural and technological disasters increased six times (Schulz 2009). A natural disaster can disrupt a global supply chain in few seconds after an outbreak (Canbolat et al. 2008; Manuj and Mentzer 2008), for example volcanic outbreak in Iceland and Earthquake in Japan in 2011 or hurricane Kathrina in 2005 (Munichre 2014; Manuj and Mentzer 2008). Or the company strategy such as global supply chain is afflicted with micro risks such as linguistic and cultural deficits and customs regulations (Cho and Kang 2001; Schniederjans and Zuckweiler 2004), transportation delays, logistics service differences (Cho and Kang 2001). All these risks can be counteracting with taking action even if these actions increase the costs. But at first moment, a **supply chain risk management** seems to be as an additional work for companies and manager as well as losses (Manuj and Sahin 2011). Supply chain disruptions cause a sales fall of 7 %, a down of an operating income of 42 % and a fall of return on assets of 35 % and an announcement of supply chain disruptions causes a shareholder return between 7 and 8 % (Hendricks and Singhal 2005). In the depth sight, it is to recognize that risk management brings profits which make the companies more efficient (Waters 2011). Therefore, efforts are done to mitigate risks for example Chapell and Peck (2006) developed risk management situations for the military supply chain by applying six-sigma method. Manuj and Mentzer (2008) developed a risk management and mitigation model for global supply chain. Manuj et al. (2009) developed an eight-step model for the design, assessment, and application of logistics and supply chain simulation model. Due to the increasing number of natural disasters and the resulting chaotic humanitarian emergencies an important area in supply chain and logistics management attracted the attention of researcher. The complex logistics environment due to emergencies put pressure on humanitarian aid agencies to deliver humanitarian aid in an appropriate and cost effective way (Thomas and Kopczak 2005; van Wassenhove 2006; Oloruntoba and Gray 2006; Kovacz and Spens 2007) along the humanitarian supply chain. The **humanitarian supply chain** encompasses the planning and management of all activities related to material, information, and financial flows in disaster relief. Importantly, it also includes coordination and collaboration with supply chain members, third party service providers and among humanitarian organizations humanitarian supply chain management is concerned with managing the efficient flow of aid materials, information, and services and aim to reduce the impact of disaster on human lives (Lijo et al. 2012). Humanitarian supply chain plays a central role in several phases of a disaster relief concept such as preparedness, immediate response, reconstruction, and recovery phase (Baumgarten et al. 2010). Each of these phases and activities require logistics support, although every phase has its requirements with regard to the duration, volume, the needed as well as the variety of supplies, urgency and procurement location.

31.3.2 Network Structure

Rapid technology development, contracting out, global markets, product dynamic, service complexity, reducing supplier and inventory practices are aspects behind commonplace complex and interlinked business environment (Deleris and Erhun 2005; Glickman and White 2006). Furthermore due to the highly competitive environment in which organizations currently operate, attention has shifted away from optimizing individual firm performance toward effective supply chain management (Bolumole 2001; Christopher 2011). Supply chain management refers to effectively managing the network of organizations to which the organization belongs (Cruijssen et al. 2007a; Christopher 2011), including both **vertical and horizontal business relations**. Therefore being part of an alliance might benefit an organization in two ways. First of all, participating firms can create a competitive advantage over their competitors (Bernal et al. 2002), which is perceived by customers as an additional value (Anslinger and Jenk 2004; Christopher 2011). Second, they might realize business growth through a net positive outcome from the project initiated by the alliance partners, which ultimately benefits the shareholders (Cruijssen et al. 2007b). The decision to engage in a horizontal cooperation is either driven by internal motives (e.g., management decisions and goals), or by external motives (e.g., evolving market conditions, the economic environment, and customer requirements) (Verstrepen et al. 2009). Due to the mentioned motives, the development of the topics **collaboration and cooperation** is growing. For example Gimenez and Ventura (2005) investigated the logistics-production and logistics-marketing interfaces and their relation with the external integration. They concluded that external collaboration among supply chain members does always contribute to improving firms' logistical performance. Soosay et al. (2008) describe how differing relationships can affect the operation of organizations and their capacities to innovate. The ability to work together with partners has enabled firms to integrate and link operations for increased effectiveness as well as embark on both radical and incremental innovation. A focal question in the area of collaboration and cooperation is how the total cost or savings should be distributed among the participants. Smaros (2007) proposed an allocation method and examined a number of sharing mechanisms based on economic models including Shapley value, the nucleolus, separable and non-separable costs, shadow prices and volume weights. Ashayeri et al. (2012) highlight that supply chains exist for more than twenty years, however, partner selection and evaluation processes are still unstructured. The right choice and evaluation of strategic partnership could bring a competitive advantage, whereas inability to establish a proper relationship would bring overwhelming problems. Research by Cousins and Spekman (2003) shows that around 60 % of partnerships fails. Incompatibility with partners is one of the most common reasons for this. Hence, choosing and evaluating the right strategic partners is highly important for the performance of the alliance among a logistics network and supply chain (Lee and Cavusgil 2006). We conclude that research on establishment of horizontal cooperation between LSPs are limited (Verstrepen et al. 2009), but extensive research has been performed on vertical business cooperation, i.e., supplier selection and service selection. Due to this fact, we pinpoint that research on collaboration and cooperation should focus on horizontal cooperation between LSPs.

31.3.3 Strategic Management

Strategy is defined as "the match an organization makes between its internal resources and skills... and the opportunities and risks created by its external environment" (Grant 1991a, b, p. 3). Even Hoover et al. (2001) pointed out that is practically difficult to deliver a high customer value while at the same time reducing cost. These trade-offs have to be resolved by developing an integrated supply chain. It targets managing the material and information flow at the strategic, tactical, and operational level by utilizing facilities, finance, people, and systems which are coordinated and harmonized as a whole (Stevens 1989). To improve customer service levels and to reduce costs a supply chain strategy has to be taken in account (Simchi-Levy et al. 2000). In the supply chain sector two well-known supply chain strategies namely **lean and agile** exist. Lean follows the idea to reduce and eliminate the waste and focus on the efficient use of resources (Ōno 1988) while the agile concept is about the ability to match supply and demand in turbulent and unpredictable market (Christopher 2000). Lean concept is usable when the demand is stable and predictable (Christopher et al. 2006). The future development of the supply chain and logistics sector essentially depends on the success of **continuing education** of new and current employees. The supply chain and logistics sector as well as the competencies and knowledge of the employees are faced with a variety of challenges, which can be subdivided in two categories. To the first category knowledge enrichment is such as global and dynamic supply chains and complex logistics chain, the second category knowledge extension is such as extensive ICT (RFID, GPS and Dynamic Routing). All these argue for the current high lack in skilled employees not only on an operational level such as truck drivers but even experts of different disciplines such as innovation management, supply chain design, ICT in logistics as well as warehouse and inventory management and order picking systems. By regarding the identified trends, employees have to increase their knowledge to tackle the tasks of logistics goods and services in a high velocity and to improve their competences. **Competence** is defined as "the ability to successfully meet complex demands in a particular context. Its manifestation, competent performance (which one may equate to effective action), depends on the mobilization of knowledge, cognitive and practical **skills**, as well as social and behavioral components such as attitudes, emotions, values and motivations" (Hakkarainen et al. 2004). Competence demonstrates also the level of student achievement in the science education context (Liu 2009). Competence is not only skills, qualification or only knowledge but all these factors are the basic for a competence of person. Nevertheless, specific further research questions in the wake of these topics could be the question of linking individual and corporate competence levels to process efficiency and corporate results (ROI and others) in logistics companies; or also the question of specific competence diversions and gaps (blue and white collar employees in logistics, gender aspects in logistics education, regional disparities in logistics education and qualification, connections and interactions of formal and informal on-the-job learning in logistics, etc.). A further element that is essential in strategic management is measuring performance. **Performance measurement** is fundamental to improve operations (Kaplan 1990), simplify communication between supply chain actors, and increase transparency of the supply chain and logistics processes (Gunasekaran and Kobu 2007). Chow et al. indicated that performance is multidimensional, because one measure is not sufficient for a logistics performance—logistics performance has to be seen as subsection of the larger conception of firm or organizational performance (Chow et al. 1994). To know the meaning of performance there are two central organizational and logistics goals which have to be defined. These are divided in two dimensions: The simplest dimension and which affect the performance—in particular logistics performance—is to differentiate between (i) efficiency and (ii) effectiveness (Gleason and Barnum 1986). In the literature they are valuable and meaningful research concept, approaches, and case studies that analyze the presented metrics and system to evaluate the performance of supply chain as well as logistics, e.g., production, distribution or inventory, and implement performance measures (Gunasekaran and Kobu 2007). Hereby is to add that the existing research in performance measurement metrics and systems focuses on analyzing current and in the practice used performance measurement systems and in studying the measures. All these argue for the complexity and difficulty in developing performance measurement metrics and systems for firms, organizations, and their logistics activities. Huang et al. present performance measures that are based on reliability, responsiveness, costs, and assets (Lai et al. 2002; Huang et al. 2005), Giannakis (2007), Simatupang and Sridharan (2005) determine performance measures to evaluate the collaboration within a supply chain; a performance management process for delivery services is set by Forslund and Jonsson (2007). Furthermore, there are general methodologies developed to measure supply chain and logistics performance, namely the balance scorecard (Kaplan and Norton 1992), supply chain council's SCOR model, logistics scoreboard, activity-based costing, and economic value added (EVA) (Lapide 2000). Nevertheless, developing and implementing a performance measurement system with appropriate key performance indicators remain as a complicated process (for example Shepherd and Gunter 2006; Lapide 2000; Chae 2009).

31.3.4 Competitive Management

Supply chain and logistics sector is facing with two themes demographic change and sustainability that need more pioneering concept to help organizations to outperform its competitor. The **demographic change** has a high impact on economy (Klumpp et al. 2012), health care service, infrastructure, mobility as well as the pension system in Germany. The working-age population (20 to 64 years old) in Germany is currently 49.8 million. In 2030, the working-age population will probably have 6.3 million fewer persons than in 2010 (Federal Minister (Bundesministerium des Inneren 2011, p. 6)) and in 2060 the working-age population will decline about 35 % compared to 2013 (Federal Statistical Office (Statistisches Bundesamt 2013). Not only Germany is faced with the demographic change even the "European Union is facing unprecedented demographic changes (an aging population, low birth rates, changing family structures and migration). In the light of these challenges it is important, both at EU and national level, to review and adapt existing policies" (European Commission 2010). However, in the future economic demand has to be adjusted due to the demographic change. Skilled and productive employee allows economic growth. To resolve the conflict due to the impact of demographic change and to mitigate the threat of a shortage of employees we should strengthen the current employees, the underutilized population in the labor system, i.e., women and disabled people and integrate the qualified immigrants in the labor system. The theme **sustainability** has been explored in a variety of settings. For example, local governments in Western Europe increasingly applied city time-access regulations to improve social sustainability. These rules significantly affect the distribution management and process of retail chain companies. Quak and de Koster (2007) studied the impact of governmental time-window pressure on retailers' logistical concepts and the consequential financial and environmental distribution performance. Carter and Dale (2008) introduce the concept of sustainability to the field of supply chain management and demonstrate the relationships among environmental, social, and economic performance within a supply chain management context. Using qualitative and quantitative survey data Walker and Brammer (2009) explored sustainable procurement in the UK public sector. Sharma et al. (2010) argued that less attention has been given to marketing's role in a green supply chain and its interface with environmentally friendly manufacturing and operations. Therefore, they identified three major strategies for achieving competitive advantage and financial performance such as reduction of surplus supply of products, reduction of reverse supply, and internal marketing. Asgari et al. (2015) presented the sustainability performance of the five major UK ports. One objective of an EU funded project (SCALE) was to improve the sustainability of food and drink supply chain logistics in the context of rising food demands, increasing energy prices, and the need to reduce emissions (Bloemhof et al. 2015). Meixell and Luoma (2015) find in which ways stakeholder pressure may affect supply chain sustainability: The three main findings includes sustainability awareness, adoption of sustainability goals, and/or implementation of sustainability practices. We conclude that further research should focus on developing innovative sustainability concepts that have to be integrated in logistics service provider network, in maritime waterways logistics, logistics hubs as well as in the air freight division.

31.4 Trend Interaction: Demographic Change

In order to visualize and analyze the specific trend of demographic change with its individual dynamic interaction with other trends, a short expert survey was used. Four individuals from an annual German logistics summit[1] were interviewed using an interaction chart. Results are depicted in the following Table 31.2.

The relationships between demographic change and the theme fields of the first column (agility to tracking and tracing) were polled. The experts were asked to indicate from strong positive (++) to strong negative (−). The given information is represented in column # 1 to # 4. As can be seen by the variance, the respondents to the areas of coordination, human resources, and inventory/warehouse are very uneven (bold print). Referring to the average strength of trend interaction, it becomes clear that for example *organizational learning* as well as *collaboration* and *information integration/industry 4.0* are strongly influenced by demographic change. This is very interesting—and for the latter two trend themes may be quite unexpected. But with a second thought, it is evident that this expert feedback is quite thorough and foresighted: Collaboration as well as information integration

---

[1]"Zukunftskongress Logistik", hosted by Fraunhofer IML in Dortmund, Germany.

Table 31.2 Demographic change and dynamic trend interaction (from 2 strong positive to -2 strong negative)

Area	#1	#2	#3	#4	⌀	Variance
	Demographic Change with …					
Agility/lean/flexibility	1	1	-2	0	0.00	2.00
Closed loop	0	2	-1	0	0.25	1.58
Collaboration	2	2	2	1	**1.75**	0.25
Cooperation	2	2	1	1	1.50	0.33
Coordination	2	2	-1	2	1.25	**2.25**
Distribution	1	-1	1	2	0.75	1.58
Green supply chain/logistics	0	1	1	1	0.75	0.25
Human resources	-1	0	-2	2	-0.25	**2.92**
Information technology/industry 4.0	1	2	2	2	**1.75**	0.25
Inventory/warehouse	0	1	-2	2	0.25	**2.92**
Operations management	1	0	-2	0	-0.25	1.58
Organizational learning (education and training)	2	2	2	2	**2.00**	0.00
Packaging	0	1	-1	1	0.25	0.92
Performance management and measurement	-1	0	0	0	-0.25	0.25
Purchasing/supply management	0	1	-1	0	0.00	0.67
Relief/humanitarian operations	1	-1	0	1	0.25	0.92
Return logistics	-1	2	0	2	0.75	**2.25**
Risk management	1	1	0	1	0.75	0.25
Skills/competences	1	0	2	2	1.25	0.92
Supply chain integration	2	1	1	0	1.00	0.67
Sustainability coordination	1	2	1	1	1.25	0.25
Tracking and tracing	1	0	0	0	0.25	0.25

and industry 4.0 have to be designed to be operable intuitively and with (almost) no knowledge of the system. Therefore, it requires highly educated people to develop these. Therefore, the conclusion to this regard should be adapted. Though automatization and artificial intelligence may be advanced and increasingly taking over human tasks, in the near future complex and conceptual tasks will still reside in the hands of capable humans who have to be found, trained and given time to accumulate experience with logistics operations. And this will be very hard facing demographic change and a "war for talents." Therefore, the logistics industry may have to drastically change their operating scheme in human resource management: From a typical provider of low or unskilled labor in large quantities (blue collar) to a hunter of rare talent in high qualification areas such as strategic management, information technology and intercultural competences agile, resilient and sustainable global cooperating supply chains of the future.

31.5 Conclusion

In outlining trend themes and connecting them to the major trend of demographic change this paper has provided three major results: (a) A bunch of important trend themes for logistics can be identified from the research literature, among them sustainability, information integration/industry 4.0, supply chain collaboration, and coordination as well as integration, operational and performance topics such as vehicle routing, reverse logistics, purchasing, and distribution, "soft themes" such as skills/competences and organizational learning as well as risk management and agile, lean and flexible logistics processes. (b) A new trend interaction approach has been tested—and shall be extended to further experts and surveys in the future—in order to gauge the interaction of demographic change with the other identified main trends in logistic, providing a "proof of concept" for this methodological approach. (c) The results of the trend interaction analysis turned out to show that besides human resources and skills/competences as obvious suspects, the experts rated a high interdependency also for collaboration and information integration/industry 4.0. This is obviously an interesting insight and can be reflected further—and will surely influence the strategic perspective on human resource and knowledge management in logistics when thinking about these important long-term trends.

Altogether the contribution has shown the value of a deep insight into trend developments for the strategic development of dynamic logistics concepts for the future. Research shall continue on this path in order to avoid just jumping to individual trends and management fads—but instead think hard about the interconnections and long-term implications of logistics trends. This will enhance the capability to design resilient, agile, and sustainable supply chains in the future.

Acknowledgments The research presented is connected to the grant "DO.WERT", NRW Department of Science, program "FH STRUKTUR" as well as to funding from "FH BASIS NRW."

References

Anslinger P, Jenk J (2004) Creating successful alliances. J Bus Strategy 25(2):18–22
Asgari N, Hassani A, Jones D, Nguye HH (2015) Sustainability ranking of the UK major ports: methodology and case study. Transp Res Part E: Logist Transp Rev 78:19–39
Ashayeri J, Tuzkaya G, Tuzkaya UR (2012) Supply chain partners and configuration selection: An intuitionistic fuzzy Choquet integral operator based approach. Expert Syst Appl 39(3): 3642–3649
Baumgarten H, Kessler M, Schwarz J (2010) Jenseits der kommerziellen Logistik–Die humanitäre Hilfe logistisch unterstützen. In: Dimensionen der Logistik–Funktionen Institutionen und Handlungsebenen. Springer, Wiesbaden
Bernal SMH, Burr C, Johnsen RE (2002) Competitor networks: international competiveness through collaboration. The case of small freight forwarders in the High-Tech ForwarderNetwork. Int J Entrep Behav Res 8(5):239–253

Bloemhof JM, van der Vorst JG, Bastl M, Allaoui H (2015) Sustainability assessment of food chain logistics. Int J Logist Res Appl 18(2):101–117

Bolumole YA (2001) The supply chain role of third-party logistics providers. The Int J Logist Manag 12(2):87–102

Buchkremer R (2015) Text Mining im Marketing- und Sales-Umfeld. In: Lang M (ed) Business Intelligence erfolgreich umsetzen. Symposion, Düsseldorf, pp 101–119

Bundesministerium des Inneren (2011) Demography Report. http://www.bmi.bund.de/SharedDocs/Downloads/DE/Themen/Gesellschaft-Verfassung/DemographEntwicklung/demografiebericht_kurz_en.pdf?__blob=publicationFile. Accessed 18 Aug 2015

Canbolat YB, Gupta G, Matera S, Chelst K (2008) Analysing risk in sourcing design and manufacture of components and sub-systems to emerging markets. Int J Prod Res 46(18):5145–5164

Carter CR, Dale SR (2008) A framework of sustainable supply chain management: moving toward new theory. Int J Phys Distrib Logist Manag 38(5):360–387

Chae B (2009) Developing key performance indicators for supply chain: an industry perspective. Supply Chain Manag Int J 14(6):422–428

Chapell A, Peck H (2006) Managing risk in the Defence supply chain: is there a role for six sigma? Int J Logist Res Appl 9(3):253–267

Charles C, Moerenhout T, Bridle D (2014) The Context of Fossil-Fuel Subsidies in the GCC Region and Their Impact on Renewable Energy Development. http://iisd.org/gsi. Accessed 12 July 2015

Chistopher M, Peck H, Towill D (2006) A taxonomy for selecting global supply chain strategies. Int J Logist Manag 17(2):277–287

Cho J, Kang J (2001) Benefits and challenges of global sourcing: perception of US Apparel retail firms. Int Mark Rev 18(5):542–561

Chow G, Heaver TD, Henriksson LE (1994) Logistics performance: definition and measurement. Int J Phys Distrib Logist Manag 24(1):17–28

Christopher M, Towill D (2001) An integrated model for the design of agile supply chains. Int J Phys Distrib Logist Manag 31(4):235–246

Christopher M (1999) Logistics and supply chain management: strategies for reducing cost and improving service. financial times. Pitman Publishing, London

Christopher M (2000) The agile supply chain: competing in volatile markets. Ind Mark Manag 29(1):37–44

Christopher M (2011) Logistics and supply chain management: strategies for reducing cost and improving service, 4th edn. Pearson Hall, London

Cooper MC, Douglas ML, Janus DP (1997a) Supply chain management: more than a new name for logistics. Int J Logist Manag 8(1):1–14

Cooper MC, Ellram LM (1993) Characteristics of supply chain management and the implications for purchasing and logistics strategy. Int J Logist Manag 4(2):13–24

Cooper MC, Lambert DM, Pagh JD (1997b) Supply chain management: more than a new name for logistics. Int J Logist Manag 8(1):1–14

Cousins P, Spekman R (2003) Strategic supply and the management of inter–and intra organisational relationships. J Purch Supply Manag 9(1):19–29

Craighead CW, Hanna JB, Gibson BJ, Meredith JR (2007) Research approaches in logistics: trends and alternative future directions. Int J Logist Manag 18(1):22–40

Cruijssen F, Cools M, Dullaert W (2007a) Horizontal cooperation in logistics: opportunities and impediments. Transp Res Part E 43(2):129–142

Cruijssen F, Dullaert W, Fleuren H (2007b) Horizontal cooperation in transport and logistics: a literature review. Transp J 46(3):22–39

Davenport TH, Short JE (2003) Information technology and business process redesign. Oper Manag Crit Perspect Bus Manag 1:97

Deleris L, Erhun F (2005) Risk management in supply networks using monte-carlo simulation. In: Simulation conference, proceedings of the winter. IEEE

Deutsche Post DHL (2014) Streetscooter. http://www.dpdhl.com/content/dpdhl/de/presse/mediathek/videos/streetscooter.html. Accessed 13 June 2015

Dong M, Chen FF (2005) Performance modeling and analysis of integrated logistic chains: an analytic framework. Eur J Oper Res 162(1):83–98

European Commission (2010) Demografic Analysis. http://ec.europa.eu/social/main.jsp?catId=502&langId=en. Accessed 23 July 2014

Fawcett SE, Vellenga DB, Truitt LJ (1995) An evaluation of logistics and transportation professional. J Bus Logist 16(1):299

Forslund H, Jonsson P (2007) The impact of forecast information quality on supply chain performance. Int J Oper Prod Manag 27(1):90–107

Giannakis M (2007) Performance measurement of supplier relationships. Supply Chain Manag Int J 12(6):400–411

Gimenez C, Ventura E (2005) Logistics-production, logistics-marketing and external integration: their impact on performance. Int J Oper Prod Manag 25(1):20–38

Gleason JM, Barnum DT (1986) Toward valid measures of public sector productivity: performance measures. Urban Transit Manag Sci 28(4):379–386

Glickman TS, White SC (2006) Security, visibility and resilience: the keys to mitigating supply chain vulnerabilities. Int J Logist Syst Manag 2(2):107–119

Grant RM (1991a) The resource-based theory of competitive advantage: implications for strategy formulation. Knowl Strategy 33(3):3–23

Grant RM (1991b) The resource-based theory of competitive advantage: implications for strategy formulation. Knowl Strategy 33(3):3–23

Grimm JH, Hofstetter JS, Sarkis J (2014) Critical factors for sub-supplier management: a sustainable food supply chains perspective. Int J Prod Econ 152:159–173

Gunasekaran A, Kobu B (2007) Performance measures and metrics in logistics and supply chain management: a review of recent literature (1995–2004) for research and applications. Int J Prod Res 45(12):2818–2840

Hakkarainen K, Palonen T, Paavola S, Lehtinen E (2004) Communities of networked expertise: professional and educational perspectives von European Association for Research on Learning and Instruction. 1 eds, Amsterdam, Oxford

Hendricks KB, Singhal VR (2005) An Empirical analysis of the effect of supply chain disruptions on long-run stock price performance and risk of the firm. Prod Oper Manag 14:35–52

Hoover W, Eloranta E, Holmstrom J, Huttunen K (2001) Managing the demand-supply chain. John Wiley & Sons, New York

Huang SH, Sheroan SK, Keskar H (2005) Computer-assisted supply chain configuration based on supply chain operations reference (SCOR) model. Comput Ind Eng 48:377–394

Kaplan RS (1990) Measures for manufacturing excellence. Harvard Business School Press, Boston

Kaplan RS, Norton DP (1992) The Balanced Scorecard: Measures that Drive Performance. Harv Bus Rev 70:71–79

Klumpp M, Bioly S, Abidi H (2012) Zur Interdependenz demographischer Entwicklungen, Urbanisierung und Logistiksystemen. In: Göke M, Heupel T (eds) Wirtschaftliche Implikationen des demographischen Wandels. Gabler, Wiesbaden

Kovacs G, Spens KM (2007) Humanitarian logistics in disaster relief operations. Int J Phys Distrib Logist Manag 37(2):99–114

Lai KH, Ngai EWT, Cheng TCE (2002) Measures for evaluating supply chain performance in transport logistics. Trans Res Part E 38:439–456

Lapide L (2000) What about measuring supply chain performance. Achiev Supply Chain Excell Through Technol 2:287–297

Lee Y, Cavusgil TS (2006) Enhancing alliance performance: the effects of contractual-based versus relational-based governance. J Bus Res 59:896–905

Lijo J, Ramesh A, Sridharan R (2012) Humanitarian supply chain management: a critical review. Int J Serv Oper Manag 13(4):498–524

Liu X (2009) Linking competence to opportunities to learn: models of competence and data mining. Springer, New York
Manuj I, Mentzer JT, Bowers MR (2009) Improving the rigor of discrete event simulation in logistics and supply chain research. Int J Phys Distrib Logist Manag 39(3):172–201
Manuj I, Mentzer JT (2008) Global supply chain risk management. J Bus Logist 29(1):133–155
Manuj I, Sahin F (2011) A model of supply chain and supply chain decision-making complexity. Int J Phys Distrib Logist Manag 41(5):511–549
McKinnon AC (2013) Starry-eyed: journal rankings and the future of logistics research. Int J Phys Distrib Logist Manag 43(1):6–17
Meixell MJ, Luoma P (2015) Stakeholder pressure in sustainable supply chain management: a systematic review. Int J Phys Distrib Logist Manag 45(1/2):69–89
Munichre (2014) Natural Disasters 2014. http://www.munichre.com/site/wrap/get/documents_E-285925502/mr/assetpool.shared/Documents/5_Touch/Natural%20Hazards/NatCatService/Annual%20Statistics/2014/mr-natcatservice-naturaldisaster-2014-Loss-events-worldwide-percentage.pdf. Accessed 16 June 2015
Oloruntoba R, Gray R (2006) Humanitarian aid: an agile supply chain? Int J Supply Chain Manag 11(2):115–120
Ōno T (1988) Toyota production system: beyond large-scale production. Productivity press
Quak HJ, De Koster MBM (2007) Exploring retailers' sensitivity to local sustainability policies. J Oper Manag 25(6):1103–1122
Schniederjans MJ, Zuckweiler KM (2004) A quantative approach to the outsourcing-insourcing decision in an international context. Manag Decis 42(8):974–986
Schulz S (2009) Disaster relief logistics: benefits of and impediments to cooperation between humanitarian organizations. Haupt Verlag, Bern et al
Scott C, Westbrook R (1991) New strategic tools for supply chain management. Int J Phys Distrib Logist Manag 21(1):23–33
Sharma A, Iyer GR, Mehrotra A, Krishnan R (2010) Sustainability and business-to-business marketing: A framework and implications. Ind Mark Manag 39(2):330–341
Shepherd C, Gunter H (2006) Measuring supply chain performance: current research and future directions. Int J Product Perform Manag 55(3/4):242–258
Simatupang TM, Sridharan R (2005) The collaboration index: a measure for supply chain collaboration. Int J Phys Distrib Logist Manag 35(1):44–62
Simchi-Levi D, Kaminsky P, Simchi-Levi E (2000) Designing and managing the supply chain, concepts, strategies, and case studies. McGraw-Hill, Boston
Småros J (2007) Forecasting collaboration in the European grocery sector: observations from a case study. J Oper Manag 25(3):702–716
Soosay CA, Hyland PW, Ferrer M (2008) Supply chain collaboration: capabilities for continuous innovation. Supply Chain Manag Int J 13(2):160–169
Sperling D (1995) Future drive: electric vehicles and sustainable transportation. Island Press, Washington D.C
Statistisches Bundesamt (2013) Bevölkerung. https://www.destatis.de/DE/ZahlenFakten/GesellschaftStaat/StaatGesellschaft.html. Accessed 13 June 2015
Stevens GC (1989) Integrating the supply chain. Int J Phys Distrib Logist Manag 19(8):3–8
Thomas AS, Kopczak LR (2005) From logistics to supply chain management: the path forward in the humanitarian sector. Fritz Institute, San Francisco, CA
Torabi SA, Hassini E, Jeihoonian M (2015) Fulfillment source allocation, inventory transshipment, and customer order transfer in e-tailing. Transp Res Part E: Logist Transp Rev 79:128–144
Van Der Aalst WMP, Ter Hofstede AHM, Weske M (2003) Business process management: a survey. Business process management. Springer, Berlin Heidelberg 1–12
Van Wassenhove LN (2006) Humanitarian aid logistics: supply chain management in high gear. J Oper Res Soc 57(5):475–489
Verstrepen S, Cools M, Cruijssen F, Dullaert W (2009) A dynamic framework for managing horizontal cooperation in logistics. Int J Logist Syst Manag 5(3/4):228–248

Walker H, Brammer S (2009) Sustainable procurement in the United Kingdom public sector. Supply Chain Manag Int J 14(2):128–137
Waters D (2011) Supply chain risk management—vulnerability and resilience in logistics. 2nd edn, London
Yao MJ, Chu WM (2008) A genetic algorithm for determining optimal replenishment cycles to minimize maximum warehouse space requirements. Omega 36(4):619–631
Zijm WMH, Klumpp M (2016) Logistics and supply chain management: trends and developments, Spinger, Berlin Heidelberg 1–20

Part IV
New Business Models

Chapter 32
Future Logistics: What to Expect, How to Adapt

Henk Zijm and Matthias Klumpp

Abstract As a result of global societal and economic as well as technological developments logistics and supply chains face unprecedented challenges. Climate change, the need for more sustainable products and processes, major political changes, the advance of internet technology in logistics and cyber-physical production systems pose challenges that require radical solutions, but also present major opportunities. The authors provide a literature review as well as a roadmap on selected issues and argue that logistics has to reinvent itself not only to address these challenges, but also to cope with mass individualization on the one hand while exploiting business applications of artificial intelligence on the other hand. An essential challenge will be to find a compromise between these two developments—in line and in combination with the known triple-bottom line for sustainability. This is presented with an extended analysis on what to do regarding worker qualification and training in logistics in order to cope with these developments.

Keywords Trends · Sustainability · Physical internet · Training · Qualification

32.1 Introduction

Logistics provides the link between subsequent stages of a supply chain, i.e., the entire production and distribution chain from raw materials to final customers for a product or a service, as well as the reverse flow of products, leading to possible reuse of materials, or components (closed loop supply chain). Almost always, such

---

H. Zijm (✉) · M. Klumpp
University of Twente, Enschede, The Netherlands
e-mail: w.h.m.zijm@utwente.nl

M. Klumpp
e-mail: matthias.klumpp@fom.de

M. Klumpp
FOM University of Applied Sciences, Essen, Germany

a production and distribution chain is not executed by one industry, but instead encompasses a number of separate and distributed companies and organizations jointly operating in a chain or network. This so-called end-to-end supply chain is represented in the Supply Chain Operations Reference (SCOR) model (Simchi-Levi et al. 2008). Current supply chains often span the entire globe and involve production, trade and logistics organizations around the world. In this paper, we review how trade, production, and logistics became the global business it is today, and what future developments can be expected (Woxenius and Sjöstedt 2003). We argue that, faced with today's challenges, a fundamental rethinking of the way we organize production and logistics as well as information management and training is needed.

This research contribution is structured as follows: Sect. 32.2 provides a methodological literature review regarding trends and development directions that can be identified in the existing research literature. Section 32.3 outlines the development from past supply chain structures and concepts to modern day structures and predicaments. Technological trends and solutions are the topic of Sect. 32.4, whereas Sect. 32.5 describes change and adaption processes of training and knowledge management in order to improve logistics in the future.

32.2 Literature Review: Logistics Trends

A large number of articles devoted to theory building, hypothesis testing, and exploratory research exist for the field of logistics and supply chain management. In order to outline this evolution, we identify trends based on a thorough literature review. While logistics and supply chain management topics are published in a broad array of journals (Fawcett et al. 1995; Grimm et al. 2014), we use academic journals as our source of information. For the analysis the data platform web of science was used in order to retrieve all supply chain management and logistics articles published from 2005 to 2015, collected with the priority journals on supply chain management and logistics as listed by McKinnon (2013). From these journals, 3,469 articles were identified which served to detect major themes (Table 32.1).

In order to identify trends, a further analysis was conducted using SAS Enterprise Miner. A list of keywords of these themes was used and their frequency in the abstract and title of all articles determined. In a third step, the trend themes were grouped into four main research areas: *business process management, competitive advantage, strategic management,* and *network structure*. The area of *business process management* deals with management of activities that produce a certain output based on customer demand (Cooper et al. 1997; Davenport and Short 2003). Furthermore, it provides methods, techniques, and tools to support the design, enactment, management, and analysis of operational business processes (Van der Aalst et al. 2003).

Table 32.1 Trend themes in logistics and supply chain management

Risk management	Sustainability	Coordination
Operations management	Green SC/LOG	Collaboration
Return logistics	Information technology/industry 4.0	Cooperation
Vehicle routing	Demographic change	Supply chain integration
Distribution	Performance management and measurement	Relief/humanitarian operations
Purchasing/supply management	Closed loop	Packaging
Skills/competences	Human resources	Tracking and tracing
Organizational learning (education and training)	Inventory/warehouse	Agility/lean/flexibility

The area *competitive advantage* is a major topic in academia and business practice, including innovative concepts and tools that support organizations to outperform competitors facing major trends. The area *strategic management* describes the identification and implementation of objectives based on assessments of internal and external factors considering efficient resource allocation. The *network structure* of the supply chain gives insight in the flow of materials and information from start to end. Furthermore, it shows the value that has been created due to cooperation with partners. Finally, the articles were classified in four time slots (2005–2007, 2008–2010, 2011–2013, 2014–2015, see Table 32.2). In order to evaluate those trends identified also for the future, a historic analysis is implemented in the next section.

32.3 Current Trends and Historic Development

The first trend category of **business process management** is rooted in the basis of modern economics and management: In his groundbreaking book *"An Inquiry into the Nature and Causes of the Wealth of Nations,"* Adam Smith developed a first scientific framework on the principles of economic production, starting with the "principle of labor division" (Smith 1776). This evolved toward the dominant philosophy of *efficiency through specialization*, worked out toward a first theory on production organizations by Charles Babbage (*"On the Economy of Machinery and Manufactures"*—Babbage 1835). Early scientific management theories were developed a bit later, e.g., by Frederick Winslow Taylor (cf. Hopp and Spearman 2008). His work on time and motion studies, best working practices and in particular the differential piece rate system was followed by Frank and Lilian Gilbreth as well as Henry Gantt. A first attempt to systemize quality management was developed by Walter Shewhart through his work at Bell Labs on Statistical Quality Control methods.

Table 32.2 Distribution of trend themes in logistics and supply chain management 2005–2015

Main areas	Key themes	2005–2015 (total)	Abstract				Title				Ranking
			2005–2007	2008–2010	2011–2013	2014–2015	2005–2007	2008–2010	2011–2013	2014-2015	
Business process management	Risk management	79	0	15	33	13	1	4	11	2	9
	Operations management	28	4	10	6	4	0	2	1	1	18
	Reverse logistics	62	2	11	13	8	1	10	10	7	14
	Vehicle routing	73	7	9	16	11	3	7	11	9	13
	Distribution	415	52	115	105	69	12	31	20	11	2
	Purchasing/supply management	188	16	55	47	23	6	15	14	12	4
	Relief operations/humanitarian	53	1	18	8	5	2	11	3	5	16
	Packaging	15	1	5	3	3	0	1	1	1	19
	Tracking and tracing	5	1	0	1	1	0	0	1	1	22
	Inventory/warehouse	442	70	108	90	47	18	48	37	24	1
Competitive advantage	Sustainability	131	3	16	35	44	2	7	8	16	6
	Green SC/LOG	80	0	6	17	21	0	5	12	19	9
	Information technology/industry 4.0	60	6	21	15	7	2	6	2	1	15
	Demographic change	4	1	3	0	0	0	0	0	0	23
Strategic management	Performance management & measurement	50	5	13	11	11	2	3	3	2	17
	Closed loop	3	0	0	1	2	0	0	0	0	24
	Human resources	13	1	6	2	3	0	0	1	0	20

(continued)

Table 32.2 (continued)

Main areas	Key themes	2005–2015 (total)	Abstract				Title				Ranking
			2005–2007	2008–2010	2011–2013	2014–2015	2005–2007	2008–2010	2011–2013	2014–2015	
	Organizational learning	11	0	3	2	2	0	2	0	2	21
	Skills/competences	75	5	19	27	16	0	3	4	1	11
	Agility/lean/flexibility	251	23	68	61	35	10	22	24	8	3
Network structure	Coordination	104	18	32	24	15	3	7	3	2	7
	Collaboration	185	17	41	42	33	8	14	12	18	5
	Cooperation	75	11	18	22	12	1	4	4	3	11
	Supply chain integration	97	7	21	19	17	4	7	13	9	8

The trend category **network structure** was introduced with increased production scales and longer supply chains as a result of globalization. Still, mass production, based on *economies of scale* and consequently limited product diversity, continued to dominate industrial production in the first decades after the Second World War. However, starting with the sixties, as prosperity grew, consumers began to demand variety and differentiation, leading to more complex products. In response, manufacturing industries introduced more versatile machines that could produce a variety of products, albeit at the cost of large setup or changeover times. Also, the transfer of production to low wage countries in the Far East and Southern America was an attempt to sustain mass production at affordable costs. Efficiency also characterized modern logistics: the introduction of the container and modern material handling systems meant an advantage in processing the growing global transport flows. Functional specialization, concentration of mass production in large factories, the shift of production to low wage countries, and as a result longer and more complex supply chains to be coordinated, marked the industrial landscape still in the seventies and eighties.

Then, first glimpses of advanced **strategic management** and **competitive advantage** aspects arrived with the sustainability question: The two oil crises of 1973 and 1979 for the first time revealed the weaknesses of the prevalent production philosophy. Raw material prices and interest rates rose sharply and industrial companies started to realize that long supply chains represented large amounts of stock and hence capital invested. Also, long supply chains make it hard to adapt to quickly changing market demands. Companies were inert, and not prepared for the flexibility that a changing society required. Publications such as "Limits to Growth" from the Club of Rome stressed the depletion of natural resources and the pollution of our natural environment as important economic limitations (Meadows et al. 1972). Concurrently, the introduction of flexible manufacturing systems, often based on computerized (CNC) machining and robotized assembly, helped to *balance efficiency and flexibility*, not only in production but also in nodes of logistics networks (material handling and distribution centers). In addition, attempts were made to synchronize and integrate supply chains by means of administrative information systems such as MRP and ERP, or by introducing new production philosophies such as Just-in-Time, or lean and agile manufacturing that focus on rigidly removing any buffer stocks. These were the heydays of the Toyota Production System and of SMED (Single Minute Exchange of Die), an engineering philosophy developed by Shigeo Shingo, who systematically sought to reduce machine setup or changeover times (Shingo 1985). Still, however, factories became more flexible, long and expensive supply chains due to functional specialization and dispersed production of parts and components continued to dominate. But more and more it is realized that current supply chains are *fundamentally unsustainable.* Current production and logistics systems cause serious environmental damage, due to for instance the emission of hazardous materials (CO_2, NO_x, particulate matter), congestion, stench, noise, and more general the high price that has to be paid in terms of infrastructural load. While the European Committee has set clear targets to reduce Greenhouse Gas Emissions

(GGE) in 2050 to 60 % as compared to 1990, the percentage of transport-related GGE has increased from 25 % in 1990 to 36 % today (ALICE 2014).

Also, future *demographics* will be a challenge and the importance of described aspects of the trend category **competitive advantage** (sustainability, demography, information integration) will increase: The current world population of 7.2 billion is projected to increase by 1 billion over the next 12 years and reach 9.6 billion by 2050 (LOG2020 2013). Though within Europe, population size is predicted to be stable; aging continues, meaning that people in general will work longer. Europe-based companies should be prepared for *scarcity of human resources* and be able to provide working conditions that extend the working life of employees. The need to further *increase productivity* while at the same time diminishing the ecological and social footprint, requires a quality upgrade of the human resource pool, e.g., by better education and training, including lifelong learning programs. In parallel, productivity can be improved by better support tools, easier access to relevant information, and finally further automation of both technical processes (i.e., robotics) and decision-making (i.e., artificial intelligence). Border crossing supply chains and logistics systems often concern high-value goods, and therefore are vulnerable to crime and illicit acts. Within the European research programs, various projects have developed roadmaps to enhance supply chain safety and security. Regarding safety, extensive attention has been paid to safe working conditions (for instance driving hour regulations) but the fight for supply chain security, abandoning crimes, and illegal activities appears to be a harder one. Economic crimes for example include: theft (robbery, larceny, hijacking, looting, etc.), organized immigration crime (human trafficking, illegal immigration), IPR violations and counterfeiting and customs law violations (tax fraud, prohibited goods). Alternatively, ideologically or politically motivated crimes occur, next to obvious vandalism (Hintsa 2011). Another aspect of supply chain security is *supply chain resilience*, which can be defined as the ability to maintain, resume, and restore operations after a major disruption (Gaonkar and Viswanadham 2007). Disruptions to supply chains can prove costly: according to research conducted by Accenture, significant supply chain disruptions have been found to cut the share price of impacted companies by 7 % on average (World Economic Forum, Accenture 2013). Finally, commercial product life cycles still tend to become even shorter; at the same time an increased reuse of products, components, and materials, both via (electronic) second markets and in so-called closed loop supply chains can be observed (cradle-to-cradle, *circular economy*). Mass customization is an important aspect of current consumer markets, enabled by fast technological developments. The rapid advance of *e-commerce* is another characteristic of today's markets, with mixed consequences: on the one hand it reduces the number of links in the supply chain, but without adequate regulation of both forward and reverse flows of packages it leads to a rapid increase of transport moves, causing more urban congestion and pollution. Finally, we mention the concept of a *sharing economy*, i.e., the notion that customers no longer buy an actual product but *only the service* the product represents (e.g., cloud computing, music streaming, car sharing). These phenomena will have a profound impact on the ecological footprint

32.4 What to Expect: Technology and the Future of Supply Chains

The outlined *four trend categories* call for a significant change when designing future-proof supply chains. Fortunately, both technological and socialeconomic innovations provide adequate tools that may help to address that challenge.

Technological innovations: The design of new lightweight (bio-)materials and their application in a variety of products pose new possibilities to diminish both costs and ecological footprints. Another manifestation of improvement is the continuous development of cleaner engines (e.g., electric, LNG). Technologies such as 3D-printing and additive manufacturing in general are based on material addition, instead of material removal as in classical machining, hence in principle have a waste avoidance potential. Smart packaging also may help to reduce volumes and to avoid waste, in particular in the case of biodegradable package materials. Modular product design allows transportation of components instead of products, resulting in a higher package density and allowing customization closer to the end-user (Zijm and Douma 2012). The same holds for 3D printing and micromachining which are not only a step forward toward mass customization but in addition have a profound logistics impact, for instance in fostering "local for local" production. When applied in small batch manufacturing, 3D printing may induce a shift from stock- to order-based production with shorter lead times and a reduction of stocks.

Automation and Robotics: The impact of robotics has already been visible for a long time, e.g., in automotive assembly lines but also in *warehouses* and distribution centers (Automatic Storage and Retrieval Systems), often consisting of high bay storage racks which are served by fully automated cranes, and equipped with automatic identification, e.g., RFID. Apart from the visible hardware, innovative warehouse management systems help to coordinate and synchronize activities, in close communication with information systems covering suppliers and customers (see also the high numbers in the trend analysis). Similar developments can be found at container terminal sites in both seaports and inland harbors. Without exception, all such systems rely heavily on smart sensor and actuator systems, where devices are equipped with sensors that *automatically* signal when actions such as ordering or replenishment have to be initiated. Additionally, materials and machinery themselves are able to communicate with each other and find solutions based on *autonomous decision making* using state-of-the-art algorithms. The world of transportation is currently innovating rapidly, as demonstrated for instance by various experiments with freight vehicle platooning, in which a convoy of freight trucks is controlled by one driver—transportation in 2050 is foreseen to be largely unmanned transportation.

Business Information Systems, new business models: Complex modern supply chains are characterized by the fact that many stakeholders are involved in shaping its ultimate manifestation. These multistakeholder and multi-decisionmaker environments require adequate mechanisms to respond to requirements, including distributed architectures, cloud computing solutions, cognitive computing, and agent-based decision support systems. Organizational innovations are indispensable to exploit the potential of advanced information and decision support architectures. The recent attention for data driven models (*big data and predictive analytics*) marks an important step toward full-blown automated decision architectures. The design and acceptance of decision models based on both horizontal and vertical cooperation in logistics networks, however, is difficult. Although many stakeholders quickly recognize the potential win-win situation arising from collaboration, they find it in general hard to give up decision autonomy. Mathematically, game-theoretical approaches have proven to provide adequate tools to handle such multistakeholder games; for instance the Shapley value calculation defines a "fair" allocation of cooperation gains to single actors. But the idea that players may give up their individual solution in order to achieve an overall equilibrium solution is hard to accept in particular for private companies that were used to concentrate on individual profits.

Circular and sharing economy, servitization: The key idea behind *servitization* is the realization that both private consumers and industrial asset owners basically need the functionality of assets and products, rather than the products themselves (Cohen et al. 2006; Neely 2008). Also, the model of a *circular economy* is based on the idea that products and systems that are disposed of can be either restored and reused or disassembled after which components and parts are given a second life in next generation equipment. Another option is to jointly use equipment in a predefined group of people (*sharing economy*). Those products or systems are either owned by individual group members or remain property of the supplier and can be leased or hired. It is clear that such developments may have important consequences for supply chain design, planning and control, because the focus will at least partly switch from delivering products to delivering services. An example is the trend toward car sharing, implying for logistics that cars may not have to be delivered to the individual end customer but toward car sharing operators.

The Physical Internet: The concept was initially proposed by Montreuil (2011) as a radical way to integrate economic, social, and ecological goals in a new sustainable logistics framework. It is defined as a logistics system in which modular packages are automatically routed from source to destination through a network of hubs and spokes. Major elements of such a network more or less exist for parcels, pallets, containers, and swap bodies. A full-fledged physical internet may be built upon all these elements with the holistic integration of existing elements and concepts as the main challenge. To arrive at such enhanced cooperation levels, shippers, manufacturers, retailers, carriers, and other providers of logistics services should take the broader sustainability goals into the economic equation. This requires new ways of decision-making on financial and market criteria but also on safety, security, and environmental aspects. In particular, transnational governance and regulation are needed to achieve such a cultural shift, and to encourage collaboration, coordination, and horizontal partnerships (see trend **network structure**). A major challenge is to design a multifaceted decision support system, with distributed automated execution by intelligent agents. Radical business models based on openness and sharing of resources are required; but this level of openness is in contrast with the core of, e.g., supply chain security. Therefore, adoption will require profound changes with respect to roles and responsibilities of stakeholders. Achieving such a combination of physical and electronic infrastructure is one step, stimulating LSPs to connect to it is a bigger challenge. Within the European Technology Platform for Logistics ALICE five roadmaps are defined on logistics innovation areas, constituting an integral roadmap toward a realization of the Physical Internet (Fig. 32.1).

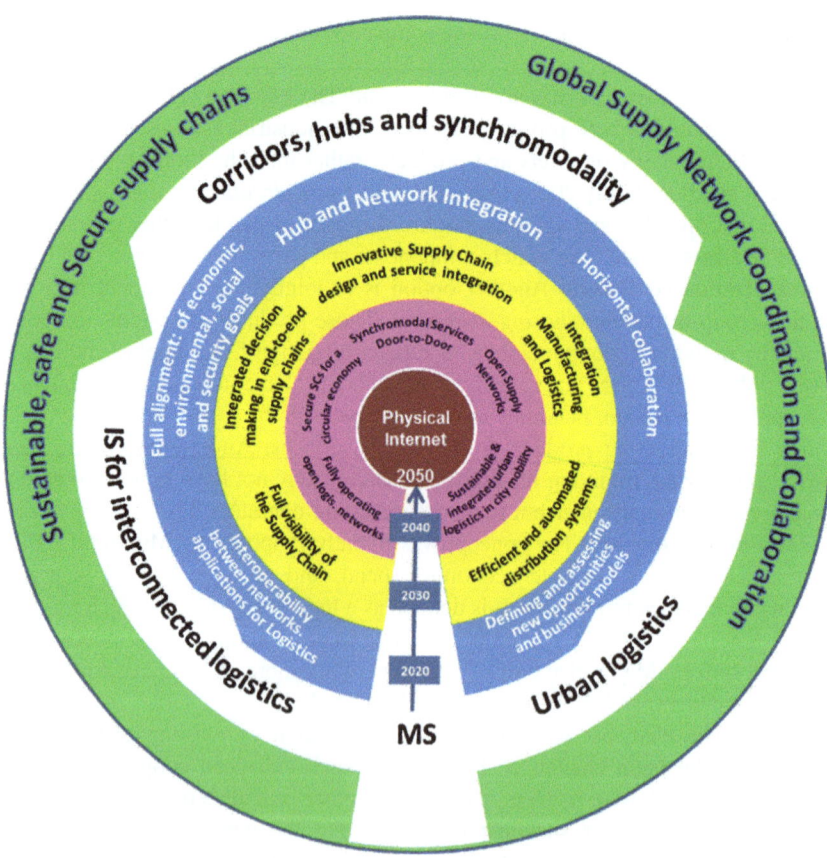

Fig. 32.1 ALICE roadmap toward the physical internet (ALICE 2014)

32.5 How to Adapt: Logistics Training and Knowledge Management

Stemming from the last trend category of **competitive advantage**, future logistics systems will innovate profoundly: Technology innovation and information systems as expected with the Physical Internet concept depend on the presence of a competent workforce at all levels in both private organizations and government institutions. Training is needed to adopt and adequately apply new technologies while the design of smart business models requires analytical skills to understand the increasing complexities of modern global supply networks. Therefore, apart from natural resources and despite the growing world population, the *quality and availability of human competencies* appear to be the most important limiting factor (Wu 2007). Regarding the question of available qualifications and competencies of the human workforce, little is known besides the general recognition that innovation requires a close match between technology, policy as well as business support, and human skills development (i.e., Aghion et al. 2009).

As depicted in Fig. 32.2, starting on a general timeline at about the industrial revolution (point "A"), the necessary or expected competence level of the workforce has increased on average (black line). For logistics processes it can be argued that this still ongoing process has a "double nature." First, existing activities such as truck driving, warehouse processes, or production processes increasingly demand higher competence levels—as demonstrated for example by new EU regulations requesting further training of drivers regarding safety, sustainability, and handling of hazardous goods as well as technology usage. Whereas only thirty years ago truck driving was a typical "unskilled" profession without any necessary training to do the job properly (apart from a driver's license), today no untrained individual can just start driving in complex transport processes as a multitude of systems (toll systems, routing systems, communication systems, auto ID systems, etc.) have to be mastered. Second, new activities arise in logistics and global supply chains, typically with a high competence requirement, such as IT systems management, logistics consulting, logistics and supply chain finance, logistics tender management, logistics controlling (Klumpp 2013). Between the ever increasing expectations and requirements regarding human competence levels, a "gap" is opening up over time as the required training for humans has for each and every person to start "anew": learning can hardly be "inherited" or automated. Longer schooling and training programs are needed in order to arrive at the required competence levels of a modern logistics and business environment. This can be termed a "knowledge accumulation gap" (gray field) that arises due to the fact that humans are not able to accumulate knowledge over generations, as opposed to machines and computers that are increasingly able to do so. Artificial intelligence applications may one day mitigate or close this gap. Besides human knowledge and competence levels, automated or artificial intelligence (nonhuman) competence levels (dotted line in Fig. 32.2) are expected to impact technology and business development. Though artificial intelligence had a somewhat "slow start" during the sixties, seventies, and eighties of the preceding century Newell (1982), it has significantly accelerated in solution contribution width and depth recently. This is connected to the trend of "deep learning," allowing computers autonomously to acquire new knowledge and to find links as well as directions of further learning and meaning themselves (Erhan et al. 2010). As outlined in Fig. 32.2 (point "B"), automated systems were initially very slowly adopted; examples in logistics include the automated gearbox in trucks, partly automated cranes and warehouse equipment as well as automated communication and transmission devices in logistics management (EDI systems, automated decision protocols). These "separate" and limited systems never really "matched" human competence levels, which is why the dotted line traverses significantly below average human competence levels between B and C. In recent years however—symbolized by point "C" in Fig. 32.2—automated systems have undergone a definite change, partly described by a "merging" as formerly separate systems now are increasingly coupled and are beginning to interact. For example, state-of-the-art automated warehouses are integrated systems of software (warehouse management systems), hardware (moving goods) and even optimization (error analysis, automated storage optimization, learning and prognosis). This integration increases the capability of such systems and accelerates their innovation speed. In some cases artificial intelligence and automated systems are already overtaking human competence levels ("D"): Regarding truck driving, the combination of the old automated gearbox and GPS-based navigation systems allows trucks to efficiently downshift *before* a steep slope of an oncoming mountain street is even visible to the human driver. This form of foresight and decision is a new capability of automated systems, which has recently reached new levels in freight platooning and automated passenger car driving experiments. What comes after point "D" can only be hypothesized—but it is not unlikely that a future point "E" lies ahead, where automated systems even exceed the expectations (set by humans) of society and business. This may sound risky, as "unintended and unforeseen behavior" of automated systems may worry humans. But as most technologies, it can easily be argued that risks and opportunities are usually embedded in any development, from the taming of fire to nuclear chain reactions. As examples, some of an unknown multitude of applications can be listed for the area beyond "E" indicated with a question mark:

- Automated production systems may increase output on single workdays (i.e., Monday) due to identified repeating sequences of increased customer demand.
- Automated trucks may for example leave the motorway without a specific order to do so—having information about a jam ahead or even a severe accident in order to make way for emergency operations.
- Automated logistics systems may decide to switch to a different supplier in another country having analyzed reports about imminent civil war hostilities or fraud in the current supplier's country.
- Automated warehouse systems may decide to use less occupied times like Fridays to prepare and rearrange storage places of specific products which are in high demand on Mondays to a place with shorter exit ways.

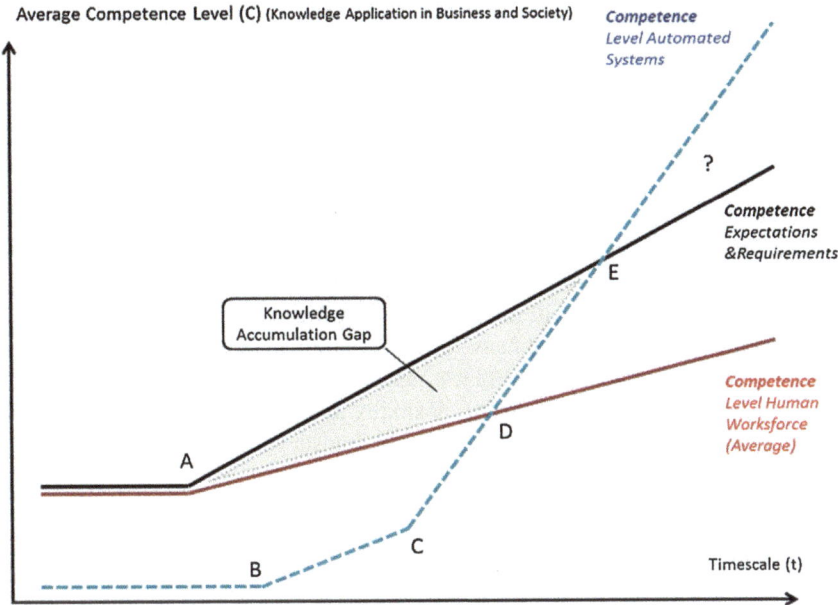

Fig. 32.2 Competence development for human beings and computers

In the light of the described qualification and training developments, the general model of innovation in a partnership of technological development and human qualification and training in implementation has to be scrutinized. In the past, often a *sequential model* was implemented. This model of technology development, followed by implementation and finally (human) training had a distinctive risk mechanism: workers were only trained for technologies already developed and implemented, "lost training" was seldom. But in a modern world, this model is outdated. Current models use a *parallel approach* for at least part of the timeline, regarding implementation and training experiences as essential input for further technology development (user involvement in research and development). In the future, it is suggested that in an environment of largely automated work innovation processes may even take place *without* any large-scale human training. In such systems, human roles may be limited to technology development as well as general oversight. Artificial intelligence and robotics appliances may take over the innovation process completely by introducing new manufacturing as well as management decision concepts without detailed human training. Such a scenario implies that technology development and implementation are two intertwined and parallel processes—as can be observed already in the smartphone app market today.

32.6 Conclusion

From the four identified trend areas, *competitive advantage* seems to be most important: Technology innovation and smart information systems (Physical Internet) are directed toward overall sustainability and heavily depend on a competent workforce. This brings about a new level of *interdependence* of technology innovation and knowledge of logistics personnel. In the future, logistics companies are in the same playing field regarding human resources as internet start-ups and other high-qualification service providers (i.e., in the financial and health industries). As dynamics in supply chains will increase, also a new form of competence will gain more importance: Not the "hard facts" of stored information which are taught today may be valuable, but more the "soft facts" and competences like adaptability (of the mind—followed by the supply chain), flexibility and creativity may be heralded in the future as the main competences needed by logisticians.

References

ALICE (2014) European technology platform for logistics. http://www.etp-logistics.eu
Aghion P, David PA, Foray D (2009) Science, technology and innovation for economic growth: Linking policy research and practice in 'STIG Systems'. Res Policy 38:681–693
Babbage C (1835) On the economy of machinery and manufactures. Charles Knight, London
Cohen M, Agrawal N, Agrawal V (2006) Winning in the aftermarket. Harvard Bus Rev 84 (5):129–138
Cooper MC, Lambert DM, Pagh JD (1997) Supply chain management: more than a new name for logistics. Int J Logist Manag 8(1):1–14
Davenport TH, Short JE (2003) Information technology and business process redesign. Operations management: critical perspectives on business and management 1:97
Erhan D, Bengio Y, Courville A, Manzagol P-A, Vinvent P, Bengio A (2010) Why does unsupervised pre-training help deep learning? J Mach Learn Res 11(1):625–660
Fawcett SE, Vellenga DB, Truitt LJ (1995) An evaluation of logistics and transportation professional. J Bus Logist 16(1):299
Grimm JH, Hofstetter JS, Sarkis J (2014) Critical factors for sub-supplier management: a sustainable food supply chains perspective. Int J Prod Econ 152:159–173
Gaonkar RS, Viswanadham N (2007) Analytical framework for the management of risk in supply chains. IEEE Trans Autom Sci Eng 4(2):265–273
Hintsa J (2011) Post-2001 Supply Chain Security—Impacts on the Private Sector. Doctoral Thesis, HEC University of Lausanne
Hopp W, Spearman M (2008) Factory physics: foundations of manufacturing management. Waveland Press, Long Grove
Klumpp M (2013) How to structure logistics education: Industry Qualifications Framework or topical structure? In: Pawar KS, Rogers H (eds): Resilient supply chains in an uncertain environment, ISL 2013 Proceedings, Vienna Nottingham, 895–902
LOG2020 (2013) Logistics and Supply Chain Management 2020: training for the future. WP2: Positioning Paper. ERASMUS Project ID: 527700-LLP-1-2012-1-NL-ERASMUS-EMCR
Meadows DH, Meadows DL, Randers J, Behrens WW III (1972) The limits to growth. Universe Books, New York

Montreuil B (2011) Towards a Physical Internet: meeting the global logistics sustainability grand challenge. Logist Res 3(2–3):71–87
Neely A (2008) Exploring the financial consequences of the servitization of manufacturing. Oper Manag Res 1(2):103–119
Newell A (1982) The knowledge level. Artif Intell 18(1):87–127
Shingo S (1985) A revolution in manufacturing: the smed system. Productivity Press, New York
Simchi-Levi D, Kaminski P, Simchi-Levi E (2008) Designing and managing the supply chain: concepts: strategies and case studie, McGraw-Hil International, Boston
Smith A (1776) An Inquiry into the nature and causes of the wealth of nations. In: Strahan W, London
Van der Aalst WMP, Arthur HM, Hofstede T, Weske M (2003) Business process management: a survey. Business process management. Springer, Berlin Heidelberg, pp 1–12
World Economic Forum, Accenture (2013) Building resilience in Supply Chains. An Initiative of the Risk Response Network In collaboration with Accenture. Geneva
Woxenius J, Sjöstedt L (2003) Logistics trends and their impact on European combined transport —services, traffic and industrial organization. Log Man 5:25–36
Wu YC-J (2007) Contemporary logistics education: an international perspective. Int J Phys Distrib Logist Manag 37(7):504–528
Zijm H, Douma A (2012) Logistics: more than transport. In Weijers S, Dullaert W (eds) Proceedings of the 2012 Freight Logistics Seminar. Venlo: 395–404

Chapter 33
Concept and Diffusion-Factors of Industry 4.0 in the Supply Chain

Hans-Christian Pfohl, Burak Yahsi and Tamer Kurnaz

Abstract The purpose of this exploratory study is to understand how technology innovations discussed with respect to the fourth industrial revolution diffuse within the supply chain. First, the main characteristic and interrelated features of the term "Industry 4.0" are identified to come up with a definition. We understand that the fourth industrial revolution is characterized by the technological trends of digitalization, autonomization, transparency, availability of real-time information and collaboration. Based on the assumption, that all technological innovations addressing these characteristics, are contributing to diffuse the concept of Industry 4.0 in the supply chain, causal hypotheses are stated within a structural model to further discuss the diffusion-factors and barriers of implementation. All results are conceptually derived based on a structured literature review.

Keywords Diffusion · Industry 4.0 · Supply chain · Innovation

33.1 Introduction

During the first three industrial revolutions, technological innovations changed the product-portfolios and the respective manufacturing processes of many industries. Steam machines, electrical drives, combustion engines, and the assembly line production system affected how products were manufactured and enabled the creation of new product functionalities (Bauernhansl et al. 2014). The aim of these innovations was to make the physical value stream, i.e., the manufacturing and delivery of a product from the producer to the end-consumer, more cost-efficient

---

H.-C. Pfohl (✉) · B. Yahsi · T. Kurnaz
TU Darmstadt, Supply Chain and Network Management, Darmstadt, Germany
e-mail: pfohl@scnm.tu-darmstadt.de

B. Yahsi
e-mail: yahsi@scnm.tu-darmstadt.de

T. Kurnaz
e-mail: kurnaz@scnm.tu-darmstadt.de

and robust. And in the end, newly available products enabled a changed lifestyle for the households. However, the currently ongoing fourth industrial revolution is different, as the demand for change in the business models is not only pushed by the availability of a new technology, but also pulled by the trend of a more socially collaborative, convenience-demanding and technology-affine way of life of the households. Because of this bilateral impact as well as the inclusion of multiple research area, the research field of "Industry 4.0" is highly complex and dynamic.

This paper aims to help affected companies in technology-driven industries which are currently investing in technological and conceptual innovations. It makes the term "Industry 4.0" more tangible by identifying its main characteristics. It is assumed that all currently available technological innovations addressing these characteristic trends are relevant with respect to their disruptive capabilities (Christensen et al. 2002). It then explains the diffusion-factors of the relevant technological innovations based on identified implementation-barriers. The analysis of diffusion-factors is based on the theories developed in Rogers (2003).

33.2 Methodology and Contribution

The basic method of the analysis is a structured literature review with respect to the diffusion-factors of technological innovations in the scope of the fourth industrial revolution. This is executed according to Baker (2000) and Cooper (1988). The review includes high-rated international journals, which were published since 2010. The following journals were selected according to the VHB-ranking (Henning-Thurau et al. 2014) and their relevancy with respect to the scientific topic: *Management Science, Operations Research, Journal of Management Studies, Organization Science, M&SOM, Transportation Science and Information Systems Research*. Other scientific databases (mainly EBSCOHOST) were added to the sources of the literature review, as currently most of the relevant published articles with respect to the topic can be found in scientific magazines and lower-rated journals. Studies published by companies (e.g., consulting firms like KPMG) or research institutes were validated as well and bring up interesting hypotheses and results, especially in the German literature. In total, 152 published articles, scientific papers and books were reviewed throughout this analysis.

The analyses showed three main results with respect to the research field of "Industry 4.0". At first, we could understand that this topic is merely discussed in internationally known journals. Most publications are studies from consulting firms (El-Darwiche et al. 2013) or research institutes (Kagermann et al. 2013). The second finding is that the relevant literature is limited to descriptive or instrumental studies discussing beneficial architectures of specific technologies on the process level (Lee et al. 2015). Other scholars perform studies to come up with recommendations for the industrial practice (Brettel et al. 2014). This paper therefore widens the horizon of the exploratory research to the explanatory-level and discusses the diffusion-factors of technological innovations as part of causal structural model.

Whilst the relevant research with respect to the fourth industrial revolution is currently limited to the key resources and key activities of companies, their business can be differentiated into further and equally important partial models, which are explained in the Canvas model proposed by Osterwalder and Pigneur (2010). This paper uses that model as a framework to enable a more holistic analysis which is including the whole end-to-end process of the value creation, i.e., from the collaboration with supply chain partners until the moment when the end-customer uses the product.

33.3 Characterizing the Term "Industry 4.0"

The term "Industry 4.0" is not ultimately defined in the relevant literature (Brettel et al. 2014). Hence, a variety of topics are discussed within the scope of this research-field. This paper states a definition based on characterizing features, that are discussed in the relevant literature: (1) Digitalization of processes and products, (2) autonomization of processes and decisions, (3) transparency of organizations and customers, (4) mobility of products and access of information, (5) modularization of products and facilities, (6) collaboration within a network, (7) socializing of machines and the decision-making processes, (8) flexibility of production facilities, (9) decentral controlling of machines and (10) the real-time availability of information.

These features are interrelated in two dimensions (see Fig. 33.2). First, some features are enabling the other feature from a technological perspective. For example, the availability of real-time information is discussed as a main enabler for the flexibility of suppliers in value chain (Chan et al. 2009). Second, some features of specific Industry 4.0-related technologies and methods can or cannot be utilized in specific parts of the Canvas business model. For example, the cloud computing technology is addressing the trend of digitalization, mobility and flexibility. This digitalization function of cloud computing may be used to offer a new value proposition over a cloud system to the customer. For each match in both dimensions a score-value is counted for each technology. If both score-values are used to evaluate the importance of a specific characteristic feature, the portfolio shown in Fig. 33.1 highlights the trends of digitalization, collaboration, real-time availability, transparency and autonomization using both values.

In the following, the relevant characteristic features are described in detail. Only features with a score-value for the usability in business-models above five, and a score-value with respect to the technological enabler-function above four (blue marked field in the portfolio) are seen as contributing to the diffusion of Industry 4.0 in the supply chain. The thresholds were chosen by using the median of both dimensions highest score-values.

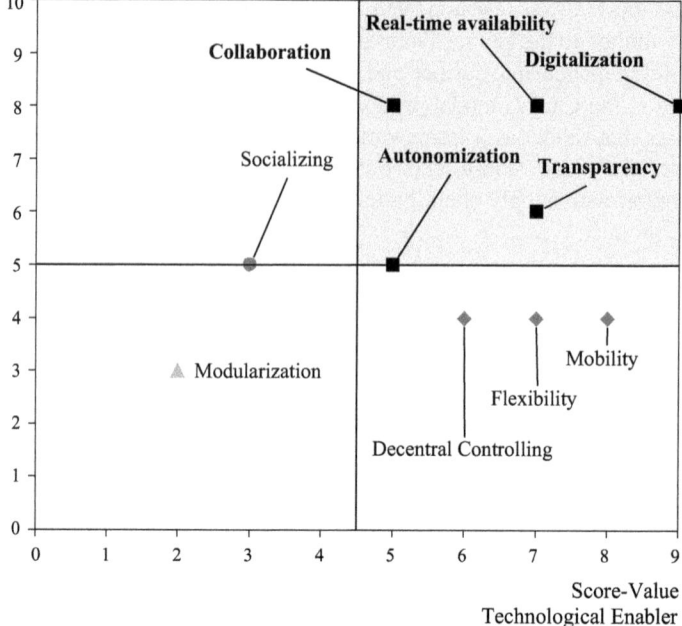

Fig. 33.1 Importance-portfolio of characteristic features

- The **digitalization** of processes and products: The companies' internal processes, product components, communication channels and other key aspects of the supply chain are undergoing an accelerated digitalization process (Geisberger and Broy 2012). The digitalization is a key enabler for the other characteristics.
- The **autonomization** of processes and decisions: Industry 4.0 technologies and concepts are enabling machines and algorithms of future companies to make decisions and perform learning-activities autonomously. This autonomous decision-making and learning is based on man-made algorithms and enables whole factories and manufacturing facilities to work with minimum human-machine-interaction (Angelov 2012).
- The **transparency** of organizations and customers: While global supply chains are characterized by highly complex structures, the available Industry 4.0 technologies are increasing the transparency of the whole value creation process. Through this increase in transparency, decision-making in the company will be more collaborative and efficient. Not only the supply chain processes, but also the behavior of corporate partners and customers will be more transparent to the company (Wang et al. 2007).

- The ability of **collaboration** within a network: Just as human beings in our society are interacting in social networks, the companies' processes will be defined and activities will be decided through the interaction of machines and human beings within specific networks in and out of the companies' organizational borders.
- The availability of **real-time information** for the operational and strategic management: Technologies providing real-time information along the value chain are key enablers for all other trends of the fourth industrial revolution and contribute to the diffusion of Industry 4.0 in many ways. For example, supply and production uncertainties may be reduced with the implementation of algorithms using real-time information (Mogre et al. 2014).

The above given results of the conceptual analysis leads to the following definition of the term Industry 4.0: *Industry 4.0 is the sum of all innovations derived and implemented in a value chain to address the trends of digitalization, autonomization, transparency, collaboration and the availability of real-time information of products and processes.*

33.4 Diffusion-Factors of Industry 4.0

The aim of this chapter is to model the relationship between diffusion-factors and the diffusion-success of digital technological innovations discussed with respect to the fourth industrial revolution. The generation of a hypothesis model requires theoretically founded assumptions about the causality between the lateral variables in the model the empirical examination of the stated hypotheses. This paper states eight hypotheses, which affect the diffusion of technological innovations within the supply chain. It defines the diffusion as the process, with which a technological innovation is spread across the supply chain network (Feldmann et al. 2010).

In the following, the causal interrelation of the specific characteristic features is firstly described and finally the causal influences of diffusion-factors on the main enabler digitalization (see Fig. 33.2). Three criteria are used to identify the relevant hypotheses with respect to their effect on the digitalization of technologies out of the many discussed in the relevant literature: The simplicity and compatibility of a hypothesis as well as the grounding on expert experiences.

Simplicity is often mentioned as a criterion to accept one hypothesis out of a set of hypotheses. It describes how simple a hypothesis is. For example, a theory might be said to be simpler than another if it assumes the existence of fewer entities, causes, or processes. Compatibility means that a hypothesis is consistent with the well-founded scientific results. The request for expert experiences is stated because of the needed practical relevancy. If the score-value is used to evaluate the rationality and plausibility of discussed hypotheses, the following structural equations model can be developed (Fig. 33.2).

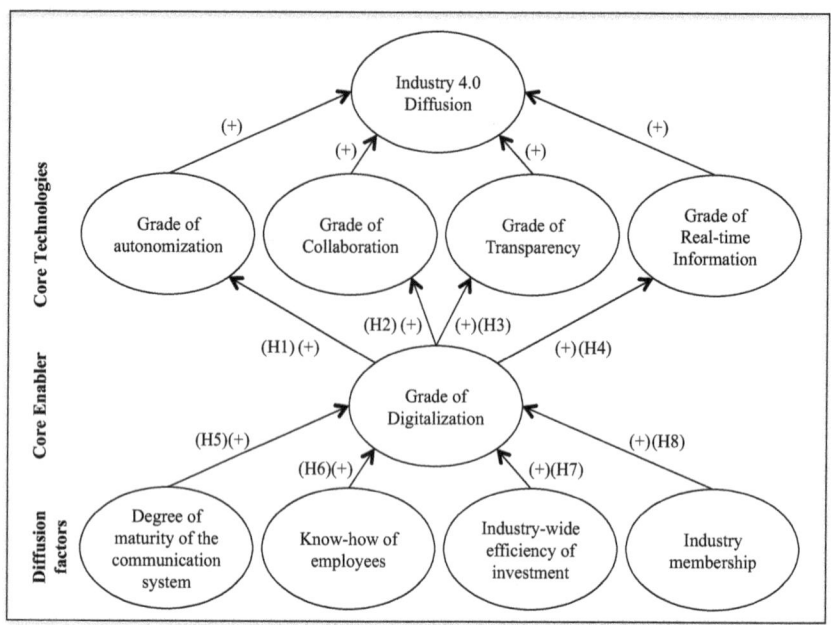

Fig. 33.2 Structural model of Industry 4.0 diffusion

The grade of autonomization of processes and the decision-making within an organization is one of the core characteristic features of the term Industry 4.0. It enables machines and algorithms of future companies to make decisions and perform learning-activities autonomously. The digitalization on the other side of this causal hypothesis means the documentation and transportation of products, processes, and infrastructure data into a digital form, so that data can be received, sent, processed and saved at any place, at any time via a mobile connection. An example technology according the previous chapter would be the concept of a smart factory. We use these definitions to state that the grade of digitalization not only enables the autonomization (Roland Berger Consulting 2014), but also could be used to improve forecasting models and decision-making processes. *This leads to the first hypothesis (H1): The higher the grade of digitalized products, processes, and infrastructures within an organization, the higher is the diffusion of autonomous decision-making and processes within the supply chain network.*

The collaboration within a supply chain was identified as a further key characteristic feature. It means the cooperation between supply chain members in a network across organizational borders. An example of a technological innovation with respect to the fourth industrial revolution is the communication between manufacturing processes of different organizations using the automatically machine-to-machine-technology (M2M). This technology allows faster and more efficient production of products (Roland Berger Consulting 2014). This technology could enable an automated ordering of new materials or an autonomous determination of the current business needs. Since these automatically collaborating machines are organized in decentralized structures, they require a digitalized network structure to operate in for an efficient communication (Horvath und Partners Management Consulting 2015). *This leads to the second hypothesis (H2): The higher the grade of digitalized infrastructures within the organizations, the higher is the diffusion of collaborative technologies within the supply chain network.*

Since supply chain structures are characterized by a huge complexity from a value stream perspective, the demand for more visibility and transparency is obvious. At the same time, the digitalization of processes, products and decisions along the whole supply chain leads to a higher standardization and transparency. However, a higher digitalization within the supply chain requires more process analysis and documentation. A technology, which would increase the transparency of customer-actions in the supply chain, would be social media. *The third hypothesis therefore is as follows (H3): The digitalization in a supply chain is increasing the transparency of the whole value creation process, and therefore the higher the grade of digitalized products and processes within the collaborating organization, the higher is the diffusion of technological innovations increasing the transparency in supply chain* (Sabbagh et al. 2012).

It could be understood that the digitalization of processes and products is the main key enabler for the availability of real-time information along the whole supply chain (Roland Berger Consulting 2014). For example, supply and production uncertainties may be reduced with the implementation of algorithms using real-time information, which were digitalized previously. For example, digitalized organizational structures make real-time information about each product and its location available at all times. *It can therefore be stated, that digitalization is an enabler of real-time information availability, and a high level of digitalization within the collaborating organization leads to a high level of real-time information availability (H4).*

The degree of maturity of the communication technology given is a precondition for the digitalization of products and processes within an organization (Sabbagh et al. 2012). So, the key requirement for intelligent applications and innovative services of the future are powerful, secure communications networks and the interoperability of technologies. According to the previous chapter, a key technology, which is to be installed to accelerate the diffusion of Industry 4.0 within the supply chain, are cyber-physical systems. *Hence, it is stated that, if the degree of technical maturity of communication systems is low, organizations will hesitate to digitalize their process and products (H5).*

Since the digitalization of products will as well change the manufacturing processes and work procedures in supply chain networks, a change in qualification requirements for employees is expected. The digitalization of whole value chains demands new requirements in terms of the existent IT-related knowhow. It is to be questioned, if or if not an industrial company should hire a supply chain manager or a data scientist for their future logistics processes—or both. We therefore see that an inadequate bundle of qualifications of employees in specific departments in the company related to supply chain issues is an important challenge for the successful diffusion of digitalized products and processes in the value chain. *It is therefore stated that if the less employee's qualifications meet the requirements specific IT- and data analysis topics in specific departments, the less will the digitalized products and processes diffuse in the specific company and its value chain (H6)* (PwC 2014; Bertschke 2015).

The implementation of technological innovations as such, but the previously explained technological innovations in the scope of the fourth industrial revolution (Fig. 33.2) in specific, requires a high amount of investment effort. At the same time, the implementation of these technologies is risky and highly complex. Low-budget value chains, with a industry-wide relatively low rate of investment, will focus on simple management of supply-buyer-relationships first and will only invest in technology increasing the digitalization in the value creation process if a short-term return is to be expected (PwC 2014). A number to measure the efficiency of investments within an industry is the ROI (return on investment). *We use this number to state the following hypothesis (H7): The industry-wide efficiency of investments (ROI) has a positive impact on the diffusion of digitalized products and processes within a supply chain.*

Many products or services are merely suitable for a digitalization process. Examples could be transportation services in the tourism-business, services in form of legal advice or the services of consulting firms. These can only be partially digitalized, e.g., consulting firms could perform data acquisition, analysis and the presentation of results using videoconferences and virtual teams only (PwC 2014). This way traveling costs would be intensively reduced. However, the services of the consulting firm itself, delivering ideas, cannot be changed. *It is therefore stated that the industry in which a company is active, is affecting the digitalization of products and processes in the way that, if an organization is active in the manufacturing industry, the level of digitalization is high (H8).*

33.5 Conclusion

In this paper, we define the term Industry 4.0 and highlight its five characteristics. Based on a conceptual analysis using the Canvas model, each characteristic feature is analyzed using its score-value in two dimensions. The most important features are the trends of autonomization, collaboration, transparency, and the availability of real-time information in decision-making processes. Above all characteristic features the trend for digitalizing processes, products and business models is dominating. Technologies addressing these trends are not only relevant, but their diffusion is highly possible and companies focusing on the implementation of these technologies can secure their competitive advantage.

The grade of digitalization itself is then affected by the maturity-degree of communication systems, the currently existent knowhow of employees with respect to the technology, the efficiency of investments made in the respective supply chain as well as the industry membership. Using the structural model with the developed hypotheses, a company may be able to understand how technological innovations addressing the trends of Industry 4.0 can be implemented in the own supply chain. As an example, it is stated that technological innovations already addressing the above mentioned trends are therefore better diffusing in the supply chain when the respectively required know-how in the affected companies is high.

Using this analyses companies are able to (1) better define and understand the characteristic features of the term "Industry 4.0", (2) better reflect on the disruptive capabilities of technological innovations available and (3) understand the diffusion-factors of the technologies addressing the trends of Industry 4.0.

As this paper focuses only on the first aspect, it leaves the empirical validation of the hypotheses for further research. This can be done within a confirmatory study defining manifest indicators for each latent variable and execute an empirical testing based on the results of a survey (structural equation modeling).

References

Angelov P (2012) Autonomous learning systems: from data streams to knowledge in real-time. Wiley, New York, US
Baker MJ (2000) Writing a literature review. Market Rev 219–247. doi:10.1362/1469347002529189
Bauernhansl T, ten Hompel M, Vogel-Heuser, B (2014) Industrie 4.0 in Produktion, Automatisierung und Logistik. Springer, Wiesbaden, Germany
Bertschke I (2015) Industrie 4.0 – Digitale Wirtschaft: Herausforderungen und Chancen für Unternehmen und die Arbeitswelt. ifo Schnelldienst 68(10):3–5
Brettel M, Friederichsen N, Keller M, Rosenberg M, (2014) How virtualization, decentralization and network building change the manufacturing landscape: an Industry 4.0 Perspective. Int J Sci Eng Technol 8(1):37–44. http://scholar.waset.org/1999.8/9997144. Accessed 14 Oct 2015
Chan FTS, Bhagwat R, Wadhwa S (2009) Study on suppliers' flexibility in supply chains: is real-time control necessary? Int J Prod Res 47(4):965–987. doi:10.1080/00207540701255917
Christensen CM et al (2002) Disruption, disintegration and the dissipation of differentiability. Ind Corp Change 11(5):955–993
Cooper HM (1988) Organizing knowlegde syntheses: a taxonomy of literature reviews. Knowl Soc 1(1):104–126. doi:10.1007/BF03177550
El-Darwiche B et al (2013) Digitization for economic growth and job creation—regional and industry perspectives. Booz & Company, New York, US
Feldmann K, Franke J, Schüßler F (2010) Development of micro assembly processes for further miniaturization in electronics production. Manufac Technol 59(1):1–4. doi:10.1016/j.cirp.2010.03.005
Geisberger E, Broy M (2012) agendaCPS – Integrierte Forschungsagenda Cyber-Physical Systems. Springer, Berlin, Germany
Hennig-Thurau T, Walsh G, Schrader U (2014) VHB-Jourqual – ein Ranking von betriebswirtschaftlich-relevanten Zeitschriften auf der Grundlage von Expertenurteilen. Zeitschrift für betriebswirtschaftliche Forschung 56:520–545
Horvath und Partners Management Consulting (2015) Auswirkungen der Digitalisierung auf die Steuerung von Banken. Steering Business in a Digital World, 6
Kagermann H et al (2013) Recommendations for implementing the strategic initiative Industrie 4.0. Federal ministry of education and research, Frankfurt/Main, Germany
Lee J, Bagheri B, Kao HA (2015) A cyber-physical systems architecture for industry 4.0-based manufacturing systems. Manuf Lett 3:18–23. doi:10.1016/j.mfglet.2014.12.001

Mogre R, Wong C, Lalwani C (2014) Mitigating supply and production uncertainties with dynamic scheduling using real-time transport information. Int J Prod Res 52(17):5223–5235. doi:10.1080/00207543.2014.900201

Osterwalder A, Pigneur Y (2010) Business model generation: a handbook for visionaries, game changers, and challengers. Wiley, New York, US

PwC (2014) Industrie 4.0—Chance und Herausforderungen der vierten industriellen Revolution. Frankfurt/ Main, Germany.

Rogers EM (2003) Diffusion of Innovations

Roland Berger Consulting (2014) Die digitale Transformation der Industrie

Sabbagh K et al (2012) Maximizing the impact of digitalization. PwC 4

Wang C, Heng M, Chau P (2007) Supply chain management—issues in the new era of collaboration and competition. Idea Group Publishing, London, UK

Chapter 34
On Upper Bound for the Bottleneck Product Rate Variation Problem

Shree Ram Khadka and Till Becker

Abstract The problem of minimizing the maximum deviation between the actual and the ideal cumulative production of a variety of models of a common base product, commonly known as the bottleneck product rate variation problem, arises as a sequencing problem in mixed-model just-in-time production systems. The problem has been extensively studied in the literature with several pseudo-polynomial exact algorithms and heuristics. In this paper, we estimate an improved largest function value of a feasible solution for the problem when the mth power of the maximum deviation between the actual and the ideal cumulative productions has to be minimized.

Keywords Bound · Product rate variation problem · Nonlinear integer programming problem

34.1 Introduction

Manufacturing systems have developed from mass production to mass customization. Customers demand a high variety of products at reasonable prices. This development forces manufacturing companies to look for approaches to cut down costs in the manufacturing process. For example, the Toyota production system Monden (2011), Ōno (1988) aims to align the sequence of manufactured model variants to

---

S.R. Khadka (✉)
Central Department of Mathematics, Tribhuvan University, Kathmandu, Nepal
e-mail: shreeramkhadka@gmail.com
URL: http://www.psls.uni-bremen.de

S.R. Khadka · T. Becker
Production Systems and Logistic Systems, Faculty of Production Engineering,
University of Bremen, Bremen, Germany
e-mail: tbe@biba.uni-bremen.de

T. Becker
BIBA - Bremer Institut Für Produktion Und Logistik GmbH,
University of Bremen, Bremen, Germany

the actual demand of the customers in order to eliminate inventories. The challenge of this just-in-time approach is to prevent shortages while keeping the inventory at a minimum level. Empirical observations have confirmed the positive effects of the implementation of just-in-time systems in manufacturing companies Huson and Nanda (1995).

The bottleneck product rate variation problem (abbreviated as BPRVP) is a sequencing problem in mixed-model just-in-time (abbreviated as MMJIT) production systems. The MMJIT production systems with negligible change-over costs between the models have been used in order to respond to the previously mentioned customer demands for a variety of models of a common base product without holding large inventories or incurring large shortages Miltenburg (1989). The BPRVP minimizes the maximum deviation between the actual cumulative productions over the observed periods from the ideal one. This problem has been widely investigated in the literature since it has a model with a strong mathematical base and wide real-world applications, see Dhamala and Khadka (2009), a survey and therein.

In Steiner and Yeomans (1993), solved the BPRVP with an absolute deviation objective in pseudo-polynomial time $O(DlogD)$, where D stands for the total demand of all the models. Moreover, the problem with different objective functions has been solved in the same complexity Dhamala et al. (2010), Khadka (2012). The problem is transformed into a perfect matching which yields a feasible solution. Then a bisection search algorithm that runs between the lower and the upper bounds is used for optimality. In this paper, we propose an improved upper bound for the BPRVP.

The remainder of the paper is structured as follows. In Sect. 34.2, we present a nonlinear integer programming formulation. In Sect. 34.3, we discuss the bounds with the level curve in Sect. 34.3.1, the lower bound in Sect. 34.3.2 and the upper bound, which is the main contribution of the paper, in Sect. 34.3.3. The last section concludes the paper.

34.2 Problem Formulation

Let given D be the total demand of $n, n \geq 2$, different models with d_i copies of model $i, i = 1, 2, \ldots, n$. The time horizon is partitioned into D equal time units under the assumption that each copy of a model $i, i = 1, 2, \ldots, n$, has equal processing time. A copy of a model is produced in a time unit k means that the copy of the model is produced during the time period from $k - 1$ to k, $k = 1, 2, \ldots, D$. Let $r_i = \frac{d_i}{D}$ be the demand rate and x_{ik} and kr_i, $i = 1, 2, \ldots, n; k = 1, 2, \ldots, D$, be the actual and the ideal cumulative productions, respectively, of model i produced during the time units 1 through k. An inventory holds if $x_{ik} - kr_i > 0$, and a shortage incurs if $kr_i - x_{ik} > 0$. We assign equal cost for both inventory and shortage. Miltenburg (1989) and Steiner and Yeomans (1993) gave a nonlinear integer programming formulation for the BPRVP as follows with m being a positive integer:

Minimize

$$F_m = \max_{ik} |x_{ik} - kr_i|^m$$

subject to

$$\sum_{i=1}^{n} x_{ik} = k, \quad k = 1, 2, \ldots, D$$
$$x_{i(k-1)} \leq x_{ik}, \quad i = 1, 2, \ldots, n; \ k = 2, 3, \ldots, D$$
$$x_{iD} = d_i, x_{i0} = 0, \quad i = 1, 2, \ldots, n$$
$$x_{ik} \geq 0, \text{integer}, \quad i = 1, 2, \ldots, n; \ k = 1, 2, \ldots, D.$$

34.3 Bounds

34.3.1 Level Curve

There exist nD deviations between the actual and the ideal cumulative productions of D copies of n models. The value of the actual cumulative production $x_{ik}, i = 1, 2, \ldots, n; k = 1, 2, \ldots, D$, is sequence-dependent integer from $\{0, 1, \ldots, d_i\}$. However, the value of the ideal cumulative production $kr_i, i = 1, 2, \ldots, n; k = 1, 2, \ldots, D$, is sequence-independent rational number such that $kr_i \in \{\frac{d_i}{D}, \frac{2d_i}{D}, \ldots, d_i\}, i = 1, \ldots, n$. Let $j, j = 1, \ldots, d_i$, be the number of copies of a model $i, i = 1, 2, \ldots, n$, and (i, j) be the jth copy of model $i, i = 1, 2, \ldots, n$. The actual cumulative production $x_{ik}, i = 1, 2, \ldots, n; k = 1, 2, \ldots, D$, has nD values with $x_{ik} \in \{j | j = 0, 1, 2, \ldots, d_i; i = 1, 2, \ldots, n\}$. There exist at most $n + D$ different values of x_{ik} for the BPRVP. Hence, one can replace x_{ik} by j with $j = 0, 1, \ldots, d_i; i = 1, 2, \ldots, n$, in the level curve (see Fig. 34.1) of the objective value of the function of the BPRVP. The level curve for copy (i, j) of the objective function of the BPRVP is defined as

$$f_{ij}^m = |j - kr_i|^m, i = 1, 2, \ldots, n; j = 0, 1, \ldots, d_i.$$

We set a horizontal line with a suitable value $B > 0$ intersecting the level curve for each copy $(i, j), i = 1, 2, \ldots, n; j = 1, 2, \ldots, d_i$, of the objective function of the BPRVP on the planning horizon $[0, D]$. The horizontal line with the value $B > 0$ is called a bound for the BPRVP. It is important to find the smallest function value, i.e., the tight lower bound $LB_m > 0$ as well as the largest function value, i.e., the tight upper bound $UB_m > 0$ with $LB_m \leq f_{ij}^m \leq UB_m$ of a feasible solution of an instance of the BPRVP so that one can minimize the maximum deviation in a reasonable time.

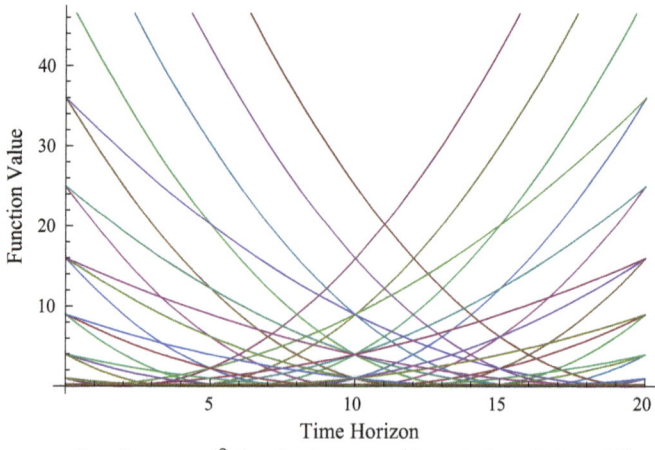

Level curves f^2_{ij} for the instance $(d_1 = 4, d_2 = 6, d_3 = 10)$

Fig. 34.1 Level curves f^2_{ij} for the instance $(d_1 = 4, d_2 = 6, d_3 = 10)$

34.3.2 Lower Bound

The importance of a lower bound LB_m on the optimal value of function F_m for the BPRVP results from the fact that a feasible sequence s of an instance $(d_1, d_2, \ldots, d_n)$ obtained with an objective function value equal to the lower bound is optimal. However, not all instances are even feasible at this value.

The lower bound LB_m has to satisfy the inequality

$$LB_m \leq |x_{ik} - kr_i|^m, \qquad (34.1)$$

for all $i = 1, 2, \ldots, n$, and $k = 1, 2, \ldots, D$, for any feasible solution $s \in \mathcal{X}$, where

$$\mathcal{X} = \{s = (x_{ij}) | i = 1, 2, \ldots, n; j = 1, 2, \ldots, d_i\}$$

is the set of all feasible solutions of an instance for the BPRVP.

One can reduce the inequality (34.1) into the inequality

$$LB_m \leq \min_i |j - kr_i|^m, \qquad (34.2)$$

for all $j = 1, 2, \ldots, d_i$. Since a copy of a model has to be processed during the first time unit, we can put $j = 1$ in the inequality (34.2).

Then, the inequality becomes

$$LB_m \leq |1 - r_{max}|^m, \qquad (34.3)$$

where

$$\max_i r_i = r_{max}.$$

Thus, the tight lower bound $LB_m = (1 - r_{max})^m$, see Dhamala et al. (2010), Khadka (2012). It is noteworthy that a lower bound on the absolute deviation objective function $F_1 = \max_{ik} |x_{ik} - kr_i|$ for the BPRVP has been established by Steiner and Yeomans (1993).

34.3.3 Upper Bound

A necessary and sufficient condition for the existence of a feasible sequence for the BPRVP with the objective of minimizing F_1 is that the value UB_1 must satisfy the two inequalities

$$\sum_{i=1}^{n} (\lfloor k_2 r_i + UB_1 \rfloor - \lceil (k_1 - 1) r_i - UB_1 \rceil) \geq k_2 - k_1 + 1 \qquad (34.4)$$

and

$$\sum_{i=1}^{n} (\lceil k_2 r_i - UB_1 \rceil - \lfloor (k_1 - 1) r_i + UB_1 \rfloor) \leq k_2 - k_1 + 1 \qquad (34.5)$$

where $k_1, k_2 \in \{1, \ldots, D\}, k_1 \leq k_2$. The interval $[k_1, k_2]$ overlaps at least to some extent the time interval within which copy (i, j) is sequenced, see Brauner and Crama (2004).

The two inequalities as the necessary and sufficient condition for the existence of a feasible sequence in order to minimize the objective function F_m of the BPRVP are

$$\sum_{i=1}^{n} (\lfloor k_2 r_i + \sqrt[m]{UB_m} \rfloor - \lceil (k_1 - 1) r_i - \sqrt[m]{UB_m} \rceil) \geq k_2 - k_1 + 1 \qquad (34.6)$$

and

$$\sum_{i=1}^{n} (\lceil k_2 r_i - \sqrt[m]{UB_m} \rceil - \lfloor (k_1 - 1) r_i + \sqrt[m]{UB_m} \rfloor) \leq k_2 - k_1 + 1 \qquad (34.7)$$

with $k_1, k_2 \in \{1, \ldots, D\}, k_1 \leq k_2$, see Khadka (2012).

Steiner and Yeomans (1993) investigated an upper bound $UB_1 = 1$ for the objective function F_1 of the BPRVP. It is not only true for the objective function F_1 but is equally true for the objective function F_m. However, the value of the upper bound can be improved to be equal to $UB_1 = 1 - \frac{1}{D}$ for the objective function F_1, Brauner and Crama (2004) and $UB_m = (1 - \frac{1}{D})^m$ for the objective function F_m, Khadka (2012).

In this paper, we show that the upper bound $UB_m = (1 - \frac{1}{D})^m$ can further be improved for the objective function F_m in the case of an instance $(d_1, d_2, \ldots, d_n)$ with $d_{min} \geq 1$, where

$$\min_i(d_i) = d_{min}, i = 1, 2, \ldots, n.$$

If UB_m is an upper bound on the largest value of the objective function F_m of a feasible solution for the BPRVP, then this bound UB_m satisfies the inequality

$$\left| x_{ik} - kr_i \right|^m \leq UB_m, \tag{34.8}$$

for all $i = 1, 2, \ldots, n; k = 1, 2, \ldots, D$, for any feasible solution $s \in \mathcal{X}$.

We claim that the improved upper bound on the largest value of the objective function F_m is

$$UB_m = (1 - r_{min})^m, \tag{34.9}$$

where

$$\min_i(r_i) = r_{min}, i = 1, 2, \ldots, n.$$

We need to show the proposed upper bound (34.9) satisfying the two inequalities (34.6) and (34.7).

We can write

$$\left\lfloor k_j r_i + \sqrt[m]{UB_m} \right\rfloor = \left\lfloor k_j r_i + 1 - r_{min} \right\rfloor .$$

$$= \left\lfloor k_j r_i + 1 - \frac{d_{min}}{D} \right\rfloor, i = 1, 2, \ldots, n; j = 1, 2 .$$

If $k_j r_i$ is an integer,

$$\left\lfloor k_j r_i + 1 - \frac{d_{min}}{D} \right\rfloor = k_j r_i, \ i = 1, 2, \ldots, n; j = 1, 2 ,$$

and if $k_j r_i$ is not an integer,

$$k_j r_i = \lfloor k_j r_i \rfloor + (k_j r_i)_{frac}, \ i = 1, 2, \ldots, n; j = 1, 2 ,$$

where $(k_j r_i)_{frac}$ is the fractional part of $k_j r_i$.

Since $\frac{d_{min}}{D} \leq (k_j r_i)_{frac} \leq 1 - \frac{d_{min}}{D}$,

$$\lfloor k_j r_i + 1 - r_{min} \rfloor = \left\lfloor \lfloor k_j r_i \rfloor + (k_j r_i)_{frac} + 1 - \frac{d_{min}}{D} \right\rfloor$$
$$\geq \lfloor k_j r_i \rfloor + 1$$
$$> k_j r_i, \; i = 1, 2, \ldots, n; j = 1, 2.$$

Therefore,

$$\left\lfloor k_j r_i + \sqrt[m]{UB_m} \right\rfloor \geq k_j r_i, \; i = 1, 2, \ldots, n; j = 1, 2. \tag{34.10}$$

Likewise,

$$\left\lceil k_j r_i - \sqrt[m]{B_m} \right\rceil = \lceil k_j r_i - 1 + r_{min} \rceil$$
$$= \left\lceil k_j r_i - 1 + \frac{d_{min}}{D} \right\rceil, \; i = 1, 2, \ldots, n; j = 1, 2.$$

If $k_j r_i$ is an integer,

$$\left\lceil k_j r_i - 1 + \frac{d_{min}}{D} \right\rceil = k_j r_i, \; i = 1, 2, \ldots, n; j = 1, 2,$$

and if $k_j r_i$ is not an integer,

$$\left\lceil k_j r_i - 1 + \frac{d_{min}}{D} \right\rceil = \left\lceil \lfloor k_j r_i \rfloor + (k_j r_i)_{frac} - 1 + \frac{d_{min}}{D} \right\rceil$$
$$\leq \lceil \lfloor k_j r_i \rfloor \rceil$$
$$< k_j r_i, \; i = 1, 2, \ldots, n; j = 1, 2.$$

Therefore,

$$\left\lceil k_j r_i - \sqrt[m]{B_m} \right\rceil \leq k_j r_i, \; i = 1, 2, \ldots, n; j = 1, 2. \tag{34.11}$$

Hence, using the inequalities (34.10) and (34.11),

$$\sum_{i=1}^{n} (\lfloor k_2 r_i + \sqrt[m]{B_m} \rfloor - \lceil (k_1 - 1) r_i - \sqrt[m]{B_m} \rceil) \geq \sum_{i=1}^{n} k_2 r_i - \sum_{i=1}^{n} (k_1 - 1) r_i$$
$$\geq k_2 - k_1 + 1.$$

And,

$$\sum_{i=1}^{n}(\left\lceil k_2 r_i - \sqrt[m]{B_m} \right\rceil - \left\lfloor (k_1-1)r_i + \sqrt[m]{B_m} \right\rfloor) \le \sum_{i=1}^{n} k_2 r_i - \sum_{i=1}^{n}(k_1-1)r_i$$
$$\le k_2 - k_1 + 1.$$

This shows that
$$UB_m = (1 - r_{min})^m$$

satisfies the two inequalities (34.6) and (34.7) proving to be the upper bound of a feasible solution for the BPRVP with the objective function

$$\max_{i,k} \left| x_{ik} - kr_i \right|^m, i = 1, 2, \ldots, n; \; k = 1, 2, \ldots, D.$$

It is noteworthy that for any instance $(d_1, d_2, \ldots, d_n)$ with $d_{min} = 1$, the upper bound,
$$UB_m = (1 - r_{min})^m,$$

is exactly the same as the upper bound,
$$UB_1 = 1 - \frac{1}{D},$$

established in Brauner and Crama (2004) for the BPRVP with the objective function F_1 and,

$$UB_m = (1 - \frac{1}{D})^m,$$

in Khadka (2012) for the objective function F_m, m being a positive integer.

It is clear that the upper bound
$$UB_m = (1 - r_{min})^m$$

is tight.

34.4 Concluding Remarks

We considered the bottleneck product rate variation problem. Literature shows that the problem has already been widely investigated. An exact solution procedure with complexity $O(DlogD)$ and a number of heuristics have been developed in the literature. A lower bound $1 - r_{max}$ for the problem with the objective function F_1, Steiner and Yeomans (1993), and $(1 - r_{max})^m$ for the objective function F_m, m being a positive integer, Khadka (2012), and an upper bound $1 - \frac{1}{D}$ for the objective function F_1,

Brauner and Crama (2004), and $(1 - \frac{1}{D})^m$, Khadka (2012), for the objective function F_m of a feasible solution for the BPRVP have been established. The lower bound is tight but is not the case for the upper bound. The upper bound could be improved. In this paper, we have proposed a tight upper bound $(1 - r_{min})^m$ of a feasible solution for the problem.

Acknowledgments The research of Shree Ram Khadka was supported by the European Commission in the framework of *Erasmus Mundus* and within the project *cLINK* and Kantipur Engineering College. The work of Till Becker has been supported by the Institutional Strategy of the University of Bremen, funded by the German Excellence Initiative.

References

Brauner N, Crama Y (2004) The maximum deviation just-in-time scheduling problem. Discrete Appl Math 134(1):25–50

Dhamala TN, Khadka SR (2009) A review on sequencing approaches for mixed-model just-in-time production system. Iran J Optim 1(3):266–290

Dhamala TN, Khadka SR, Lee MH (2010) A note on bottleneck product rate variation problem with square-deviation objective. Int J Oper Res 7(1):1–10

Huson M, Nanda D (1995) The impact of just-in-time manufacturing on firm performance in the US. J Oper Manage 12(3–4):297–310. http://www.sciencedirect.com/science/article/pii/027269639500011G

Khadka SR (2012) Mixed-model just-in-time sequencing problem

Miltenburg J (1989) Level schedules for mixed-model assembly lines in just-in-time production systems. Manage Sci 35(2):192–207

Monden Y (2011) Toyota production system: an integrated approach to just-in-time. CRC Press

Ōno T (1988) Toyota production system: beyond large-scale production. Productivity Press

Steiner G, Yeomans S (1993) Level schedules for mixed-model, just-in-time processes. Manage Sci 39(6):728–735

Chapter 35
Crowdsourcing in Logistics: An Evaluation Scheme

Matthias Klumpp

Abstract Crowdsourcing concepts are a major development in the context of social media and Industry 4.0 as well as a possible enabler for dynamic logistics processes and concepts. In the light of increasing sustainability, flexibility, and security expectations towards global supply chains, solutions for logistics based on crowdsourcing may be promising—but still the proof of concept is missing for logistics applications in practice. Combined with a theoretical framework for crowdsourcing in logistics, this contribution outlines two case studies for regarding e-mobility, knowledge management in logistics, as well as disaster applications. This could be a supportive approach in developing new business models for real-life intelligent transport systems including customers and multiple stakeholder perspectives.

Keywords Intelligent transport systems · Crowdsourcing · Logistics e-mobility · Logistics qualification · Logistics disaster relief management

35.1 Introduction

Crowdsourcing applications in logistics are increasingly discussed in the context of smartphone and Internet of things dissemination success (e.g., Berrang-Ford and Garton 2013; Bowen 2012). Whereas the "natural" application areas in information technology are already established (Gefen et al. 2015), the application frontier is now moving in the direction of further—especially service—industries such as the health and banking industries (i.e., Prpić 2015; Smith and Merchant 2015; Lisha and Hongjie 2015).

M. Klumpp (✉)
University of Twente, Enschede, The Netherlands
e-mail: matthias.klumpp@fom.de

M. Klumpp
FOM University of Applied Sciences, Essen, Germany

It can be assumed that this technology application can also improve logistics concepts and processes to become more efficient and effective in the context of major new developments and requirements, e.g., sustainability, flexibility, security (e.g., Barbier et al. 2012; Pitt et al. 2011; Heipke 2010; Munro 2013) and especially facing the challenge of demographic change and the increasingly limited access towards human resources and knowledge for logistics operators (e.g., Zijm and Douma 2012; Fugate et al. 2012; Barbier et al. 2012; Woxenius and Sjöstedt 2003).

The *research question* for this contribution is as follows:

- RQ: "Can the assumption be sustained that there will be significant and suitable application areas for crowdsourcing in logistics?"

The specific research method applied in this contribution is the case study method in order to describe specific logistics situations and value added by crowdsourcing implementations in logistics. This may trigger further discussion and also further innovative first-mover projects in these directions. Therefore, the contribution is structured as follows: Sect. 35.2 will provide a theoretical framework regarding the basic concept and technology requirements of crowdsourcing as well as general requirements for logistics applications. In Sect. 35.3, a case study regarding specific applications for vehicles and especially e-vehicles is outlined. Section 35.4 describes theoretical further applications within the field of logistics education and knowledge management, whereas Sect. 35.5 is giving an innovation outlook for the field of humanitarian logistics. The last section is providing a short outlook regarding further developments and research question.

35.2 Conceptual Framework

Based on modern information and communication technologies and their broad dissemination of multiple users the idea of crowdsourcing appeals to the *coordinated but autonomous action* of these user groups in contexts of software development (open source projects), innovation, and other pluralistic value-creation processes (cp. Reichwald and Piller 2006; Schulze and Hoegl 2008; Schulz 2009; Whitla 2009; Schenk and Guittard 2009; Estellés-Arolas and González-Ladrón-de-Guevara 2012). Usually higher innovation rates and lower costs are connected to these concepts as the advantages of crowdsourcing are manifold. But especially for logistics the characteristics of *long distance and distributed search* for contributors, data or solutions (Afuah and Tucci 2012) may be interesting due to the already global and distributed nature of today's supply chains.

Logistics *specific characteristics* have to be taken into account for crowdsourcing applications, namely

(a) flexibility,
(b) security,
(c) sustainability,

(d) scalability, and[1]
(e) economic viability.

A discussion regarding these criteria may highlight the suitability of crowdsourcing applications from a theoretical point of view

(a) As logistics is providing just-in-time availability of goods, it is very important due to the real-world nature of transportation processes and probable physical interruptions—i.e., due to traffic situations, weather, or security incidents—to have a certain degree of *flexibility* (Bai and Sarkis 2013; Naim et al. 2010; Tofighi et al. 2016). Crowdsourcing adheres to this by incorporating a very decentralized approach based on individual and small-scale contributors, usually private individuals—this makes the preferred setup with private smartphone applications for example very flexible per definition.
(b) *Security* concerns are an important consideration within logistics processes (Lam et al. 2015; Govindan and Chaudhuri 2016). This may be harder on crowdsourcing applications due to the open nature of contributors' networks and numbers, especially the dynamic nature of contributors. Therefore, a *minimal feasibility hurdle* has to be observed in terms of admittance and checking of participants as, i.e., lately the crowdsourcing peer-to-peer platform *Uber* has experienced regarding the participating drivers (security check, social security integration; BBC 2015).
(c) Furthermore, sustainability considerations are also of high importance to logistics companies and their customers (Colicchia et al. 2013; Ugarte et al. 2015). Therein, especially with the *social* dimension as for example with *data protection* and labor standards specific requirements and caution have to be maintained. This can be combined with the security issue mentioned above regarding participation checks and security assessments for data protection. Regarding *ecological* efficiency, crowdsourcing peer-to-peer applications are usually rated very positive due to the low-invest and low-resource concept of using, i.e., private assets. This avoids major resource waste due to idle assets on a global or regional scale.
(d) A specific logistics requirement—but sure nothing new either on Internet e-commerce applications—is the question of *scalability* (Knapen et al. 2015; Choi et al. 2015). Very volatile as well as high numbers of users and providers have to be incorporated without process failures into crowdsourcing networks. Usually this is not a major problem due to the easy-to-use as well as easy-to-build smartphone applications used for crowdsourcing.
(e) For the last requirement of *economic viability*, one can, i.e., historically compare crowdsourcing applications to the implementation hurdles for RFID due to high unit and production as well as implementation costs (Bunduchi et al. 2011; Véronneau and Roy 2009). Therefore, this may be a severe hurdle

---

[1]Suitability for large-scale applications e.g. in global supply chains and for millions of standardized processes and shipments in logistics settings.

for crowdsourcing applications in logistics as the implementation costs, i.e., in changing standard transportation and logistics processes may be quite high.

Altogether, the theoretical requirements posed by logistics applications as outlined are *not* to be evaluated as prohibitive for crowdsourcing applications. Some requirements like security and sustainability (data protection) issues as well as the business investment case have to be recognized and addressed professionally—but those should not completely prevent this concept from being applied in logistics.

35.3 Case Study E-Mobility

Electric mobility is one of the major new developments in personal as well as cargo transport in the context of *green logistics* (Davis and Figliozzi 2013; Augenstei 2015). This is mainly due to reductions in raw energy consumption and (local) emissions—depending on the production scheme of the used electricity—as well as significant noise reduction especially in urban areas (Pallas et al. 2014; Hurley et al. 2012). In this context of *electric* vehicles used in logistics applications for an efficient use of these new propulsion technologies for sustainability and also long-term cost reasons are in high demand. An example for private cars—but also feasible for trucks—is the "SignalGuru" application developed at MIT for a communication among smartphones in cars identifying green waves and optimal driving strategies in cities in order to save energy (Koukoumidis et al. 2011).

The solution also outlined in Fig. 35.1 is based on the communication and interaction of several smartphones in different cars recognizing signals (red, green) and rhythms. Especially for e-mobility this can be important and lower logistics costs as especially the starting phases during driving are consuming most of the energy in city delivery trucks. If "flow driving" with a reduced number of traffic light stops can be implemented, the total range of electric trucks would increase significantly, fostering their application in urban distribution and last-mile logistics.

Additional further implementation applications for crowdsourcing could be the question of ad hoc peer-to-peer delivery networks (Roumboutsos 2014; Fatnassi et al. 2015) within city transportation as, i.e., applied by Uber for private passenger transport. If this would be applied for cargo delivery—as Amazon is planning to do (Reuters 2015)—a combination with (private) electric vehicles would provide a very sustainable[2] as well as efficient solution for future urban logistics.

---

[2]Again, depending on the specific production mix of electricity in the region and country of application.

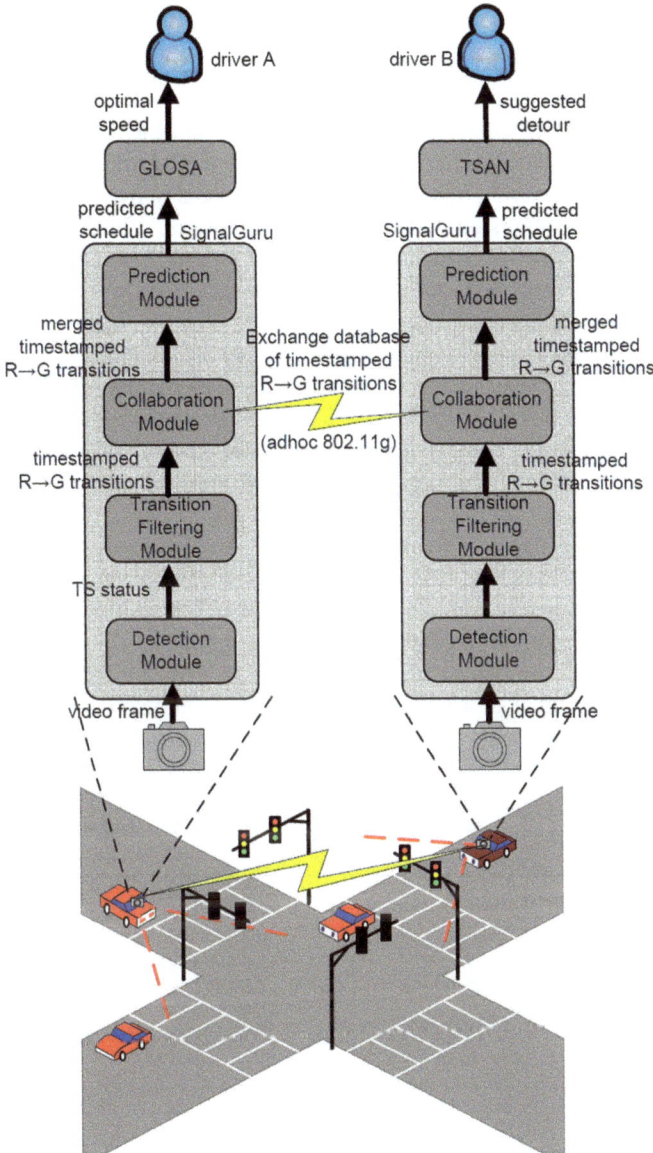

Fig. 35.1 SignalGuru interactive model framework (Koukoumidis et al. 2011)

35.4 Application Area Logistics Training and Knowledge Management

Creation of knowledge and cataloging are the most valuable things the crowd can do; because the crowd is *fast, efficient, and effective* (e.g., Lithium 2012). Consider Wikipedia: The roughly 30 million articles of Wikipedia in over 280 languages are designed in multi-authorship of free working volunteers, written and continually corrected jointly by the principle of collaborative writing, expanded, and updated. Web 2.0 and 3.0 technologies are transforming the landscape of learning and enable new learning environments.

To assimilate many small contributions—known as crowdsourcing—can be an aspect of knowledge management (KM) and open learning platforms (e.g., Corneli and Mikroyannidis 2010). They described crowdsourcing through the collaborative KM theory of "shared context in motion," called basho in Japanese, formally known as "ba". Nonaka and Toyama developed the ba theory of KM in 2003 (Nonaka and Toyama 2003). They suggest that knowledge is created as people interact in a shared context. Figure 35.2 depicts repeated phases of socialization, externalization, combination, and internalization (SECI) (Nonaka et al. 2000).

In view of the rapid technological progress, the importance of lifelong learning will increase in the future. In combination with the imminent demographic change, knowledge transfer will be a fundamental part; to gain the knowledge and for further expand. In order to accommodate logistics specific knowledge the form of the above-mentioned wiki is useful. It is conceivable to establish knowledge

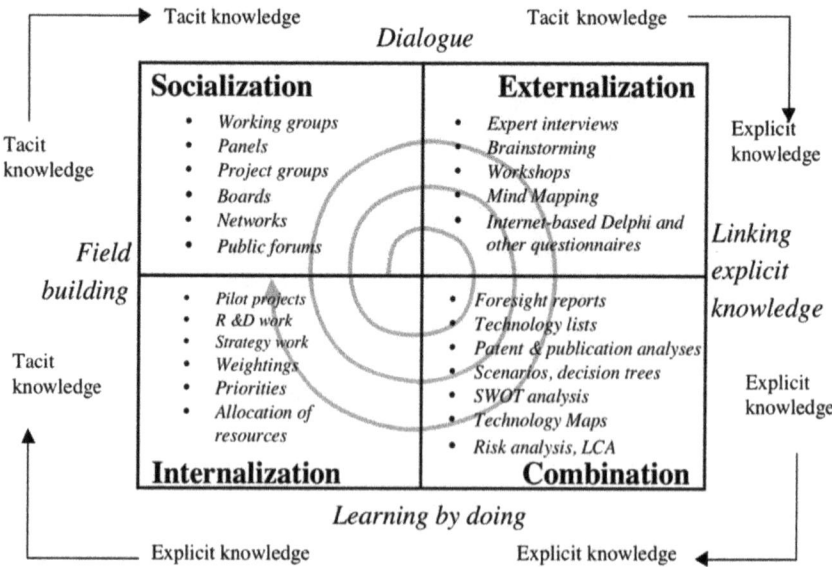

Fig. 35.2 SECI model by Eerola and Jorgensen (Eerola and Joergensen 2002), p. 12

management for specific topics, for example, to establish a logistics wiki. For this, some things have to be clarified, for example, how to ensure the quality of this open platform? It regulates the possibility here to write something too strong, it will be the bottleneck here. The hurdle may write, a post should not be too large. On the other hand, the question is to clarify who administrates this wiki, who will be the hosting institution. The success of Wikipedia has demonstrated that it is quite possible to establish such applications with a profound and far-reaching impact.

35.5 Outlook: Humanitarian Logistics

For the application area imminent *disaster relief management* also the mobile and smartphone data connections as crowdsourcing element could be used due to the specific situations and needs in humanitarian logistics

(a) In many cases there is no established fixed-line communications infrastructure as, i.e., after earthquakes only mobile phone connections are available if any.
(b) Furthermore, the short text and information messages in app communications can be a lighter load on communications infrastructure compared to other coordination mechanisms requiring spoken language among individuals.
(c) Many people may be available as supporters and help for humanitarian processes, but usually these people have low experience and competence levels regarding logistics processes—in these cases, easy-to-use smartphone applications may alleviate the situation significantly.
(d) A high level of detailed information is necessary in order to coordinate the required processes of disaster relief in the imminent time after such an event—i.e., rescue, support, material handling as well as supplied transport processes have to be based on hands-on information on the ground near the disaster site.
(e) In many cases the exact region and geographical dimensions of impacts of disasters are unknown—a field where especially the multiuser help of, i.e., smartphone users throughout the region may help as those are distributed evenly and accidentally in the whole area or country.

Some indicative *possible application areas* may be the following, necessitating further details in future discussions as (cp., i.e., Gao et al. 2011; Özdamar and Ertem 2015):

(i) Health measurement and monitoring via smartphone applications in order to organize and steer optimized medical relief transports (i.e., detecting contagions like cholera in disaster areas earlier);
(ii) "Snapshot" evaluation of (smart) phone networks before and after disaster timeframes in order to locate missing persons by "missing mobile phone signatures";

(iii) Using smartphone QR scanners instead of fixed barcode or RFID scanners for logistics operations in disaster areas by a standardized application (app).

For the first application option, an application with a simple setup which could potentially improve overall transport performance in disaster and humanitarian situations as from the "aggregated" data of individual users and humans a automated information and supply flow (e.g. of medicine items) could be organized by logistics systems. In essence, automated distribution and tour planning tools could supply the necessary medicine to the affected people on the basis of online available and real-time health information provided and aggregated by such systems.

35.6 Conclusions

The outlined case study regarding electric vehicles and city traffic flow as well as the two further application areas may show the important potential of crowdsourcing solutions and concepts for cooperative intelligent transport systems. Therefore, the initial research question regarding "Can the assumption be sustained that there will be significant and suitable application areas for crowdsourcing in logistics?" has to be answered *positive*. Yes, there will be future application areas in logistics, especially in *special fields* as outlined, i.e., city and last-mile logistics (cp. also O'Reilly 2014; Crainic and Montreuil 2015).

Further research is needed and suggested here in order to outline the further specific requirements and application potentials of such crowdsourcing apps towards the different field of logistics and transport applications. For example such research questions may entail but are not restricted to

- Which factors and restrictions apply to an investment calculation regarding crowdsourcing applications in logistics?
- How can mass payment systems and processes be established in order to support the very detailed and sometimes low-value activities in logistics?
- Which interaction areas can be established for e-commerce and city logistics applications as "natural habitat" areas of crowdsourcing?
- How can possible contributors for logistics crowdsourcing concepts be identified and motivated?
- By which concepts and applications can the important procedures of data and privacy protection be enabled in crowdsourcing applications in logistics?

This shows that the field of crowdsourcing applications in logistics is characterized by a high importance due to expected applications areas as well as still open and necessary research questions.

References

Afuah A, Tucci CL (2012) Crowdsourcing as a solution to distant search. Acad Manag Rev 37(3):355–375

Augenstein K (2015) Analysing the potential for sustainable e-mobility—the case of Germany. Environ Innov Societal Transitions 14:101–115

Bai C, Sarkis J (2013) Flexibility in reverse logistics: a framework and evaluation approach. J Clean Prod 47:306–318

Barbier G, Zafarani R, Gao H, Fung G, Liu H (2012) Maximizing benefits from crowdsourced data. Comput Math Organ Theory 18:257–279

BBC (2015) Uber driver background checks 'not good enough'. Online available: http://www.bbc.com/news/technology-34002051. Accessed 04 Jan 2016

Berrang-Ford L, Garton K (2013) Expert knowledge sourcing for public health surveillance: national tsetse mapping in Uganda. Soc Sci Med 91:246–255

Bowen JT (2012) A spatial analysis of FedEx and UPS: hubs, spokes and network structure. J Transp Geogr 24:419–431

Bunduchi R, Weisshaar C, Smart AU (2011) Mapping the benefits and costs associated with process innovation: The case of RFID adoption. Technovation 31:505–521

Choi YJ, Lee SH, Kim JH, Kim YJ, Pak HG, Moon GY, Ra JH, Jung Y (2015) The method to secure scalability and high density in cloud data-center. Inf Syst 48:274–278

Colicchia C, Marchet G, Melacini M, Perotti S (2013) Building environmental sustainability: empirical evidence from Logistics Service Providers. J Clean Prod 59:197–209

Corneli J, Mikroyannidis A (2010) Crowdsourcing Education: A Collaborative Knowledge Managemet Approach, an extended abstract centering on a SECI analysis of PlanetMath. http://metameso.org/~joe/docs/pm-seci.pdf. Accessed 04 Jan 2016

Crainic TG, Montreuil B (2015) Physical Internet Enabled Interconnected City Logistics. CIRRELT Research Paper No 2015–13, Montreal

Davis BA, Figliozzi MA (2013) A methodology to evaluate the competitiveness of electric delivery trucks. Transp Res Part E 49:8–23

Eerola A, Joergensen BH (2002) Technology Foresight in the Nordic Countries, Report to the Nordic Industrial Fund, Center for Innovation and Commercial Development. Risoe-R-1362 (EN), ISBN:87-550-310992 & ISBN: 87-550-3110-2. http://www.risoe.dk/rispubl/sys/syspdf/ris-r-1362.pdf. Accessed 04 Jan 2016

Estellés-Arolas E, González-Ladrón-de-Guevara F (2012) Towards an integrated crowdsourcing definition. J Inf Sci 38(2):189–200. doi:10.1177/0165551512437638

Fatnassi E, Chaouachi J, Klibi W (2015) Planning and operating a shared goods and passengers on-demand rapid transit system for sustainable city-logistics. Transp Res Part B 81:440–460

Fugate BS, Autry CW, Davis-Sramek B, Germain RN (2012) Does knowledge management facilitate logistics-base differentiation? The effect of global manufacturing reach. Int J Prod Econ 139:496–509

Gao H, Barbier G, Goolsby R (2011) Harnessing the crowdsourcing power of social media for disaster relief. IEEE Intell Syst 26(3):10–14

Gefen D, Gefen G, Carmel E (2015) How project description length and expected duration affect bidding and project success in crowdsourcing software development. J Syst Softw (in press). doi:10.1016/j.jss.2015.03.039

Govindan K, Chaudhuri A (2016) Interrelationships of risks faced by third party logistics service providers: a DEMATEL based approach. Transp Res Part E (in press). doi:http://dx.doi.org/10.1016/j.tre.2015.11.010

Heipke C (2010) Crowdsourcing geospital data. ISPRS J Photogrammetry Remote Sens 65:550–557

Hurley K, Marshall J, Hogan K, Wells R (2012) A comparison of productivity and physical demands during parcel delivery using a standard and a prototype electric courier truck. Int J Ind Ergon 42:384–391

Knapen L, Hartman IB-A, Keren D, Yasar A, Choa S, Bellemans T, Janssens D, Wets G (2015) Scalability issues in optimal assignment for carpooling. J Comput Syst Sci 81:568–584

Koukoumidis E, Peh L-S, Martonosi M (2011) SignalGuru: leveraging mobile phones for collaborative traffic signal schedule advisory

Lam HY, Choy KL, Ho GTS, Cheng SWY, Lee CKM (2015) A knowledge-based logistics operations planning system for mitigating risk in warehouse order fulfillment. Int J Prod Econ 170:763–779

Lisha Z, Hongjie L (2015) Business model innovation of logistics enterprise based on the pattern of banking business. In: International conference on logistics, informatics and service sciences (LISS). doi:10.1109/LISS.2015.7369787

Lithium (2012) From crowdsourcing to knowledge management. http://www.lithium.com/pdfs/whitepapers/Lithium-From-Crowdsourcing-to-Knowledge-Management_zl3GHD8m.pdf. Accessed 04 Jan 2016

Munro R (2013) Crowdsourcing and the crisis-affected community: lessons learned and looking forward from Mission 4636. Inf Retrieval 16:210–266

Naim N, Aryee G, Potter A (2010) Determining a logistics provider's flexibility capability. Int J Prod Econ 127:39–45

Nonaka I, Toyama R (2003) The knowledge-creating theory revisited: knowledge creation as a synthesizing process. Knowl Manage Res Pract 1(1):2–10

Nonaka I, Toyama R, Konno N (2000) SECI, Ba and Leadership: a unified model of dynamic knowledge creation. Long Range Plan 33(1):5–34

Özdamar L, Ertem MA (2015) Models, solutions and enabling technologies in humanitarian logistics. Eur J Oper Res 244(1):55–65

O'Reilly J (2014) Same-day delivery: the amazing race—the same-day delivery race puts crowdsourcing, bicycles, and delivery vans on the same team. Inbound Logistics 34(2):63–65

Pallas MA, Chatagnon R, Lelong J (2014) Noise emission assessment of a hybrid electric mid-size truck. Appl Acoust 76:280–290

Pitt LF, Parent M, Junglas I, Chan A, Spyropoulou S (2011) Integration the smartphone into a sound environmental information systems strategy: Principles, practices and a research agenda. J Strateg Inf Syst 20:27–37

Prpić J (2015) Health care crowds: collective intelligence in public health. collective intelligence 2015. Center for the Study of Complex Systems, University of Michigan

Reichwald R, Piller F (2006) Interaktive Wertschöpfung – Open Innovation. Individualisierung und neue Formen der Arbeitsteilung, Gabler, Wiesbaden

Reuters (2015) Amazon develops new delivery app: report (Tuesday, June 16, 2015). http://www.reuters.com/video/2015/06/16/amazon-develops-new-delivery-app-report?videoId=364608960. Accessed 04 Jan 2016

Roumboutsos A, Kapros S, Vanelslander T (2014) Green city logistics: systems of Innovation to assess the potential of E-vehicles. Res Transp Bus Manage 11:43–52

Schenk E, Guittard C (2009) Crowdsourcing: what can be outsourced to the crowd, and why? https://hal.archives-ouvertes.fr/file/index/docid/439256/filename/Crowdsourcing_eng.pdf. Accessed 04 Jan 2016

Schulz C (2009) Organizing user communities for innovation management. Gabler, Wiesbaden

Schulze A, Hoegl M (2008) Organizational knowledge creation and the generation of new product ideas: a behavioral approach. Res Policy 37(10):1742–1750

Smith RJ, Merchant RM (2015) Harnessing the crowd to accelerate molecular medicine research. Trends Mol Med 21(7):403–405

Tofighi S, Torabi SA, Mansouri SA (2016) Humanitarian logistics network design under mixed uncertainty. Eur J Oper Res 250:239–250

Ugarte GM, Golden JS, Dooley KJ (2015) Lean versus green: The impact of lean logistics on greenhouse gas emissions in consumer goods supply chains. J Purchasing Supply Manage (in press). doi:http://dx.doi.org/10.1016/j.pursup.2015.09.002i

Véronneau S, Roy J (2009) RFID benefits, costs, and possibilities: The economical analysis of RFID deployment in a cruise corporation global service supply chain. Int J Prod Econ 122:692–702

Whitla P (2009) Crowdsourcing and its application in marketing activities. Contemp Manage Res 5(1). doi:http://dx.doi.org/10.7903/cmr.1145

Woxenius J, Sjöstedt L (2003) Logistics trends and their impact on European combined transport —services, traffic and industrial organization. Log Man 5:25–36

Zijm H, Douma A (2012) Logistics: more than transport. In: Weijers S, Dullaert W (eds) Proceedings of the 2012 freight logistics seminar. Venlo, pp 395–404

Chapter 36
Modularization of Logistics Services—An Investigation of the Status Quo

Aleksander Lubarski and Jens Pöppelbuß

Abstract Providers of logistics services are confronted with multivariant customer needs and high competitive pressures. The concept of service modularization has been proposed as a strategy to allow for both standardization and individualization at the same time so that customer-specific demands can be fulfilled efficiently. However, the discussions on applying the concept of modularization to the service sector have largely remained academic. Little empirical evidence is available on how service modularization strategies can be applied. In this paper, we synthesize the status quo of service modularization in logistics by reviewing existing contributions from the literature and conducting expert interviews. The contribution is an overview of the current state of modularization in logistics and an outlook toward the potential for future adoption of modular strategies in this industry.

Keywords Service modularization · Logistics services · Expert interviews

36.1 Introduction

In advanced economies, the service sector has continuously gained in importance during the last decades. In Germany, especially industrial services like logistics create a high proportion of the economic output (Eickelpasch 2012). Both, the service sector in general and the specific area of industrial services are increasingly characterized by strong competition and cost pressures of providers, requiring an increasing degree of standardization and efficiency in the sales and provision of services (Böttcher and Klingner 2011). Simultaneously, the heterogeneity and complexity of customer demands are growing (Bask et al. 2011b) as customers are looking for holistic solutions for their individual problems. Hence, the

A. Lubarski (✉) · J. Pöppelbuß
University of Bremen, Bremen, Germany
e-mail: lubarski@is.uni-bremen.de

J. Pöppelbuß
e-mail: jepo@is.uni-bremen.de

standardization of services is only possible as far as the flexibility to meet individual customer demands is not diminished (Burr 2005; Pekkarinen and Ulkuniemi 2008).

Modularization has been proposed as a strategy to allow for both standardization and individualization at the same time. This strategy has been widely adopted in manufacturing, especially in the automotive industry. Recently, first attempts have also been made to transfer this strategy to the service sector (Dörbecker and Böhmann 2013) and to the area of logistics services in particular (Bask et al. 2011a). However, the discussion on using modularization concepts and methods in the service sector has largely remained academic so far, while their practical applicability is still in question. Little empirical evidence is available on how service modularization strategies can be applied or are possibly already applied in practice. One of the few studies points to possible growth of operating profit by 30 % in the field of legal services (Giannikis et al. 2015). Some first case studies have also been undertaken in logistics, identifying modularity as a helpful mindset that can improve the value cocreation of providers and customers (Bask et al. 2011a; Hautamäki et al. 2015). We agree with Dörbecker and Böhmann (2013) that more of this kind of research is needed to examine whether the concept of modularity can be actually useful for practical purposes in the service sector in general and especially logistics.

The objective of this article is therefore to synthesize the status quo of service modularization in logistics. We review existing contributions from the literature and further conduct a small qualitative empirical study using expert interviews. The contribution is an overview of the current state of modularization in logistics and an outlook for the potential future adoption of modularization practices in this industry.

36.2 Research Background

36.2.1 Concept of Modularity

The concept of modularity is considered as a possible solution to enable the trade-off between growing customer requirements for individualization and a company's necessity to standardize products or processes for cost efficiency reasons. Modularization can be seen as a process of building a complex system from smaller parts that can be designed and improved independently, yet function as a whole (Baldwin and Clarc 1997). The idea of loose coupling results in interchangeability and flexibility, as long as the interfaces between separate modules are well-defined and standardized and a clear one-to-one matching of modules and functions exists (Arnheiter and Harren 2005). Essential benefits of providing modularized products and services include, among others, the reuse of components for different products (Carlborg and Kindström 2014), faster development processes

(Böttcher and Klingner 2011), increased variety of products (Yang and Shan 2009), economies of scope and scale (Tuunanen et al. 2012) as well as cost-efficient operations (Bask et al. 2011b). Furthermore, modular product structures enable product configuration and the use of so-called Configure Price Quote (CPQ) software support, thereby improving the efficiency of quotation processes in sales (Hvam et al. 2006). Hence, benefits of modularity can be expected for both the sales process and the actual operations, i.e., the production and provision processes of products and services.

Despite its popularity and evident positive impacts, the concept of modularity has been adopted almost only in the context of products (Voss and Hsuan 2009). However, since it is an abstract architecture concept, it can be applied to different dimensions of value creation. For example, Bask et al. (2010) discuss the impacts of modularization and differentiate between (i) product modularity, (ii) modularity of production process, (iii) modularity of organization and supply chain management, and (iv) modularity of services.

36.2.2 Service Modularization

Being still a recent topic in the academic discussion, service modularization took over most of the upper mentioned architecture principles from product modularization. Some service providers have begun to imitate the automotive industry by introducing modular services (Yang and Shan 2009). However, according to Bask et al. (2010), the modularization of services has more in common with process modularity than with product modularity. Similarly, de Blok et al. (2010) point to a difference in the provision of modular products and services, while it is common practice to postpone the finalization of modular product assembly into distribution centers or even customers in manufacturing; early customer participation in the value creation is essential for service modularization in order to achieve customization. Finally, Pekkarinen and Ulkuniemi (2008) argue that due to the intangibility of services, their process character as well as the interdependency and close interaction between supplier and customers, the modularization of services may be not that intuitive and may result in implementation problems.

Carlborg and Kindström (2014) analyze different types of service process and recommend different modular strategies depending on the types. They present their typology as a matrix of two dimensions. The first dimension distinguishes service processes as being either rigid or fluid. Rigid service processes are characterized by a high level of formalization, standardization, and low task variety, and a relatively low level of information exchange between service provider and customer. On the other hand, fluid service processes require a high level of technical skill and information exchange and exhibit a high task variety. While the activities of rigid service processes are mainly directed to the customer's possessions (e.g., equipment or material), the activities of fluid service processes are mainly directed toward the customer processes. The second dimension distinguishes between the modes of

		role of customer	
		passive	*active*
service process	*rigid*	**Type 1** • Key issue: internal efficiency through a high degree of standardized modules (sub-processes) • High level of process formalization • Primarily provider-driven resources • Low level of technical skills • Modular strategy: **Bundled**, based on standardized modules.	**Type 2** • Key issue: Modularizing the customer's own experience (esp. in customer self-service) • High level of process formalization • Customer-driven resources • Low level of technical skills • Modular strategy: **Pre-defined bundled offerings**, customer chooses package.
	fluid	**Type 3** • Key issue: Interaction of modules • Provider-driven resources • High level of technical skills • Modular strategy: **Flexible bundling** based on a wide range of available service modules and supporting resources in order to respond to different customer needs.	**Type 4** • Key issue: Interaction of modules and supporting resources • Provider- and customer-driven resources • High level of technical skills • Modular strategy: **Unbundled modules** that can be combined flexibly during the coordinated, interactive service process with the customer.

Fig. 36.1 Service types, associated resources, and modular strategies (Carlborg and Kindström 2014)

customer participation in the service process, which can vary from passive (i.e., employees of the service firm produce the service) or active (i.e., customer action is required). The four cells of the matrix (see Fig. 36.1) represent four different service types that are each characterized by different key issues in modularization, supporting resources, and modular strategies (Carlborg and Kindström 2014).

36.2.3 Modularization in Logistics

Within the service sector, the area of logistics services is considered to be one of relatively high potential to benefit from the modularization concept. This is due to the complexity of the supply chain: the large number of relatively similar services and the increased use of IT for the purpose of automation of service processes (Bask et al. 2011a). The increased scale of operations and higher service specialization leads to the creation of new business models, so that third-party logistic (3PL) service companies become increasingly popular. With the help of modularity, it is made possible to integrate various functions within a company in order to substantially simplify the supply network (Arnheiter and Harren 2005), while at the same time make it more responsive to satisfy customized needs and provide service variation without higher production and inventory costs (Tu et al. 2004). In this way, logistic operations are typically provided by a large network of service providers (Hsuan Mikkola and Skjøtt-Larsen 2004), so that the final logistics value chain is separated into modules, each of which is assigned to the specific module supplier (Voordijk et al. 2006). It is believed that applying modular logic is a cost-effective way to design logistic services (Lin et al. 2010) as intermodal transport chains consists of modular elements by their very nature (Bask 2001).

Most academic papers on modularization in logistics have so far focused on this perspective, i.e., the organizational modularization of the supply chain. The actual modularization of logistic services largely remains a theoretical discussion, with no clear advice on what practical potential it actually offers. Some first anecdotal evidence, however, points to the potentials of modularity. Bask et al. (2011a) claim that many services of Finnish logistic companies were in fact based on mainly repetitive and routine standard processes that could be divided into separate reusable modules. The main finding of their study is that most of the differences are actually created in the customer interfaces, indicating that with the help of modularity, back office processes such as actual transportation can achieve considerable economies of scale, while simultaneously simplifying and enriching the customer experience via an actual customization process. In their recent study on logistics in healthcare, Hautamäki et al. (2015) come to the conclusion that modular logistics services are beneficial to customers in both monetary (increasing process and space efficiency at the customer's site) and nonmonetary ways (increasing staff's well-being and decreasing stress).

36.3 Data Collection and Analysis

In this study, we employed methods of qualitative research. Qualitative research is especially suitable when the research area is still emerging and not controllable by the investigators (Yin 2013). Our field of service modularization is such an area. We applied expert interviews as our approach to data collection. These interviews followed a semi-structured interview guideline. The guiding questions addressed the sales process of logistics firms as well as the characteristics of the actual logistic service processes. In line with the subject area of this research, interviewees from organizations that provide logistics services were selected. The interviews were conducted in September and October 2015 and lasted from 20 to 45 min each. All interview partner requested to remain anonymous. We used MAXQDA 12 as software support for analyzing the interview data.

The four organizations and the range of logistic services provided by them can be described as follows. LOCAL is a medium-sized logistics company with a more than 100 years of tradition in the market. They offer a wide range of services including transport, warehousing, heavy goods logistics, and removals. GLOBAL is a subunit of a large international company that offers air and sea freight services. HARBOR provides services related to the loading, buffering, and maintenance of freight containers as well as export packaging. Their locations are mainly close to harbors. PRODUCTION is a subunit within a large logistics company that mainly provides production logistics services including warehousing, just-in-time and just-in-sequence services, preassembly, supplier management, as well as handling and cleaning of packaging and empties. Table 36.1 gives an overview about the interviewees' organizations and their positions in these organizations.

Table 36.1 Organizations and interviewees

Organization	Number of employees	Turnover per year	Position of interviewee
Local	~100	~15 million	Sales manager
Global	~4.000	~1,500 million	Senior sales manager
Harbor	~250	~25 million	Chief executive
Production	~3.500	~250 million	Director sales

36.4 Results

36.4.1 Status Quo of Modularization in Service Sales

All interviewees point to the challenge of addressing highly individual customer demands. They also describe a high level of competitive and pricing pressure in the logistics industry. Sales employees are confronted with a high number of request for quotations (RFQs) and invitations to tender that cause a high and ever-increasing workload, even for a growing number of low-priced and rather standardized services (like, e.g., road freight). The Senior Sales Manager at GLOBAL correspondingly complains about an *"enormous RFQ mentality"* leading to the preparation of numerous quotations from which only very few are actually won. The Sales Manager at LOCAL distinguishes between two main customer groups, which are either focused on the lowest price or on well working and error-free processes.

None of the four organizations has a standardized catalog of service modules that was publicly accessible to (potential) customers (e.g., in terms of an online configurator). The Director Sales at PRODUCTION sees a key problem of such a configuration approach in the many parameters that actually determine the pricing process so that resulting bundles of service modules would always be inaccurate. He points to the problem that even within one and the same country like Germany prices can vary much, e.g., due to different levels of wages and rents. A further important aspect mentioned by LOCAL and GLOBAL is that RFQs and invitations to tender usually go along with structures for quotations that are predetermined by the customer and that the service providers have to adhere to. Consequently, they do not put much effort in defining service modules themselves, as they mostly have to adjust their offerings and description in accordance to the RFQ. The Sales Manager at LOCAL refers to what he calls additional services like data exchange interfaces and tracking and tracing that can be additionally booked, but he refrains from calling that an actual modularization approach.

The Director Sales at PRODUCTION emphasizes that the sales process of logistics services requires specialists with a high level of expertise and experience. Quotations for their contract logistics services require a considerable amount of preparation as the target processes are planned and calculated in detail using the Methods-Time Measurement (MTM) approach. This method in turn requires the sales personnel to be trained. It is even possible that processes at the customer's site are surveyed and documented or that joint workshops take place. Furthermore, he emphasizes that the sales personnel must be able to fully understand the actual demand of the customer, which has to be perfectly reflected in the quotation document in the end. Regarding the expertise required in sales, the approach of HARBOR is specific as they refrain from having dedicated sales personnel. Sales activities are performed by the personnel that also provides the actual logistics services in order to *"provide one face to the customer."* The Chief Executive at HARBOR is confident that this ensures that sales activities are performed by knowledgeable people.

As for the software support for the quotation process, all respondents exclusively rely on Microsoft Office tools like Excel spreadsheets, PowerPoint presentations, and Word text documents. They also mention that the spreadsheets for price calculations typically exhibit a high complexity. Specialized CPQ software is not used. The Director Sales at PRODUCTION mentions that quotations can be 80–120 pages long. The reuse of text is focused on blocks like sustainability and quality management that hardly change from quotation to the other. The Chief Executive at HARBOR also considers their quotations to be very *"text heavy"* as the offerings have to be explained in detail. In order to reduce the interaction required between the sales personnel and customers, GLOBAL provides steady customers with Excel-based calculation tools, so that they can calculate freight rates themselves.

36.4.2 Status Quo of Modularization in Service Provision

Several interviewees have reported on efforts to standardize processes or process parts within the logistics value chain and that they develop so-called process modules. The Chief Executive at HARBOR considers their process of container loading as a process that is standardized in all its details. At LOCAL, the Sales Manager considers processes like goods receiving, warehousing, inventory management, and outgoing goods as standardized processes. Exemptions only occur when handling very large and heavy pieces. The services that HARBOR provides can also be understood as a module for its own as they are typically integrated as one step between further logistic processes. The Director Sales at PRODUCTION for instance refers to their "goods receiving module" that defines standard services and a standardized set and order of activities. However, at the same time, he refers to the high specificity of customer demands in the 3PL business that puts narrow limits to standardization efforts, e.g., as the systematics of picking and packing vary strongly from customer to customer.

Reflecting on the interview data through the lens of the service process typology by Carlborg and Kindström (2014), a general classification of logistics services is not easy. On the one hand, complex 3PL services are certainly directed toward the customer process and require high levels of technical skills and information exchange between customer and provider. This is highlighted by LOCAL, who report on lead times of up to 2 years for establishing a contract logistics arrangements with customers. More simple freight services, on the other hand, can be considered rather rigid as they are highly standardized and exhibit a low task variety as GLOBAL, LOCAL, and HARBOR describe. They are directed toward the customer's possessions, i.e., the transportation of these possessions from one place to the other. Looking at the role of the customer as the other dimension of Carlborg and Kindström's (2014) framework, this can be considered as rather passive as the intention of customers typically is to outsource the logistics activities. As a result, logistics services can be considered as either Type 1 (passive/rigid) or Type 3 (passive/fluid) services (cf. Fig. 36.1). This classification of the logistics services offered by the interviewed organizations would suggest that they follow either a "bundled" or a "flexible bundling" modular strategy (Carlborg and Kindström 2014). The first strategy focuses on the standardization of subprocesses. The expected benefits of this strategy is to "achieve a degree of (cost-efficient) customization through the combination of different process modules" (Carlborg and Kindström 2014, p. 317). The latter strategy requires "the provider to have a wide range of service modules available (and the supporting resources) in order to respond to different customer needs" (Carlborg and Kindström 2014, p. 320). The adoption of both strategies is well observable from our interview data.

36.5 Discussion and Conclusions

Concluding from the interview data, service modularization still remains a rather theoretical concept than a widely adopted practical approach. Attempts to apply the concept of modularization can be mainly identified when looking at the internal processes of logistics companies where increasing process standardization appears to be a clear objective of all organizations that we interviewed (as proposed by the modular strategy for Type 1 services in Fig. 36.1). Looking at the sales activities, however, the idea of modularization of services appears to be completely absent, which is a strong contrast to the domain of tangible products. The modularization of complex, multivariant, and customizable products like cars and machinery offers great benefits for sales processes, as CPQ activities can be supported by dedicated software tools. Such software tools are nonexisting in the service domain, where customer-specific RFQs are followed by complex and tedious calculations mainly using basic spreadsheet tools. The mere number of customer inquiries in the logistics industry, however, actually calls for adopting the idea of modularization and CPQ software in order to make sales processes more efficient.

The interview data also indicates that the logistics industry is heterogeneous. The service portfolios of the different organizations include a wide and ever growing range of services that partly can be well standardized (reflecting the Type 1) and that partly include complex, customer-specific, and long-term projects (rather reflecting the Type 3). Therefore, there is no clear answer regarding the potentials of modularization for the complete industry, and a combination of different modular strategies might be appropriate. Hence, further research should concentrate on defining a set of evaluation criteria that could help predicting, whether and in what scope a company could benefit from the concept of modularization and specific modular strategies. Here, the typology of service process types by Carlborg and Kindström (2014) offers a promising starting point, although the modular strategies they describe might still be too vague to actually guide measures for implementing the modularization concept in organizations.

Another main obstacle in applying the concept of modularization lies in the specifics of services compared to tangible products. Services often rely on a close interaction between provider and customer throughout the whole value-creating process. This makes both the decomposition of services and the actual building of modules more complicated than in case of product modularization. Therefore, the existing methods for product modularization have to be adjusted to the service context or complete new methods need to be created (e.g., like the TM3 approach for telemedical services, Peters and Leimeister 2013). Due to the nature of service provision, special attention has to be addressed on the topic of interfaces between the modules as well as service quality to satisfy customers (Lin and Pekkarinen 2011).

In addition, more research on the possible negative effects of service modularization in practice has to be conducted since the academic literature has outlined purely advantages so far, which can perhaps be achieved only in theory. For example, more empirical data is needed on whether companies actually benefit from service module and price transparency for their customers, instead of rather having more freedom on price calculation of the final service bundle and offering inseparable service packages. Similarly, it is important to test whether a systematically established modular system is stimulating, or, on the contrary, limits radical service innovations by falling in the so-called modularization trap (Ernst 2005).

Acknowledgments This paper is written in the context of the research project BakerStreet (promotional reference FUE0576B) funded by the WFB Wirtschaftsförderung Bremen GmbH. The project is realized in cooperation with and with further financial support by encoway GmbH.

References

Arnheiter ED, Harren H (2005) A typology to unleash the potential of modularity. J Manuf Technol Manag 16:699–711. doi:10.1108/17410380510619923
Baldwin C (1997) Design Rules Volume 1, The Power of Modularity.pdf
Bask AH (2001) Relationships among TPL providers and members of supply chains-a strategic perspective. J Bus Ind Mark 16(6):470–486
Bask A, Lipponen M, Rajahonka M, Tinnilä M (2010) The concept of modularity: diffusion from manufacturing to service production. J Manuf Technol Manag 21:355–375. doi:10.1108/17410381011024331
Bask A, Lipponen M, Rajahonka M, Tinnilä M (2011a) Modularity in logistics services: a business model and process view. Int J Serv Oper Manag 10:379–399

Bask A, Lipponen M, Rajahonka M, Tinnilä M (2011b) Framework for modularity and customization: service perspective. J Bus Ind Mark 26:306–319. doi:10.1108/08858621111144370

Böttcher M, Klingner S (2011) Providing a method for composing modular B2B services. J Bus Ind Mark 26:320–331. doi:10.1108/08858621111144389

Burr W (2005) Chancen und Risiken der Modularisierung von Dienstleistungen aus betriebswirtschaftlicher Sicht. In: Herrmann PD-IT, Kleinbeck PDU, Krcmar PDH (eds) Konzepte für das Service Engineering. Physica-Verlag HD, pp 17–44

Carlborg P, Kindström D (2014) Service process modularization and modular strategies. J Bus Ind Mark 29:313–323. doi:10.1108/JBIM-08-2013-0170

De Blok C, Luijkx K, Meijboom B, Schols J (2010) Modular care and service packages for independently living elderly. Int J Oper Prod Manag 30:75–97. doi:10.1108/01443571011012389

Dörbecker R, Böhmann T (2013) The concept and effects of service modularity—a literature review. IEEE, pp 1357–1366

Eickelpasch A (2012) Industrienahe Dienstleistungen Bedeutung und Entwicklungspotenziale. Abt. Wirtschafts- und Sozialpolitik der Friedrich-Ebert-Stiftung, Bonn

Ernst D (2005) Limits to modularity: reflections on recent developments in chip design. Ind Innov 12:303–335. doi:10.1080/13662710500195918

Giannikis M, Mee D, Doran D, Papadopulos T (2015) The design and delivery of modular professional services: implications for operations strategy. In: Proceedings of the 6th international seminar on service modularity. Helsinki, Finland

Hautamäki M, Pohjosenperä T, Pekkarinen S, Juga J (2015) Co-creating value for public healthcare customer through modularity of logistics services. In: Proceedings of the 22nd EurOMA conference. Neuchâtel, Switzerland

Hsuan Mikkola J, Skjøtt-Larsen T (2004) Supply-chain integration: implications for mass customization, modularization and postponement strategies. Prod Plan Control 15:352–361. doi:10.1080/0953728042000238845

Hvam L, Pape S, Nielsen MK (2006) Improving the quotation process with product configuration. Comput Ind 57:607–621. doi:10.1016/j.compind.2005.10.001

Lin Y, Pekkarinen S (2011) QFD-based modular logistics service design. J Bus Ind Mark 26:344–356. doi:10.1108/08858621111144406

Lin Y, Luo J, Zhou L (2010) Modular logistics service platform. In: 2010 IEEE international conference on service operations and logistics and informatics (SOLI), IEEE, pp 200–204

Pekkarinen S, Ulkuniemi P (2008) Modularity in developing business services by platform approach. Int J Logist Manag 19:84–103. doi:10.1108/09574090810872613

Peters C, Leimeister JM (2013) TM3-A modularization method for telemedical services: design and evaluation. In: Proceedings of 21st European conference on information systems (ECIS)

Tu Q, Vonderembse MA, Ragu-Nathan TS, Ragu-Nathan B (2004) Measuring modularity-based manufacturing practices and their impact on mass customization capability: a customer-driven perspective. Decis Sci 35:147–168. doi:10.1111/j.00117315.2004.02663.x

Tuunanen T, Bask A, Merisalo-Rantanen H (2012) Typology for modular service design: review of literature. Int J Serv Sci Manag Eng Technol 3:99–112. doi:10.4018/jssmet.2012070107

Voordijk H, Meijboom B, de Haan J (2006) Modularity in supply chains: a multiple case study in the construction industry. Int J Oper Prod Manag 26:600–618. doi:10.1108/01443570610666966

Voss CA, Hsuan J (2009) Service architecture and modularity*. Decis Sci 40:541–569. doi:10.1111/j.1540-5915.2009.00241.x

Yang L, Shan M (2009) Process analysis of service modularization based on cluster arithmetic. IEEE, pp 263–266

Yin RK (2013) Case study research: design and methods. SAGE Publications

Part V
Frameworks, Methodologies and Tools

Chapter 37
Toward a Unified Logistics Modeling Language: Constraints and Objectives

Michael Freitag, Martin Gogolla, Hans-Jörg Kreowski,
Michael Lütjen, Robert Porzel and Klaus-Dieter Thoben

Abstract In this paper, we discuss the basic constraints and objectives of a unified logistics modeling language that allows to model dynamic logistic systems on various levels of concern from the requirements definition to their design and realization. It should, furthermore, support the compositionality and interoperability among all processes and other components even if they are modeled by means of different methods. Lastly, it should enable domain-specific views on logistic systems. A key concept, therefore, is that the model transformation allows one to bridge the gaps between different data formats and different modeling methods.

37.1 Introduction

Current logistic systems are comprised of a multitude of joined systemic components that employ individual and frequently changing technologies. Common visions for future dynamic logistic systems assume capabilities for reacting robustly to such

M. Freitag (✉) · M. Gogolla · H.-J. Kreowski · R. Porzel · K.-D. Thoben
Universität Bremen, Bremen, Germany
e-mail: fre@biba.uni-bremen.de

M. Gogolla
e-mail: gogolla@informatik.uni-bremen.de

H.-J. Kreowski
e-mail: kreo@informatik.uni-bremen.de

R. Porzel
e-mail: porzel@informatik.uni-bremen.de

K.-D. Thoben
e-mail: tho@biba.uni-bremen.de

M. Freitag · M. Lütjen · K.-D. Thoben
BIBA – Bremer Institut für Produktion und Logistik, Bremen, Germany
e-mail: ltj@biba.uni-bremen.de

changes and for adapting their structure and behavior accordingly (Gosling et al. 2010; Swafford et al. 2008). While in the domain of production, many communication technology oriented efforts are under way toward developing networked production, e.g., within the Industry 4.0, Internet of Things or Cyber-Physical Systems paradigms, analogous efforts are lacking in process-oriented modeling of such systems (Khaitan McCalley 2014). Seiger et al. define six requirements for process-oriented modeling concepts of smart cyber-physical production and logistic environments (Seiger et al. 2015). Such concepts have to reflect and to deal with the following:

- dynamics,
- heterogeneity,
- complexity,
- parallelism,
- evolution, and
- distribution.

Formal descriptions of such systems are not fully realizable given current modeling methods. At the moment, only few attempts are made in order to adapt *BPMN 2.0*, UML, or *Petri Nets* to the domain of production and logistics (Seiger et al. 2015; Khabbazi et al. 2013a, 2013b; Koniewski, et al. 2006). All approaches are inspired by the development of information systems and focused to data manipulation. The explicit consideration of material flows, logistic control rules, as well as simulation aspects, to name a few, is missing.

In the light of modeling processes and systems that overarch individual businesses, a need arises to find sustainable modeling approaches that can cope with such dynamic logistic systems. Furthermore, the dynamic nature of logistic systems requires the potential to create domain-specific models and views. Additionally, procedural, material, informational, and organizational flows have to be included in a scalable and simulatable manner. Despite their heterogeneity and multilayered nature, transitions between these layers have to work seamlessly and robustly.

Against this backdrop, we propose a set of constraints and objectives for a corresponding logistic modeling and transformation language. Our long-term goal is the development of an overarching framework that includes graphical, mathematical, data- and language-specific modeling languages. Using model transformations, the individual elements of these languages can be linked to become an open logistic modeling language. We, therefore, set out to create an analoge to the *Unified Modeling Language* (*UML*) in combination with the Object Constraint Language *(OCL)* and the Query/View/Transformation concept *(QVT)* (OMG 2003, 2008, 2011a, b, c, 2012), which serves as a quasi-standard in software engineering and subsumes several modeling techniques. This analoge is called *LogisticsML* in the following. In this paper, we discuss its constraints and objectives with respect to the levels of concern in Sect. 37.2, the compositionality and interoperability are discussed in Sect. 37.3 followed by a look into domain-specific views in Sect. 37.4. The key concept of model transformation is described in Sect. 37.5.

37.2 Levels of Concern

Today's logistic system cannot be modeled as a "monolith" in a single effort once and for all. Analogously to the development of information processing systems, one can distinguish various levels of concern including at least the following:

- The *requirements definition* describes the problems to be solved and the intentions the system should meet leading to an abstract model.
- The *design* specifies the system as a conceptual model that solves the problems and meets the intentions, but may not focus on proper and efficient executability.
- The *realization* yields an executable platform-specific model that runs robustly and efficiently.

There may be further levels mixing requirements definition, design, and realization or distinguishing various steps and aspects within the three levels. This may depend on the particular intentions one has in mind such as, for example, simulation, optimization, or verification. In any case, the levels must be compatible with each other meaning that the platform-specific model realizes the conceptual model that, in turn, fulfils the requirements correctly.

Consequently, we get the first objective for *LogisticsML*:

(O1) The language should allow one to model dynamic logistic systems on various levels of concern including at least the requirement definition and the design. Moreover, means should be provided to assure the compatibility of the models of a system on different levels.

Whereas, programming platforms such as *JAVA* and *C* ++ are used on the level of realization; they are not suitable for the requirement definition and the design. To define requirements for logistic systems, goal modeling languages like *i** and *KAOS* are in use. For the design, quite a spectrum of modeling techniques are available including *AMPL* (A Mathematical Programming Language), *EPC* (Event-driven Process Chains), *BPMN* (Business Process Model and Notation), *BPEL* (Process Execution Languages), *CMSD* (Core Manufacturing Simulation Data), *UML* (Unified Modeling Language), and *OWL* (Web Ontology Language).

In addition to the levels of concern and the multitude of modeling methods, one faces the need for domain-specific views on logistic systems further discussed in Sect. 37.4. Figure 37.1 illustrates the challenges of modeling logistic systems on a more general level.

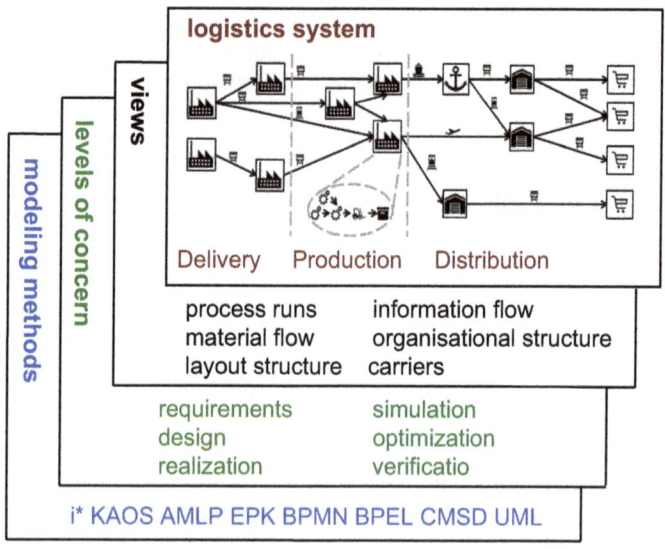

Fig. 37.1 Challenges of modeling logistic systems

37.3 Compositionality and Interoperability

As logistic systems are intrinsically complex, the models on some level of abstraction are composite entities usually consisting of various processes and other components. For their modeling, one encounters quite a spectrum of methods in the literature like *BPMN, EPC, AMPL, Petri Nets, UML,* and *OWL*. It is mandatory that the components can be combined and that they interoperate properly. This goal is hard to reach as there is not the one and only modeling method, but one must take into account that the components are specified by different methods which must be compositional and interoperable with each other.

Consequently, we get the second objective for *LogisticsML*:

(O2) The language should allow for modeling a logistic system in a structured and modular way such that the components can be composed to interoperate properly with each other even if their specifications employ different modeling methods. In particular, the language should comprise a spectrum of modeling methods.

The ensuing problems with compositionality and interoperability are manifold. For example, the sequential composition of two input/output processes requires that the output of the first process can serve as input into the second process. To enable this—despite incongruent data formats or the processes that are specified by different methods—adaptations between the data formats are necessary. The situation is similar with respect to parallel composition in particular if two processes or other components must share data, but require different formats. While several modeling methods offer such operations, none of them allows a composition if the components are specified by different methods. Hence, in meeting our second objective a multitude of challenges has to be met.

37.4 Domain-Specific Views

The proposed modeling approach aims at the development of logistic systems and their dynamics in a holistic way, so that all significant aspects can be covered. On one hand, it does not make sense to deal with a very large system on each level if only certain features are of interest in certain situations. Domain-specific views of a system such as the process structure, the material flow, the information flow, the layout structure, and the organizational flow are of great importance. This can be seen in a Thapa's study (Thapa 2013) on the interaction of the material and money flows with the information flow in just-in-time production sequencing and supply chain systems.

Consequently, we get the third objective for *LogisticsML*:

(O3) The language should make it possible to look at a logistic system from different domain-specific points of view in such a way that the further development of a specific view is properly reflected in the overall system.

In addition to this kind of projection of the overall systems to specific aspects, there may be a direct approach to domain-specific views. *LogisticsML* may be regarded as a domain-specific language that serves to formulate and solve verification tasks crucial in the logistics realm. Such a language will offer domain-specific constructs for logistics experts. The constructs will in principle be close to *UML* and *OCL* with units for structural and behavioral entities; however, as mentioned above, supported by particular logistics constructs. For example, there could be support for trade items, locations, and shipping containers; their relationships, the processes, and the properties of the processes in which these entities are involved. On the other hand, the language view will offer only a restricted set of constructs, so that they can be handled by automatic proving techniques. No interference between the logistics expert and the underlying prover engine is expected to take place. It will be possible to formulate structural as well as behavioural properties such as reachability of particular system states, redundancy freeness, deadlock freeness, or liveness of dynamic logistic processes. The prover engine will rely on a combination of testing and automatic proving techniques. It could depend on the well-known theory and practice of relational logic and its implementation in form of *Alloy* and *Kodkod*. *OCL* can be employed for the description of structural and temporal properties; however, special constructs will offer shortcuts for verification tasks typical in logistics.

37.5 Model Transformation

While there are various methods available and well established for the modeling of single logistic processes and other components of a logistic system, it is not quite clear how the compatibility of models of different levels of concern and abstraction, the compositionality and interoperability among the components as well as different views on a system can be supported and guaranteed.

The key idea in all three respects in model transformation is the following:

1. Models of the same system or the same component that differ in the degree of abstraction may be related by transforming one into the other in such a way that the requirements are preserved.
2. Compositionality requires a modeling method with composition of models. But in a heterogeneous environment, one may need to compose models of processes or other components that are specified by different modeling methods. Then the transformation into a method with composition constitutes a viable solution. Similarly, interoperability may requires the exchange of information. If the information of the interoperating components has different formats (as it occurs quite often), then again transformations between the formats can solve the problem.
3. Various views of a logistic model are obtained by projections that abstract from all features of the model that are not of interest in a respective view. In this way, domain-specific views are results of particular model transformations.

Consequently, we get the fourth objective for *LogisticsML*:

(O4) The language should provide means of model transformation between different levels of concern, between models on the same level of abstraction, but specified by different modeling methods, and from the whole system to domain-specific views.

In the literature, one encounters quite a lot of singular model transformations each quite specific for certain models, but there are only very few general approaches. In Kreowski et al. (2010); Lütjen et al. (2014), and Kreowski et al. (2014), one can find some first sketches of a framework for model transformation that applies particularly to logistic models.

The basic idea is that input models are mapped into output models as, for example, *BPMN* models into Petri nets. As the models may be very large, this cannot be done by a global operation, but the transformation must be performed step-by-step transforming only small portions in each step. Consequently, one needs intermediate models in addition to input and output. Moreover, it may happen that one needs further auxiliary information or must store some information during the transformation. This consideration leads to the concept of model transformation that is sequentially composed of three phases: The initialization where the input model is mapped into an initial working model, the internal running transformation where the initial working model is stepwise transformed into a terminal working model by means of rules and elementary operations, and finally the terminalization where the terminal working model is mapped to the output.

37.6 Conclusion

In this paper, we have sketched some basic constraints and objectives for a unified logistic modeling language which is called *LogisticsML* in analogy to the *Unified Modeling Language (UML)* that is very successful in the area of modeling information processing systems. We have formulated four objectives concerning the levels of concern, the compositionality and interoperarability, the domain-specific views, and model transformation. The latter serves as the "glue" between the other objectives that allows one to relate different levels of concern and abstraction, to bridge the gaps between different data formats and different modeling methods, and to project the entire system to domain-specific views of current interest.

The next step will be the proper development of *LogisticsML*. For this purpose, a group of researchers within the Bremen Research Cluster for Dynamics in Logistics (Log*Dynamics*) at the University of Bremen plans to cooperate in this enterprise for the next 4 years. The project will focus on the conception of model transformation based on the preliminary studies as referred in Sect. 37.5 making sure that a spectrum of modeling techniques can coexist and can be used in a compatible way. The aim is a logistic modeling and transformation language with a formal semantics that allows one to prove correctness properties. The theoretical challenge is to transform logistic models of one type into models of another type in such a way that the semantics is preserved if both involved modeling methods have got a formal semantics or that the source model inherits the semantics of the target model if only the latter has got a formal semantics. *LogisticsML* promises some improvements for practitioners including (1) a variety of modeling methods and tools so that they can choose their favorite ones and (2) the development of a dynamic logistic system in a modular and integrated way at all levels of concern.

References

Gosling J, Purvis L, Naim MM (2010) Supply chain flexibility as a determinant of supplier selection. Int. J Prod Econ 128(1):11–21

Khabbazi MR et al (2013a) Business process modelling in production logistics: complementary use of BPMN and UML. Middle-East J Sci Res 15(4):516–529

Khabbazi MR et al (2013b) Business process modeling for domain outbound logistics system: analytic perspective with BPMN 2.0. World Appl Sci J 28(3):367–377

Khaitan SK, McCalley JD (2014) Design techniques and applications of cyberphysical systems: a survey

Koniewski R, Dzielinski A, Amborski K (2006) Use of Petri nets and business processes management notation in: modelling and simulation of multimodal logistics chains. In:

Proceedings 20th European Conference on Modeling and Simulation, Institute of Control and Industrial Electronics, Warsaw University of Technology

Kreowski H-J, Kuske S, von Totth C (2012) Combining graph transformation and algebraic specification into model transformation. In: Mossakowski T, Kreowski H-J (eds) Proceedings of International Workshop on Algebraic Development Techniques (WADT 2010). Lecture Notes in Computer Science, vol 713, pp 193–208

Kreowski H-J, Franke, Hribernik K, Kuske S, Thoben K-D, von Totth C (2015) Towards a comprehensive approach to the transformation of logistic models. In: Herbert Kotzab H, Pannek J, Thoben K-D (eds) Proc. 4th International Conference on Dynamics in Logistics (LDIC 2014). Springer

Lütjen M, Kreowski H-J, Franke M, Thoben K-D, Freitag M (2014) Model-driven logistics engineering—Challenges of model and object transformation. In: Thoben K-D, Busse M, Denkena B, Gausemeier J (eds) Proc. 2nd International Conference on System-Integrated Intelligence: Challenges for Product and Production Engineering, pp. 301–310

OMG (2003) Model Driven Architecture (MDA) Guide 1.0.1. omg/03-06-01

OMG (2008) Model to text transformation language 1.0 (MOFM2T). OMG Document Number: formal/08-01-16.pdf

OMG (2011a) Meta Object Facility 2.4.1 (MOF). OMG Document Number: formal/2011-08-07

OMG (2011b) Query/View/Transformation 1.1 (QVT). OMG Document Number: formal/2011-01-01

OMG (2011c) Unified Modeling Language 2.4.1 (UML) Infrastructure and Superstructure. UML Infrastructure OMG Document Number: formal/2011-08-05; UML Superstructure OMG Document Number: formal/2011-08-06

OMG (2012) Object Constraint Language 2.3.1 (OCL). OMG Document Number: formal/2012-01-01

Seiger R, Keller C, Niebling F, Schlegel T (2015) Modelling complex and flexible processes for smart cyber-physical environments. J Comput Sci 10:137–148. ISSN 1877-7503, http://dx.doi.org/10.1016/j.jocs.2014.07.001

Swafford PM, Ghosh S, Murthy N (2008) Achieving supply chain agility through IT integration and flexibility. Int. J Prod Econ 116(2):288–297

Thapa GB (2013) Basics of informed logistics in just-in-time production sequencing and supply chain systems. J Inst Eng 9(1):54–64

Chapter 38
Potential of Improving Truck-Based Drayage Operations of Marine Terminals Through Street Turns

Niklas Nordsieck, Tobias Buer and Jörn Schönberger

Abstract Drayage operations at a marine container terminal are considered. The idea is to increase efficiency of truck-based drayage operations by reusing empty import containers (*street turns*). Different sizes of containers, i.e., 20-foot and 40-foot containers, are considered. The transport planning problem is identified as a generalization of the well-known pickup and delivery problem. However, in extension, the transport requests are only partially specified. A heuristic is developed to solve this problem. The effect of different mixtures of 20-foot and 40-foot container requests on the possibility of street turns is tested. This is based on total operation time and other performance criteria for an example instance.

38.1 Introduction

Maritime intermodal supply chains are often organized as a hub-and-spoke system. The goods to be transported are loaded in standardized containers to ease the intermodal switch. The hubs are represented by ports and the long distance transport between ports is done via container vessels. The spokes are represented by *drayage operations*. Drayage operations take place in the urban area of a port, also referred to as hinterland. Short distance transport of containers during drayage operations is usually done by trucks.

N. Nordsieck (✉) · T. Buer
Computational Logistics – Cooperative Junior Research Group of University of Bremen,
ISL -Institute of Shipping Economics and Logistics,
Box 33 04 40, 28334 Bremen, Germany
e-mail: niklas.nordsieck@uni-bremen.de

T. Buer
e-mail: tobias.buer@uni-bremen.de

J. Schönberger
Chair of Business Administration, Transport and Logistics,
TU Dresden, 01187 Dresden, Germany
e-mail: joern.schoenberger@tu-dresden.de

As in many other hub-and-spoke systems, the performance of an intermodal supply chain is highly affected by operations on the spoke level, i.e., the drayage operations. According to The Tioga Group (2011), truck-based drayage operations face several challenges: asymmetric information about bottlenecks and job status among port operators, drayage operators (i.e., trucking companies), ocean carriers, consignees of containers (i.e., information sharing and coordination problems); high waiting times of trucks to enter a terminal (i.e., queuing problems); high idle times of trucks within a terminal and unproductive moves outside the terminal often associated with the repositioning of empty containers (i.e., inefficient transport and relocation decisions). This paper focuses on the latter problem. It contributes to improving the repositioning of empty containers as well as the transport of loaded and empty containers in the urban area of a marine terminal. Therefore, it focuses on the ratio of 20-foot and 40-foot containers (*container share*) and its impact on the overall solution.

In order to improve these decisions, the study focuses on the idea to ease the *reuse* of empty containers. In particular, if a loaded import container (transported from the port into the hinterland) is reused to export goods (reloaded in the hinterland and transported to the port), this is referred to as a *street turn* (Jula et al. 2006). The reuse of means of transport is at least in some areas of application considered as best practice, e.g. the share of wooden stillages (Bernan 2008). However, reuse of empty containers is uncommon, because there are high administrative and technical barriers (The Tioga Group 2011). To reduce the technical barriers and to improve the efficiency of container drayage transport, the option of reusing empty containers is integrated into a well-known vehicle routing problem. In contrast to many earlier approaches, different containers sizes (i.e., 20-foot and 40-foot containers) are explicitly considered, because heterogeneous container sizes increase the difficulty of the repositioning problem as well as the routing problem significantly.

The container drayage problem at hand is studied as a generalization of the well-known pickup and delivery problem with time windows (PDPTW). In contrast to the PDPTW, not all of the transport requests can be considered as given. That is, some requests are only partially specified before planning begins. For these requests, either the delivery location or the pickup location has to be determined during planning. Therefore, a sequential algorithm is proposed. To estimate effects of reusing empty containers as well as consequences from the composition of the container pool, computational experiments are performed on some instances of the problem.

The remaining paper is organized as follows. Section 38.2 introduces the problem at hand and reviews the literature on vehicle routing problems in the context of container drayage. Section 38.3 presents an iterative three-phase heuristic to solve this problem. Section 38.4 studies the effects of reusing empty containers by means of a computational test. Section 38.5 concludes the paper.

38.2 Vehicle Routing with Empty and Loaded Containers

The pickup and delivery problem (PDP) is an extension of the vehicle routing problem (VRP). For the VRP, either all goods are picked up at the depot or delivered to the depot. However, as the name of the PDP indicates, for this problem the goods have to be picked up at a customer and delivered to another customer. Since the VRP is NP-hard, the PDP is NP-hard, too. There are many studies on the PDP, an introductive overview is given by Savelsbergh and Sol (1995), for example. First, this section discusses the PDP with partially specified requests, an extension of the standard PDP, which is abbreviated as PDP-PSR. Second, the literature on truck-based drayage transport is discussed.

38.2.1 The Pickup and Delivery Problem with Partially Specified Requests

In the PDPTW we are given a fleet of homogeneous vehicles and a set R of less-than-truckload requests. Each request $r := (r^+, r^-, l)$ with $r \in R$ is specified by a pickup location r^+, a delivery location r^- and a load l which is usually smaller than the capacity of a vehicle. Furthermore, a time window is given for each location. A vehicle tour is called feasible, if the tour starts and ends at a given depot location, and for each assigned request the pickup location is visited before the corresponding delivery location, time windows are met, and the capacity of the vehicle is never exceeded. The task of the PDPTW is to find a set of feasible tours such that all requests are fulfilled and the total operation time t^o of all vehicles n is minimized.

In comparison to the constraints for the PDPTW, the presented PDP-PSR has some special characteristics. Here, means of transport are containers. Their size is measured in *20-foot equivalent unit* (TEU) and they have a capacity of either 20-foot (i.e., 1 TEU) or 40-foot (i.e., 2 TEU). Each truck has a capacity of 2 TEU. The following four request types are considered (Schönberger et al. 2013):

- Import request: transport a *loaded* container from a given pickup location (e.g., a customer or a marine terminal) to given customer location in the hinterland.
- Export request: transport a *loaded* container from a given hinterland location to a given location (e.g. another customer or a marine terminal).
- Provide request: transport an *empty* container from an *a priori unknown pickup* location to a given customer location in the hinterland.
- Store request: transport an *empty* container from a given pickup location to an *a priori unknown delivery* location.

For requests which include empty container transportation, either the pickup or the delivery location has to be determined during planning. These requests are denoted as *partially specified requests*. It is assumed that the pickup location of a provide request may always be the terminal. Furthermore, the delivery location of store request may also always be a terminal. That is, a terminal is able to provide and

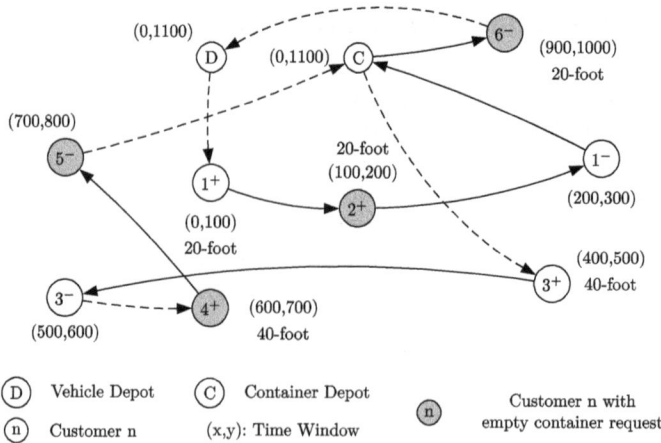

Fig. 38.1 Possible routing plan for the PDP-PSR, including standard, provide and store requests

store an unlimited number of empty containers. However, if the delivery location of a store request is used as a pickup location of a provide request, a *street turn* is present. Street turns are only feasible, if the sizes of the stored and provided containers match. Note, if an empty 20-foot container is required, it is not feasible to provide a 40-foot container.

Figure 38.1 shows a feasible tour plan for an example instance of the PDP-PSR. The dashed lines represent a vehicle carrying no container and the solid lines indicate that at least one container is carried. Gray vertices were originally part of partially specified requests which were assigned to other vertices or the container depot by the solution approach. The vehicle starts at Depot D and picks up a 20-foot container at 1^+ for a standard request. Before this container is unloaded at 1^-, an empty 20-foot container is picked up at 2^+ for a store request. This container has no particular consignee, here it is delivered to the container depot C. Afterwards a standard request including the nodes 3^+ and 3^- is fulfilled. Due to Customer 4^+ having a store request and 5^- requiring an empty container because of a provide request, the empty container is shipped from 4^+ to 5^-. Finally, 6^- has a provide request and requires an empty container, which is provided by the container depot.

38.2.2 Literature Review

The topic of optimizing container movements in the hinterland receives more and more attention in the literature. An overview is given by Table 38.1 which shows papers dealing with truck-based container drayage optimization problems. These include fundamental as well as very recent approaches. The initial studies on this problem focused on full truckload (FT) transportation, i.e., a vehicle is only able to transport one container at a time. The first papers to study less-than-truckload transportation were Vidovic (2011) and (Schönberger et al. 2013). Additionally, most

Table 38.1 Comparison of different models

Author	Truckload	Incomplete requests	No. of customers	TW
Jula et al. (2005)	FT	no	100	Yes
Sterzik and Kopfer (2013)	FT	Yes	75	Yes
Braekers et al. (2014)	FT	Yes	200	Yes
Vidović et al. (2011)	LTL	Yes	75	No
Schönberger et al. (2013)	LTL	Yes	6	No
Deidda et al. (2008)	LTL	Yes	50	No
Zhang et al. (2015)	LTL	Yes	150	No
Funke and Kopfer (2015)	LTL	Yes	75	No

presented approaches are able to deal with incomplete requests (see column 3). This means that either the pickup or the delivery location of empty containers is not known in advance and must be determined by a solution approach. The fourth column shows the size of the test instances measured as number of customers used for the computational experiments. The rightmost column of Table 38.1 indicates, whether time windows on the pickup or delivery locations are considered.

Jula et al. (2005) is one of the first papers to study drayage operations. The focus is on container drayage between port terminals, intermodal facilities and customers. Sterzik and Kopfer (2013) present the inland container transportation problem. Some requests are incomplete, however, only full truckload requests are considered.

A vehicle routing problem for drayage operations is proposed by Braekers et al. (2014). Incomplete requests and a bi-objective function, simultaneously minimizing the number of vehicles as well as the total distance, are used. However, only full truckloads are included.

In extension of the previous papers, Vidović et al. (2011) introduces less-than-truckload requests. The proposed model considers loaded and empty containers of heterogeneous sizes simultaneously. However, time windows are not regarded and all empty containers are either picked up or delivered to the container terminal, i.e., street turns are not used.

Schönberger et al. (2013) present a mathematical model for a LTL-PDP with incomplete requests. Although it is quite comprehensive compared to previous models, the commercial mixed integer programming solver CPLEX was not able to solve small test instances in a reasonable time. This is due to considering incomplete requests for heterogeneous container sizes and incorporating street turns. Deidda et al. (2008) present a similar model explicitly focusing on street turns. CPLEX was able to solve test instances based on historical data of a trucking company optimally. However, only 50 customers and no time windows are regarded.

Zhang et al. (2015) solve the container truck transportation problem for full and empty container drayage consisting of 20-foot and 40-foot containers. A state-transition model is formulated. The different types of requests are similar to those used in this paper, but there is a stronger coupling between requests: A loaded container is discharged immediately after arrival at a customer location. The container becomes available as an empty container which has to be serviced by the same vehicle, i.e., import requests and store requests are coupled. Furthermore, the approach makes a *full-twin assumption*. This means, that a truck handling two 20-foot container requests at the same time has to finish both requests before fulfilling a new request. A problem with a similar coupling of requests is studied by Funke and Kopfer (2015) and solved by a matheuristic. For the present problem there is a relaxed coupling of requests and the full-twin assumption is not necessary.

38.3 A Three-Phase Heuristic

An overview of the heuristic is given by Algorithm 1. There are three phases, namely completing requests, solving a pickup and delivery problem, and weight adjustment. In the first phase, decisions for the assignment of customers with partially specified requests are made. The second phase solves the resulting less-than-truckload PDPTW. The third phase changes parameters which serve as assignment criteria of the request completion phase for the following iteration. The goal of the heuristic is to fulfill all requests and minimize the total operation time t^o. This includes the time for driving of the trucks, handling of containers and waiting due to time windows.

The Request Completion matches nodes of partially specified requests in the best possible way. The assignment of nodes is modeled as an assignment problem and solved via commercial software package CPLEX. It is possible to forbid the involvement of the container depot or match all incomplete requests with the container depot. Hence, a pure street turn or depot direct strategy can be studied.

Input: test instance, no. of iterations
while *Weight adjustment not completed* **do**
　　Request Completion;
　　Pickup and Delivery Problem;
　　Weight Adjustment;
end
return feasible tour plan;

Algorithm 1: Structure of the heuristic

$$\min \sum_{i=1}^{m} \sum_{j=1}^{n} \left(\delta \cdot d_{ij} + \tau \cdot t_{ij}^w \right) \cdot x_{ij} \ . \tag{38.1}$$

The objective (1) is to minimize the overall costs. They consist of the weighted distance d_{ij} summed with the weighted minimum waiting time t_{ij}^w for every route. If $x_{ij} = 1$, an empty container at customer i (pickup location) shall be transported to customer j (delivery location), $x_{ij} = 0$ otherwise. δ represents the distance weight, τ represents the time weight. It holds that $\delta + \tau \equiv 1$. The problem is solved twice, once for all 20-foot container requests and once for the 40-foot container requests. Hence,

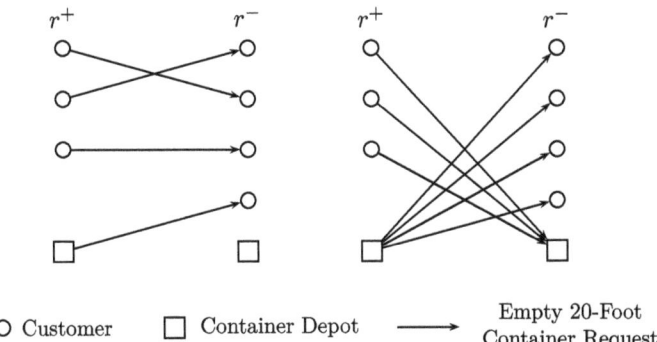

Fig. 38.2 Two possible completions of partially specified requests, based on different distance weights (*left* $\delta_{20} = 1$, *right* $\delta_{20} = 0$)

there are weights $\delta_{20}, \delta_{40}, \tau_{20}$ and τ_{40}, two for each container size. Figure 38.2 shows two possible completions for partially specified requests, each based on its distance weight δ_{20}.

The weighting parameters determine if the assignment of partially specified requests to one another is based on the geographical distance of two customers or on the necessary minimum waiting time when serving them successively. The distance weights are changed automatically in the weight adjustment phase to increase matching variety and eventually provide a higher solution diversity. In the following, this is explained in detail.

Initially, the distance weight for 20-foot containers is set to $\delta_{20} = 1$ and distance weight for 40-foot containers is set to $\delta_{40} = 0$. Based on experience from a computational study, these values seem to be good starting points. The distance weights are changed throughout the run of the heuristic. For every distance weight combination, a pickup and delivery problem is solved to generate a feasible routing plan for the test instance. Thereafter, the performance values of this solution are compared to the best found solution and if they constitute an improvement, they become the new best solution.

To systematically change the distance weights, first the heuristic determines if there is a majority of 40-foot containers or 20-foot containers. If both number of requests are equal, a majority of 40-foot containers is assumed for the following process. Now the distance weight for the majority of containers is changed while the other distance weight is fixed. More precisely, δ_{20} is decreased by 0.25 per iteration until it reaches $\delta_{20} = 0$ or δ_{40} is increased by 0.25 until it reaches the value $\delta_{40} = 1$. The distance weight that led to the best solution is fixed then and the distance weight for the other container size is changed in equidistant steps. When this procedure terminates, the best combination of distance weights δ_{20} and δ_{40} is used for one more iteration of the algorithm with five times more iterations for the route generation phase to improve solution quality.

When solving the PDP, an insertion heuristic generates a sufficient number of feasible routes. Subsequently, CPLEX solves a set covering problem and selects a subset of routes that serves all customers with the lowest total operation time. After that, the routes are improved by well-known local search operators *rearrange, shift* and *exchange*. They try to change the position of nodes within a tour or shift nodes to other tours. For further explanation of these operators, see Li and Lim (2003).

38.4 Discussion and Results of a Computational Experiment

In order to study the effects of the share of 20-foot containers on the total number of containers, a computational experiment based on a modified test instance from the literature was performed. First, the setup of the study is described in Sect. 38.4.1. Second, the results of the actual study are presented and discussed in Sect. 38.4.2.

38.4.1 Used Test Instances and Parameters of the Heuristic

38.4.1.1 Used Test Instance

For the PDP-PSR with 20-foot and 40-foot containers, there are no benchmark instances available in the literature. Recently Funke and Kopfer (2015) introduced test instances which are related to PDP-PSR, but still assume a different request model. Therefore, two well-known instances of the standard PDPTW instances of Li and Lim (2003) are extended. In the instances, either the pickup location or the delivery location for a given request has been deleted in order to generate partially specified requests. In addition, the load of all PDP requests was updated to either 1 TEU or 2 TEU.

38.4.1.2 Computer and Implementation

The heuristic was implemented in Java 8 and tested on an Intel Core I5, 2.5 Ghz CPU with 8 GB system memory. The number of iterations is set to 5,000 for all test instances.

38.4.1.3 Effects of Distance and Time Weights on the Solution Quality

The effects of different values for the weights δ_{20} and δ_{40} on the solution of a PDP-PSR are studied. These weights are used in the objective function (1) of the request completion problem and are updated in final phase of the heuristic. The test is based on the generated instance *plc205* with 50 percent partially specified requests and 20 percent 20-foot containers. It is characterized by a clustered distribution of 200 customers and a long planning horizon. Therefore, bad assignments of partially specified requests will result in long waiting times and the solution value is highly dependent on the choice of δ_{20} and δ_{40}.

For different distance weights, the results per iteration vary significantly. The weights clearly affect the outcome of the request completion phase. For every empty container that is reused, the number of requests is reduced by one: a store request and provide request can be combined to a single pickup and delivery request with a pickup of an empty container at the storage location and a delivery of an empty container at the provide location. Depending on the matching of requests, the number of requests to be fulfilled varies between 25 and 45. Consequently, the outcome of the pickup and delivery problem phase varies as well. The total operation time t^o varies between 23,074 time units and 32,848 time units. The total waiting t^w time varies between 9,914 time units and 16,911 time units. The number of required vehicles varies between 13 and 17. The total travel distance d varies between 2,665 distance units and 3,254 distance units.

38.4.2 Effect of Different Mixtures of Container Sizes

For this computational test, the percentage of 20-foot containers was changed and its effect on the overall solution was studied. Everything equal, more 20-foot containers implies more comprehensive options for routing which should increase the difficulty of an instance of the PDP-PSR. This test used instance *plc105*, containing 100 clustered customers, with ratios from 0 % to 100 % of 20-foot container requests, increasing by 5 % per step. Table 38.2 presents the total operation time t^o, the total travel distance d and the waiting time t^w. It shows the minimum, maximum and quantiles that were found during the solution process and the division of the former two values.

Table 38.2 Effect of different ratios of 20-foot containers on the solution for instance *plc105*

	t^o	d	t^w
25-quantile	12600	2210	813
50-quantile	13536	2347	1527
75-quantile	14692	2450	2429
min	11166	1796	361
max	15179	2561	3240
min/max [%]	74	70	11

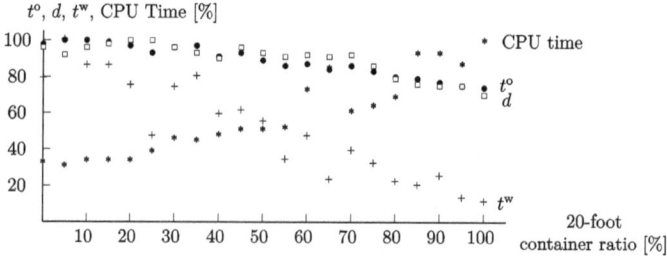

Fig. 38.3 Effect of different container ratios on performance values

Figure 38.3 presents four performance measures for an increasing percentage of 20-foot containers. A significant decrease is observable, especially for the waiting time t^w. The deviations are caused by the different solution approaches based on the distance weights. In some cases, reducing the waiting time led to the best solution, e.g., for a 25 % share of 20-foot containers. However, this results in the fact that the total driven distance is significantly high. This means that for this solution, the distance weights had a low value and therefore, the assignment of customers with partially specified requests was based on temporal closeness. Most likely, many customers were assigned to the container depot. For 35 % 20-foot containers, it is the other way round. The waiting time is long, but the driven distance is relatively small. Probably, customers with partially specified requests have been assigned to each other in most cases. Apart from these extreme cases, a balanced approach often led to the best solutions.

The main reason for the improvement in solution values as the ratio of 20-foot containers increases can be explained by the vehicle capacity. Carrying a 20-foot container, a vehicle can fulfill another request with a 20-foot container simultaneously. In contrast, a vehicle transporting a 40-foot container has no other option but to deliver said container. This results in inefficiency if bad assignments in the request completion are made because the vehicle has to wait until the consignee starts its service. In some routes of the test instances, vehicles were waiting longer than actually servicing customers due to big differences in time windows.

Therefore, it is highly important to put emphasis on the assignment of customers with partially specified requests. This is especially true for 40-foot container movements, otherwise it will result in unnecessary waiting times. Hence, 40-foot container requests should be assigned to each other based on their temporal closeness. In contrast, 20-foot containers should be assigned on spatial distance. As two small containers can be carried simultaneously by a truck, decreasing the waiting time is not as important as it is for 40-foot containers, because they can generally be avoided better. This is underlined by Fig. 38.3 when taking a closer look at the difference in waiting time for 0 % 20-foot containers and 100 % 20-foot containers. In general, a simultaneous approach of routing 20-foot containers and 40-foot containers might increase the efficiency of a routing plan tremendously.

38.5 Conclusion and Outlook

Empty container repositioning is of increasing importance in nowadays global trade as well as regional trade. Therefore, this paper introduced a heuristic for the simultaneous routing of loaded and empty 20-foot and 40-foot containers. A small computational test studied different compositions of the set of requests. The results strongly depend on the share of 20-foot container requests in all requests (see Fig. 38.3). Future research should focus on improving these solutions by incorporating additional types of containers (e.g., reefer) or include modal shifts to further close the gap between model and reality.

References

Bernan (2008) Waste Reduction: 6th Report of Session 2007-08, vol. 2 Evidence: House of Lords Paper 163-II Session 2007-08. House of Lords Papers, Stationery Office

Braekers K, Caris A, Janssens GK (2014) Bi-objective optimization of drayage operations in the service area of intermodal terminals. Transp. Res. Part E: Logist Transp Rev 65:50–69. special Issue on: Modeling, Optimization and Simulation of Logistics Systems

Deidda L, Francesco MD, Olivo A, Zuddas P (2008) Implementing the street-turn strategy by an optimization model. Maritime Policy & Manage 35(5):503–516

Funke J, Kopfer H (2015) A neighborhood search for a multi-size container transportation problem. In: 15th IFAC Symposium on Information Control Problems in ManufacturingINCOM, IFAC-PapersOnLine, vol 48, no. 3, pp 2041–2046

Jula H, Dessouky M, Ioannou P, Chassiakos A (2005) Container movement by trucks in metropolitan networks: modeling and optimization. Transp Res Part E: Log Transp Rev 41(3):235–259

Jula H, Chassiakos A, Ioannou P (2006) Port dynamic empty container reuse. Transp Res Part E: Log Transp Rev 42(1):43–60

Li H, Lim A (2003) A metaheuristic for the pickup and delivery problem with time windows. Int J Artif Intell Tools 12(2):173–186

Sol M Transp Sci (1995) Savelsbergh, MWP. 29:17–29

Schönberger J, Buer T, Kopfer H (2013) A model for the coordination of 20-foot and 40-foot container movements in the hinterland of a container terminal. In: Pacino D, Voß S, Jensen R (eds) Computational logistics, lecture notes in computer science, vol 8197. Springer, Berlin, Heidelberg, pp 113–127

Sterzik S, Kopfer H (2013) A tabu search heuristic for the inland container transportation problem. Comput Oper Res 40(4):953–962

The Tioga Group (2011) University of Texas at Austin Center for Transportation Research, University of South Carolina, Department of Civil & Environmental Engineering: Truck drayage productivity guide. In: NCFRP Report 11, Transportation Research Board, Washington

Vidović M, Radivojević G, Raković B (2011) Vehicle routing in containers pickup up and delivery processes. Proc Soc Behav Sci 20:335–343

Zhang R, Yun W, Kopfer H (2015) Multi-size container transportation by truck: modeling and optimization. Flexible Serv Manuf J 27(2–3):403–430

Chapter 39
Inventory Routing Problem for Perishable Products by Considering Customer Satisfaction and Green Criteria

Mohammad Rahimi, Armand Baboli and Yacine Rekik

Abstract This paper presents a new model for Inventory Routing Problem (IRP) considering simultaneously economic criteria, customer satisfaction level and environmental aspect for perishable products with expiration date. For this consideration, a multi-objective mathematical model has been developed. The first objective focuses on traditional inventory and distribution costs as well as recycling cost of perished products. The second objective concerns in customer satisfaction by minimization of three criteria, such as the number of delays (deliver after time windows), the quantity of backordered, and the frequency of backorders. The third objective considers Greenhouse Gas (GHG) emission, produced by different IRP activities. The proposed model is also enabled to investigate the possibility of using diesel and electrical vehicles in urban transportation. In order to cope with complexity of proposed model, Non-dominated Sorting Genetic Algorithm-II (NSGA-II) is tuned and applied. Finally, sensitivity analysis is performed to investigate the effects of variation of customer satisfaction and green aspects in economic side.

Keywords Inventory routing problem · Perishable product · Green supply chain · Customer satisfaction level · Multi-objective optimization · Evolutionary algorithm

---

M. Rahimi (✉) · A. Baboli
DISP Laboratory EA4570, INSA-Lyon, Université de Lyon, Villeurbanne, France
e-mail: Mohammad.Rahimi@insa-lyon.fr

A. Baboli
e-mail: Armand.Baboli@insa-lyon.fr

Y. Rekik
DISP Laboratory EA4570, EMLYON Business School, Écully, France
e-mail: rekik@em-lyon.com

39.1 Introduction

Classical models of Inventory Routing Problems (IRPs) try to minimize the total inventory and transportation costs and find the best strategy of inventory control on the one hand, and on the other hand determine the best routing for vehicles and quantity of product to be delivered in each period (Madadi et al. 2010). However, this approach overlooks other important criteria as Customer Satisfaction Level (CSL), environmental issues, social aspects, etc.

Other way, in real world, managers have to make their decisions by taking into account different aspects that could have conflictual impacts. Furthermore, integrating of these criteria and constraints such as respecting time windows, considering the possibility of using diesel and electrical vehicles in urban transportation, customer satisfaction level and environmental criteria make IRPs complex. Moreover, this problem becomes more complex for perishable product. Despite the importance of this subject, there are few studies which considered mentioned criteria (Sazvar et al. 2014; Al Shamsi et al. 2014).

The main aim of this study is to find the best trade-off between economic aspect, customer satisfaction, and environmental criteria in IRPs for the distribution network of perishable products.

The remainder of this paper is organized as follow. Section 39.2 reviews the relevant literature about perishable inventory routing problems. The problem under study and proposed mathematical model is addressed in Sect. 39.3. The solution approach for the presented mathematical model and experimental results are provided in Sect. 39.4. Finally, Sect. 39.5 concludes the paper and presents some perspectives.

39.2 Literature Review

There is a large number of research works related to IRPs. Some recent papers propose models and methods for economic and environmental aspect of perishable products. Custódio and Oliveira (2006) worked on proposing a new model to support negotiations between a logistics operator and retailers for calculating safety stock for distribution frozen products. Hsu et al. (2007) studied vehicle routing problem for distribution of perishable foods which need special vehicle with cold storage equipment. The authors, considered the deterioration rate of foods which is depended on its temperature. Their objective is to minimize the total cost including inventory, transportation, energy, and penalty cost due to violation time windows. Nagurney et al. (2011) focused on distribution of human blood while demand is stochastic. The authors concerned the additional cost due to produced waste of perished blood. Mirzapour Al-e-hashem and Rekik (2013) studied IRP with taking into account the transshipment option motivated by greener supply chain. The authors added a constraint about the total produced GHG emission. Al Shamsi et al.

(2014) proposed a multi-objective mathematical model to find trade-off between costs of emission and costs of inventory and distribution in order to deliver perishable products to customers. Sazvar et al. (2014) proposed a new multi-objective mathematical model to show the effect of environmental criteria in total costs of supply chain in downstream pharmaceutical industry. Coelho (2014) investigated perishable products in order to maximize the system profit. They assumed that product can have different ages and the supplier decides about the number of delivered products to each customer based on different ages of the products and by delivering fresher first or older first.

According to the literature and to the best of our knowledge, this is not any study which attempted to show the simultaneous effect of economic, CSL and green criteria in perishable IRPs. Other way, perishable products and possibility of using different kinds of vehicles (electrical and diesel) in the urban transportation are also concerned.

39.3 Problem Description

As presented previously, the first objective of proposed model concern the maximization of profit while the second objective tries to maximize CSL and the last one to minimize the environmental aspect.

The first objective has been calculated based on revenue from each product minus the inventory (holding and backorder) costs, distribution (loading, unloading, fixed and variable transportation) costs, ordering cost and recycling cost.

For the second objective function, customer satisfaction level can be improved by sending more products to customers and reduce amount of backordered products. However, from inventory side, sending more products, could cause an extra expired products and generate additional recycling cost and GHG emission. From distribution side, instead of sending more products to customers, supplier could increase the number of transportation could also cause improvement in CSL, in one hand and increasing the total costs and GHG emission, in the other hand.

The concept of CSL is integrated in the proposed model by defining three criteria as, number of delays which may occur if vehicle arrives after the allowed time windows, number of backordered products and also frequency of backorder. The analysis of transport efficiency in food supply chain in 2002 in UK (McCormick 1976) shows that 29 % of retailers visiting were occurred with delay which has direct effect in CSL while 31 % delay causes were related to traffic. For this consideration, three indexes of satisfaction level are aggregated as one customer service level objective function, based on normalized weighting factors (β_n, β_B, β_r). It should be noted that this aggregation could be presented as the importance of each component. It is noteworthy that, in this paper, we focus only on these indicators for CSL, but proposed model is "open model" and others economic, green and CSL criteria could be easily added to this propose.

In the third objective function, environmental considerations are introduced to measure the carbon footprint of the IRP (measured as an aggregate GHG emission). In order to decrease amount of produced GHG emission, the possibility of using both diesel and electrical vehicles in urban transportation has been considered. This possibility has direct impact in environmental issues in third objective function. Electrical vehicles have higher fixed and variable transportation costs compared to traditional diesel vehicles but they produce a very low level of GHG emission.

In this model, one supplier distributes food products with expiration date to different retailers under following specific assumptions (in addition of traditional assumption of IRPs):

- Supplier use a heterogeneous fleet with different capacities (small, medium and large), different modes of technology for vehicles (diesel and electric), and consequently different fixed and variable cost as well as different amount of GHG emission.
- The manager faces with backorder penalty in the case of out of stock.
- The supplier and retailers have two types of loading/unloading equipment (gas and electric) with differ in costs and amount of GHG emission.
- Each product has the specified expiration date (L_f) and it is considered to be not available for sale after this shelf life time.
- An expired product should be recycled with an additional cost. This activity produces also GHG emission.

Set of variables:

M set of retailer, index for retailer $(1, 2, \ldots, m)$
M' set of retailer and supplier, $M \cup \{0\}$
f index for product $(1, 2, \ldots, F)$
k index for vehicle type $(1, 2, \ldots, K)$
u index for vehicle technology $(1, 2, \ldots, U)$
d index for loading equipment type $(1, 2, \ldots, D)$
d' index for unloading equipment type $(1, 2, \ldots, D')$
t, t' index for period $(1, 2, \ldots, T)$

Parameters:

I_{if0} initial inventory level of product f in retailer i
h_{if} inventory holding cost in retailer i per unit of product f
IC_{if} maximum inventory capacity of retailer i for product f
γ_{if} backordering cost of one unit of product f in retailer i
O_{if} ordering cost of product f for retailer i
R_{ft} revenue from one unit product f at period t
c_{ij} distance between retailer i and j
v_{ku} variable transportation cost per unit distance for vehicle type k, technology u
fc_{ku} fixed transportation cost for vehicle type k with technology u per trip
cap_{ku} capacity of vehicle type k with technology u

d_{ift} demand of retailer i for product f at period t
L_f shelf life of product f
sp_{ku} the required time to unload one unit product from vehicle k with technology u
$[a_i, b_i]$ earliest and latest possible arriving time to retailer i
s_{ijku} average speed of vehicle type k with technology u in the rout of $i - j$
LC_d loading cost of one unit product through equipment type d
$UC_{d'}$ unloading cost of one unit product through equipment type d'
GL_d produced GHG emission of equipment type d for loading 1 kg of product
$GU_{d'}$ produced GHG emission of equipment type d' for unloading 1 kg of product
MA_f mass of one unit product f
GT_{ku} GHG emission produced by vehicle type k, technology u in one unit distance
GE average of produced emission by recycling one unit of expired product
θ_f recycling cost of one unit expired product f
G a large value number

Variables:

x_{ijkut} 1 if retailer j is visited exactly after retailer i by vehicle type k and technology u at period t, otherwise 0
I_{ift} inventory level of product f in retailer i at the end of period t
y_{kut} number of transportation type k with technology u at period t
$q_{ifkutt'}^{dd'}$ quantity of received product f by retailer i through vehicle k with technology u at period t for being used at period t' which is loaded by equipment d and is unloaded by equipment d'
$Q_{ifkutt'}^{dd'}$ total quantity of received product f by retailer i through vehicle k with technology u at period t which is loaded by equipment d and is unloaded by equipment d'
B_{ift} amount of backordered product f of retailer i at the end of period t
EX_{ift} number of expired product f in inventory of retailer i at the end of period t
n_{it} 1 if delay is occurred in visiting retailer i at period t, otherwise 0
r_{it} 1 if backorder is occurred in visiting retailer i at period t, otherwise 0
z_{ikut} 1 if retailer i is served at period t by vehicle k, technology u, otherwise 0
T_{ikut} arriving time to retailer i by vehicle k with technology u at period t
C_{ift} 1 if product f is delivered to retailer i at period t, otherwise 0
w_{ikut} a dummy variable to eliminate sub-tours

Objective functions:

$$\text{Max } f_1 = \sum_{i \in M}\sum_{f \in F}\sum_{k \in K}\sum_{u \in U}\sum_{d \in D}\sum_{d' \in D'}\sum_{t \in T} R_{ft} Q^{dd'}_{ifkut}$$
$$- \sum_{i \in M}\sum_{f \in F}\sum_{t \in T} h_{if} I_{ift} - \sum_{i \in M}\sum_{f \in F}\sum_{t \in T} \gamma_{if} B_{ift} - \sum_{i \in M}\sum_{f \in F}\sum_{t \in T} \theta_f EX_{ift}$$
$$- \sum_{i \in M}\sum_{f \in F}\sum_{t \in T} O_{if} C_{ift} - \sum_{k \in K}\sum_{u \in U}\sum_{t \in T} fc_{ku} z_{0kut} - \sum_{(i,j) \in M'}\sum_{k \in K}\sum_{u \in U}\sum_{t \in T} v_{ku} y_{kut} c_{ij} x_{ijkut}$$
$$- \sum_{i \in M}\sum_{f \in F}\sum_{k \in K}\sum_{u \in U}\sum_{d \in D}\sum_{d' \in D'}\sum_{t \in T} LC_d Q^{dd'}_{ifkut} - \sum_{i \in M}\sum_{f \in F}\sum_{k \in K}\sum_{u \in U}\sum_{d \in D}\sum_{d' \in D'}\sum_{t \in T} UC_{id'} Q^{dd'}_{ifkut}$$
(39.1)

$$\text{Min } f_2 = \beta_n (\sum_{i \in M}\sum_{t \in T} n_{it}) + \beta_B (\sum_{i \in M}\sum_{f \in F}\sum_{t \in T} B_{ift}) + \beta_r (\sum_{i \in M}\sum_{t \in T} r_{it}) \qquad (39.2)$$

$$\text{Min } f_3 = \sum_{i \in M}\sum_{f \in F}\sum_{k \in K}\sum_{u \in U}\sum_{d \in D}\sum_{d' \in D'}\sum_{t \in T} MA_f GL_d Q^{dd'}_{ifkut}$$
$$+ \sum_{i \in M}\sum_{f \in F}\sum_{k \in K}\sum_{u \in U}\sum_{d \in D}\sum_{d' \in D'}\sum_{t \in T} MA_f GU_{id'} Q^{dd'}_{ifkut} \qquad (39.3)$$
$$+ \sum_{(i,j) \in M'}\sum_{k \in K}\sum_{u \in U}\sum_{t \in T} GT_{ku} y_{kut} c_{ij} x_{ijkut} + \sum_{i \in M}\sum_{f \in F}\sum_{t \in T} GEEX_{ift}$$

Constraints:

$$I_{ift} - B_{ift} = I_{if(t-1)} - d_{ift} + \sum_{k,u,d,d'} Q^{dd'}_{ifkut} - EX_{ift} - B_{if(t-1)} \qquad \forall i \in M, f, t \quad (39.4)$$

$$B_{ift} r_{it} \geq \sum_{k,u,d,d'\, t > t'} q^{dd'}_{ifkutt'} \qquad \forall i \in M, f, t \quad (39.5)$$

$$I_{ift} \times B_{ift} = 0 \qquad \forall i \in M, f, t \quad (39.6)$$

$$\sum_{k,u,d,d',t'} q^{dd'}_{ifkutt'} \leq IC_{if} - I_{if(t-1)} \qquad \forall i \in M, f, t \quad (39.7)$$

$$\sum_{k,u,d,d'}\sum_{t' > (t+L_f)} q^{dd'}_{ifkutt'} \leq EX_{ift} \qquad \forall i \in M, f, t \quad (39.8)$$

$$\sum_{t' \in T} q^{dd'}_{ifkutt'} \leq Q^{dd'}_{ifkut} \qquad \forall i \in M, f, k, u, d, d't \quad (39.9)$$

$$\frac{C_{ift}}{G} \leq \sum_{k,u,d,d'} Q^{dd'}_{ifkut} \leq C_{ift} G \qquad \forall i \in M, f, t \quad (39.10)$$

$$T_{ikut} + \sum_{f,d,d'} Q^{dd'}_{ifkut} \cdot sp_{ku} + \frac{c_{ij}}{s_{ijku}} \leq T_{jkut} + G(1 - x_{ijkut}) \qquad \forall (i,j) \in M', k, u, t \quad (39.11)$$

$$a_i \le \sum_{k \in K} \sum_{u \in U} T_{ikut} \le (b_i + G.n_{it}) \quad \forall i \in M, t \quad (39.12)$$

$$\sum_{i \in Mf,d,d',t'} q^{dd'}_{ifkutt'} \le cap_{ku} y_{kut} z_{0kut} \quad \forall k, u, t \quad (39.13)$$

$$\sum_{j \in M'} x_{ijkut} = \sum_{j \in M'} x_{jikut} = z_{ikut} \quad \forall i \in M', k, u, t \quad (39.14)$$

$$w_{ikut} - w_{jkut} + (|G|+1)x_{ijkut} \le |G| \quad \forall (i,j) \in M, k, u, t \quad (39.15)$$

Equation (39.1) maximizes the expected profit which is equal to revenue from each product minus total cost function; which includes holding cost, backordering cost, recycling of expired products, ordering cost, fixed and variable transportation cost, loading and unloading cost. Equation (39.2) minimizes the total delays which might occur when visiting the customers, number of backordered products and number of frequency of backorders. Equation (39.3) minimizes GHG emissions produced due to loading and unloading products, transportation and also produced emission due to recycling the expired products. Constraint (39.4) balances the inventory level in the periods. Constraint (39.5) calculates amount of backordered product. Equation (39.6) guarantees value of I_{ift} and B_{ift} cannot be positive simultaneously. Equation (39.7) verifies the capacity constraint of the retailer inventory for each product. Constraint (39.8) calculates number of expired products. Constraint (39.9) calculates total amount of received products for using in different periods by the retailer. Constraint (39.10) determines that if retailer received any product. Equation (39.11) determines arriving time to next retailer which should be after the time of visiting recent retailer plus the unloading time of products plus the traveling time between two retailers. Equation (39.12) verifies the time windows constraint and calculates also number of delays. Constraint (39.13) guarantees that the capacity of each vehicle should not be exceeded. Moreover, this constraint eliminates tour making without visiting the supplier. Equations (39.14) and (39.15) are also developed for making tours and eliminating sub-tours. Due to the existence of nonlinear equations in the objective functions and constraints (39.5), (39.6) and (39.13), linearization methods are applied to convert the model to its equivalent linear form.

39.4 Resolution Approach

In order to check validity of proposed model, a small size instance problem is generated randomly where supplier distributes 2 types of products to 2 retailers in 2 periods by different vehicles in term of capacity and technology. To solve this problem, we used the "GAMS v22.2" optimization software and solver CPLEX v10.1. The CPU time implies that it is needed about 12 min to solve the small size of problem. The previous studies show the IRPs are NP-hard optimization problems (Shukla et al. 2013) and we used a meta-heuristic to solve this problem. There are different algorithms as heuristic and meta-heuristic to solve the NP-hard multi-objective problems. Based on previous studies (Amorim and Almada-Lobo 2014), Non-dominated Sorting Genetic Algorithm-II (NSGA-II) (Deb et al. 2002) is an efficient algorithm for solving the IRPs and we selected and used this algorithm to cope with complexity and solve the proposed model. This algorithm starts by an initial solution (population) that is randomly generated. In each iteration, using crossover and mutation as two operators, new solutions (children) are produced using existing solutions. It is noteworthy that in this paper a binary tournament selection is used for selection of parents and swap is selected as crossover and mutation operators. In each iteration, the objective value of each solution is compared against two strategies, ranking and crowding distance.

The structure of the solutions contains 13 matrices. For example, the matric $[X]_{i \times j}^{k \times u \times t}$ determines genes which assign vehicle type k and technology u to route between retailer i and j at period t.

39.4.1 Experimental Results

In order to study the performance of proposed model, two instance problems (Table 39.1) are generated randomly (Table 39.2) according to our past experience in a real case. In the Problem No. 1, food industry supplier distributes 2 types of products to 5 retailers in a planning horizon composed of 2 periods (periods are weekly). In the Problem No. 2, the supplier distributes 8 types of products to 15 retailers in a planning horizon with 6 periods. In the both instance problems, supplier uses 3 types of vehicles (small, medium, large) and 2 types of different vehicles in technology.

In order to improve the efficiency of the proposed NSGA-II algorithm and find the best results, a Taguchi design is applied as a statistical technique (DOE method) to find the best combination of the parameters of NSGA-II. Taguchi proposes two main categories of factors: controllable and noise. The noise factors are out of control and their elimination is often impossible. The method attempts to find the optimal level of controllable factors by minimizing the effect of noise factors (Chatsirirungruang 2009). Using this method, the best configuration of four critical parameters of NSGA-II is: Number of Population = 100, Maximum Iteration = 120, Crossover Rate = 0.7 and Mutation Rate = 0.2 (Fig. 39.1). It is noteworthy that both problems are solved by NSGA-II algorithm which is coded

Table 39.1 Size of instance problem

	Problem No. 1	Problem No. 2
|M| × |F| × |K| × |U| × |T|	5 × 2 × 3 × 2×2	15 × 8 × 3 × 2×6

Fig. 39.1 Signal-to-noise ratio for each level of the parameters

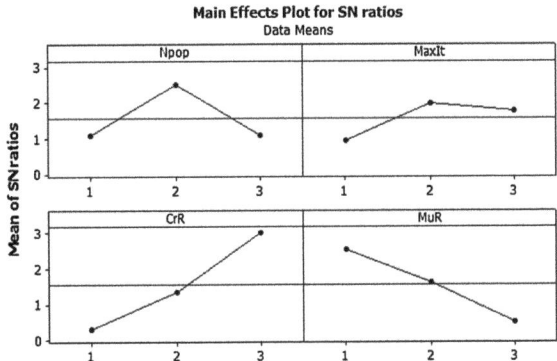

Table 39.2 Random value of parameters

Parameter	Value/Uniform distribution	Parameter	Value/Uniform distribution
I_{if0} (Pallet)	$\sim U[0,3]$	c_{ij} (Km)	$\sim U[5,35]$
h_{if} (€)	$\sim U[10,20]$	d_{ift} (Pallet)	$\sim U[2,12]$
IC_{if} (Pallet)	$\sim U[1,2] \times \max\{d_{ift}\}$	L_f (Period)	$\sim U[0.05, 0.25] \times |T|$
γ_{if} (€)	$\sim U[10,20]$	MA_f (Kg)	$\sim U[50, 300]$
O_{if} (€)	$\sim U[40,60]$	a_i	0
θ_f (€)	$\sim U[5,30]$	b_i (Hour)	$\sim U[0.5, 8]$
R_{ft} (€)	$\sim U[100,200]$	GE(kg/pallet)	3.7

using MATLAB 2014 and run on a personal computer Intel Corei7, 2.27 GHZ with 12.0 GB RAM.

Table 39.3, shows the value of best solutions and computational times for each problem. The reported CPU time that could be as a performance indicator, implies about 15 min is necessary to solve the problem No. 2 as a problem of large size.

Figure 39.2 shows also the obtained optimal Pareto frontier which consist a set of solutions which could be obtained from trade-off between economic, CSL and green criteria for two instance problems.

Table 39.3 The best value of each objective in final Pareto

Pro.	OF	The best solutions			CPU time (s)
		Solution A	Solution B	Solution C	
1	Z_1	**1,799.00**	835.00	709.00	27.04
	Z_2	10.30	**3.00**	3.90	
	Z_3	43.53	73.03	**5.51**	
2	Z_1	**34,762.00**	17,296.00	13,941.00	896.52
	Z_2	381.40	**139.90**	203.70	
	Z_3	1,646.19	1,742.80	**325.62**	

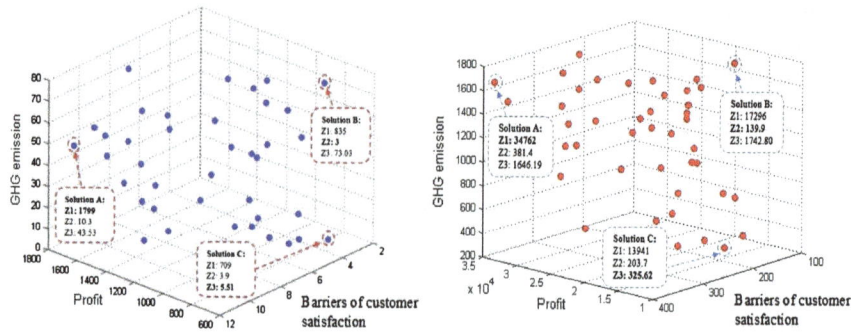

Fig. 39.2 Optimal Pareto frontier for problem No. 1 (*Left*) and problem No. 2 (*Right*)

The analysis of optimal Pareto frontier implies that the maximum of the customer satisfaction (Z2) could be obtained by decreasing the profit (Z1) by about 50.2 % in problem No. 2 with regard to the best economical solution.

Moreover, another analysis of obtained results of problem No. 2 shows that the decision maker can reach the lowest level of produced GHG emission (Z3) by 59.9 % reduction in profit (Z1). Since, in this study the profit of product is 10 % of the product price, this analysis shows that the retailer can decrease its product price by only 5.99 % to reach best level of green criteria. It means, the firm could achieve the best green solution and sell the product for only 5.99 % of increasing of price. This increasing can be justified by a green label, if the manager decides that customer has to pay this increasing of price, if not, a part of this increasing can be supported by producer. The analysis of two intermediate solutions in problem No. 2 shows that there is a possibility that decision maker increase CSL and Profit, at the same time without being aware of the GHG emission. Based on this analysis, decision maker can increase 23.1 % in CSL (Z2) and 26.7 % in profit (Z1) while amount of GHG emission (Z3) will be increased 62.4 %. It is noteworthy that, obtained Pareto enables the manager to make the compromise decisions between conflicting issues of profit, CSL and green.

39.5 Conclusion

This study proposes a new multi-objective mathematical model to handle the effect of customer satisfaction and green issues on the economic aspect of perishable IRPs. This study integrates the concept of CSL by defining three new indicators such as number of delays in visiting the retailers, amount, and number of backordered products. Due to the complexity of the problem, NSGA-II algorithm is developed to solve the model and obtain optimum Pareto frontier for the generated instance problems. Obtained results show the manager could motivate the green-friendly customers by a green label and sell their products only 5.99 % more expensive. Moreover, the best amount of CSL can be reached by 50.2 % reduction in profit. However, by the proposed model, decision makers are able to find the best compromise between these three aspects.

As future research, age and price strategy, dynamic routing and stochastic parameters for demand, several costs and delivery time will be studied.

Acknowledgment The authors gratefully acknowledge the scientific and financial supports of Region Rhône-Alpes, France, ARC7 in developing this paper.

References

Al Shamsi A, Al Raisi A, Aftab M (2014) Pollution-Inventory routing problem with perishable goods. In: Golinska P. (ed) Logistics operations, supply chain management and sustainability, vol 42, EcoProduction. Springer International Publishing, pp. 585–596. doi:10.1007/978-3-319-07287-6_42

Amorim P, Almada-Lobo B (2014) The impact of food perishability issues in the vehicle routing problem. Comput Ind Eng 67:223–233

Chatsirirungruang P (2009) Application of genetic algorithm and Taguchi method in dynamic robust parameter design for unknown problems. Int J Adv Manuf Technol 47:993–1002. doi:10.1007/s00170-009-2248-8

Coelho GL (2014) Optimal joint replenishment, delivery and inventory management policies for perishable products. Comput Oper Res 47:42–52

Custódio A, Oliveira R (2006) Redesigning distribution operations: a case study on integrating inventory management and vehicle routes design. Int. J. Logist. 9:169–187

Deb K, Member A, Pratap A, Agarwal S, Meyarivan T (2002) A fast and elitist multiobjective genetic algorithm. IEEE Trans Evol Comput 6:182–197

Hsu C-I, Hung S-F, Li H-C (2007) Vehicle routing problem with time-windows for perishable food delivery. J Food Eng 80:465–475

Madadi A, Kurz ME, Ashayeri J (2010) Multi-level inventory management decisions with transportation cost consideration. Transp. Res. Part E Logist. Transp. Rev. 46:719–734

McCormick GP (1976) Computability of global solutions to factorable nonconvex programs: Part I —Convex underestimating problems. Math Program 10:147–175. doi:10.1007/BF01580665

Mirzapour Al-e-hashem SMJ, Rekik Y (2013) Multi-product multi-period Inventory Routing Problem with a transshipment option: a green approach. Int. J. Prod, Econ

Nagurney A, Masoumi AH, Yu M (2011) Supply chain network operations management of a blood banking system with cost and risk minimization. Comput. Manag. Sci. 9:205–231. doi:10.1007/s10287-011-0133-z

Sazvar Z, Mirzapour Al-e-hashem SMJ, Baboli A, Akbari Jokar MR (2014) A bi-objective stochastic programming model for a centralized green supply chain with deteriorating products. Int J Prod Econ 150:140–154

Shukla N, Tiwari MK, Ceglarek D (2013) Genetic-algorithms-based algorithm portfolio for inventory routing problem with stochastic demand. Int J Prod Res 51:118–137. doi:10.1080/00207543.2011.653010

Chapter 40
Decentralized Routing of Automated Guided Vehicles by Means of Graph-Transformational Swarms

Larbi Abdenebaoui and Hans-Jörg Kreowski

Abstract The use of decentralized control systems in the field of intralogistics offers high promises, but presents also various technical challenges. One of them is the development of adequate modeling approaches. Based on the problem of routing of automated guided vehicles (AGVs), this paper demonstrates how graph-transformational swarms can be used to model and develop decentralized solutions. Representing the environment as a graph, every AGV is modeled as a swarm member. They act locally and simultaneously like swarms members do in nature. The movement of the AGVs is regulated in such a way that they reach their targets on shortest paths without collisions.

Keywords Intralogistics · Automated guided vehicles · Swarm computation · Graph transformation

40.1 Introduction

The logistic networks grow increasingly, becoming more and more difficult to be controlled. Nevertheless, the nowadays applied controls are mostly central and often turn out to be inflexible to deal with large and dynamic networks. One of the most significant current paradigm that faces this complexity is the so-called autonomous control approach (cf. Hülsmann et al. 2011). This approach proposes that each logistics object, such as a container or an automated guided vehicle, receives its own computing processor and make its decision autonomously. Therefore, the components can react locally and quickly to changes in the environment. However, a major challenge within this kind of decentralized approach is how the individuals act and cooperate with each other to reach a desired global goal. In this paper, we introduce and discuss graph-transformational swarms as a formal modeling approach to

---

L. Abdenebaoui (✉) · H.-J. Kreowski
University of Bremen, P.O. Box 330440, 28334 Bremen, Germany
e-mail: larbi@informatik.uni-bremen.de

H.-J. Kreowski
e-mail: kreo@informatik.uni-bremen.de

decentralized control using as case study the routing problem of automated guided vehicles.

Automated guided vehicles (AGVs) are driverless transportation engines that follow traditionally guide-paths like lines on the ground. Their use is expanding rapidly in the last decades. Beside the classical application in small manufacturing systems, nowadays, the tendency is to use AGVs more and more for transport in high complex systems including external areas like container terminals (for a general overview, see, e.g., Vis 2006).

One of the important problems that a designer of an AGV system faces in complex areas is the collision-free routing problem. The classical way to solve this problem is the central time windows planning (see, e.g., Smolic-Rocak et al. 2010; Taghabonidutta and Tanchoco 1995; ter Mors et al. 2010). The tendency in the last years is to explore more decentralized approaches (e.g., Schwarz and Sauer 2012; Weyns et al. 2008). This paper contributes to this research, by proposing graph-transformational swarms as a rule-based modeling and solution generator approach.

A graph-transformational swarm consists generally of members that act and interact simultaneously in an environment, which is represented by a graph. The members are all of the same kind or of different kinds. Kinds and members are modeled as graph transformation units (Kreowski et al. 2008) each consisting of a set of graph transformation rules specifying the capability of members and a control condition which regulates the application of rules. The basic framework is introduced in Abdenebaoui et al. (2013), where a simple ant colony, cellular automata and discrete particle systems are modeled to demonstrate the usefulness and flexibility of the approach. The graph-transformational swarms in this paper is composed of three kinds. *preparator*, *assigner* and *navigator*. Every node in the graph gets a member of kind *preparator* and every AGV gets a member of kind *assigner* and a member of kind *navigator* reflecting in this way the ideas of autonomous control approach. The movement of the AGVs is regulated in such a way that they reach their targets on shortest paths without collisions.

This paper is organized as follows. In Sect. 40.2, graph-transformational swarms is recalled starting with the basic concepts of graph transformation. Section 40.3 presents the main contribution of this paper. It proposes a routing solution as a graph-transformational swarms and discuss some computational characteristics. The conclusion is in the Sect. 40.4.

40.2 Basics of Graph-Transformational Swarms

This section recalls the concept of graph-transformational swarms starting with the basic components of the chosen graph transformation approach as far as needed in this paper (for more details, see, e.g., Ehrig et al. 2006; Kreowski et al. 2006, 2008; Rozenberg 1997).

40.2.1 Basic Concepts of Graph Transformation

A *(directed edge-labeled) graph* consists of a set of nodes and a set of labeled edges such that every edge is directed. If the target is equal to the source, then the edge is called a *loop*.

If G is a subgraph of H, we use the notation $G \subseteq H$. A match $g(G)$ of a graph G in a graph H is a subgraph of H obtained by renaming the nodes and edges of G.

A *rule* $r = (N, L, K, R)$ consists of four graphs, the *negative context* N, the *left-hand side* L, the *gluing graph* K, and the *right-hand side* R such that $N \supseteq L \supseteq K \subseteq R$. We depict a rule as $N \longrightarrow R$ dashing the elements in N that do not belong to L. In this paper, we consider rules that manipulate only edges (i.e., the nodes are neither deleted nor added). We use the same relative positions of nodes in N and R. K can be identified in this way as the identical parts of L and R. Given the rules $r_i = (N_i, L_i, K_i, R_i)$ for $i = 1, \ldots, n$, the *parallel rule* $p = \sum_{i=1}^{n} r_i$ is given by the disjoint unions of the respective components N_i, L_i, K_i, R_i.

The application of a rule $r = (N, L, K, R)$ to a graph G replaces a match of L in G by R such that the match of K is kept. If L is a proper subset of N the match of L must not be extendable to a match of N. A rule application is denoted by $G \underset{r}{\Longrightarrow} H$ where H is the resulting graph and called a *direct derivation* from G to H. A sequence $G = G_0 \underset{r_1}{\Longrightarrow} G_1 \underset{r_2}{\Longrightarrow} \cdots \underset{r_n}{\Longrightarrow} G_n = H$ is called a *derivation* from G to H. Two direct derivations $G \underset{r}{\Longrightarrow} H_1$ and $G \underset{r'}{\Longrightarrow} H_2$ of two rules r and r' are *(parallel) independent* if the corresponding matches intersect only in gluing items.

The following considerations are based on the parallelization theorem in Kreowski (1977). A parallel rule $p = \sum_{i=1}^{n} r_i$ can be applied to G if and only if the rules r_i for $i = 1 \ldots n$ can be applied to G and the matches are pairwise independent. Moreover, the r_i can be applied one after the other in arbitrary order deriving in each case the same graph as the application of p to G.

This allows the use of massive parallelism in the context of graph transformation based on local matches of component rules which are much easier to find then matches of parallel rules.

A *control condition* C is defined over a finite set P of rules and specifies a set $SEM(C)$ of derivations. Typical control conditions are priorities among the rules. Other examples of control conditions that are used in this paper are the expression $\|r\|$ and $[r]$. $\|r\|$ requires that a maximum number of rule r be applied in parallel. $[r]$ requires that the rule r may be applied or not.

A *graph class expression* X specifies a set of graphs denoted by $SEM(X)$. The class of all graph class expressions is denoted by $\mathcal{X}$. We use the graph class expressions *distance* and *id-looped*. $SEM(distance)$ contains all loop-free graphs where each edge is labeled with a distance (i.e., a value $d \in \mathbb{N}$). $SEM(id\text{-}looped(distance))$ contains all graphs G where the nodes are numbered from 1 to n and every node has an additional loop labeled with the corresponding number. In the following, if a node has an i-loop we call it i-node or simply i.

A *graph transformation unit* is a pair $gtu = (P, C)$ where P is a set of rules and C is a control condition over P. The *semantics* of gtu consists of all derivations of the rules in P allowed by C. A unit gtu_0 is *related* to a unit gtu if gtu_0 is obtained from gtu by relabeling and renaming. The set of units related to gtu is denoted by $RU(gtu)$.

40.2.2 Graph-Transformational Swarms

A graph-transformational swarm consists of members of the same kind or of different kinds to distinguish between different roles members can play. All members act simultaneously in a common environment represented by a graph. The number of members of each kind is given by the size of the kind. While a kind is a graph transformation unit, the members of this kind are modeled as units related to the kind so that all members of some kind are alike.

A swarm computation starts with an initial environment and consists of iterated rule applications requiring massive parallelism meaning that each member of the swarm applies one of its rules in every step. The choice of rules depends on their applicability and the control conditions of the members as well as on a cooperation condition. Moreover, a swarm may have a goal given by a graph class expression. A computation is considered to be successful if an environment is reached that meets the goal.

Definition 1 (*swarm*) A *swarm* is a system $S = (I, K, s, m, c, g)$ where I is a graph class expression specifying the set of *initial environments*, K is a finite set of graph transformation units, called *kinds*, s associates a *size* $s(k) \in \mathbb{N}$ with each kind $k \in K$, m associates a family of *members* $(m(k)_i)_{i \in [s(k)]}$ with each kind $k \in K$ with $m(k)_i \in RU(k)$ for all $i \in [s(k)]$,[1] c is a control condition called *cooperation condition*, and g is a graph class expression specifying the *goal*.

A swarm may be represented schematically depicting the components *initial*, *kinds*, *size*, *members*, *cooperation* and *goal* followed by their respective values.

Definition 2 (*swarm computation*) A *swarm computation* is a derivation $G_0 \underset{p_1}{\Longrightarrow} G_1 \underset{p_2}{\Longrightarrow} \cdots \underset{p_q}{\Longrightarrow} G_q$ such that $G_0 \in SEM(I)$, $p_j = \sum_{k \in K} \sum_{i \in [s(k)]} r_{j_{ki}}$ with a rule $r_{j_{ki}}$ of $m(k)_i$ for each $j \in [q]$, $k \in K$ and $i \in [s(k)]$, and c and the control conditions of all members are satisfied.

That all members must provide a rule to a computational step is a strong requirement because graph transformation rules may not be applicable. In particular, if no rule of a swarm member is applicable to some environment, no further computational step would be possible and the inability of a single member stops the whole

---

[1] $[n] = \{1, \ldots, n\}$.

swarm. To avoid this global effect of a local situation, we assume that each member has the empty rule $(\emptyset, \emptyset, \emptyset, \emptyset)$ in addition to its other rules. The empty rule gets the lowest priority. The empty rule is called *sleeping rule*.

40.3 Routing of AGVs as Graph-Transformational Swarms

In this section, we propose a solution to the routing problem of the AGVs using the notion of graph-transformational swarms. We model the infrastructure where the AGVs operate as an *id-looped distance* graph. Usually, in a graphical representation the nodes correspond to the ends or intersections of paths including important stations like pick-up and delivery locations. The edges represent the path or segments of paths in the infrastructure depending on their lengths. The distance of an edge can correspond to the distance of the corresponding path or to some cost of traversing it.

In this paper, the task assignment have the most simple form serving solely the simulation purposes of the computational steps. A transport mission is specified by a start node and a target node. We represent the state of an AGV operating in the system and having as target the node T by an edge e labeled by a, T. We call such an edge an *AGV-edge*. Every AGV tries to reach the assigned target using the minimal distance possible and avoiding conflicts with other AGVs.

40.3.1 The Routing Swarm

We propose a solution based on two stages. The first one consists of the preparation of the layout in a such way that the AGVs follow later only local information. The second one consists of the navigation process of the AGVs depending on an arbitrary task assignment.

The parameter m is the number of AGVs and can be chosen freely. The swarm has three kinds: *preparator*, *assigner* and *navigator*. Their sizes are n, m, and m, respectively where n is the number of nodes in the underlying graph $G \in SEM(id\text{-}looped (distance))$. The members are obtained by relabeling in such a way that every node in the graph gets assigned a member of kind *preparator*, and every AGV gets assigned a member of kind *assigner* and a member of kind *navigator*. How relabeling is achieved, is described below in the detailed introduction of the kinds. The cooperation condition requires that *preparator* is applied as long as possible realizing the layout preparation followed by an arbitrary repetition of assignments each followed by an arbitrary number of navigations. The swarm is schematically presented in Fig. 40.1.

Fig. 40.1 The schematic representation of the graph-transformational swarm *routing*

$$
\begin{aligned}
&\text{routing}(m) \\
&\quad \texttt{initial:} \quad \textit{id-looped}(distance) \\
&\quad \texttt{kinds :} \quad \textit{preparator,assigner,navigator} \\
&\quad \texttt{size :} \quad n = \#nodes, m, m \\
&\quad \texttt{members:} \quad preparator_i \text{ for } i \in [n] \\
&\quad\quad\quad\quad\quad\quad\ assigner_j \text{ for } j \in [m] \\
&\quad\quad\quad\quad\quad\quad\ navigator_k \text{ for } k \in [m] \\
&\quad \texttt{coop:} \quad preparator!; (assigner; navigator^*)^* \\
&\quad \texttt{goal:} \quad \textit{all vehicles arrived}
\end{aligned}
$$

40.3.2 Layout Preparation

The goal of the layout preparation is that every node in the graph can indicate to an AGV having the target T which next node can be visited to reach T with the minimal distance possible. Given an i-node, we code such an indicator as an outgoing edge e labeled with a pair T, D. We say that i has an indicator to T with the distance D using the successor s, where s is the destination of e. If D is minimal considering simple paths up to the maximal lengths l, we say that the indicator is l-minimal. If D is minimal considering all possible paths, then the indicator is optimal. A path composed from indicators to a target T is called an indicator path to T. If every node in the graph has only optimal indicators to every reachable node, then the graph is called *fully indicated*.

The members of kind *preparator* realize the layout preparation process. The kind *preparator* specified in Fig. 40.2 initializes this process with rule *init*. It adds an indicator in a X-node to a direct successor s provided that such indicator does not yet exist. The rule *connect* connects an X-node with an existing indicator path to T. It is applied if a direct successor s of X exists having an indicator to T with a distance D_1 provided that there is no other direct successor of X having the same target T with a distance D_2 such that $D_2 < D_1 + d$. The rule *connect* generates an indicator to T with the new distance $D_1 + d$ using the successor s. If an X-node has two indicators to a target T with different distances, the rule *select* deletes the one with the larger distance selecting in this way the best one to be kept. The control condition requires that the rule *init* is applied with maximum parallelism. Afterward, The rule *connect* is applied followed by *select* both with maximum parallelism. Because of the negative application condition of *init*, for every successor node *init* is applied only once in the whole swarm computation while *connect* and *select* are iterated with the repetition of *preparator*.

In the swarm, there are n members of kind *preparator*. A member $preparator_i$ for $i \in [n]$ is obtained from *preparator* by relabeling all occurring X with i. In the following, we describe how the members work together using the computation in Fig. 40.3 as illustrating example. In the first step, all members apply the rule $\|init\|$ generating in every node an indicator to all successor nodes (see the result of the derivation p_1 in the example). In the second step, the parallel application of the rule

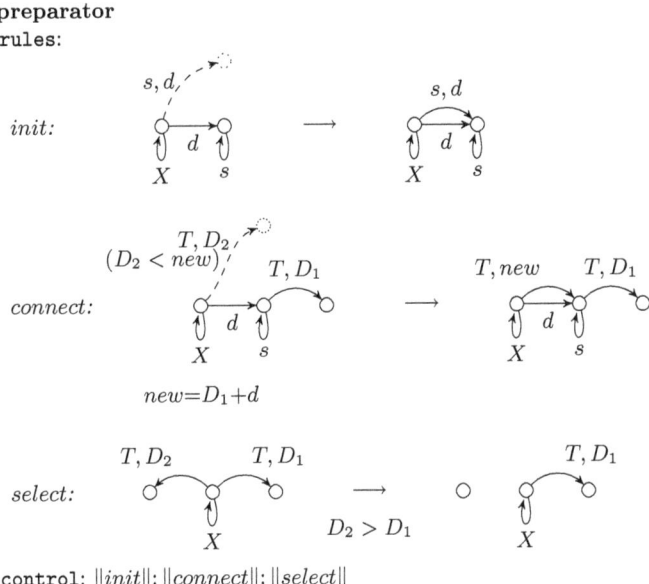

Fig. 40.2 The unit *preparator*

‖*connect*‖ connects all nodes to construct indicator paths of length 2. It connects also those that are already connected to indicator path of length 1 if the new distance is smaller than the old (in the example p_2 adds indicators in the nodes 2 and 3). In the third step all members apply ‖*select*‖ deleting all indicators using path of length 2 and 1 that are not 2-minimal (p_3 deletes the indicator in 2 to 1 with distance 3 keeping the minimal indicator to 1 with distance 2). Note that a node can have more than one minimal indicator to the same target (in the example, the node 3 gets two indicators to 1 with the same distance 4). By induction, one can proof that in $2L - 1$ steps all L-minimal indicators are constructed. If the longest path with a minimal distance is constructed, then the *preparator*-members can not apply any rule anymore (except the sleeping rule). And the constructed indicators are optimal. Because the length of such a path is shorter or equal $n - 1$, the number of steps is bounded by $2n - 3$. Summarizing, the following correctness result holds:

Theorem 1 *Given an id-looped distance graph G, the swarm routing transforms it in a fully indicated graph in a number of steps bounded by $2n - 3$ where n is the number of nodes in G.*

Note that, the layout preparation process can be considered as a distributed version of the Dijkstra's shortest path algorithm (cf. Dijkstra 1959).

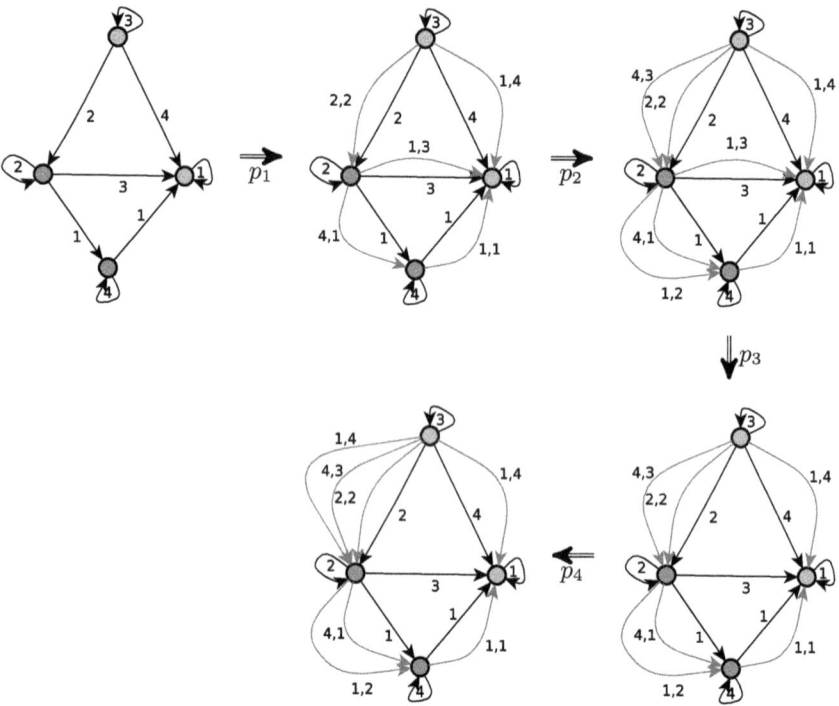

Fig. 40.3 A sample computation of the swarm *routing* illustrating the layout preparation process

40.3.3 Assignment and Navigation

The kinds *assigner* and *navigator* model the task assignment and navigation process from the point of view of the AGVs (Figs. 40.4 and 40.5). The kind *assigner* has just a single rule *assign* that creates a vehicle edge labeled with a, T between two arbitrary nodes provided that this position is free and that the vehicle edge is not yet present in the whole graph. The control condition [*assign*] requires that the rule may be applied or not so that not every vehicle must be present any time. The member *assigner*$_j$ for $j \in [m]$ are obtained from the kind *assigner* by relabeling all occurring a with a_j and the a' by a_k for $j \neq k$.

The kind *navigator* contains two rules *move* and *arrive*. The rule *move* is responsible of the forward movement of the AGV until the target is reached. It moves the AGV a with the target T forward following a next indicator to T provided that there is no other AGV there. If the target node is reached the rule *arrive* can be applied. The rule *arrive* deletes the AGV-edge signaling in this way to the task assigner that the AGV a is free for a new assignment. The control condition requires that *move* or *arrive* is applied. The member *navigator*$_k$ for $k \in [m]$ are obtained from *navigator* by relabeling all occurring a with a_k.

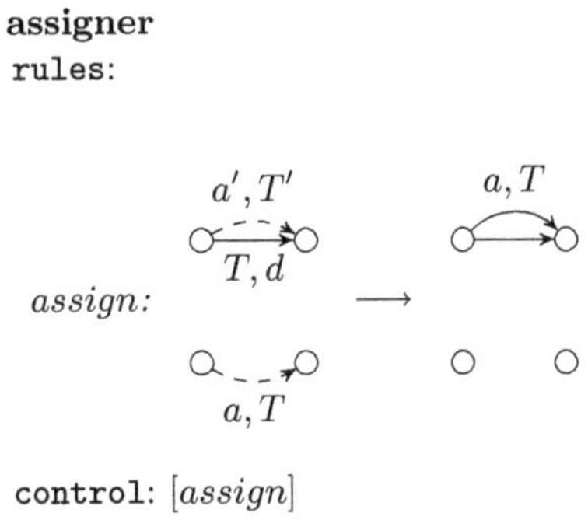

Fig. 40.4 The unit *assigner*

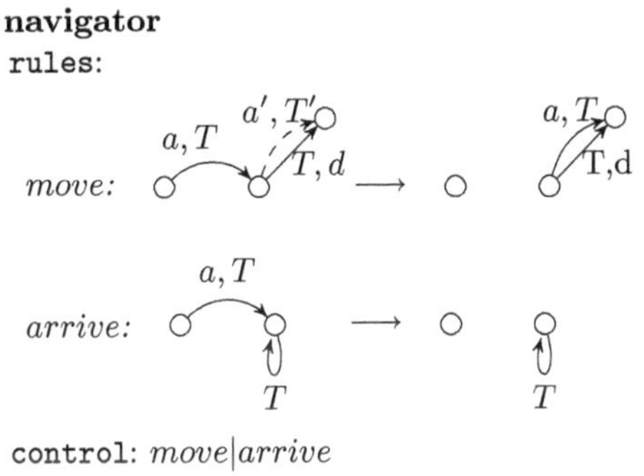

```
assigner
rules:

           a',T'              a,T
assign:  o--→•            o→→o
           T,d
                    ⟶
           o----→o         o    o
              a,T

control: [assign]
```

Fig. 40.5 The unit *navigator*

```
navigator
rules:

              a',T'                a,T
         a,T  o                    o
move:    o   /T,d      ⟶    o    /T,d
         o--o                     o

         a,T
arrive:  o→o           ⟶    o    o
           ↕                      ↕
           T                      T

control: move|arrive
```

After the layout preparation, only members of kind *assigner* and *navigator* are active. The *assigner* members create an arbitrary number lesser or equal m of AGV-edges in parallel. Afterwards, all created AGV-edges move forward in parallel. If two or more AGVs want to follow the same indicator, then one of them is selected randomly due to the negative application condition in the rule *move*. If an AGV arrives to its goal, the corresponding edge will be deleted. In this way the number of active AGVs decreases when the application of *navigator* is repeated. If the number of repetition is high enough, the swarm reaches its goal otherwise the process start again by assigning new tasks to inactive vehicles. The swarm repeats this process until the goal is reached. Especially, we have the following result.

Theorem 2 *If the swarm routing reaches its goal, each AGV that has been assigned to a target reaches this target collision-free.*

The random selection by concurring vehicles is obviously not fair. Because it is possible that a vehicle waits for a very long time while new coming vehicles are chosen to traverse. To make it fairer, one may add priorities to the vehicles that augment with waiting time as for example in Schwarz and Sauer (2012).

40.4 Conclusion

In this paper, we have proposed a decentralized solution to the automated guided vehicles routing problem using graph-transformational swarms. We have demonstrated how it is possible to generate a complex behavior of the whole swarm based only on simple rules on the individual level. The swarm solves the problem in two phases. The first one consists of the preparation of the graph in such a way that every node can indicate which node can be visited next in order to reach a target node in the shortest distance. The second phase consists of the navigation process following the best paths and avoiding conflicts in form of collisions.

In order to present the case study within the scope of this paper, it has been necessary to make some simplifications. In the future, more sophisticated versions will be investigated. For example, a more realistic task assignment process should be simulated and negotiations based on priorities should be integrated in the solution. However, we have shown how graph-transformational swarms can be used to model intralogistics processes, providing visualization and analysis capabilities as advantages. In particular, the formal modeling concept provided by graph-transformational swarms allows the verification of correctness properties as demonstrated with the two given theorems.

We are convinced that this kind of concepts applies well to all modeling problems within the scope of cyber-physical systems where a variety of machines, vehicles, and other components cooperate in a common environment acting and interacting simultaneously and to a good part autonomously (cf. The National Science Foundation 2008).

References

Abdenebaoui L, Kreowski HJ, Kuske S (2013) Graph-transformational swarms. In: Bensch S, Drewes F, Freund R, Otto F (eds) Proceedings of fifth workshop on Non-Classical Models for Automata and Applications—NCMA 2013. Umeå, Sweden, 13–14 Aug 2013. Österreichische Computer Gesellschaft, pp 35–50

Cyber-physical systems. Program Announcements & Information. The National Science Foundation, 30 Sept 2008

Dijkstra EW (1959) A note on two problems in connection with graphs. Numer Math 1(5):269–271

Ehrig H, Ehrig K, Prange U, Taentzer G (2006) Fundamentals of algebraic graph transformation (Monographs in theoretical computer science. An EATCS Series). Springer, Berlin, Heidelberg

Hülsmann M, Scholz-Reiter B, Windt K (2011) Autonomous cooperation and control in logistics. Springer, Berlin Heidelberg

Kreowski HJ (1977) Manipulationen von Graphmanipulationen. Ph.D. thesis, Technische Universität, Berlin

Kreowski HJ, Klempien-Hinrichs R, Kuske S (2006) Some essentials of graph transformation. In: Ésik Z, Martín-Vide C, Mitrana V (eds) Recent advances in formal languages and applications, Studies in Computational Intelligence, vol 25. Springer, pp 229–254

Kreowski HJ, Kuske S, Rozenberg G (2008) Graph transformation units—an overview. In: Degano P, Nicola RD, Meseguer J (eds) Concurrency graphs and models, essays dedicated to ugo montanari on the occasion of his 65th Birthday. Lecture Notes in Computer Science (LNCS), vol 5065. Springer-Verlag, New York, pp 57–75

Kreowski HJ, Kuske S, Rozenberg G (2008) Graph transformation units—an overview. In: Degano P, Nicola RD, Meseguer JD (eds) Concurrency, graphs and models. Lecture notes in computer science, vol 5065. Springer, pp 57–75

Rozenberg G (ed) (1997) Handbook of graph grammars and computing by graph transformation, vol 1. Foundations. World Scientific, Singapore

Schwarz C, Sauer J (2012) Towards decentralised agv control with negotiations. In: Kersting K, Toussaint M (eds) Proceedings of the sixth starting ai researchers symposium. Frontiers in artificial intelligence and applications, vol 241. IOS Press

Smolic-Rocak N, Bogdan S, Kovacic Z, Petrovic T (2010) Time windows based dynamic routing in multi-AGV systems. IEEE T Autom Sci Eng 7(1):151–155

Taghaboni-dutta F, Tanchoco JMA (1995) Comparison of dynamic routeing techniques for automated guided vehicle system. Int J Prod Res 33(10):2653–2669

ter Mors A, Witteveen C, Zutt J, Kuipers FA (2010) Context-aware route planning. In: Dix J, Witteveen C, (eds) Multiagent system technologies, 8th German Conference, MATES 2010, Leipzig, Germany. Lecture notes in computer science, vol 6251. Springer, pp 138–149

Vis IF (2006) Survey of research in the design and control of automated guided vehicle systems. Eur J Oper Res 170(3):677–709

Weyns D, Holvoet T, Schelfthout K, Wielemans J (2008) Decentralized control of automatic guided vehicles: applying multi-agent systems in practice. In: Companion to the 23rd ACM SIGPLAN conference on object-oriented programming systems languages and applications, OOPSLA Companion'08, New York, NY, USA. ACM, pp 663–674,

Chapter 41
Interoperability of Logistics Artifacts: An Approach for Information Exchange Through Transformation Mechanisms

Marco Franke, Till Becker, Martin Gogolla, Karl A. Hribernik and Klaus-Dieter Thoben

Abstract The information flow in logistics is covered by a growing set of systems and standards. To handle the increasing heterogeneity, the exchange of information rather than the exchange of data is favored because unlike data information contains a precise meaning. This precise meaning would reduce the false interpretation between stakeholders within information flows. The exchange of information on the system level requires the application of transformations on the communication interface levels. In this paper, a corresponding transformation and communication approach is presented. Finally, the impact of the conceptual approach is shown on a hypothetical application scenario.

Keywords Intralogistic systems · Heterogeneity · Interoperability for information exchange · Transformations for integration · Logistics artifacts

---

M. Franke (✉) · T. Becker · K.A. Hribernik · K.-D. Thoben
BIBA – Bremer Institut Für Produktion Und Logistik, Hochschulring 20, 28359 Bremen, Germany
e-mail: tbe@biba.uni-bremen.de

K.A. Hribernik
e-mail: hri@biba.uni-bremen.de

K.-D. Thoben
e-mail: tho@biba.uni-bremen.de

M. Gogolla
Faculty of Mathematics and Computer Science, University of Bremen, Bibliothekstraße 1, 28359 Bremen, Germany
e-mail: gogolla@informatik.uni-bremen.de

T. Becker · K.-D. Thoben
Faculty of Production Engineering, University of Bremen, Bibliothekstraße 1, 28359 Bremen, Germany

41.1 Introduction

Current megatrends such as globalization, growth of product variants, shorter product life cycles, and quick pervasion of new technologies require companies to become more agile and flexible. This megatrends directly influence the logistics activities along the whole supply chain, from the costumer via the manufacturer to the suppliers. The logistic processes are challenged according to the planning and realization of both the transport and storage of goods, which take into account time, quality, and cost constraints (Kumar 2007). The significance of these constraints increases rapidly due to the growing popularity of online shopping and the customers' desire for transparency and same-day delivery which have impacts to established production and logistic processes. These trends cause a reduction of delivery lot size, while in turn increasing the overall amount of deliveries (Ickert et al. 2007). To handle the logistics effort, automation through IT systems is placed on a global scale in which all systems are interconnected massively. The count of involved systems additionally increases through modular designed production and logistics systems, which are capable of interacting with each other through networked components by design. The higher count of systems also results in a higher number of interfaces and complex networked topologies, which increases the necessary effort to control these systems (Becker et al. 2015). The development of future production and logistic systems is supported by the development and implementation of cyber-physical systems (CPS) in production and logistic systems (Veigt et al. 2013). A possible application scenario for these kind of logistic systems is, for example, ubiquitous logistics (Wellsandt et al. 2013) or the ad hoc sharing of logistic infrastructure (Freitag et al. 2015).

The interfaces between interweaved IT logistic systems are currently implemented in a variety of proprietary data formats and de facto standards to transfer electronic business data across sectors. Hereby, each communication flow must overcome the data specific heterogeneity (Wache 2003) which is implemented through pragmatically implemented data integration solutions leaking flexibility and adaptability for dynamics in logistics. Apart from this heterogeneity, the essential logistic process of the unambiguous identification of goods is also provided by several standards at this time. The existence of numerous standards challenges the frictionless information exchange between interfaces of logistics systems to handle the above mentioned constraints of logistics.

This paper proposes an approach to achieve a robust communication in logistics according to both the increasing amount of involved IT systems and provided standards and de facto standards. For that purpose, the objective of the proposed approach it to resolve the heterogeneity of data formats through the exchange of information and not any longer of heterogeneous data between stakeholders. The implementation of this approach would prevent that a stakeholder especially an IT system must be capable of interpreting all used standards and data formats between him and all communication partners correctly. To demonstrate the necessity of resolving the heterogeneity, the authors present the current heterogeneity of common standards in logistics. Then, an approach is presented to overcome the data integration conflicts in logistics. Subsequently, the approach is discussed on basis of a job shop scenario, which contains intra logistic processes which are influenced by external events from suppliers. Finally, a summary and outlook is given.

41.2 Current Initial Situation for Interoperability in Logistics

The interoperability between enterprises is considered by ATHENA Interoperability Framework at four levels: "data/information (for information interoperability), services (for flexible execution and composition of services), processes (for cross-organizational processes) and enterprise/business (for collaborative enterprise operations)" (Vernadat 2010). This article focuses on the interoperability level data/information which is an enabler for the other three interoperability levels according to the wide involvement of IT systems in logistic processes and services. The mentioned requirements toward interoperability in logistics include the ability to mediate data between heterogeneous IT systems distributed over stakeholders spreading over different networks to enable a flexible and adaptable information flow. The interoperability has not achieved mostly on the application layer according to the OSI model yet. The logistics relevant applications define syntax and semantics of the applied terminology. Hereby, the interoperability could only be achieved between terminologies if the schematic conflicts, semantic conflicts and the intensional conflicts can be resolved (Goh 1997). The overall terminology includes both exchangeable commands and status information of logistics entities. One of the most common standards, which is available in more than 80 countries (GS1 Austria 2015) is EDIFACT EANCOM. The syntax and semantics of a command's payload define, e.g., the trade item, the location, and the legal entity. The representation of such items is standardized, too. Global Location Number (GLN), Global Trade Item Number (GTIN), etc. are examples of standards for unambiguous identification of subjects and objects in logistics processes (EANCOM 2012). The EDIFACT standard including identification standards creates a common language and terminology to automate the communication processes. Alternative standards are also available. Common combinations are (EDIFACT <-> ANSI ASC X12), (GLN <-> ISO/IEC) 6523 or (GTIN <-> ISO/IEC 15459). The availability and usage of different standards for trading relies also on historical reasons in which the markets were more separated geographically. While the EDIFACT standard has been more present in Europe, ANSI ASC X12 has been more applied in America. Apart from the covered and transferred information of a logistic object, new trends are going to increase the amount of information and its heterogeneity. In the following, two exemplary trends are presented.

Firstly, the integration of sensors in goods or in packages enables the recording of item level specific information. This information can be used to monitor the state of a good and to influence the logistic chain. This kind of information is product type specific. The representation and meaning of contained data is available through different standards. Common standards are the Open Data Format (O-DF) and the Open Messaging Interface (O-MI) standard. The objective of O-MI/O-DF is to transfer information on the product item level. The syntax and semantics differ to other standards such as EANCON.

Secondly, the creation of product specific services from the perspective of the logistic provider and product owner includes the usage of social media such as Tweeter, fora and Facebook as data sources as well as commercialization platforms.

The above mentioned trends result in information flows, which require the aggregation of data over a wide range of heterogeneous sources. The representation of the available information beyond the data varies according to its syntax and semantics. To get access to the information the application of semantic data integration methods are necessary. To capture the available information beyond the data to be visible for communication partners, an information model is required, which could define the terminology and the restrictions of all relevant standards. In such an information model the terms such as owner, good, or order could be specified and used precisely by stakeholders including employees as well as software agents. This information model should be extendable through each standard that could be relevant in the future or in an industry specific logistic process. For the modeling purpose, ontologies as the foundation/ground information model are a common approach to model the terminology of a specific application domain, which has been resulted in different foundation ontologies such as DOLCE (Masalo et al. 2002) and web sites such as http://schema.org/ for offering ontology concepts. Each application domain could create its own ontology to define the terminology and the exchanged information in the corresponding information flows. The creation of an ontology to cover all aspects of logistics would result in a monolithic ontology. However, the creation and maintenance of a monolithic ontology is not convincing either in many respects (Masalo et al. 2002).

The availability of different standards, the expected integration of product specific information, the availability of Product Service Systems, Product Avatars, and other information driven product services demand the application of robust information flows in logistics including a clear understanding of the exchanged information over the boundaries of standards.

41.3 Approach

This paper proposes an approach for a robust communication in logistics. To enable a flexible and adaptable information flow, one of the key challenges is to exchange information and not only data any longer. Bellinger (2004) defines information as "data that has been given meaning by way of relational connection." In this article the combination of information models and transformation definitions is applied to transform the data stored in the different data sources of a logistic system into information.

The proposed procedure of exchanging information within a logistic system is defined as an activity in which one transmitter sends information to 1... n receivers at a specific timestamp. The information exchange represents decentralized processes in which no central entity such as a mediator or another server based information system is contained in the transmission process. In that case the information exchange is located directly between the interfaces of systems. Each system should take the responsibility of providing information and not just data for the receiver to achieve the interoperability. The step from data to information could be managed through the general approach of transformation to resolve the schematic, semantic and intensional data integration conflicts (Goh 1997; Wache 2003). A transformation, in particular a model transformation approach includes according to Kleppe et al. (2003) three components, namely the transformation, transformation definition, and transformation rules. (Kleppe et al. 2003) defines these as follows: "A transformation is the automatic generation of a target model from a source model, according to a transformation definition. A transformation definition is a set of transformation rules that together describe how a model in the source language can be transformed into a model in the target language. A transformation rule is a description of how one or more constructs in the source language can be transformed into one or more constructs in the target language."

On basis of the above mentioned definitions for the terms 'information' and 'transformation' the authors propose an extended Logistics Artifact, which is capable of exchanging information including the responsibility for transforming his data to information. Without this mandatory capability, a central node has to consider the interoperability in a central manner for all contained Logistics Artifacts in a logistic system. In this paper a Logistics Artifact is defined as any kind of object which provides relevant logistics information. Logistics relevant information could be for example ranges from an ID as barcode to an offer in an Enterprise Resource Planning (ERP) system. In all cases, the data is necessary for the logistics process and should be available as information. The principle of an extended logistics Artifact is illustrated in Fig. 41.1.

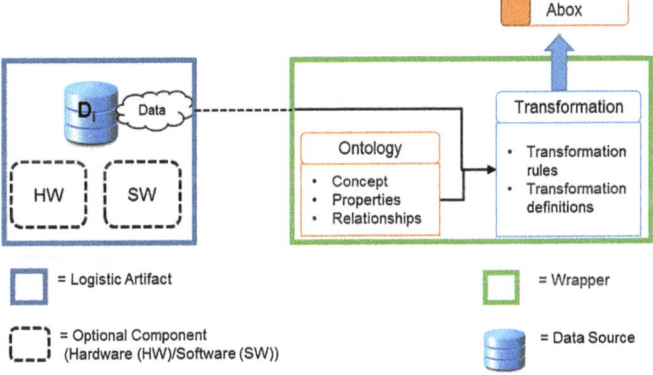

Fig. 41.1 Extended Logistics Artifact capable of information exchange

The extended Logistics Artifact of Fig. 41.1 consists of two parts, namely the Logistics Artifact and the Wrapper. The transformation of data into a set of information entities is realized within a Wrapper. A Wrapper as proposed by Doan et al. (2012) is a component to transform the data from the source view into the target view. It is a component, which enables the information provision through a loose coupling to the Logistics Artifact. The Wrapper is responsible for the transformation of data into information through the application of the contained and data source specific transformation definitions. The representation of information will be realized through ontologies. Guarino (1998) defines ontologies as formal, partial specifications of an agreement over the description of a domain. Therefore, a semantic understanding of the exchanged information is possible as well as the solving of data integration conflicts between Extended Logistics Objects in an information flow.

The beginning of a concrete information flow between more than one extended Logistics Artifact can be realized through a Peer-to-Peer (P2P) topology, which is going to satisfy the decentralized character of the communication. In particular, unstructured peer-to-peer networks are easy to set up and could be adapted to different regions (Chervenak et al. 2008). In addition, unstructured networks are highly robust in the case of joining and leaving nodes (Jin and Chan 2010). The possibility of joining and leaving nodes is a mandatory requirement for moving Logistics Artifacts in logistics. In such a topology not all extended Logistics Artifacts would be involved in an instantiated information flow but rather the set of extended Logistics Artifacts, which could contribute information to a query and the ones, which would need the requested information. The starting point of an information flow would be the need to publish new information (such as a status update) or to request the latest information. In both cases, the addressed extended Logistics Artifact resolves a look up command to identify the possible communication partners. The look up command would send an intensional query to all available extended Logistics Artifacts to check which ones both understand the information to be traded and whether the communication partners are useful for the intended information flow. After the synchronization phase is finished, an extensional query would be sent to all relevant extended Logistics Artifacts to collect information.

The aggregation of the collected information is not carried out on one specific extended Logistics Artifact but rather on all involved extended Logistics Artifacts during the information flow. For that purpose a container is created by the first extended Logistics Artifact and is forwarded to each involved extended Logistics Artifacts. The container is instantiated only once and is used to store the information, which will be filled by each extended Logistics Artifacts, which is illustrated in Fig. 41.2. It implicates an ordering regarding the time stamps when an extended Logical Artifact could add its information into the container. The ordering correlates both to the underlying logistics processes and the internal structure of the applied information models. For example, an entity could insert the information, which includes a foreign key once the information providing the primary key has been already inserted.

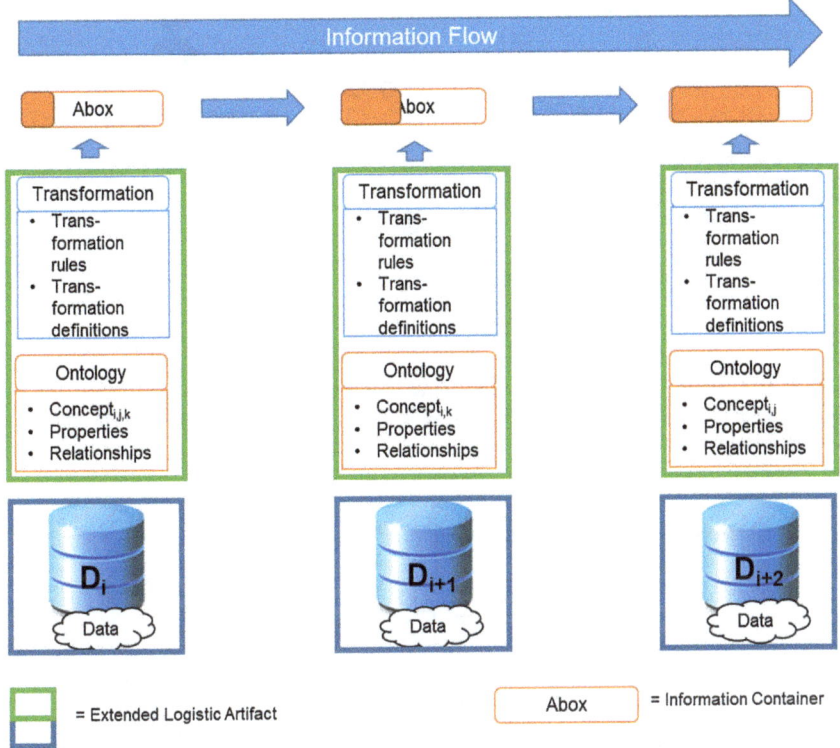

Fig. 41.2 Usage of an information container in information flow to aggregate all data specific information to one logical view

In the following, an example of the different kinds of information flows is presented in an exemplary logistic system.

41.4 Application Example

The following example illustrates the differences between a logistic system using the existing data exchange standards and a more flexible system, which includes an information model for the information flow. In the latter case, information and not data are exchanged.

Figure 41.3 shows a shop floor scenario with two machines M_1 and M_2, a container with three semi-finished products waiting in front of M_1, a forklift truck (F), a storage for finished goods, and an external customer (C). Several applications for data exchange are possible. Case 1 can be the recognition of the container by the forklift truck, which could be realized by an attached barcode or RFID tag and a corresponding sensor at the forklift truck. The data flow is unidirectional toward the

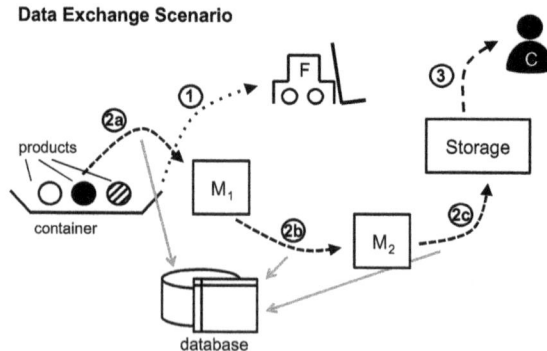

Fig. 41.3 A shop floor scenario showing three exemplary data transmission cases: (*1*) the identification of a container by a forklift truck, (*2*) the recording of processing events of a product, and (*3*) the submission of data

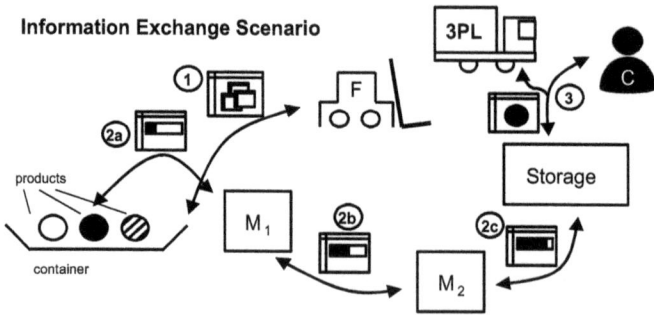

Fig. 41.4 The same shop floor scenario with increased information exchange capability: (*1*) Information objects can be extended in a modular fashion (*2*), the development of a physical object over time can be depicted in the corresponding information object

forklift truck. Similarly, Case 2 shows the movement of a part from the container to M_1 (2a), then to M_2 (2b), and finally to the storage (2c). During this process, events such as arrival or departure at a machine or the storage are stored centrally in an event log, usually as a row in table, containing product ID and a timestamp. Although the flow of data virtually accompanies the physical material flow, the real data flow is directed toward the central database (indicated by the gray arrows). Other processes that need to access this data need to have predefined interfaces to the database containing the table, and they need to be aware of how to interpret the data. The third case is the transmission of data to the customer. This can be billing data, which is submitted by, e.g., a predefined XML message. Again, the interface between the internal and the customer's IT systems needs to be specified in advance. In the following Fig. 41.4 the same scenario is presented, but now including the idea of applying information exchange and not simply data exchange.

Case 1 shows the exchange of information between the forklift truck and the container via an information artifact. The underlying information model is able to convey additional information, e.g., about the products included in the container. Moreover, the information flow is bidirectional by nature, because the information about the container is decentral stored, so that not only information about the container can be provided to other entities on the shop floor, but also the information describing the container and its state can be updated by other entities. E.g., the forklift truck can reposition the container and update the position information in the container artifact. The information exchange in the manufacturing process of the product in Case 2 does no longer depend on a central database. The information object carries all relevant information about the product. With the progress of the manufacturing process, the information content in the information object grows as the product moves from M_1 via M_2 to the storage. Again, a bidirectional exchange of information is possible at any time. In the data exchange scenario, Case 3 described the transmission of billing data to the customer. In the enhanced information exchange scenario, the information object of the finished product carries all necessary information, including billing information. But the information exchange is now more flexible, as the information can be made available to other parties as well without the need for the design of additional interfaces. In this example, a logistics service provider can access the information and is able to retrieve and to update the product information.

41.5 Summary and Outlook

The illustrated syntactical and semantic heterogeneity in the communication within logistic processes results from the application of different systems, standards, and data repositories, which codify equal or similar information. A piece of information usually passes through different systems and possible different standards during an information flow in a global context. The set of relevant and to be covered systems and standards appears likely to increase due to the trends of Internet-of-Things, Industry 4.0, and Item Level Product Lifecycle Management. To enable an information flow over the boundaries of standards and pragmatically implementations, the exchange of information and corresponding application of transformations on the interfaces is required. The presented approach resides on a conceptual level. The next steps are the exploration of requirements for the necessary transformation capabilities on the interfaces. Then, an adequate transformation mechanism including different transformation methods could be defined.

Apart from the technical development, the necessary transformations of the data at the interfaces must be considered in all lifecycle phases of logistic systems. The design, the simulation and the execution requires models of the logistics process considering the above mentioned information flows. To consider the information flow, the data model, the information model and the transformations of a logistics artifact in the design phase of a logistic system, a modeling approach is necessary.

Furthermore, the integration of logistics artifact into a logistic system requires the considering of more aspects. To capture and model all aspects a modeling approach is required.

Acknowledgments The work of Till Becker has been supported by the Institutional Strategy of the University of Bremen, funded by the German Excellence Initiative.

References

Becker T, Weimer D, Pannek J (2015) Network structures and decentralized control in logistics: topology, interfaces, and dynamics. Int J Adv Log 4:1–8. doi:10.1080/2287108X.2015.1012329

Bellinger G, Castro D, Mills A (2004) Data, information, knowledge, and wisdom. http://www.systems-thinking.org/dikw/dikw.htm, Accessed 23 Oct 2015

Chervenak A, Bharathi S (2008) Peer-to-peer approaches to grid resource discovery. In: Danelutto M et al (eds) Making grids work: proceedings of the CoreGRID workshop on programming models grid and P2P system architecture grid systems, tools and environments, vols 12–13, Heraklion, Crete, Greece. Springer, pp 59–76. doi:10.1007/978-0-387-78448-9_5

Doan A, Halevy A, Ives Z (2012) Principles of data integration. Elsevier

EANCOM (2012) S4, EANCOM® MESSAGES. http://www.gs1au.org/products/gs1_system/emessaging/eancom_2002_v2012/ean02s4/part1/part1_01.htm#Introduction. Accessed 17 August 2014

Freitag M, Becker T, Duffie NA (2015) Dynamics of resource sharing in production networks. CIRP Ann Manuf Technol 64:435–438

Guarino N (1998) Formal ontology in information systems. In: Guarino N (ed) Proceedings of the international conference on Formal Ontology in Information Sys-tems (FOIS'98): 3-15. IOS Press, Trento, Italy

Goh CH (1997) Representing and reasoning about semantic conflicts in heterogeneous information sources. Massachusetts

GS1 Austria (2015) EANCOM. http://www.gs1.at/gs1-leistungen-a-standards/ecom/eancom. Accessed 22 Oct 2015

Ickert L, Matthes U, Rommerskirchen S, Weyand E, Schlesinger M, Limbers J (2007) Abschätzung der langfristigen Entwicklung des Güterverkehrs in Deutschland bis 2050 (Study for the German Ministry of Traffic, Construction and Urban Development). Prognosen und Strategieberatung für Transport und Verkehr

Jin X, Chan S-HG (2010) Unstructured peer-to-peer network architectures. In: Shen et al (eds) Handbook of peer-to-peer networking, vol 119. Springer. ISBN 978-0-387-09750-3

Kleppe A, Warmer J, Bast W (2003) MDA Explained: the model-driven architecture: practice and promise. Addison Wesley, Boston

Kumar A (2007) From mass customization to mass personalization: a strategic transformation. Int J Flex Manuf Syst 19:533–547

Masolo C, Borgo S, Gangemi A, Guarino N, Oltramari A, Schneider L (2002) The WonderWeb library of foundational ontologies. WonderWeb Deliverable 17

Veigt M, Lappe D, Hribernik KA, Scholz-Reiter B (2013) Entwicklung eines Cyber-Physischen Logistiksystems. Ind Manage 29(1):15–18

Vernadat FB (2010) Technical, semantic and organizational issues of enterprise interoperability and networking. Annu Rev Control 34(1):139–144

Wache H (2003) Semantische Mediation für heterogene Informationsquellen. Universität Bremen

Wellsandt S, Werthmann D, Hribernik K, Thoben K-D (2013) Ubiquitous logistics: a business and technology concept based on shared resources. In: Cunningham P, Cunningham M (eds) International Information Management Cooperation (IIMC), Proceedings of the eChallenges Conference (e-2013), Dublin, Ireland, 8 pp. ISBN: 978-1-905824-40-3

Chapter 42
Airflow Behavior Under Different Loading Schemes and Its Correspondence to Temperature in Perishables Transported in Refrigerated Containers

Chanaka Lloyd, Reiner Jedermann and Walter Lang

Abstract Supply chains are a highly evolving line of trading. The cool chains responsible for transportation of perishables are one subcategory that is demanding technological support to reduce the quality-related losses that they suffer due to temperature variations, among other reasons. Even distribution and ventilation of refrigerated air inside containers is imperative to maintain the perishables at the desired temperature range, avoiding degradation and spoilage. However, lack of research on airflow movement behavior—and convenient means of measuring spatial airflow speed—within packed containers makes it difficult to determine the hotspot scenarios, which is a prime cause of the said degradation. This paper presents a methodology to parametrically measure spatial airflow and analyzes the airflow behavior under different container loading schemes and how the airflow affects the internal pallet temperature.

42.1 Introduction

The preservation of food quality and avoidance of losses in food supply chains is an important target, both at present and in future (Saguy et al. 2013). Moreover, transportation of perishable food, such as meat, fish, fresh vegetables, and fruits, introduces an additional element to the latter challenge: maintaining the desired temperature set point in the refrigerated containers. In consideration of the

---

C. Lloyd (✉) · R. Jedermann · W. Lang
Institute for Microsensors, -Actuators and -Systems (IMSAS),
University of Bremen, Bremen, Germany
e-mail: clloyd|rjedermann|wlang@imsas.uni-bremen.de

C. Lloyd
International Graduate School for Dynamics in Logistics (IGS),
University of Bremen, Bremen, Germany

transportation of, for example, climacteric fruits, supply chains face further challenges in the need to avoid hotspots created by maturing fruits that produce heat (Snowdon 2010). This paper concentrates mostly on banana transportation. They are transported ideally between 13 and 14 °C. However, even if it is assumed that the desired temperature is maintained by the cooling unit of the container, bananas could already be at a higher temperature before loading. Therefore, maturation process may have already started which becomes a cascading event and spread throughout a section of the container due to the presence of ethylene, thereby producing heat. This creates isolated hotspots (Moureh et al. 2009). If the produced ethylene and the warm air are not removed, it causes unwanted degradation of the bananas.

The Intelligent Container project (Lang et al. 2011; Grunow and Piramuthu 2013) shows the methods for and importance of the usage of RFID technology in perishable food transportation. In terms of this paper, the focus is on research performed in refrigerated containers packed with bananas. More specifically, the research here is limited to airflow behavior in the refrigerated containers. It helps researchers understand the ventilation process in the container and enables the adoption of correctional measures—changing the banana-pallet loading scheme in the container, changing banana box design, increasing the cooling load, etc.

The measurement of airflow within a container is imperative to get a direct measure of the ventilation within a container. However, temperature profiling (Mai et al. 2012; Jedermann et al. 2010) of a container is the direct measure of the condition of the transported product and is of utmost importance. Therefore, it is highly relevant to correlate airflow and temperature. In measuring the airflow within a container packed with bananas, only the airflow through gaps can be measured. Since measurement of flow (in m^3/h) is not possible, airflow speed (in m/s) is measured at various, spatially distributed locations in real-time using wireless airflow sensors specifically designed for this task.

This paper presents three test cases based on airflow measurements: two comparison tests involving two different banana-pallet loading schemes and a study case on correlation between the airflow around a pallet and its internal temperature. All tests were carried out in a full scale 40-foot container prototype integrated with remote sensing capabilities (Jedermann et al. 2010). In addition, short section on the wireless airflow sensor is also presented. Apart from few research activities (Laniel et al. 2011; Ruiz-Garcia et al. 2009), wireless sensors are not heavily used in food transport containers. Therefore, this sensor stands as a good candidate for future research, especially in airflow mapping.

Section 42.2 illustrates the field test conditions and equipment, airflow sensor (Sect. 42.2.1), banana loading schemes, and the test cases (Sect. 42.2.2) mentioned above. Section 42.3 details the test cases in depth and discusses all the results. Section 42.4 summarizes and concludes the results.

42.2 Field Tests

The field tests were performed in the container prototype under a scaled down scenario, i.e., using only 11 and 12 banana pallets under two different pallet loading schemes. In commercial transportation 20 banana pallets are required to fill a 40-foot refrigerated container. The test cases (Sect. 42.2.2) presented in this paper compares the two loading schemes for their airflow behavior. Although two cases have different number of pallets, it is the only possibility, as loading/unloading a full scale container of 20 pallets takes long time and effort. Such logistic difficulties prevented the researchers from using a fully loaded container at will.

The aforementioned two loading schemes—*conventional* loading and *chimney* loading—differ only by the way the banana pallets are arranged inside the container. Figure 42.1a shows the two loading schemes side by side. It also shows the partition wall, which is raised inside the container right after the last two pallets in order to achieve the scaled down container. Figure 42.1b shows a single pallet that is made up of 48 individual banana boxes. The pallet consists of eight banana box levels, also called *tiers*, called by tier 1, 2, 3 … 8. Three empty vertical spaces are created in the chimney scheme, as seen on Fig. 42.1a, due to the loading mechanism.

The main aim of the field tests was to measure the airflow speed in-between the gaps between adjacent pallets and the spaces between pallets and the container walls, thereby analyze the spatial airflow. In addition, the airflow speed of the input air—pumped in by the cooling unit—in the ducts below the pallets and the return air—absorbed back into the cooling unit—over the pallets were measured. All these types of airflow speed in different gaps describe all major airflow paths in the container. These are depicted in Fig. 42.2a, which is a cross-section of a banana container as shown in Fig. 42.1a. The horizontal airflow (input and return air) is not mentioned in this paper but in [Lloyd et al. 2013]. In addition, in order to assess how the airflow affects the temperature inside a banana pallet, two pallets were equipped with 4 temperature sensors and 4 airflow sensors each. This test is further explained in Sect. 42.2.2.

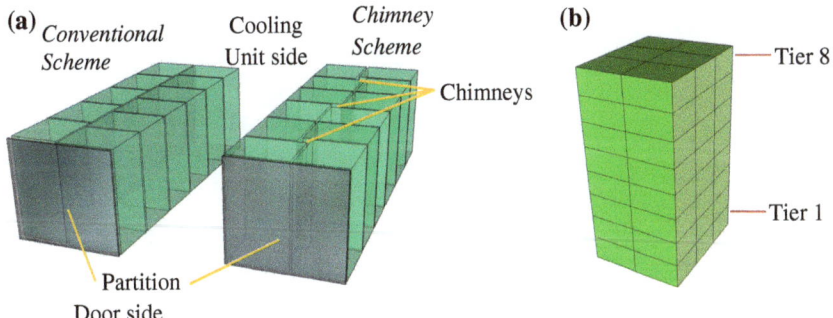

Fig. 42.1 Banana pallet loading schemes (**a**); a single banana pallet with 48 banana boxes (**b**)

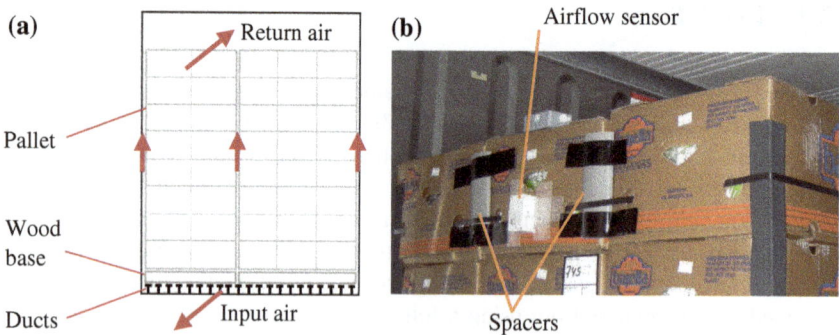

Fig. 42.2 Cross section of a packed banana container (**a**); a banana pallet with an airflow sensor and spacers mounted on tier 8 (**b**)

Figure 42.2b shows a pallet affixed with an airflow sensor. The spacers, approx. 1–2 cm thick plastic blocks in place to force a gap—where required—in-between two adjacent pallets, are also shown on either side of the airflow sensor. The sensor is mounted vertically, meaning that it measures the speed of the air that flows from bottom to top, which is the natural direction for air in a refrigerated banana container (Lloyd et al. 2013). Most of the tests were conducted using 15 wireless airflow sensors, networked using the BananaHop protocol (Jedermann et al. 2011). Sensor data was read out in real-time using the remote connection integrated on the intelligent container.

42.2.1 Wireless Airflow Sensor

The airflow sensor is based on a thermal flow sensor (Buchner et al. 2006). This miniature sensor (Fig. 42.3) works on the principle of voltage difference between two thermopiles. A heater element in the sensor is powered up and calibrated at 5 mW for the sensors used in the experiments in this paper. This power is

Fig. 42.3 Wireless airflow sensor enclosure with its air channel (*left*) and its internal components (*right*)

maintained throughout the measurement period. The method of sampling the sensor is called the Constant Power (CP) method. The heat dissipation profile of the heater element is disturbed when a flow of air passes over the sensor membrane, which is directly above the heater element. This change is measurable by the difference of the voltage between the two thermopiles. This specially designed airflow sensor (Lloyd et al. 2013), with its circuit, thermal flow sensor and battery, is integrated onto a wireless TelosB platform that is based on IEEE 802.15.4 standard.

42.2.2 Test Cases

The field tests using a packed banana container were conducted in January and May of 2013, with the support of Dole GmbH, Germany. In general, each test was conducted—with the cooling unit running—for approx. 45 min. Test case 3 was run for 15 h. For the purpose of this paper, three test cases are formed:

- **Test case 1**
  This is a study of airflow behavior in the gap between the banana pallets and the left container wall along the length of the container from the cooling unit to the partition wall. Airflow movement in this gap for the conventional loading scheme is described in (Lloyd et al. 2013). This study compares the airflow behavior between the two loading schemes as depicted in Fig. 42.4a. The airflow sensors were vertically mounted (as in Fig. 42.2b) on the pallets on tiers 2, 5, and 8. Both the tests—conventional and chimney—were conducted with spacers attached on all aforementioned tiers (also as shown in Fig. 42.2b), thereby forcing a slight space between the pallets and the container wall along the length of the container.

- **Test case 2**
  This test case considers the airflow behavior in the mid-section of the container as shown in Fig. 42.4b. The spacers were used in this test as well. In conventional setup (left of Fig. 42.4b), the airflow sensors were mounted only on the tier 8. However, in the chimney setup (right of Fig. 42.4b) the sensors were mounted on tiers 2, 3, and 8 at the locations indicated on the cross-section figure. Therefore, only the tier 8 sensor values were considered for the comparison of data with the conventional setup.

- **Test case 3**
  The main aim of this test (Fig. 42.5) was to assess how the airflow affects the internal temperature of banana pallets. Therefore, the scenario of a pallet with and without spacers was considered. Ideally, the airflow data around the pallet should be obtained from the same pallet—with and without the spacers—in

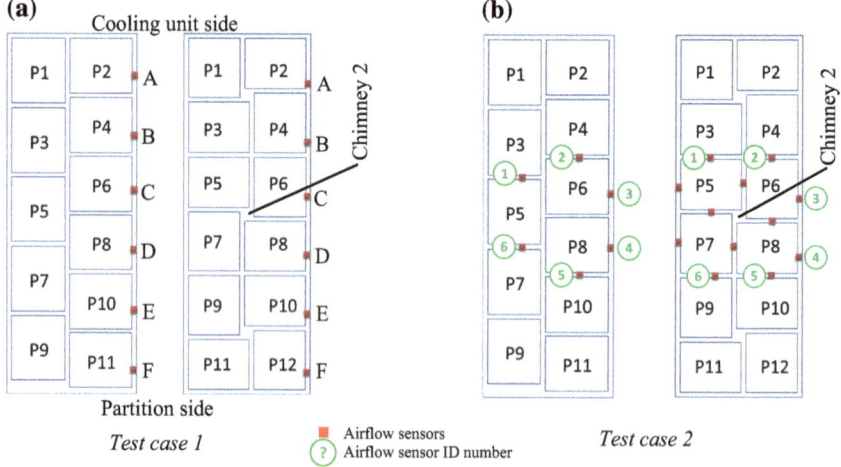

Fig. 42.4 Airflow sensors in the vertical gap between the pallets and the right-side container wall for both loading schemes (test case 1) (**a**); airflow sensors in the vertical gaps in-between adjacent pallets in the mid-section of the container under both loading schemes (test case 2) (**b**)

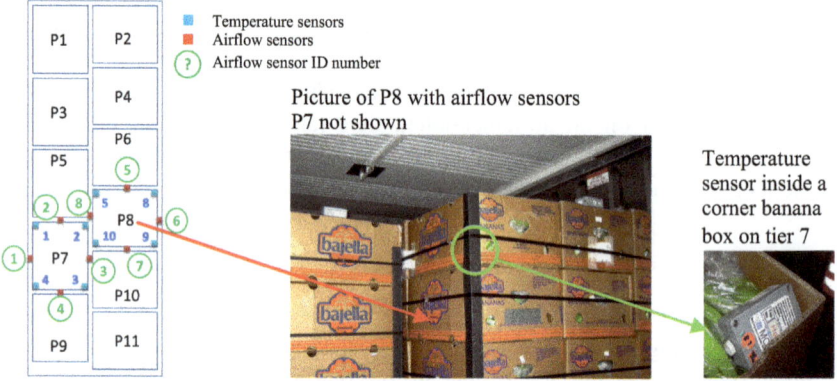

Fig. 42.5 Test case 3: Airflow sensors in the vertical gaps in-between adjacent pallets and temperature sensors inside corner banana boxes on tier 7, under the conventional scheme. P7 is with spacers and P8 without

order to compare the two cases. To implement that, the container needs to be loaded/unloaded two times. The conditions, such as the gaps created when loading the pallets in the second loading, would be different from the first loading. Therefore, in order to monitor the airflow behavior around pallets with/without spacers, two adjacent pallets were chosen, P7 and P8 (P—acronym for **P**allet. Pallets are numbered from 1 to 12). P7 was with the spacers around the entire pallet and P8 without. Additionally, temperature sensors, 4 per each pallet, were inserted in the corner boxes on tier 7 of both pallets as shown in Fig. 42.5. The test was run for 15 h from start to finish where the temperature set point of the cooling unit was set to 13 °C (as bananas were precooled to 13 °C), then to 16 °C for 6.5 h and then back to 13 °C until the sensor batteries died (approximately 6.5 h).

42.3 Results

42.3.1 Test Case 1: Pallet–Wall Gap

The setup of test case 1 is shown in Fig. 42.4a below. A, B, C…F are position identifiers for the sensors that are distributed along the length of the container. At each identifier position, the red square symbolizes three airflow sensors that are placed on 3 different tiers. There were no sensors placed at position identifier D.

Table 42.1 lists the airflow speed values of both the conventional and chimney schemes of test case 1. "-" means there was no sensor there, data unavailable or data is too erroneous. Three chimneys are formed inside the container under this scaled down scenario. All chimneys recorded high airflow speed values than any other vertical airflow speeds in other gaps. For example, chimney 2 airflow speed was approximately 4.5 m/s. The conventional setup does not have any such large, vertical open shafts. The difference of the different loading schemes is clearly evident when comparing the values in Table 42.1.

There are only four sensor positions in the chimney setup that exceed the value of the conventional setup by 0.20 m/s or more. In {tier 2, E, conventional} case, it seems the pallet bottom was completely blocking the air flow upwards, hence zero. In all other sensor positions the conventional setup recorded much higher values. Obvious explanation is the presence of the chimneys.

Table 42.1 Comparison of airflow speed values (at tier 8, 5, and 2—top to bottom of the pallet) of both the conventional and chimney schemes in test case 1

Tier	Scheme	A	B	C	E	F
Tier 8	Conventional	0.59	0.60	0.97	1.31	1.18
	Chimney	0.29	0.88	0.27	–	0.83
Tier 5	Conventional	0.35	1.24	0.95	0.61	–
	Chimney	0.55	0.54	1.21	0.28	–
Tier 2	Conventional	0.48	0.99	0.24	0.00	1.09
	Chimney	0.57	–	0.16	0.27	0.20

42.3.2 Test Case 2: Pallet–Pallet Gap Around Container Mid-Section

The comparison of airflow in the mid-section of the container for both loading schemes is not straight forward. However, the airflow sensors (ID numbers from 1 to 6) shown in Fig. 42.4b are strategically placed in the conventional scheme in such way that it best resembles the airflow sensor locations of the chimney scheme. Table 42.2 lists the airflow speed values of both the schemes of test case 2. Again, what is very clear is the vast difference of the airflow behavior between the two loading schemes in the container mid-section as well. With majority of the air escaping at high speed through chimney 2, the values on other vertical gaps around the pallets are far less than that of the conventional case. The airflow values of the 4 vertical gaps that connect to the chimney (values not in Table 42.2) were even lower, at 0.17, 0.10, 0.28, and 0.52 m/s, which further enforces the theory that air is, in fact, sucked into the shaft creating a low pressure volume around chimney 2.

42.3.3 Test Case 3: Airflow Versus Temperature

Test case 3 considers the airflow around two adjacent pallets under the conventional loading scheme, which is how most supply chains transport the banana pallets. The intent of this test was to determine how the airflow around a pallet is different when installed with spacers around it to enable better ventilation, and to assess if such increased ventilation would cause the temperature inside a banana pallet to drop faster. Figure 42.5 shows the test setup and how the temperature sensors are inserted in banana boxes.

At a single glance of Fig. 42.6b, which shows the temperature rise (13 to 16 °C) and fall (16 to 13 °C) of all temperature sensors, curves do not coincide or rise together in unison. In order to write off the possibility of sensor error a post-test calibration was run on all eight sensors (Fig. 42.6a). It shows how the temperature of all sensors rise together in the red boxes marked on the figure. Temperature offsets were found—maximum being 0.75 °C—and Fig. 42.6b was compensated accordingly. The temperature sensor numbers (1 to 5 and 8 to 10) are marked in the

Table 42.2 Comparison of airflow speed values (at tier 8) of both loading schemes in test case 2

Sensor ID	Conventional scheme (m/s)	Chimney scheme (m/s)
1	1.16	0.33
2	0.82	0.69
3	1.93	0.20
4	–	1.23
5	0.82	0.21
6	2.44	–

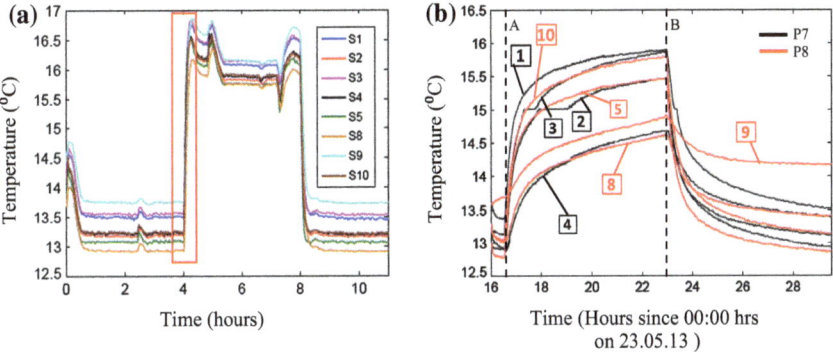

Fig. 42.6 Calibration of temperature sensors (**a**); temperature behavior around two pallets with and without spacers (**b**) over the entire test

rectangular boxes in Fig. 42.6b. An interesting airflow versus temperature behavior is found when comparing the airflow values of the eight airflow sensors (1 to 8 in Fig. 42.5) distributed in the gaps around the two pallets.

If a tangent line is drawn at the point where the temperature curves meet the dashed line A on Fig. 42.6b, it is easily discernible that sensor 1, 10, 3, 5, and 2 are having high ascend rates, which indicates that internal temperature of the boxes increased rapidly in comparison to other sensors. The same is true for the cooling cycle beyond dashed line B. The order of descent of temperature was similar to that of the order of ascent. Airflow sensors 1, 2, and 8 recorded values of 1.66, 0.78 and 1.49 m/s. Temperature sensors 1, 2, 5, and 10 are in the neighborhood of these fast airflow paths, which explain the high ascent and descent of temperature. Airflow sensors 3 and 4 recorded values of 0.71 and 0.52 m/s, also much higher compared to other three airflow values on P8 (0.31, 0.69, and 0.47 m/s for airflow sensors 5, 6 and 7, respectively).

Evidently, the airflow values around P7 (with spacers) were much higher than that of P8 (without spacers). Lack of airflow around P8 caused the slow ascent and descent of temperature on temperature sensors 8 and 9, which are sandwiched between the aforementioned three, slow moving drafts around P8. Sensor 4 is special case, where its immediate surroundings are thought to have blocked the airways, hence the slow ascent/descent.

42.4 Conclusion

Airflow behavior and its effects on pallet temperature were tested. For the purpose of airflow behavior, two banana loading schemes were compared. The chimney scheme clearly exhibits different characteristics to that of the conventional banana loading, where the chimneys funnel most of the upward airflow movement. These comparatively high speed drafts seem to create a low pressure region around the chimney that possibly enhance horizontal airflow movement—towards the chimney shaft—inside the pallets, which is expected to keep the bananas at the desired temperature during transport. The internal temperature analysis, versus airflow, is not fully deterministic of how the pallet core temperature changes with and without more space around the pallets. However, it shows how the internal contour region located just before the outer surface of the pallet responds to varying airflow speed. These results are encouraging and promising enough to investigate further.

Acknowledgments The research project "The Intelligent Container" was supported by the Federal Ministry of Education and Research, Germany, under the reference number 01IA10001. We thank the support of Dole Fresh Fruit Europe for the provision of test facilities.

References

Buchner R, Maiwald M, Sosna C, Schary T, Benecke W, Lang W (2006) Miniaturized thermal flow sensors for rough environments. In: 19th IEEE international conference on MEMS, Istanbul, pp 582–585. doi:10.1109/MEMSYS.2006.1627866

Grunow M, Piramuthu S (2013) RFID in highly perishable food supply chains—Remaining shelf life to supplant expiry date? Int J Prod Econ 146(2):717–727. doi:10.1016/j.ijpe.2013.08.028

Jedermann R, Moehrke A, Lang W (2010) Supervision of banana transport by the intelligent container. In: Kreyenschmidt J (ed) CoolChain-Management—4th International Workshop. University of Bonn, Germany

Jedermann R, Becker M, Görg C, Lang W (2011) Testing network protocols and signal attenuation in packed food transports. Int J Sensor Netw (IJSNet) 9(3/4):170–181. doi:10.1504/IJSNET.2011.040238

Lang W, Jedermann R, Mrugala D, Jabbari A, Krieg-Brückner B, Schill K (2011) The intelligent container – a cognitive sensor network for transport management. Sensors J IEEE 11(3):688–698. doi:10.1109/JSEN.2010.2060480

Laniel M, Émond JP, Altunbas AE (2011) Effects of antenna position on readability of RFID tags in a refrigerated sea container of frozen bread at 433 and 915 MHz. Transp Res Part C: Emerg Technol 19(6):1071–1077. doi:10.1016/j.trc.2011.06.008

Lloyd C, Issa S, Lang W, Jedermann R (2013) Empirical airflow pattern determination of refrigerated banana containers using thermal flow sensors. In: 5th international workshop on cool chain-management. http://ccm.ytally.com/index.php?id=176

Mai NTT, Margeirsson B, Margeirsson S, Bogason SG, Sigurgísladottír S, Arason S (2012) Temperature mapping of fresh fish supply chains—air and sea transport. J Food Process Eng 35 (4):622–656. doi:10.1111/j.1745-4530.2010.00611.x

Moureh J, Tapsoba S, Derens E, Flick D (2009) Air velocity characteristics within vented pallets loaded in a refrigerated vehicle with and without air ducts. Int J Refrig 32(2):220–234. doi:10.1016/j.ijrefrig.2008.06.006

Ruiz-Garcia L, Lunadei L, Barreiro P, Robla I (2009) A review of wireless sensor technologies and applications in agriculture and food industry: state of the art and current trends. Sensors 9 (6):4728–4750. doi:10.3390/s90604728

Saguy SI, Singh RP, Johnson T, Fryer PJ, Sastry SK (2013) Challenges facing food engineering. J Food Eng 119(2):332–342. doi:10.1016/j.jfoodeng.2013.05.031

Snowdon AL (2010) Carriage of bananas (musa spp.) in refrigerated ships and containers: preshipment and shipboard factors influencing cargo out-turn condition. Acta Hort. (ISHS) 879:375–383. http://www.actahort.org/books/879/879_40.htm

Chapter 43
VANET Security Analysis on the Basis of Attacks in Authentication

Nimra Rehman Siddiqui, Kishwer Abdul Khaliq and Jürgen Pannek

Abstract Elevation of technology in hardware, software and communication have lead to an idea of Intelligent Vehicles, which communicate via Vehicular Ad hoc Networks (VANET) to provide road safety and improved driving conditions. In VANETs, both multi-cast and uni-cast routing is applicable. These protocols can be used by an attacker to breach and endanger the network by accessing and manipulating information of the vehicle and communicating data. As authentication allows to trust both user and data, it requires considerable attention in the security framework of VANETs. Here, we provide a classification of attacks on the basis of security requirements to narrow down the parameters of consideration and comparison of previously proposed protocols in VANET. Pointing out security gaps regarding distinct threats, the classification will allow to designed new secure network control methods.

Keywords VANET · IEEE 802.11p · Authentication · Security · Attacks

---

N.R. Siddiqui (✉) · K.A. Khaliq
International Graduate School for Dynamics in Logistics,
Universität Bremen, Bremen, Germany
e-mail: reh@biba.uni-bremen.de

K.A. Khaliq
e-mail: kai@biba.uni-bremen.de

J. Pannek
Department of Production Engineering, Universität Bremen,
Bremen, Germany
e-mail: pan@biba.uni-bremen.de

N.R. Siddiqui · K.A. Khaliq · J. Pannek
BIBA – Bremer Institut für Produktion und Logistik GmbH,
Bremen, Germany

43.1 Introduction

Statistics shows that 1.24 million people died in 2014 due to lack of driver cooperation and poor road safety concerns (Toroyan 2013). This raises a question of human life security and safety, that has lead to the idea of 'Smart/Intelligent vehicles' and the concept of Vehicular Ad hoc Network (VANET) as an enabler for the latter. VANET is a wireless ad hoc network, which provides communication between vehicles and infrastructure units in ad hoc mode, cf. (Zeadally et al. 2012; Hartenstein and Laberteaux 2009). Moreover, it provides support for a number of wireless products (i.e., remote key-less entry, personal digital assistants, laptops, and mobile telephones) presently used in vehicles. There is an increasing range of VANET applications, which can be split into safety and non-safety applications, and which induce increased communication requirements in terms of Vehicles-to-Vehicles (V2V) and Vehicle-to-Infrastructure (V2I), see e.g., Harsch et al. (2007). Safety applications include life-critical, safety warning, traffic signal violation, curve speed, emergency braking light, pre-crash sensing, forward collision, left turn assist, lane change warning and stop sign assist (Hartenstein and Laberteaux 2009). Non-safety application includes remote vehicle personalization/diagnostics, internet access, digital map downloading, real-time video relay (Akbar et al. 2014), value-added advertisement, route diversions, electronic toll collection, parking availability, and active prediction Kumar et al. (2013); Akbar et al. (2015). While applications can be classified even further, see e.g., Engoulou et al. (2014), user and data are considered trustworthy as the serious actions may be taken based on data received from other vehicles. Hence, security of communicating channel is a necessity of a protected network deployment.

To design a protected vehicular network, data integrity, access control, authentication, privacy, and availability parameters of the network needs considerable attention, cf. Engoulou et al. (2014); Zeadally et al. (2012); Mokhtar and Azab (2015). Authentication of user and data is considered as a fundamental security requirement of VANET (Raya et al. 2006) and lies on the upper hierarchical level of VANET security. This is due to the fact that a network user is the most involved component of the system. A network user may violate the security by acting as a malicious attacker, prankster, greedy driver and or eavesdropper (Engoulou et al. 2014; Zeadally et al. 2012), which may impaired us to solve the increasing number of road accidents and injuries (Nantulya and Reich 2002; Sethi and Mitis 2013) witnessed in the last years. Hence, an effective authentication mechanism is of utmost importance. Here, we present an idea to improve a security framework by classifying attacks, which allows to analyze proposed protocols and to choose a proper solution in the concept of authentication.

The rest of the paper is structured as follows. Section 43.2 presents the state of the art and gives an overview of previous research. We also present classifications and proposed ideas to solve respective attacks. Section 43.3 discusses major security requirements of VANET along with the attacks classes on the basis of VANET security parameters as well as our own classification of attacks. In Sect. 43.4, we discuss proposed solutions for attacks and further explain gaps within these solutions. In the same section, we placed all attacks into a simple classification to find an understandable solution for approximately all attacks. Finally, Sect. 43.5 concludes the paper and outlines of future research directions.

43.2 State of the Art

Intelligent Transportation System (ITS) came up with an idea of VANET to cut down the traffic congestion and the number of road accidents. According to security standards, VANET must meet some security requirements before it is deployed in real time and—by virtue of an Ad hoc network—is subject to a variety of internal as well as external attacks from different types of entities (Mobile, Laptop, and vehicles). Here, we present a classification of attacks on the basis of security requirements and respective solution methods to identify gaps in the framework.

Within the literature, we find different classifications, which were proposed on the basis of various parameters. In 2004, Golle et al. (2004) identified the attacks based on their nature (communicate with wrong information about itself or any other node), target (local or remote targets), impact (undetected, detected, or corrected), and scope (limited or extended on the basis of victims covering area). A related classification was proposed by Mokhtar and Azab (2015), who illustrate attacks on different operating layers (application layer, transport layer, network layer, link layer, and physical layer). Combined, both classifications provide a critical analysis of attacks, yet do not support security analysis of attacks. In particular, several attacks such as *Sybil attack* and *DoS attack* exhibit similar characteristics yet different security requirements. Hence, it becomes a time-consuming task for researchers to find solution, though the classification gives well descriptive knowledge of attacks.

In 2006, Raya et al. (2006) specified an attacker profile according to three norms for classification, active versus passive, malicious versus rational and outsider versus insider. Active attackers have the authority to operate the network while passive attacker can attack by eavesdropping only. Rational attacker differ from malicious attackers by specifying of their target. The idea of outsider and insider attacks are somewhat alike to active and passive attackers, yet insider attackers posses two subtypes named authenticated node and industrial node. In Raya and Hubaux (2007), the authors extend this classification by local versus extended attacker. An attacker belongs to the extended class when it can able to control several entities whereas a local attacker can only work in a limited control area. A relative classification of above mentioned both categorization is presented by Papadimitratos et al. (2008), who proposed Vehicular Communication (VC) adversaries as: active and passive adversaries analogous to Raya et al. (2006) and internal and external attacker. Active adversaries are able to alter message data during broadcasting while passive adversaries can only collect data from the network. In this classification, an attack can be within all three of classes, i.e., Sybil attack is an active, rational and insider attack of the network. Hence, three solutions for the same attack exist, which increases the number of options but also the complexity.

Table 43.1 Classification of attacks on the basis of VANET security requirements

Authentication	Availability	Confidentiality	Integrity and data trust	Privacy
GPS spoofing/Movement tracking	DoS	Non-repudiation	Message suppression	Third party
Sybill	DDoS	Eaves dropping	Illusion	
Node impersonation	Jamming	Traffic analysis	Replay	
Tunneling	Greedy behavior		Fabrication	
Key/Certificate replication	Black hole			
Message tempering	Gray hole			
Message suppression	Sink hole			
Masquerading	Worm hole			
Replay	Malware			
Unauthorized preemption	Broadcast tempering			
Non-repudiation	Spamming			

Despite increased complexity, none of the above classifications explicitly mentioned attacks or security requirements, which is preeminent before deploying VANET. Classification of this survey enlist all possible attacks in every security requirement to establish a more secure VANET for deployment.

Incorporating the above, security requirements for VANET can be structured into seven different requirements: authentication, availability, confidentiality, integrity and data trust, accountability, privacy, and access control. As accountability provides verification of sending and receiving nodes, we observe that this point is covered by authentication and can therefore be dropped. Additionally, access control provides vehicles the capability of accessing various services offered by remote nodes Mokhtar and Azab (2015). This can considered a feature of VANET, but it is not a security requirement. Therefore, we narrowed down to five essential requirements in VANET, cf. Table 43.1.

43.3 Security Requirements of VANET

VANET provides both safety and non-safety applications and it completely relies on the facts and figures provided by vehicular nodes (Qian and Moayeri 2008). Hence, security is an important factor and a major challenge for the communication between vehicle-to-vehicle (V2V) and vehicle-to-infrastructure (V2I) upon deployment of VANET. To handle this issue efficiently, we identified five required critical parameters from Engoulou et al. (2014); Zeadally et al. (2012); Mokhtar and Azab (2015); Lin et al. (2007) specifically for a secure VANET framework design in Table 43.1 and listed all attainable attacks of respective parameters in a nutshell. Here, we details these factors *Authentication, Availability, Confidentiality, Integrity, and Data Trust* and *Privacy*.

Authentication in a vehicular network involves protecting legitimate nodes from inside and/or outside attackers. In VANET, the latter may infiltrate the network by use of a false identity while suppressing, fabricating, altering or replaying legitimate messages, revealing spoofed GPS signals, and impeding the introduction of misinformation into the vehicular network (Zeadally et al. 2012). Authentication covers the entity authentication, entity identification and attribute authentication Fuentes et al. (2010).

Availability assures that the network is available all the time, even if it is under an attack using alternative mechanisms without affecting and compromising its performance and parameters (Al-Kahtani 2012). The availability requirement implies that every node should be capable of sending any information at any time. As most interchanged messages in VANET affect the road traffic safety, this requirement is critical in this environment.

Confidentiality is required when certain nodes want to communicate in private. Upon implementation of VANET, this option is reserved for law enforcement authority vehicles to communicate with each other to convey private information (Sakib and Reza 2010).

Integrity and Data Trust ensures that a sent message should reach the destination on the chosen path without any alteration. The received messages must verify at the destination using an appropriate algorithm for integrity. Then, the receiver will be able to corroborate the sender's identity during the transaction.

Privacy is achieved in the sense that user related information must be confidential, and no other node can acquire these facts. In VANET, these data include the drivers name, the license plate, speed, position, and traveling routes.

Table 43.1 indicates that the majority of attack options is in the areas of authentication and availability. Yet, if authentication is broken, all other categories (availability, confidentiality, privacy and data trust, integrity) are at risk as well. Hence, although every requirement needs significant attention, authentication is of particular importance. As it represents the primary door to the network, a number of solutions have been proposed to reduced attacks due to lack of network authenticity, cf. Table 43.2.

In the area of VANET, these solutions can be classified into two main classes: ID-based authentication and message-based authentication (Lu et al. 2012). In Sect. 43.4 below, we present a clear picture of both these methods and thoroughly explains the importance of these two classes.

Table 43.2 Authentication attacks detailed

Attacks	Description	Solution
GPS Spoofing/Movement Tracking/Position Faking (Agrawal et al. 2013; Zeadally et al. 2012)	Attacker generate the signals using GPS simulator that are stronger than signals generated by genuine satellites. This also affects routing in VANETs scope, and attacker can easily fool vehicles into thinking that they are in a different location	GPS free position system (Caruso et al. 2005; Čapkun et al. 2002)
Sybill (Zeadally et al. 2012; Newsome et al. 2004; Douceur 2002)	The attacker forge multiple identifications in a certain region by broadcasting messages with multiple identifications. It is a severe and ubiquitous problem in VANET authentication	Node validation of neighboring nodes (Newsome et al. 2004)
Node impersonation (Lin and Lu 2015)	Adversary change its identity to another vehicle or to RSU, phone, laptop, to pretend to be another vehicle after maliciously behaving in the network	Bloom filter system (Liu et al. 2007)
Tunneling (Zeadally et al. 2012; Agrawal et al. 2013; Sanzgiri et al. 2002)	Attacker exploit the momentary loss of positioning information when a vehicle enters a tunnel. Before it receives the authentic positioning information, the attacker injects false data into the on-board unit	Hop-distance and time delay calculation (Juwad and Al-Raweshidy 2008)
Key/Certificate replication (Zeadally et al. 2012)	Attacker impaired the system by duplicating a vehicle's key/certificate across the network. This attack would confuse authorities and prevent identification of vehicles in hit-and-run event	Shared Authentication token (Zeadally et al. 2012)
Message tempering (Zeadally et al. 2012)	Attacker modify the messages (routing information) exchanged in V2V or V2I communication modes, in order to falsify transaction application requests or to forge responses	Hop count and time delay analysis (Juwad and Al-Raweshidy 2008)
Message suppression (Zeadally et al. 2012; Lin et al. 2007)	Attacker alter the message during or after transmission	System of direct/indirect/piggybacking trust (Niranjan et al. 2012)

(continued)

Table 43.2 (continued)

Attacks	Description	Solution
Masquerading (Zeadally et al. 2012)	By posing as a legitimate vehicle in the network, outsider attackers can conduct a variety of attacks such as forming black holes or producing false data	Bloom filter system (Liu et al. 2007)
Replay (Zeadally et al. 2012)	In a replay attack, the attacker re-inject previously received packets back into the network, poisoning a node's location table by replaying beacons	Shared authentication token (Zeadally et al. 2012)
Unauthorized preemption (Lin and Lu 2015)	Attacker can control RSU and traffic lights, to provide special traffic priority for emergency vehicles. The adversary may illegally interrupt traffic lights by manipulating the traffic light preemptive system in order to get a better traffic condition for themselves	Group Signature and Identity-based Signature techniques (GSIS) (Lin et al. 2007)
Non-repudiation (Qian and Moayeri 2008)	The sender of a message cannot deny having sent the message, and cannot deny communication participation in the network	ARAN and PKI (Sanzgiri et al. 2002; Wasef et al. 2010; Wu et al. 2007)

43.4 Protocols Reasoning of Attacks in Authentication

Authentication is known as a 'fundamental security functions of Vehicular Communication (VC)', since it provides authenticity of both the data origin and also verifies the sender of the data (Raya et al. 2006). To design a secure framework with an effective authentication system, a better analysis of attacks in authentication is required. In Table 43.2 we listed all possible attacks in authentication and their complete definitions. For better security, identification of previously proposed solutions and deficiency in the latter is required to overcome the attacks. This section discusses attacks, respective solutions, and simpler classification with respect to authentication.

Sybil attacker adapt multiple identities to communicate in the network. In Newsome et al. (2004) some validation methods were proposed by using radio source testing, key validation for random key pre-distribution, position verification, registration and code attestation to overcome the Sybil attack. In all of these methods, neighboring nodes identify whether the communicating node is a Sybil attacker or not. The major flaw in the radio sources testing is the consumption of bandwidth, as a node is assigning a separate channel to every node to validate. A rivalry in random key distribution is the analysis of every node as a Sybil attacker for $1/p$ times, where p indicates the number of activated nodes. Registration is a good approach for initial level verification, yet it is conveniently accessible for attackers to modify this information. Therefore, we cannot guarantee the accuracy of information from this method. The methods proposed for position verification concept can be easily attacked by GPS spoofing. And last, code attestation is an agreeable approach for remote verification, but it is not yet available for wireless sensor networks. Above all methods, random key pre-distribution method is likely to have the most promising among all of them because it is based on easily analyzed cryptographic principles.

GPS spoofing attack can be reduced by using GPS-free position systems. In Caruso et al. (2005) the Virtual Coordinate assignment protocol (VCap) was introduced for constructing a sensor network to find location information based on hop distance calculation. Another method proposed a distributed infrastructure-free positioning algorithm without relying on GPS uses the distances between the nodes to build a relative coordinate system in which the node positions are computed in two dimensions (Čapkun et al. 2002). The aforementioned approaches were proposed for MANET specifically. However, they can be applicable in the scope of VANET if they can be rendered to be more accurate.

In Liu et al. (2007), the authors used bloom filters to securely evaluate the honesty of a node itself with the help of Temper Proof Device (TPD) to reduce the node impersonation attack and masquerading attack. The idea was basically proposed for nodes possessing a number of pseudonyms for privacy issues. As node impersonation and masquerading attacks are based on acquiring a number of certificates and keys, this filter allows to reduce these attacks.

Tunneling attacker interfere with the route establishment process between the network entities and try to create a secret route to connect two nodes placed at distinct regions. Therefore, the idea of hop count and time delay analysis Secure Adaptive On-demand Distance Vector (SAODV) was presented in Juwad and Al-Raweshidy (2008). The method uses hash chains to authenticate hop count of Route Request (RREQ) and Route Replay (RREP) messages in such a way that allows every node that receives the messages (either an intermediate node or the final destination) to verify that the hop count has not been decremented by an attacker. This method also prevents network from message tampering attack.

In Dötzer et al. (2005), the authors proposed the Vehicle ad hoc Reputation System (VARS), a distributed approach based on reputation to overcome message suppression attack. The system presented an idea of event-related messages, forwarded in such a way that the information can be trusted by receiving nodes by using methods of direct and indirect trust as well as an opinion of piggybacking (Niranjan et al. (2012)).

A protocol of shared authentication token is proposed against the Replay attack (Sung-Ming and Kuo-Hong 1997). In the protocol, the generated hash value is evidently one-time if a secure one-way hash function is employed. The counter value INCRi of the hash value is incremented after each session. If replay attacker re-injects previously received packets, this is prohibited as the INCRi token value is changed.

The Authenticated Routing for Ad hoc Network (ARAN) implements an idea for non-repudiation defense mechanism by using cryptographic certificates to offer routing security (Sanzgiri et al. 2002).

Unauthorized attacker may gain access to RSU or traffic signals illegally because of low authenticity. To compensate, a secure protocol GSIS proposed in Lin et al. (2007) applies an ID-based and group-based authentication scheme. Within this method, group-based authentication reduces the overhead and ID-based authentication uses to reduced keys and certificates complications.

All of the above mentioned attacks have distinct solutions. Together with time delays, costs and complexity of the network itself, the design of a security framework is a complex task. To overcome this complexity, Engoulou et al. 2014 proposed two types of authentication approaches named as ID authentication and Property Authentication. To adapt this idea to the VANET setting, we consider Vehicle Authentication as ID-based, and Data/Massage Authentication as Message-based method. As a result, we can specify an elementary classification of attacks in Table 43.3, which elaborates clearly that we can solve these attacks by certain type of authentication methods and their respective solutions. Sybil, node impersonation, tunneling, masquerading, replay, non-repudiation, and unauthorized preemption attacks, according to their definitions can be solved by ID-based authentication solutions, which were proposed in Guo et al. (2007); Kamat et al. (2006). On the other hand, to solve message authentication problems solutions were presented in Wasef et al. (2010), Raya et al. (2006) and Perrig et al. (2001).

We like to note that different approaches for ID-based authentication have been presented, for instance, group-based, nongroup-based, symmetric key-based, and asymmetric key-based methods of authentication. Group-based authentication was first proposed in Guo et al. (2007), and the authors suggested that nearby vehicles should be structured into a group and a public key is allocated to every group. Each

Table 43.3 Elementary categorization and solutions

Authentication methods	Attacks	Solutions
ID-based	Sybil Node impersonation Tunneling Masquerading Replay Unauthorized preemption Non-repudiation	Group and probabilistic-based signature verifications (Guo et al. 2007) Identity verification with public and private key (Kamat et al. 2006)
Message-based	Message tempering Message suppression	Verification of digital signatures (Wasef et al. 2010) sparabreak Need of a vehicular public key infrastructure (Raya et al. 2006) Verification with TELSA (Perrig et al. 2001)

vehicle poses its own private key along with group public key, but it group public key to authenticate itself. A different infrastructure-based approach is based on mutual handshaking of RSU and network vehicle. Moreover, in Kamat et al. (2006) the authors presented an identity-based framework for authentication, which uses an identity-based cryptography scheme. This method includes a setup phase and also pseudonym generation by Trusted Authority (TA), and only TA is allowed to provide vehicles with legal identities. This approach also supports two other security requirements of VANET, integrity and confidentiality.

Similar to alternatives for ID-based authentication, data or message authentication can be achieved using PKI. A node can use its unique private key to generate a unique digital signature for an outgoing message. When a signed message is received to the recipient, the recipient uses the sender's public key to verify the digital signature of the sender of the message (Wasef et al. 2010). Research summarizes that there are also broadcast message authentication methods such as the Elliptic Curve Digital Signature Algorithm (ECDSA), which uses signature verification for message authentication, but is comparably slow (Raya et al. 2006). In contrast to ECDSA, Timed Efficient Stream Loss-tolerant Authentication (TELSA) using symmetric cryptography provides a fast signature verification, but suffers from delayed key disclosure. Additionally, receivers store data until the key is disclosed, which provides an opportunity to degrade the network efficiency to malicious attackers by flooding receivers memory with junk data (Perrig et al. 2001).

Finally, some solutions can support both ID-based authentication and message authentication. In Biswas et al. (2011), an ID-based proxy signature technique with ECDSA was proposed to combine the identity-based features with the ECDSA proxy signature.

43.5 Conclusion and Future Work

In this paper, we presented a featured and structured survey of attacks in particular to authentication of VANET. We discussed existing solutions to independent threads to illustrate the complexity in designing a security framework. To reduce the complexity of the latter, we proposed a classification of attacks on the basis of network security requirements. Based on literature, we identified authentication is the most important parameter and reduced the number of classes present in VANETs. After classification, we discussed all attacks experienced in the scope of authentication. Then, we summarized the most significant methods from literature to reduce attacks and render a network secure. We categorized these methods, which use distinct protocols and methods, to use synergies and simplify the framework design. In a follow-up of this paper, we will present a more detailed analysis of authentication and a structured framework to make network authenticity more secure both in terms of vehicle and data.

Acknowledgments This research was supported by the European Commission in the framework of Erasmus MUNDUS and within the project FUSION.

References

Agrawal A, Garg A, Chaudhiri N, Gupta S, Pandey D, Roy T (2013) Security on Vehicular Ad-hoc Networks (VANET): a review paper. Int J Emerg Technol Adv Eng 3(1)

Akbar MS, Khaliq KA, Qayyum A (2014) Vehicular MAC Protocol Data Unit (V-MPDU): IEEE 802.11 p MAC protocol extension to support bandwidth hungry applications. In: Vehicular ad-hoc networks for smart cities. Springer, pp 31–39

Akbar MS, Khan MS, Khaliq KA, Qayyum A, Yousaf M (2014) Evaluation of IEEE 802.11n for multimedia application in VANET. Proc Comput Sci 32:953–958

Al-Kahtani MS (2012) Survey on security attacks in Vehicular Ad hoc Networks (VANETs). In: Proceedings of the 6th International Conference on Signal Processing and Communication Systems (ICSPCS), pp 1–9

Biswas S, Mišić J, Mišić V (2011) ID-based safety message authentication for security and trust in vehicular networks. In: *Proceedings of the 31st International Conference On Distributed Computing Systems Workshops (ICDCSW)*, pp 323–331

Čapkun S, Hamdi M, Hubaux J-P (2002) GPS-Free positioning in mobile ad hoc networks. Cluster Comput 5(2):157–167

Caruso A, Chessa A, De S, Urpi A (2005) GPS free coordinate assignment and routing in wireless sensor networks. In: Proceedings of the 24th annual joint conference of the IEEE Computer and Communications Societies (INFOCOM), pp 150–160

Dötzer F, Fischer L, Magiera P (2005) Vars: a vehicle ad-hoc network reputation system. In: Proceedings of the 6th IEEE international symposium on a world of wireless mobile and multimedia networks, pp 454–456

Douceur JR (2002) The Sybil attack. In: Peer-to-peer systems, pp 251–260

Engoulou RG, Bellaïche M, Pierre S, Quintero A (2014) VANET security surveys. Comput Commun 44:1–13

Fuentes JMD, González-Tablas AI, Ribagorda A (2010) Overview of security issues in vehicular ad-hoc networks

Golle P, Greene D, Staddon J (2004) Detecting and correcting malicious data in VANETs. In: Proceedings of the 1st ACM international workshop on vehicular ad hoc networks, pp 29–37

Guo J, Baugh JP, Wang S (2007) A group signature based secure and privacy-preserving vehicular communication framework. Mob Netw Veh Environ 2007:103–108

Harsch C, Festag A, Papadimitratos P (2007) Secure position-based routing for VANETs. In: Proceedings of the 66th IEEE Vehicular Technology Conference (VTC-2007), pp 26–30

Hartenstein H, Laberteaux K (2009) VANET Vehicular Applications and Inter-networking Technologies

Juwad M, Al-Raweshidy H (2008) Experimental performance comparisons between SAODV & AODV. In: Second Asia International Conference on Modeling & Simulation (AICMS 2008), pp 247–252

Kamat P, Baliga A, Trappe W (2006) An identity-based security framework for VANETs. In: Proceedings of the 3rd international workshop on vehicular ad hoc networks, pp 94–95

Kumar V, Mishra S, Chand N (2013) Applications of VANETs: present & future. Commun Netw 5(1):12

Lin X, Sun X, Ho P-H, Shen X (2007) GSIS: a secure and privacy-preserving protocol for vehicular communications. IEEE Trans Veh Technol 56(6):3442–3456

Lin X, Lu R (2015) Vehicular ad hoc network security and privacy

Liu B, Zhong Y, Zhang S (2007) Probabilistic isolation of malicious vehicles in pseudonym changing VANETs. In: Proceedings of the 7th IEEE international conference on Computer and Information Technology (CIT-2007), pp 967–972

Lu H, Li J, Guizani M (2012) A novel id-based authentication framework with adaptive privacy preservation for VANETs. In: Proceedings of the Computing, Communications and Applications Conference (ComComAp), pp 345–350

Mokhtar B, Azab M (2015) Survey on security issues in vehicular ad-hoc networks. Alexandria Eng J 54(4):1115–1126

Nantulya VM, Reich MR (2002) The neglected epidemic: road traffic injuries in developing countries. Br Med J 324(7346):1139

Newsome J, Shi E, Song D, Perrig A (2004) The sybil attack in sensor nnetworks: analysis & defenses. In: Proceedings of the 3rd international symposium on information processing in sensor networks, pp 259–268

Niranjan P, Srivastava P, Soni RK, Pratap R (2012) Detection of wormhole attack using hop-count and time delay analysis. Int J Sci Res Publ 2(4):1

Papadimitratos P, Buttyan L, Holczer TS, Schoch E, Freudiger J, Raya M, Ma Z, Kargl F, Kung A, Hubaux J-P (2008) Secure vehicular communication systems: design and architecture. IEEE Commun Mag 46(11):100–109

Perrig A, Canetti R, Song D, Tygar JD (2001) Efficient and secure source authentication for multicast. In: Network and Distributed System Security Symposium. NDSS, vol 1, pp 35–46

Qian Y, Moayeri N (2008) Design of secure and application-oriented VANETs. In: Proceedings of the 67th IEEE vehicular technology conference (VTC-2008), pp 2794–2799

Raya M, Papadimitratos P, Hubaux J-P (2006) Securing vehicular communications. IEEE Wirel Commun Mag 13:8–15 Special Issue on Inter-Vehicular Communications

Raya M, Hubaux J-P (2007) Securing vehicular ad hoc networks. J Comput Secur 15(1):39–68

Sakib RK, Reza B (2010) Security issues in VANET. Ph.D. dissertation, Department of Electronics and Communication Engineering, BRAC University

Sanzgiri K, Dahill B, Levine BN, Shields C, Royer EMB (2002) A secure routing protocol for ad hoc Networks. In: Proceedings of 10th IEEE international conference on network protocols, 2002, pp 78–87

Sethi D, Mitis F (2013) Road traffic injuries. In: Successes and failures of health policy in Europe: four decades of divergent trends and converging challenges, p 215

Sung-Ming Y, Kuo-Hong L (1997) Shared authentication token secure against replay and weak key attacks. Inf Process Lett 62(2):77–80

Toroyan T (2013) Global status report on road safety 2015 Supporting a decade of action. World Health Organization, Department of Violence and Injury Prevention and Disability, Geneva

Wasef A, Lu R, Lin X, Shen X (2010) Complementing public key infrastructure to secure vehicular ad hoc networks. IEEE J Wirel Commun 17(5):22–28

Wu B, Chen J, Wu J, Cardei M (2007) A survey of attacks and countermeasures in mobile ad-hoc networks. In: Wireless network security, pp 103–135

Zeadally S, Hunt R, Chen Y-S, Irwin A, Hassan A (2012) Vehicular ad hoc Networks (VANETS): Status, results, and challenges. Telecommun Syst 50(4):217–241

Lightning Source UK Ltd.
Milton Keynes UK
UKHW02f1157051018
330062UK00003B/36/P